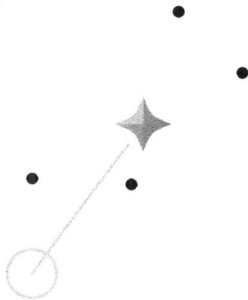

100 YEAR
STARSHIP®

canopus   310 LY

100 YEAR STARSHIP™

canopus 310 LY

finding **earth 2.0**

100YSS PUBLIC SYMPOSIUM 2015

100 YEAR STARSHIP®

2015 Conference Proceedings

Finding Earth 2.0

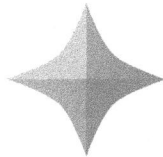

Published by 100 Year Starship®
www.100yss.org

The papers included in this edition reflect the opinions of the original authors only and do not
represent those of the 100 Year Starship®.

ISBN-13: 978-0-9903840-2-1
ISBN-10: 0-9903840-2-0

⋅⋆ 100 YEAR STARSHIP ®
canopus **310 LY**

The 100 Year Starship® exists to make the capability of human travel beyond our solar system a
reality within the next 100 years. We unreservedly dedicate ourselves to identifying and pushing the radical leaps in knowledge and technology needed to achieve interstellar flight, while
pioneering and transforming breakthrough applications that enhance the quality of life for all
on Earth. We actively seek to include the broadest swath of people and human experience in
understanding, shaping and implementing this global aspiration.

For more information, visit www.100yss.org

Edited by Mae Jemison, M.D., Jason D. Batt, and Alires J. Almon
Design and Layout by Jason D. Batt

Statements of fact and opinion in the papers, presentations, and proceedings are those of
the respective authors and contributors and not of the editors or sponsor(s) of 100 Year
Starship®.

Cover Logo Credit:
the barbarian group

PRINTED IN THE UNITED STATES OF AMERICA

100 YEAR STARSHIP®
Table of Contents

TECHNICAL TRACKS

finding earth 2.0
100YSS PUBLIC SYMPOSIUM 2015
★ 100 YEAR STARSHIP™
canopus 310 LY

Finding Earth 2.0 explores the technologies and processes to definitively find a planet outside our solar system capable of supporting Earth-evolved or terran life, as well as the impact of the indisputable discovery of an Earth analogue or Earth 2.0, so to speak.

Presenters, speakers, and attendees at the Symposium asked questions and sought answers to what specific capabilities and systems—scientific, technical, and societal—will be needed over the next five to twenty-five years, not to merely suggest or catalog Earth analogue candidate exoplanets, but to identify at least one definitive Earth 2.0.

Finding Earth 2.0 has profound implications for technology, knowledge, and systems across the spectrum of human experience, skills, capacities, perspectives, and ambitions with each new milestone reached. All are a true part of a bold, audacious journey.

"Finding Earth 2.0 has profound implications across the spectrum of human experience..."

– Dr. Mae Jemison

100 YEAR STARSHIP®
2015 Public Symposium
Finding Earth 2.0

Symposium Overview

Wow! In 2015, the 100 Year Starship Public Symposium tackled a number significant milestones on the way to interstellar capabilities. A couple, relatively down to Earth and the other, has consequences that could "shake the ground" we stand upon!

(continued on next page)

The 100 Year Starship 2015 Public Symposium to took on the challenge of "Finding Earth 2.0." By summer 2015, it was becoming increasingly clear that ours is not the only planet in the universe capable of supporting life, as life exists here on Earth. In fact, scientists and researchers around the world knew or were pretty sure that:

- More than 1,800 planets outside of our solar system—exoplanets—had been confirmed by Kepler Telescope and other astronomical data;
- Between 8.8 and 40 billion earth-sized planets were estimated to orbit stars within the so-called "habitable zone" in just our galaxy, the Milky Way.
- Estmations were that 132-160 habitable worlds existed in our stellar neighborhood—that is, within 10 parsecs (33.6 light years) of our star, the Sun.

100 Year Starship's (100YSS) purpose is to ensure that the capabilities for human interstellar flight exists within 100 years. Such capabilities include not only propulsion, sustainability, health and investment, but also the ability to look beyond our solar system and discover what is out there. And, to understand how such discoveries impact life here on Earth.

Both the process of definitively finding and the actual indisputable discovery of an exoplanet capable of supporting Earth-evolved or "terran life" —an Earth analog or Earth 2.0 so to speak—are game changers. That is, we considered the process and the discovery itself integral parts of the phenomenon Finding Earth 2.0; and each has profound implications for technology, knowledge and systems across the spectrum of human skills and capacities as well as experiences and perspectives on the future.

Atmosphere, geology and weather. • Design and propulsion of probes. • Definitions of life and intelligence. • Telescopes and remote sensing. • Data storage, compression, transmission and communication. • Economics and investment. • Education. Individual, national and global ambition. • Ethics. • Materials. • Law. • Politics. • Religion. • Resources.

2015 attendees were tasked to consider what specific capabilities and systems—scientific, technical and societal—would be needed over the next 5-25 years; not to merely suggest or catalog earth analogue candidate exoplanets, but to identify at least one definitive Earth 2.0—and to consider how such a discovery itself will impact our world and space exploration.

100YSS 2015 Public Symposium Changed Spatial Coordinates. Here on Earth, we moved the Starship Public Symposium city venue from Houston to Santa Clara in the heart of Silicon Valley. The lure was what happens at the intersection of audacious space exploration and the bold smarts, high stakes, cutting edge innovation of Silicon Valley. Here's a sampling of what existed.

Real time, precise directions from your phone. Ultra-light weight high temperature insulation. Insight into the first seconds of the universe using particle accelerators. A wholly connected world.

100YSS challenged attendees to think even bigger; to create the advances in science, engineering, bold social, economic and cultural commitments necessary to explore deep space beyond our solar system and improve life on Earth—today.

100YSS launched the Canopus Awards for Excellence in Interstellar Writing, an annual prize to recognize the finest fiction and non-fiction literary works that contribute to the excitement, knowledge and understanding of interstellar space exploration and travel. It debuted during Science Fiction Stories Night. We wanted to celebrate the authors, ideas and well-told stories, fiction and non-fiction, that result in broader public engagement, commitment and advocacy around the technological, cultural, political, economic, scientific and innovative required to achieve an Inclusive Audacious Journey that Transforms Life Here on Earth and beyond—the objective of 100YSS.

Often, the most powerful ideas start with the simple question, "What if?" Writing based on this query—science fiction and exploratory non-fiction, challenges us to look beyond the world in front of us, to imagine what could be, and envision distant worlds and evolved realities. Such speculative tales inspire scientists and engineers, artists and teachers, politicians and cultural icons toward discovery, invention, and exploration; and in so doing, have transformed our world.

The human interstellar journey, travelling beyond our solar system to another star, is a grand challenge that has capture our imagination across the globe. It is also extraordinarily difficult. And it cannot be successfully accomplished without mustering the full of human experience, skills and talent. We believe it critical to broaden the participants in this journey by integrating the artistic expression of the interstellar exploration.

Winners were:
Previously Published Long-Form Fiction (40,000 words or more)
InterstellarNet: Enigma, Edward M. Lerner (FoxAcre)

Previously Published Short-Form Fiction (40,000 words or less)
"The Waves", Ken Liu (Asimov's 12/12)

Original Fiction (1,000-5,000 words)
"Everett's Awakening", Yelcho

Original Non-Fiction (1,000-5,000 words)
"Finding Earth 2.0 from the Focus of the Solar Gravitational Lens", Louis Friedman & Slava Turyshev

Attendees at SFS enjoyed discussion by Pat Murphy, Juliette Wade, Brenda Cooper and Jacob Weisman led by Jason Batt.

Classes & Technical Tracks
Participants in the 100YSS classes augmented their knowledge over range of topics taught by experts. Discussions in the Technical Tracks and Plenary Sessions were robust and far reaching. And fun.

Plenary sessions were far reaching around the "Finding Earth 2.0" theme.

State of the Universe hosted by Jill Tarter, Ph.D. was defined with presentations by Pete Worden, Ph.D., Executive Director of Breakthrough Listen; and the exploration of biosignatures, new telescopes, exoplanets the evolution of life here by Niki Parenteau, Ph.D., Chris Burke and Natalie Batalha, Ph.D.

Hakeem Olluseyi, Ph.D. hosted a new formulation of "Trending Now" with a panel of experts who answered audience questions, ad hoc.

Amy Millman hosted a powerful luncheon panel that explored the need for "Purposeful Inclusion" with Mar-ianne Caponnetto, Ronke Olabisi, Ph.D., and Gwen Artis.

Pat Murphy, Nebula Award winning author luncheon keynote informed on Playing with the Stars: Realities of Human and Childhood Development in Finding Another Earth, An exploration of human creativity, achievement, play and mental health.

Morgan Cable urged us not to just follow the water when trying to understand "What's Just Right for Life?" Scott Howe, Ph.D. considered 3D Space Construction with In Situ Materials. Radical Leaps with Peter Swan, Ph.D., Seth Shostak, Ph.D. and Phillip Lubin, Ph.D. considered mining at the Next Star, techniques to search for life in exotic destinations and methods of using beamed propulsion to achieve relativistic speeds.

I had the pleasure of moderating the "Who's Got Next" plenary discussion between the Virgin Galatic CEO George Whitesides and Mmboneni Muofhe, south Africa's Deputy Director General for Science and Technology. The final plenary moderated by Karl Aspelund, Ph.D. addressed "Who Goes and Who Decides" with Lt. General Patrick O'Reilly, Louis Freidman, Ph.D amd Derk Austin, Ph.D.

Accelerating Creativity featured creator of television series Extant, Mickey Fisher, and the short film, Project Kronos followed by an All Hallows Eve Party. Yes, there were costumes and Star Lord emceed!

For Thought...
How will knowing, with ever-increasing certainty that the conditions for the existence of life outside our planet and solar system exist, impact our view of ourselves, our planet and our destiny?

The answers to this question and whether they propel us forward or send us scurrying to the closet to hide behind a security blanket will depend upon who is with us on the journey of discovery and our ability to look with the full range of talent, experience, knowledge and wisdom of this planet.

The content and energy of 100 Year Starship 2015 Public Symposium was palpable. I left with a even stronger sense that the pursuit of an extraordinary tomorrow will create a better world today.

100 YEAR STARSHIP™
canopus 310 LY

finding earth 2.0
100YSS PUBLIC SYMPOSIUM 2015

THURSDAY, OCTOBER 29

REGISTRATION *California Ballroom Foyer* 8 AM–7 PM	**100YSS MEMBER GOVERNANCE &** **PLANNING MEETINGS** *Salon 1-3* 1:30 PM–4:30 PM
2015 SYMPOSIUM OPENING RECEPTION *Sedona Room* 630 - 8:30 PM	**100YSS LOUNGE** *Sedona Room* 8:30 - 11 PM

100YSS Classes-100YSS

Friday & Saturday Morning

Classes focus on fundamental principles of disciplines that underlie the capabilities of interstellar flight. The classes help attendees build a greater understanding and appreciation of the 100YSS mission and challenges. Classes range from space vehicle design and astronomy to social sciences, policy, space life sciences, and education.

THE WALL

TAKE A BREAK : Salon 4

Networking, engagement, solutions, questions. Work with Symposium attendees, speakers & special guests.

Pick a color, get your time slot and throw your ideas against The WALL and see what sticks.

FRIDAY, OCTOBER 30

REGISTRATION & 100YSS STORE	OPENING PLENARY & KEYNOTE LUNCHEON	SCIENCE FICTION STORIES NIGHT & CANOPUS AWARDS
California Ballroom Foyer 7:30 AM – 9 PM	*Plenary Room* 11:30 AM – 12:45 PM	*Plenary Room* 7:30 – 10 PM

7:15–9 AM	Continental Breakfast	*California Ballroom Foyer*		

100YSS CLASSES

8:00–8:50 AM	*Neighborhood Watch: An Advanced look at our Space Neighborhood* Bobby Farlice- Rubio Fairfield Museum	*Science Fiction: The History of the Future* J. Daniel Batt Author, 100YSS Creative and Editorial	*Small Satellites and Cubesats: A Huge New Industry* Randa and Roderick Milliron Interorbital Systems

TECHNICAL TRACKS: SESSIONS – A

9:15– 11:15 AM	**Becoming an Interstellar Civilization** *Salon 1* Derek Austin, Ph.D. Kathleen Toerpe, Ph.D. *Co-Chairs*	**Designing for Interstellar** *Salon 2* Karl Aspelund, Ph.D. *Chair*	Data Communications & IT *Salon 3* Ron Cole *Chair*	**Life Sciences:** *Salon 9* Terry Mulligan, Ph.D. *Chair*

11:30-11:50	**Opening Plenary** *Plenary Room*	*Finding Earth 2.0: Evolutionary Step or Societal Imperative* , Mae Jemison, M.D., Principal 100YSS *Technical Track Show & Tell*, Pam Contag, Ph.D., 100YSS Technical Track Chair; CEO and Founder , Molecular Sciences Institute
12-12:45 PM	Keynote Luncheon *Plenary Room*	*Playing with the Stars: Realities of Human and Childhood Education Development in Finding Another Earth, An exploration of human learning, creativity, achievement, play and mental health.* Keynote: Pat Murphy, Nebula Award-winning science fiction author, San Francisco Exploratorium and Klutz Science Books for Kids Moderator, Kathleen Toerpe, Ph.D.

TECHNICAL TRACKS: SESSIONS – B

1:15-3:15 PM	**Becoming an Interstellar Civilization** *Salon 1* Derek Austin, Ph.D. Kathleen Toerpe, Ph.D. *Co-Chair*	**Propulsion & Energy** *Salon 2* Hakeem Oluseyi, Ph.D. *Chair*	**Life Sciences:** *Salon 9* Terry Mulligan, Ph.D. *Chair*

3:30-4:30 PM	Destinations & Astronomy *Salon 3* Margaret Turnbull, Ph.D. *Glen*	
4:45-5:45 PM	PLENARY *Plenary Room*	*The Data Must Flow:" Really Big Data, IT and Communications in Finding Earth 2.0: Handling Massive Data* Ron Cole, Moderator, Michael Flynn,, Ph.D., Fellow, Stanford Univ. & Maxeler Dataflow, and Mmboneni Muofhe (Deputy Director General, Dept. of Science & Tech- South Africa
6:00 – 7:00 PM	Posters *Sedona Room*	**Poster Session & Reception** Hosted Technical Track Poster . Chair: Tim Meehan, Ph.D.
7:30-10 PM		**Science Fiction Stories Night** Moderator: Jason Batt Pat Murphy, Juliette Wade, Brenda Cooper, and, Jacob Weisman, Publisher Tachyon **Inaugural Canopus Awards for Excellence in Interstellar Writing** **Author Book Signing** *Hosted by* Borderlands Bookstore 100YSS Lounge Open Until 11:30 PM

finding earth 2.0
100YSS PUBLIC SYMPOSIUM 2015
100 YEAR STARSHIP™
canopus 310 LY

SATURDAY, OCTOBER 31

REGISTRATION & 100YSS STORE	KEYNOTE LUNCHEON	ACCELERATING CREATIVITY & ALL HALLOWS EVE PARTY
California Ballroom Foyer 7:30 AM–9 PM	*Plenary Room* 11:45 AM –1 PM	*California Ballroom, Foyer &* *Sedona*

7:30-8:30	**Continental Breakfast**	*California Ballroom Foyer*
9:00– PM	**PLENARY** *California Ballroom*	**STATE OF THE UNIVERSE: Where are we on the road to Earth 2.0?** *Facilitated by* Jill Tarter, PhD, Bernard Oliver Chair, & Co-founder, SETI Institute *A conversation with* Dr. Pete Worden, Chair Breakthrough Listen & former NASA Ames Center Director; *biosignatures, exoplanets, new telescopes, and geological formation of Earth 1.0, our planet, ties the evolution of life here.* Niki Parenteau, Ph.D.; NASA Ames; Chris Burke, NASA Ames ; Natalie Batalha, Ph.D. NASA Ames.
1:50-11:50 AM	**PLENARY** *California Ballroom*	HAKEEM Oluseyi, Ph.D., *Chief Science Officer, Discovery Channel, Astrophysicist and Professor Florida of Technology, TED Fellow* . **Ask Hakeem and his designated pinch hitters what's TRENDING NOW**
12:15-1:15 PM		**Keynote Luncheon** *California Ballroom* *Purposeful Inclusion: The Key to Achieving Radical Leaps in Technology and Enhancing Life on Earth. A discussion between* Amy Millman, CEO & Co-Founder Springboard Enterprises; Marianne Caponnetto, Marketing, Branding & Startups; Rinke Olabisi, Ph.D., Asst. Prof. Biomedical Engineering, Rutgers University.
1:30– 2 PM		**Plenary Session:** *California Ballroom* **What's Just Right for Life?** *Don't (Just) Follow the Water,* Morgan L. Cable, Ph.D. Report from Keck Institute *Study.*
2-2:30 PM		**Plenary Session:** *California Ballroom* **Design 2.0:** *3 D Space Construction with In Situ Materials.* Scott Howe, PhD, Jet Propulsion Laboratory
2:30-3:30 PM		**Plenary: Radical Leaps** *California Ballroom* *These are the researchers, technologies and systems pushing the very sharp edge of knowledge, engineering and capabilities. Stroll through an overview of the tech and then take a Deep Dive in breakout sessions!* *Radical Leaps Facilitator*: Peter Swan, Ph.D., President International Space Elevator Consortium. *"Critical Resource Replenishment at Exotic Destinations - Mining at the Next Star"* Swan, *"Technological Approaches that can search for Life Close Up in Exotic Destinations"* Seth Shostak, Ph.D., SETI Institute *"Achieving Relativistic Speeds,"* Phillip Lubin, Ph.D., Professor, Physics, Univ. California Santa Barbara
3:30-5:30 PM		**DEEP DIVES** Salons 1-3 *Ninety minute immersion into specific topics with experts in specific subject matter covered during the Symposium. Specifically, Achieving Relativistic Velocities; Communicating Science; and Mining the Next Star & 3D In Situ Construction; Approaches to the Search for Life & Don't Just Follow the Water.*

6-6:45 PM	**ACCELERATING CREATIVITY** Reception *California Ballroom Foyer; PRIME in Sedona Room*
7:00-11:30 PM	**ACCELERATING CREATIVITY** Dinner and Dancing *California Ballroom*

Artificial Intelligence & Seeking New Worlds:

Mickey Fisher, *creator of the CBS space drama series* EXTANT, *in conversation with* Dr. Mae Jemison and Dr. Karl Aspelund *considers the potential of artificial intelligence (AI)—pro and con— in deep space exploration and the enhancement of life on Earth. What are the ways in which the interplay between the sciences and art push ideas and advancements in innovation, society and space exploration. The short film, Project Kronos, an imaginative documentary poses an unusual solution to the overwhelming data and computational challenges of interstellar exploration. And additional context for discussion.*

Halloween Party, with Miss Q Space Cat, *cosmic trick or treat, Space Bingo, Interstellar Trivia, scavenger hunt, DJ and music and dancing.*

SUNDAY, NOVEMBER 1

8-9:30 AM	**Continental Breakfast**
8:30-9:15	**Plenary Session:** *California Ballroom* **Track Chair Summary & THE WALL Results** *Discussion of the papers and ideas that emerged from the Technical Track presentations. Slide show of the results of THE WALL.*
9:30-10:45 AM	**PLENARY:** *California Ballroom* **Who's Got Next:** *The Ideas, people and organizations set to continue the journey* Virgin Galactic CEO George Whiteside's and South Africa's *Deputy Director General for Science and Technology* Mmbonene Muofhe *will discuss the next people, organizations, and technologies in space. South Africa is the home of the new Pan Africa University Space Sciences Institute.*
10:55–11:55 AM	**PLENARY:** *California Ballroom* **Who's Goes, Who Decides, Why?** Moderator: Karl Aspelund, Ph.D., Anthropologist, Industrial Design, Asst. Prof Textile & Design, Univ. of Rhode Island; **Lt. General Patrick O'Reilly,** Senior Fellow, Atlantic Council; **Louis Friedman, PhD,** Author and Co-founder, The Planetary Society; **Derek Austin, PhD,** Director, New Clergy Program, The Chautauqua Institute. This panel considers the dimensions of the influence of economics, religion and various national priorities on audacious projects, exploration and investment in audacious projects.
12 –12:30	**SYMPOSIUM CLOSE & 100YSS The Next Year**

21

Science Fiction Stories Night Authors

PAT MURPHY

Science Fiction Stories Night
Luncheon: Playing With the Stars

Pat is a writer, a scientist, and sometimes a toy maker. Her fiction writing has won a number of awards, including the Nebula Award for Science Fiction, the World Fantasy Award, the Philip K. Dick Award for best paperback original, the Christopher Award, and the Theodore Sturgeon Memorial Award. She also co-founded the James Tiptree Memorial Award. When she is not writing fiction, she writes books about science. For upwards of 20 years, she was a writer at the Exploratorium, San Francisco's hands-on museum of science, art, and human perception. She also teaches writing. She has taught in Stanford University's Creative Writing Program, at the University of California at Santa Cruz, and at the Clarion Speculative Fiction Workshops in Michigan and Seattle.

BRENDA COOPER

Science Fiction Stories Night

Brenda Cooper is the author of eight science fiction and fantasy books. Her most recent novel is Edge of Dark (Pyr, 2015). Also releasing this year is a short fantasy collection, Beyond the Waterfall Door (self-published via Kickstarter, 2015), and a science fiction collection, Cracking the Sky (Fairwood Press, 2015). Her other works include The Creative Fire (Pyr, 2012) and The Diamond Deep (Pyr, 2013) as well as the Silver Ship and the Sea series (Tor Books) and Building Harlequin's Moon, with Larry Niven (Tor, 2005). Her most recent short fiction includes Elephant Angels (Heiroglyph, 2014) and Biology at the End of the World (Asimov's, August 2015). Brenda blogs frequently on environmental and futurist topics, and her non-fiction has appeared in Slate and Crosscut. Winner of the 2007 Endeavor Award for a distinguished science fiction or fantasy book written by a Pacific Northwest author or authors Brenda lives in Bellevue, Washington with her wife and three dogs. Brenda also serves as the Chief Information Officer for the City of Kirkland, which is a Seattle suburb. Brenda was educated at California State University, Fullerton, where she earned a BA in Management Information Systems.

JACOB WEISSMAN

Science Fiction Stories Night

Jacob Weisman is the editor and publisher at Tachyon Publications, which he founded in 1995. He has been nominated for the World Fantasy Award three times for his work at Tachyon and is the series editor of Tachyon's Hugo, Nebula, Sturgeon, and Shirley Jackson Award-winning novella line. His writing has appeared in The Nation, Realms of Fantasy, the Louisville Courier-Journal, The Seattle Weekly, and The Cooper Point Journal. Jacob is rumored to be hard at work on his first novel.

JULIETTE WADE, PHD

Science Fiction Stories Night

Juliette grew up in the Monterey Bay area of California, but her family always maintained close relationships overseas as well as locally, so she had the chance to travel and to learn French at an early age. She has always been very interested in science, and toyed with becoming a Biology major in college, but after suffering in Organic Chemistry I decided that language was the way to go. When she showed up at the Linguistics department talking about the fascinating links between language and culture, they sent her over to Anthropology. In the end, she majored in Anthropology and Japanese as an undergraduate, and then turned to theoretical Linguistics for her Master's degree. She has lived in Japan three times. She taught Japanese for two years before beginning her Ph.D. in Education. In her dissertation she combined everything she had done before, looking at cultural clashes in classrooms where American students were learning Japanese, and investigating how those clashes influenced the students' use of politeness grammar. Her research took me back to Japan and involved lots of listening to people talk, and breaking down how they expressed information about social position. She started writing fiction while studying for her Ph.D., and by the time her degree was finished and her son was born, she realized she'd never stop writing. At this point she works at home, caring for her kids and writing.

Plenary Speakers

MICKEY FISHER

Accelerating Creativity

Originally from a small town in Southern Ohio, Mickey Fisher is the now Los Angeles based creator of the CBS sci-fi event series EXTANT, Executive Produced by Steven Spielberg and starring Academy Award Winner Halle Berry. Prior to that (his 20-year "overnight" breakthrough) he wrote and directed a number of independent feature films and plays for various theaters around the country. He's a proud alum of The University of Cincinnati's College Conservatory of Music Musical Theatre Program, and a member of Actor's Equity and The Writer's Guild of America.

GEORGE WHITESIDES

Who's Got Next Panelist

George is responsible for guiding all aspects of building the world's first commercial spaceline including our spaceflight program as well as our small satellite launch capability. This includes oversight of our sister company, The Spaceship Company (TSC), to manufacture a fleet of WhiteKnightTwo and SpaceShipTwo space vehicles. Prior to Virgin Galactic, George served as Chief of Staff for NASA, where he provided policy and staff support to the agency's Administrator. Upon departure from NASA, he received the Distinguished Service Medal, the highest award the agency confers. Prior to his role at NASA, George served as Executive Director of the National Space Society (NSS), a space policy and advocacy group that was founded by Apollo program leader Wernher von Braun and the journalist Hugh Downs. He is a member of the World Economic Forum's Global

Agenda Council on Space Security; an advisory board member of the Rotary National Award for Space Achievement Foundation; a fellow of the UK's Royal Aeronautical Society; and an associate fellow of AIAA. He previously chaired the Reusable Launch Vehicle Working Group for the FAA's Commercial Space Transportation Advisory Committee. He served for four years on the Board of Trustees of Princeton University. George has testified on American space policy before the United States Senate, the United States House of Representatives, and the President's Commission on Implementation of United States Space Exploration Policy. Space News selected him as one of 12 "People to Watch" in the space industry. An honors graduate of Princeton University's Woodrow Wilson School, George earned his undergraduate degree in public and international affairs. He later earned a master's degree in geographic information systems and remote sensing from the University of Cambridge in England, and was a Fulbright Scholar to Tunisia. George is a licensed private pilot and certified parabolic flight coach.

MMBONENI MUOFHE
Who's Got Next? Panelist
The Data Must Flow Panelist

Mr. Mmboneni Muofhe is the Deputy Director-General (DDG) for Technology Innovation at the Department of Science and Technology in South Africa. He was previously DDG for International Cooperation and Resources and has held several positions at the Department, including being Chief Director: International Resources, Director: Strategic Partnerships and Director: Global Projects. During this period he oversaw South Africa's growing participation in EU Research Funding programmes, led the mobilization of Official Development Assistance funds to support South Africa science system and the Department's partnership with Multinational Companies. Mr Muofhe's earlier working and post-graduate years were mainly in agricultural biotechnology. He was a UNESCO Biotechnology Fellow at the Agricultural Research Council in 1997 before joining the Foundation for Research Development (now National Research Foundation) as Coordinator and then Manager for Technology and Human Resource for Industry Programme (THRIP). Mr Muofhe's current responsibilities include overseeing the implementation of South Africa's Space Science and Technology Strategy, Bioeconomy Strategy, Energy Research and Technology Strategy, Innovation Priorities Instruments (including new Research Areas) and oversight of the IP act.

LOUIS FRIEDMAN, PHD
Who Goes, Who Decides, Why? Panelist
Closing & 100YSS: The Year Ahead

Dr. Friedman lectures in the U.S. and abroad about planetary missions and space exploration programs, has written many popular articles about planetary exploration and space policy as well as op-eds in major newspapers. He has frequently testified to the U.S. Congress about programs and policies in space exploration. He has traveled on field expeditions to Kamchatka, the Arctic and Antarctic, tours to observe Halley's Comet, Belize and to several places in the former Soviet Union. Asteroid 3651 was named for Louis and Connie Friedman by its discoverer, Eleanor Helin. Recently he was co-leader of the Keck Institute for Space Studies (KISS) Asteroid Retrieval Mission Study at Caltech. He is also co-leader of a new KISS study: Science and Technology to Explore the Interstellar Medium. In March 2015 he was reappointed (after and earlier stint in 2011)

to the NASA Innovative Advanced Concepts (NIAC) External Council. His book examining the implications of robotic interstellar precursor missions using nano-spacecraft and solar sails on future human space flight is expected to be published in the fall of 2015 by the University of Arizona Press. It is tentatively titled: From Mars to the Stars: The Future of Human Space Flight. Dr. Friedman is a native of New York City. He received a B.S. in Applied Mathematics and Engineering Physics at the University of Wisconsin in 1961, an M.S. in Engineering Mechanics at Cornell University in 1963, and a Ph.D. from the Aeronautics and Astronautics Department at M.I.T. in 1971. His Ph.D. thesis was on Extracting Scientific Information from Spacecraft Tracking Data.

PETER SWAN, PHD
Radical Leaps Moderator

President of the International Space Elevator Consortium. As such, he leads a team who further the concept with incremental studies and yearly conferences. Over the last ten years he has published five books on the topic as co-author and/or co-editor. They are: Space Elevators: An Assessment of the Technological Feasibility and the Way Forward [2013], Design Considerations for Space Elevator Tether Climbers [2014], Space Elevator Concept of Operations [2013], Space Elevator Survivability – Space Debris Mitigation [2012], and Space Elevators Systems Architecture [2005]. He graduated from the US Military Academy in 1968 with a Bachelor of Science degree and served 20 years in the Air Force with a variety of research and development positions in the space arena. He taught at the Air Force Academy and retired as a Lieutenant Colonel. Upon retirement in 1988, he joined Motorola on the Iridium satellite program. He lead the team responsible for the development of the Iridium spacecraft bus. Pete received his Ph.D. from the University of California at Los Angeles in Mechanical Engineering with a specialty in space systems. He has published many papers and a few books; two of which are on preparing for SCUBA trips.

MICHAEL FLYNN, PHD
The Data Must Flow Panelist

Michael Flynn received his Ph.D. from Purdue University in 1961. He joined IBM in 1955 and for ten years worked in the areas of computer organization and design. He was design manager of prototype versions of the IBM 7090 and 7094/II, and later for the System 360 Model 91 Central Processing Unit. Between 1966 and 1974 Prof. Flynn was a faculty member of Northwestern University and the Johns Hopkins University. In 1975 he became Professor of Electrical Engineering at Stanford University, and was Director of the Computer Systems Laboratory from 1977 to 1983. He was founding chairman of both the ACM Special Interest Group on Computer Architecture and the IEEE Computer Society's Technical Committee on Computer Architecture. Prof. Flynn was the 1992 recipient of the ACM/IEEE Eckert-Mauchley Award for his technical contributions to computer and digital systems architecture. He was the 1995 recipient of the IEEE-CS a href="http://www.pgg.com/mjf-award.html" target="_top"></a Harry Goode Memorial Award in recognition of his outstanding contribution to the design and classification of computer architecture. In 1998 he received the Tesla Medal from the International Tesla Society (Belgrade), and an honorary Doctor of Science from Trinity College (University of Dublin), Ireland. He is the author of three books and over 250 technical papers.

CLAUDIO MACCONE, PHD
SETI Astronomer

Claudio Maccone (born 6 February 1948, Torino, Italy) is an Italian SETI astronomer, space scientist and mathematician. In 2002 he was awarded the "Giordano Bruno Award" by the SETI League, "for his efforts to establish a radio observatory on the far side of the Moon." In 2010 he was appointed Technical Director for Scientific Space exploration by the International Academy of Astronautics.[Since 2012, he has chaired the SETI Permanent Committee of the International Academy of Astronautics, succeeding Seth Shostak of the SETI Institute, who held that position from 2002 to 2012. Maccone's two vice-chairs are his fellow Academician H. Paul Shuch and Mike Garrett of Astron (the Netherlands Institute for Radio Astronomy). He obtained his PhD at the Department of Mathematics of King's College London in 1980. He then joined the Space Systems Group of Aeritalia (later called Alenia Spazio S.p.A. and now Thales Alenia Space Italia S.p.A.) in Turin as a technical expert for the design of artificial satellites, and got involved in the design of space missions. In 2000 he was elected as Co-Vice Chair of the SETI Committee of the IAA. He has published over 100 scientific and technical papers, most of them in "Acta Astronautica." In 2010, Maccone was appointed Technical Director of Scientific Space Missions for the International Academy of Astronautics. In 2012, he became a founding member of the Advisory Council of the Institute for Interstellar Studies.

JILL TARTER, PHD
State of the Universe Moderator, SETI, Institute

Jill Tarter holds the Bernard M. Oliver Chair for SETI Research at the SETI Institute in Mountain View, California. Tarter received her Bachelor of Engineering Physics Degree with Distinction from Cornell University and her Master's Degree and a Ph.D. in Astronomy from the University of California, Berkeley. She served as Project Scientist for NASA's SETI program, the High Resolution Microwave Survey, and has conducted numerous observational programs at radio observatories worldwide. She is a Fellow of the AAAS and the California Academy of Sciences, she was named one of the Time 100 in 2004, and one of the Time 25 in Space in 2012, received a TED prize in 2009, public service awards from NASA, multiple awards for communicating science to the public, and has been honored as a woman in technology. Since the termination of funding for NASA's SETI program in 1993, she has served in a leadership role to design and build the Allen Telescope Array and to secure private funding to continue the exploratory science of SETI. Dr. Tarter was recently awarded the prestigious Jansky Lectureship, which honors outstanding contributions to the field of Radio Astronomy. Many people are now familiar with her work as portrayed by Jodie Foster in the movie Contact.

MARIANNE CAPPONNETTO
Lunchson: Purposeful Inclusion

Marianne Caponnetto. As Founder & President of MCW Group, Inc., a New York based strategic consulting firm, Ms. Caponnetto works with start-ups, mid-market and Fortune 500 purpose-driven companies defining and executing significant transformational and growth objectives. Her experience in leading and optimizing purpose-driven change has been honed at a diverse range of corporate and entrepreneurial companies. In her role at MCW Group, she is a trusted advisor and hands on operator for a select group of global clients. From 2006 to 2008, Ms. Caponnetto was the Chief Sales and

Marketing Officer and a key member of the management team that redefined digital marketing leadership at DoubleClick, Inc., which led to its acquisition by Google Inc. in 2008. From 1994 to 2005, she held executive positions at IBM in the Global Media Entertainment Industry, where she enabled the transition from a product to a solutions focus with responsibility for a billion dollar market. Ms. Caponnetto joined IBM in 1994 as a key member of the marketing leadership team that reinvented the IBM brand and repositioned the Company. She pioneered IBM's adoption of the internet as IBM's defining marketing medium and critical proof point of e-business transformation. Prior to IBM, she led Strategic and Corporate Marketing for Dow Jones & Co., after a successful career in the advertising business. Ms. Caponnetto has served on digital technology, media, and advertising industry boards and is active on several private and non-profit company boards. Ms. Caponnetto is a graduate of the University of California, Berkeley.

SETH SHOSTAK, PHD
Radical Leaps Panelist

Seth claims to have developed an interest in extraterrestrial life at the tender age of ten, when he first picked up a book about the solar system. This innocent beginning eventually led to a degree in radio astronomy, and now, as Senior Astronomer, Seth is an enthusiastic participant in the Institute's SETI observing programs. He also heads up the International Academy of Astronautics' SETI Permanent Study Group. In addition, Seth is keen on outreach activities: interesting the public – and especially young people – in science in general, and astrobiology in particular. He's co-authored a college textbook on astrobiology, and continues to write trade books on SETI. In addition, he's published nearly 300 popular articles on science, gives many dozens of talks annually, and is the host of the SETI Institute's weekly science radio show, "Big Picture Science" And, as might be evident from this overly effusive bio, he is also editor of the Institute's Explorer magazine.

PETE WORDEN, PHD
Director, NASA Ames Research Center

Dr. S. Pete Worden (Brig. Gen., USAF, ret.) is the NASA Ames Research Center Director. Prior to becoming Director, Dr. Worden was a Research Professor of Astronomy, Optical Sciences and Planetary Sciences at the University of Arizona where his primary research direction was the development of large space optics for national security and scientific purposes and near-earth asteroids. Additionally he worked on topics related to space exploration and solar-type activity in nearby stars. He is a recognized expert on space issues—both civil and military. Dr. Worden has authored or co-authored more than 150 scientific technical papers in astrophysics, space sciences, and strategic studies. Moreover, he served as a scientific co-investigator for two NASA space science missions. In addition to his former position with the University of Arizona, Dr. Worden served as a consultant to the Defense Advanced Research Projects Agency (DARPA) on space-related issues. During the 2004 Congressional Session Dr. Worden worked as a Congressional Fellow with the Office of Senator Sam Brownback (R-KS), where he served as Senator Brownback's chief advisor on NASA and space issues. Dr. Worden retired in 2004 after 29 years of active service in the United States Air Force. His final position was Director of Development and Transformation, Space and Missile Systems Center, Air Force Space Command, Los Angeles Air Force Base, CA. In this position he was responsible for developing new directions for Air Force Space Command programs and was instrumental in initiating a major Responsive Space Program designed to produce space systems and launchers capable of tailored military effects on timescales of hours. Dr. Worden was commissioned in 1971 after receiving a Bachelor of Science degree from the University of Michigan. He entered the Air Force in 1975 after graduating from the University of Arizona with a doctorate in astronomy. Throughout the 1980s and early 1990s, Dr. Worden served in every phase of development, international negotiations and implementation of the Strategic Defense Initiative, a primary component in ending the Cold War. He twice served in the Executive Office of the President. As the staff officer for initiatives in the George Bush administration's National Space Council, Dr. Worden spearheaded efforts to revitalize U.S. civil space exploration and earth monitoring programs. Dr. Worden commanded the 50th Space Wing that is responsible for more than 60 Department of Defense satellites and more than 6,000 people at 23 worldwide locations. He then served as Deputy Director for Requirements at Headquarters Air Force Space Command, as well as the Deputy Director for Command and Control with the Office of the Deputy Chief of Staff for Air and Space Operations at Air Force headquarters. Prior to assuming his current position, Dr. Worden was responsible for policy and direction of five mission areas: force enhancement, space support, space control, force application and computer network defense. Dr. Worden has written or co-written more than 150 scientific technical papers in astrophysics, space sciences and strategic studies. He was a scientific co-investigator for two NASA space science missions. He and his wife Nancy reside in Placitas, New Mexico.

MORGAN CABLE, PHD
What's Just Right for Life? Moderator
Dr. Morgan L. Cable is a Research Scientist in the Instrument Systems Implementation and Concepts Section at the NASA Jet Propulsion Laboratory (JPL) in Pasadena, California. She is also a Project Science Systems Engineer for the Cassini Mission, which has been exploring the Saturn system for over 10 years. Morgan's research focuses on organic and biomarker detection strategies, through both in situ and remote sensing techniques. While earning her Ph.D. in Chemistry at the California Institute of Technology, she designed receptor sites for the detection of bacterial spores, the toughest form of life. As a NASA Postdoctoral Fellow at JPL, Morgan developed novel protocols to analyze organics such as amines and fatty acids using small, portable microfluidic sensors. She is currently working as a Collaborator on the Mapping Imaging Spectrometer for Europa (MISE), an instrument selected for NASA's next mission to Jupiter's icy moon Europa; this spectrometer will map Europa's surface and search for organics, salts and minerals. Dr. Cable's research interests also include 'weird' life. She has performed laboratory experiments to study the liquid hydrocarbon lakes of Titan, a moon of Saturn. She has been involved in several studies led by the Keck Institute for Space Studies, the most recent of which was to explore what kinds of life could survive or even thrive in exotic solvents (other than liquid water). In addition to biomarker sensor design and the search for 'weird' life, Morgan has also explored several extreme environments on Earth that serve as analogs for other places in the solar system, such as Mars. She was involved in research expeditions to the driest desert in the world, the Atacama Desert in Chile, and to the summit of Mt. Kilimanjaro in Tanzania. Morgan has also co-led a team of young researchers on multiple expeditions to Iceland to study how

life colonizes a fresh lava field. The goal of this work is to inform future Mars sample return missions in terms of sample selection, preservation and analysis.

RONKE OLABISI, PHD
Luncheon: Purposeful Inclusion
Dr. Ronke Olabisi is a member of the 100YSS Research team, focused in Life Sciences. Dr. Ronke Olabisi is a member of the Biomedical Engineering Department of Rutgers University. Her PhD research centered on limb lengthening (1) muscle and joint function; (2) the elastic and viscoelastic properties of tendon; and (3) the lengthened tissue histomorphology. As a postdoctoral fellow Dr. Olabisi studied the biophysics of bone and seashell at Wisconsin's Synchrotron Radiation Center, bone tissue engineering at Rice University using hydrogel microencapsulated cells genetically modified to express bone morphogenetic protein, and differentiating mesenchymal stem cells down an osteogenic lineage by varying microencapsulation parameters. Dr. Olabisi's current research interests include orthopedic tissue engineering and regenerative medicine for injury, aging, disease, and space flight. Dr. Olabisi received her B.S. degree in Mechanical Engineering from Massachusetts Institute of Technology, an M.S. Degree in Mechanical Engineering and an M.S. in Aerospace Engineering from University of Michigan, and her Ph.D. in Biomedical Engineering from University of Wisconsin-Madison.

AMY MILLMAN
Luncheon: Purposeful Inclusion
Amy Millman is a passionate advocate for women entrepreneurs building Big Businesses Starting Small. In 2000, she co-founded Springboard Enterprises, a non-profit venture catalyst which sources, coaches, showcases and supports women-led companies seeking equity capital for product development and expansion. Springboard has assisted hundreds of women entrepreneurs in raising billions in investments and connecting with thousands of expert resources. The successes of Springboard entrepreneurs include 10 IPOs, legions of high value M&As and a community of accomplished serial entrepreneurs. During her career in Washington, DC, she served as a representative for several industry groups and was appointed as Executive Director of the National Women's Business Council during the Clinton Administration. She served on the boards of many organizations including her current service with JumpStart Inc. and Enterprising Women Magazine.

PATRICK O'REILLY
Lieutenant General, US Army (retired)
Who Goes, Who Decides, Why? Panelist
Closing & 100YSS: The Year Ahead
Pat is a Nonresident Senior Fellow at the Atlantic Council with expertise in recoverable energy, cyber-security, aerospace and missile defense. He also is the Senior Vice President at Alphabet Energy, Inc. leading the application of thermoelectric technology to convert waste heat directly into electricity for the U.S. government and military. Additionally, Pat is a member of National Advisory Committee for Spark 101 (a non-profit organization to stimulate students' interest in Science, Technology, Engineering and Math). During his Army career, Pat was the Director of the US Missile Defense Agency and held many positions of responsibility for the procurement of missile defense systems, combat support vehicles and power generation equipment. He holds functional and technical expertise in aerospace,

electronics, and physics and was an Associate Professor of Physics at West Point. He is a West Point graduate with master's degrees in Physics, National Security and Strategic Studies, and Management.

Technical Track Chairs
PAMELA CONTAG, PHD
Overall Technical Track Chair
Pamela R. Contag, PhD, has founded four early stage technology companies, most recently ConcentRx, a cell-based immune therapy company and Cygnet Inc. commercializing a platform technology to discover beneficial microbes for applications in food, renewable fuels, low cost therapeutics, industrial enzymes, and carbon dioxide capture. With more than 25 years of microbiology research experience, Dr. Contag is widely published in the field of Microbiology and Optical imaging and has over 35 patents in Biotechnology. Dr. Contag received her Ph.D. in Microbiology.

REV. DEREK AUSTIN, PHD
Becoming an Interstellar Civilization Track Co-Chair
Rev. Dr. Derek Austin currently resides in Lakeville, CT, and is the Director of Clergy Development Programming for Chautauqua Institution in Western N.Y. Prior to accepting this position, he served several churches in New England as an Intentional Transitional Minister for ten years. Although being raised in upstate NY, Derek went to school on the West coast and lived most of his adult life in San Diego, CA, until relocating to New England in 2005. With a Bachelor degree in Music and a Master of Arts in Education, Derek taught vocal music at a performing arts magnet school in southern California in the 1990's, and was the Conductor of a professional San Diego civic Chorus, before pursuing his Doctoral studies at San Francisco Theological Seminary in San Anselmo, a part of the Graduate Theological Union of Berkeley, CA. Derek has also served on several boards in New England that both resource and explore the changing nature of ministry, church life, and non-profit dynamics. Leadership development has been a cornerstone of Derek's ministry and professional life, honing an expertise in change management, transitional dynamics, and corporate goals assessment. Enjoying an active life, Derek can be found during his free time hiking a mountain, kayaking a lake, attending a staged play, or watching old episodes of West Wing.

MAGGIE TURNBULL, PHD
Interstellar Space, Stars, & Destination Track Chair
Dr. Turnbull is an astrobiologist whose expertise is in identifying planetary systems that are capable of supporting life as we know it. She developed a Catalog of Habitable Stellar Systems for use in the search for extraterrestrial intelligence (SETI) and she has studied the spectrum of the Earth to identify telltale signatures of life. She is currently leading two teams to prepare for NASA's WFIRST mission, slated for launch in 2025. WFIRST will be the first mission to directly image planetary systems orbiting the nearest sunlike stars, and the first mission with the hope of determining the atmospheric composition and surface characteristics of those planets. When not thinking about alien worlds and missions to get us there, Maggie can be found keeping honey bees, raising monarchs, tapping sugar maples, and cross country skiing across the north woods with her dogs.

TERRY MULLIGAN, PHD
Life in Space Track Chair

HAKEEM OLUSEYI, PHD
Propulsion & Energy Track Chair
Hakeem M. Oluseyi, PhD is an internationally recognized astrophysicist, science TV personality, and global science education activist. His research interests span the fields of astrophysics, cosmology, and technology development. He currently has 7 U.S. patents, 4 EU patents and over 60 scholarly publications in the areas of astrophysics, optics and detector technologies development; nanotechnology manufacturing; observational cosmology; and the history of astronomy. Dr. Oluseyi leads a group studying processes by which electromagnetic fields and plasmas interact in order to understand solar atmospheric heating and acceleration, which has resulted in a new in-space propulsion technology.

DAN HANSON
Interstellar Innovations Enhance Life on Earth Track Co-Chair
Dan is a principal with Technology Innovation Group, Inc. (TIG). Dan has a keen interest in leveraging art, science, and education infrastructure to promote economic development in regional economies across the globe. TIG pursues it mission through two primary service offerings: advising governments, foundations, and communities wanting to build technology-based economies; and serving as translational consultants with institutions and private companies to commercialize specific technologies, primarily those with public health or economic development benefits. TIG works with clients that are developing products and services based on complex technologies, and universities and research institutions that desire to move discoveries from the laboratory to businesses.

RON COLE
Data, Communications, and Information Technology Track Chair
Ron finished a 50-year career with the US Intelligence Community in May 2013. Ron received the Civilian Defense Meritorious Award upon leaving the NSA to begin working with Scitor Corporation as a systems engineer technical advisor to the NSA and the National Reconnaissance Office (NRO) on system development and data processing. Ron left Scitor to go to work for Riverside Research as a senior advisor for five years to the National Geospatial-Intelligence Agency on technology developments for mission execution and then moved to support the NRO on policy and management issues.

KARL ASPELUND, PHD
Designing for Interstellar Chair
Karl Aspelund, PhD, is an anthropologist with a design background. He is assistant professor at the Department of Textiles, Fashion Merchandising and Design at the University of Rhode Island and visiting professor of ethnography at the University of Iceland. His interests lie in examining the role of textiles and design in identity-creation, the environmental impact of the textile life-cycle, and how designers may contribute to environmental sustainability. He is currently investigating the design and cultural needs and constraints of apparel in long-term space exploration. Dr. Aspelund was recently a speaker at TEDx Reykjavik in Iceland.

KATHLEEN TOERPE, PHD
Becoming an Interstellar Civilization Co-Chair

Kathleen D. Toerpe, PhD, is a social and cultural historian who researches the human dimension of outer space through an emerging field called "astrosociology." She is the Deputy CEO for Programs and Special Projects with the Astrosociology Research Institute, and volunteers as a NASA/JPL Solar System Ambassador. She has served as a Historian-in-Residence and a museum Educational Curator and has provided local outreach programming, oral history program management and exhibit curation. As a research and applied astrosociologist, she investigates how individuals and societies react and respond to space exploration and astrobiological discoveries, and how those responses can reflect, predict, inform or mitigate social and cultural conflict here on earth.

TIMOTHY MEEHAN, PHD
Poster Sessions Chair

Timothy Meehan, PhD has extensive expertise in biomolecular analytical science and diagnostics. He has applied nanotechnology approaches to bioengineering through a collaborative effort with the Australian Stem Cell Centre in order to achieve large scale human synthetic whole blood production. Tim is a seasoned small business leader with experience in genetic diagnostics, microbial parasite detection and commercial New Space startup ventures. At Saber Astronautics he is working with NASA performing reduced gravity hardware flight tests and developing the next generation of autonomous fault detection and recovery solutions for spacecraft.

BOBBY FARLICE-RUBIO
Fairbanks Museum and Planetarium

100YSS Class Organizer

Bobby Farlice-Rubio has been a Science Educator at the Fairbanks Museum & Planetarium in St. Johnsbury, Vermont since 2003. There he teaches classes, to visiting students and the public at large, on a wide variety of subjects ranging from Astronomy and Natural Sciences to History and Culture. Mr. Farlice-Rubio may also be seen in his monthly "Star Struck" segments on WCAX-TV's news show "The :30," on which he presents the latest happenings in the field of Astronomy. Raised in Hialeah, Florida from Cuban and African-American roots, Bobby is also an avid musician who plays in a local band called Tritium Well, as well as his solo musical endeavor, Bobby & The Isotopes. He currently resides in Barnet, Vermont with his partner and their four children.

Event Moderators
ALIRES ALMON *Luncheon, Moderator*

KARL ASPELUND, PHD *Accelerating Creativity, Moderator*

JASON D. BATT *Science Fiction Stories Night, Canopus Awards, Moderator*

KATHLEEN TOERPE, PHD *Luncheon: Playing with the Stars Moderator*

Sponsors
MOSAIC GLOBAL TRANSPORTATION

NICHE MARKETING

SCHOLASTIC, INC.

THE JEMISON GROUP, INC.

CARWAY COMMUNICATIONS

PREFERRED DOCUMENT SOLUTIONS

PSAV

SANTA CLARA MARRIOTT

Special Thanks
ALIRES ALMON

JASON D. BATT

PAM CONTAG, PHD

SHADIA SADAQA

JULIEA ROBINSON-NELSON

YEABAN GOMPAH

MARK KETTLES

JOANN BROWN

MAE JEMISON, MD

RODNEY JOHNSON

TONY FADDOUL

TECHNICAL TRACKS

Session A
9:15-11:15 am

Becoming an Interstellar Civilization

Pushing the Outside of the Envelope: Thought Experiments as a Didactic for Becoming an Interstellar Civilization
Michael Waltemathe, Ruhr-University Bochum, Germany

Interstellar Embryos through Interstellar Teens: Negotiating Human Development through Education in an Off-world Context
Janet A. de Vigne, United Kingdom

Reconciling Science and Religion through Eternity
Rex Pay, USA

Engaging Religious Opposition to Human Space Exploration
Michael Waltemathe, Ruhr-University Bochum, Germany

Theology Re-imagined in Light of Extraterrestrial Intelligence
Peter Michael Hess, Theology for a Sustainable Future, USA

An Ethical Framework for Robotic Probe Design
David Burke, Galois, Inc.,USA

Data Communications and Information Technology

Not Enough Bandwidth: Communication on the Edge:
Jaym Gates, Uplift Aeronautics, USA

Communications' Requirements on a Generational Starship:
Alexander Sweetman, Metropolitan State University of Denver, USA

Managing Technical Debt in Space
Jesse Warden, Accenture,USA

Recent Progress in Interstellar Links Exploiting the FOCAL Space MissionKLT;
Claudio Maccone International Academy of Astronautics and INAF, Italy

Designing for Interstellar

Design Considerations for Interstellar Spacecraft
Thomas M Hancock, SAIC, USA

Solid Ground in Interstellar Space: Mission and Metaphor
Hank Hine, The Salvador Dali Museum, & Charles Hine. Hine Inc. Consulting, USA

Advanced Design Methodology for Deep Space Systems
Christopher Andrew Corner, Steersman Technology Ltd, UK

Leapfrog Across the Galaxy, to Intergalactic Travel
Peter Swan and Bruce Mackenzie, Mars Foundation, USA

We Are All Astronauts: Net Zero Design on Earth and Beyond
Tom Hootman, MKK Consulting Engineers, USA

Designing for Earth 2.0: New Design Philosophies and Processes
Karl Aspelund, University of Rhode Island, USA

Life in Space: Health, Astrobiology, Earth Biology, and Bioengineering

Virtual Human
Kurt Zatloukal, Medicine University of Graz

Nanomedicine in Space Applications
Frank Josef Boeh, NanoApps Medical, Inc., Canada

3D Printed Sharkskin for Enhanced Interstellar Wound Healing
Chelsea Magin, Michael Drinker, MiKayla Henry, Dylan Neale, Bradley Willenberg, Shravanthi Reddy, Gregory Schultz, Anthony Brenna

Neurocognitive Behavioral Health Monitoring During Space Missions
Curtis Cripe, USA

Astrosociology: Deviance Aboard a Long-Duration Spaceflight
Jim Pass, USA

Session B
1:15-3:15 pm

Becoming an Interstellar Civilization

A Roadmap to Interstellar Travel: Societal Challenges
Abigail Sherriff, International Space University, USA

The More Things Change . . . A 13th Century Model for Space-based Communities
Joelle Rollo-Koster and Karl Aspelund, University of Rhode Island, USA

Memory Issues for 100YSS Missions: Lifelogging, Social Media, and Virtual Time Capsules
Jo Ann Oravec, University of Wisconsin at Whitewater, USA

Tomorrowland: Recreating the Cultural Ambition for Interstellar Travel
Jaym Gates, Uplift Aeronautics,USA

Propulsion and Energy

Solar Flare in a Can: Magnetic Reconnection Propulsion
Hakeem Oluseyi

Photonic Laser Thruster Laboratory Demonstration towards Interstellar Photonic Railways
Young Bae, Y.K. Bae Corporation, USA

A Casimir-like Quantum Sail
Scott Smith, USA

Status of Solar Sail Propulsion within NASA: Moving toward Interstellar Travel
Les Johnson, NASA, USA

Electric Sail Propulsion for Exploring Nearby Interstellar Space
Les Johnson, Bruce Wiegmann, Mike Bangham, NASA, USA

Life in Space: Health, Astrobiology, Earth Biology, and Bioengineering

The Contributions of Occupational Science to the Readiness of Long Duration Space Exploration
Howard Koh

Starship Alpha: The Case for an Earth-based Protostarship
Manuel Richey, Honeywell International, Inc., USA

Who Gets to Go? On the Ethics of Sending Life on an Interstellar Journey
Michael Waltemathe, Ruhr-University Bochum, Germany

Finding Life on Earth 2.0: Philosophical and Theological Perspectives on an Anthropocentric Biocentrism
Michael Waltemathe, Ruhr-University Bochum, Germany

Evo-SETI Mathematical Theory about the Evolution of Life on Earth and Exoplanets
Claudio Maccone, International Academy of Astronautics (IAA) and INAF, Italy, Italy

Session C
3:30-4:30pm

Interstellar Space, Stars, and Destinations

Will There Ever Be Another Earth?
Dr. Margaret Turnbull, SETI Institute, USA

Testing the Boundaries of Planetary Habitability
Stephen Kane, San Francisco State University, USA

Goldilocks: A Find-Grained Exoplanet Taxonomy
Patrick Talbot and Dennis Ellis, Talbot Consulting, USA

Poster Session
6:00-7:00 pm

Gateway Space Station
John Blincow, GatewaySpaceStation.com

Asteroids R Us
James Baldini, USA

Commentary about Angelo Baymasecchi, S.J. (2015-21030), First Jesuit that Went on a Pilgrimage to the Dyson Sphere Surrounding Rigil Kentaurus
Oscar Garrido Gonzalez, Spain

Tennessee Valley Interstellar Workshop's Virtual Science Competition
Les Johnson and Joe Meany, NASA, USA

A Roadmap for Interstellar Travel
Abigail Sherriff and Chris Welch, International Space University, USA

The Alternative Star
Amalie Helen Sincalir, Lifeboat Foundation, USA

Plenary
Sessions

100 YEAR STARSHIP™

The Data Must Flow

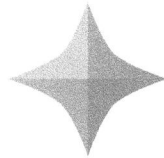

Computers Now and Ahead

Michael Flynn, Ph.D.

Fellow Stanford University and Maexeler Dataflow,

Silicon scaling

- 50 years of Moore's law
- Silicon wafer 30 cm in Diameter represents 700 1 cm squared chips
- Photolithography improves 2x every 2-3 years
- Transistor density improves as square and speed improves linearly
- Power density (and heat removal) set limit to frequency scaling in about 2005, and future interest went to architecture improvement and multiprocessors

Now

- 14 nm feature size
- 1-3 cm squared chip
- Memory chip 4Gb
- Processor chip 5 Billion transistors
- Currently move to 3D, stacking chips multiple memory or processor + memory
- Approaching physical limits but 50-100x improvement predicted by 2030.

Compute capability goes beyond technology

- 10x every 5 years
- So for we've strayed with conventional computing models (linear, 1D, control flow)
- More is possible with 2D spatial programming using dataflow.
- Results go to destination action (+,*, etc) not registers/ memory.
- Create a custom dataflow graph for each application and implement the graph in hardware.

Dataflow computing

- Easy to create hardware for an array of actions, custom interconnections are more difficult.
- But large FPGA enables flexible interconnections.
- With 10,000 actions simultaneously executing significant speedup results
- Open SPL openspl.org

Example: $X^2 + 30$

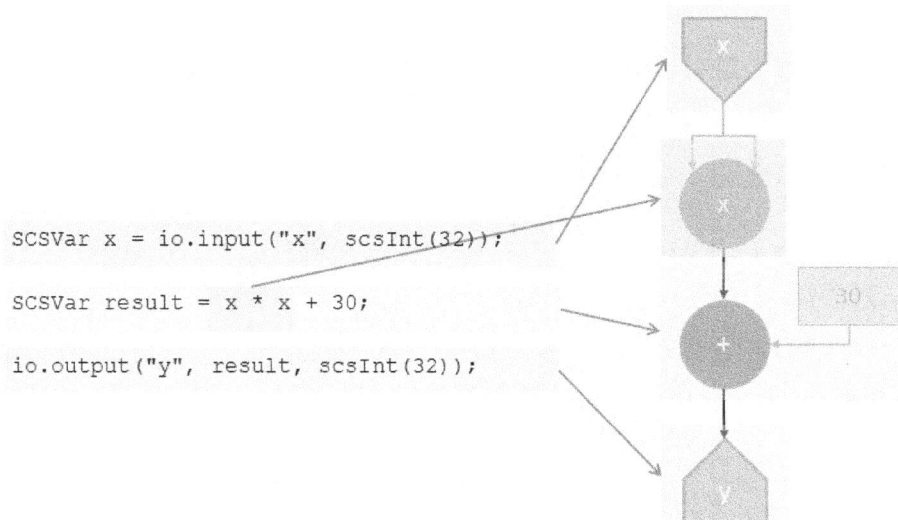

```
SCSVar x = io.input("x", scsInt(32));

SCSVar result = x * x + 30;

io.output("y", result, scsInt(32));
```

Intelligent computing is emerging

- Learning / reasoning is improving
- Robotics /language translation / speech recognition illustrate the emergence of intelligence.
- Still a way to go

The more distant promise: Quantum logic/computers

- Superposition of quantum states can provide exponential improvement over classical computation
- Qubits (superposition of "1" and "0") are to basic element with entangled states: N entangled qubits can potentially perform 2^N simultaneous operations
- Quantum states communicate faster than the speed of light

Beyond that is the fruit fly

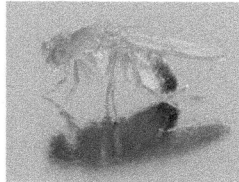

Fruit fly

- Length 2.5 mm; volume 2 mm^3

- 20 milligram; 1 month lifetime

- Vision: 800 units each w 8 photoreceptors for colors thru the UV (200k neurons); 10x better than human in temporal vision

- Also olfaction, audition, learning/memory

- Flight: wings beat 220x /sec; move 10 cm/sec; rotate 90^0 in 50 ms;

The Square Kilometer Array

Mmboneni Muofhe

Deputy Director Genral, Department of Sicence and Technology, South Africa

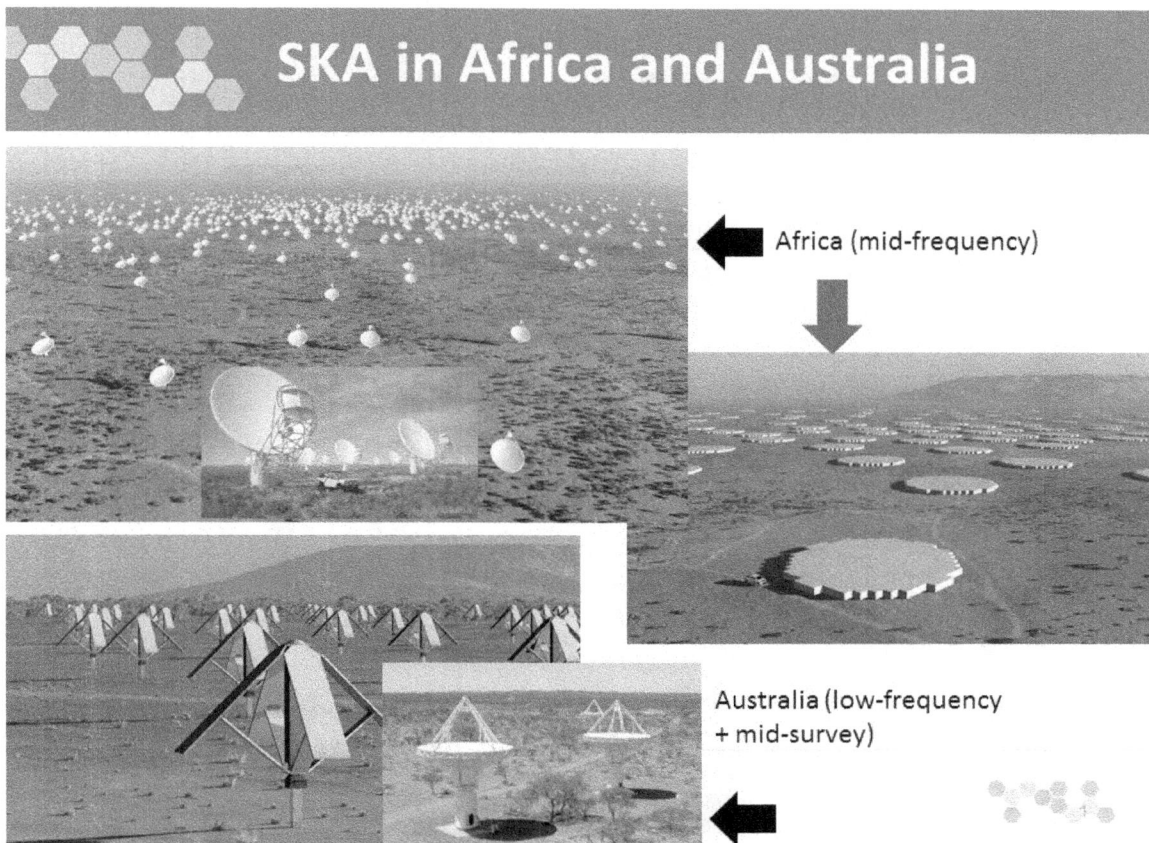

SKA in Africa and Australia

Africa (mid-frequency)

Australia (low-frequency + mid-survey)

Big Data Africa Programme

- The SKA will be a Big Data machine (exascale) – the ultimate prize will be the discoveries from the data

- Big Data phenomenon –collection of <u>data sets</u> so large and complex that it is difficult to process using traditional methods & technologies.

- Complexity = Function (volume, velocity, variety and veracity)

- Data has become a natural resource (data mining) – value addition is critical and therefore the new set of skills to do this

MeerKAT Data Processing

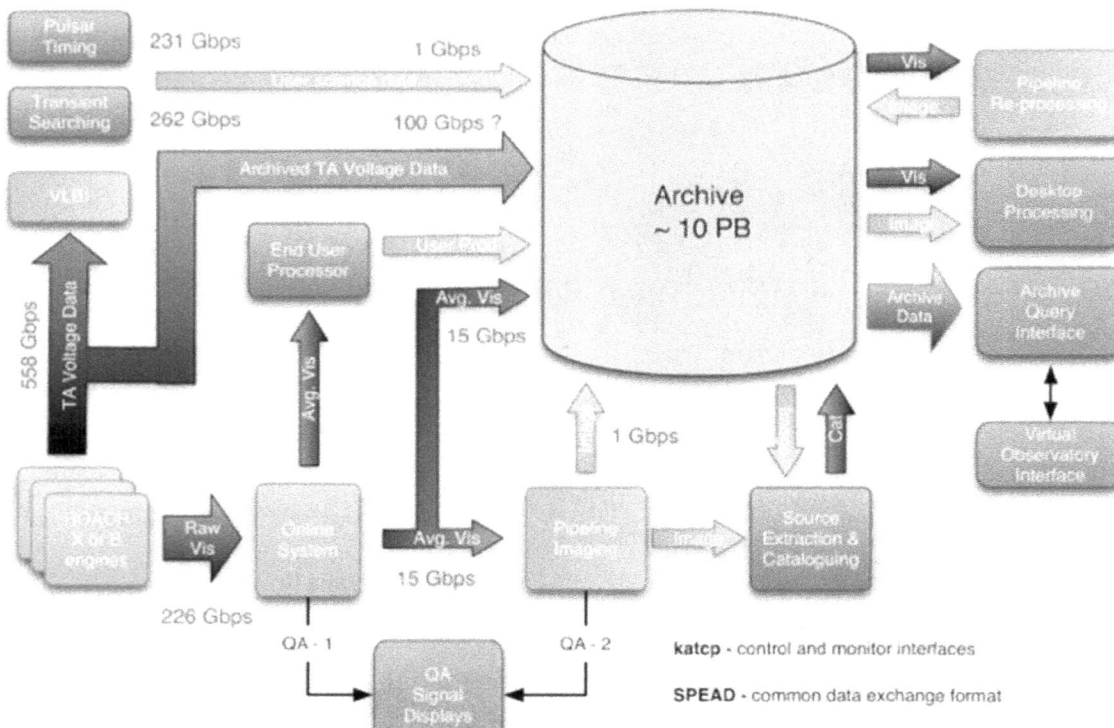

Big Data Initiatives

- Research Chair in Big Data at UCT

- Recently launched a R50m Inter-varsity Institute for Data-Instensive Astronomy (IDIA) on 3 September 2015 – Universities involved include UCT, UCT and NWU

- IBM established a R 700m Research Lab with a big focus on Big Data at Wits University.

- CISCO – set up a R 50m optic fibre research centre at Nelson Mandela Metropolitan University – to find cost-effective solutions for data transport

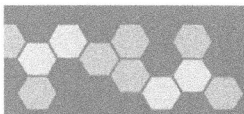

100 YEAR STARSHIP™

Radical Leaps

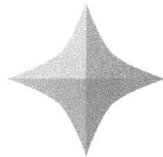

Space Mineral Resources:
Prior to, During, and After Interstellar Flights

Peter A. Swan, PhD

Member, International Academy of Astronautics

President, Member BofD's, International Space Elevator Consortium

Director, Space Way Research Institute

Space Mineral Resources:
Prior to, During & After Interstellar Flights

Peter A. Swan, Ph.D.   Member, International Academy of Astronautics
          President, Member BofD's, International Space Elevator Consortium
          Director, Space Way Research Institute

Cathy Swan, Ph.D. Member, International Academy of Astronautics, President, SouthWest Analytic Network, Docent, Phoenix Art Museum.

100 Year StarShip Symposium

Note: many images from Heinlein Prize Trust and Excalibur Exploration

10/31/2015

Image from IAA Study

49

Vision -- *Leverage the Phenomenal Resources – "Along the Way".*

Hypothesis:
For Interstellar travel, mining and
processing raw resources is essential:

- Develop Tools, Practice and Grow Skills in Earth-Moon Ecosystem
- Logistics support for Interstellar Flight will include processed SMRs
- Logistics re-supply along the way will require special designs
- Once arrival in target Solar System, will need resource replenishment

Today's Topics

- Introduction
- IAA Study
- Logistics – Develop Tools

- Logistics – Store Supplies
- Logistics – Examples
- Conclusions & Questions

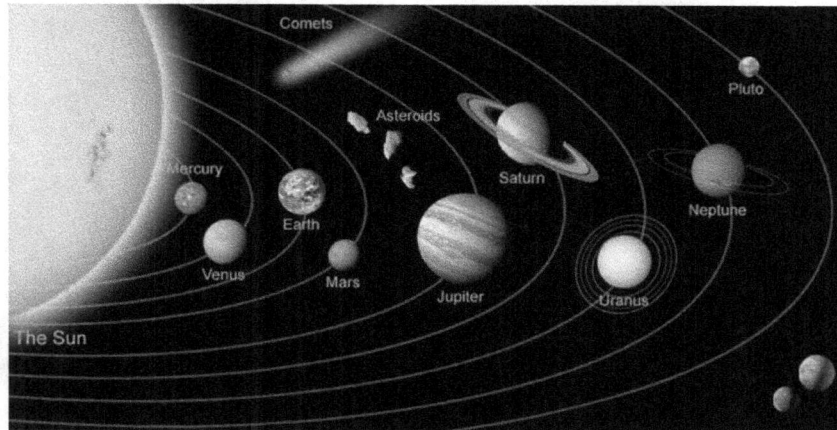

Image from 100 YSS

10/31/2015   5

> Any sufficiently advanced technology is indistinguishable from magic.
>
> Arthur C. Clarke
>
> One man's 'magic' is another man's engineering. 'Supernatural' is a null word.
>
> Robert A. Heinlein

Opening Interstellar Flights Starts at Earth Moon L-1 – A Vision

Images by chasedesignstudios.com

Looking to the Future

Today's Topics

- Introduction
- IAA Study
- Logistics – Develop Tools
- Logistics – Store Supplies
- Logistics – Examples
- Conclusions & Questions

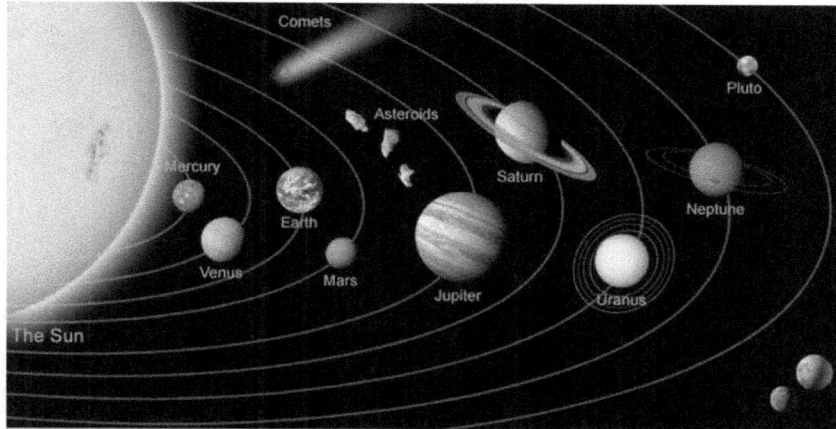

Image from 100 YSS

10/31/2015

9

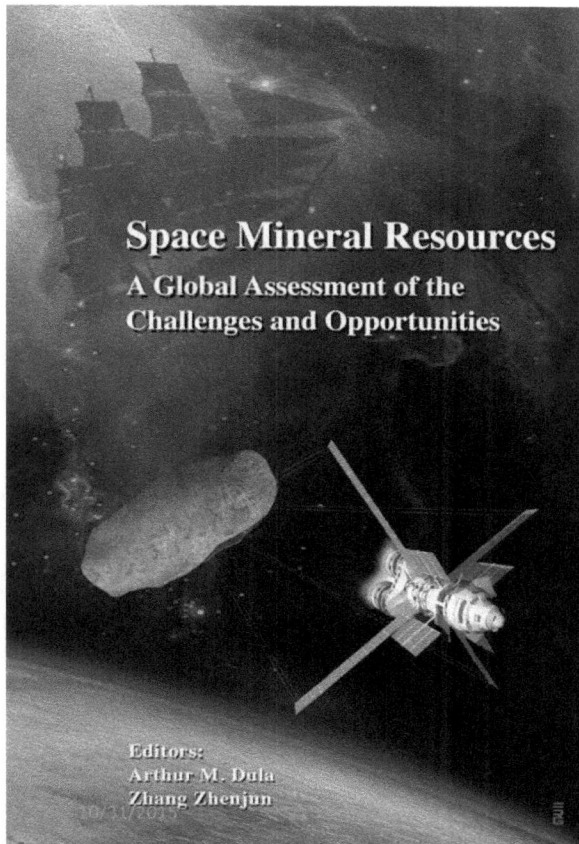

International Academy of Astronautics

Editors

Arthur M . Dula
Zhang Zhenjun

Research and
Editorial Staff:

Dr. Peter Swan
Mr. Roger Lenard
Dr. Cathy Swan
Mr. Brad Blair
Ms. Anat Friedman
Ms. Shirazi Jaleel-Khan
Mr. Jason Juren

Authorized by the
IAA Scientific Commission
October 2012

Published (385 pages)
September 2015

10

Purpose of the SMR Study

- To provide, in one document, the current state of the art of the technology, economics, law & policy related to Space Mineral Resource opportunities and to make recommendations for moving forward.
- To provide a logical, systematic and practical road map to promote and encourage near term evaluation, development and use of space mineral resources.
- No comprehensive summary of the current literature on this subject is now publicly available. This IAA study is the first comprehensive study of the subject; and, thus it should be of significant value to its development for the benefit of humanity.

From Robert Heinlein Trust

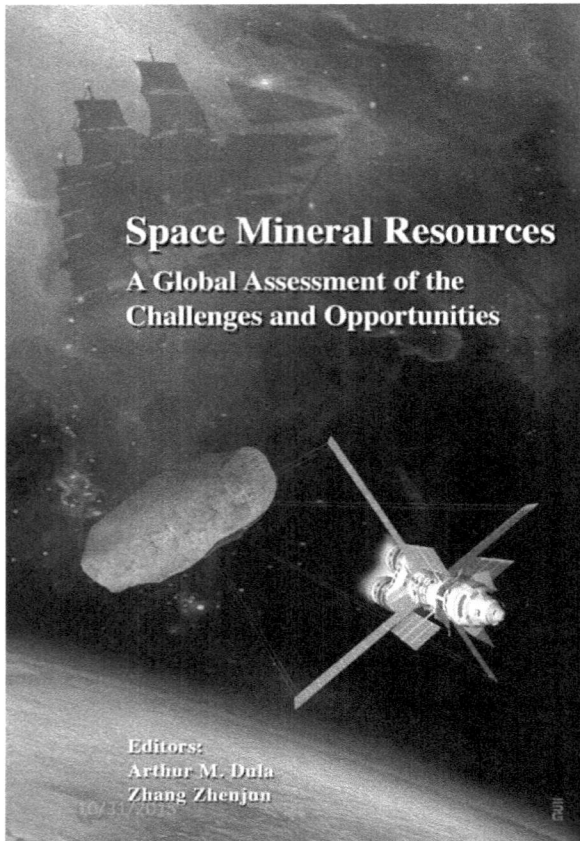

Space Mineral Resources
A Global Assessment of the Challenges and Opportunities

Editors:
Arthur M. Dula
Zhang Zhenjun

International Academy of Astronautics

Recommendation

Develop technologies, corporations and government relationships to support the following action plan.

Phase One: Initiate the business infrastructure on Earth   2014-2020

Phase Two: Execute prototype flights to potential asteroids as well as testing hardware in LEO   2015-2022

Phase Three: Initiate mining operations with sale of product   2018-2029

Expected Results: Selling water at the Earth-Moon Lagrangian Point #1.

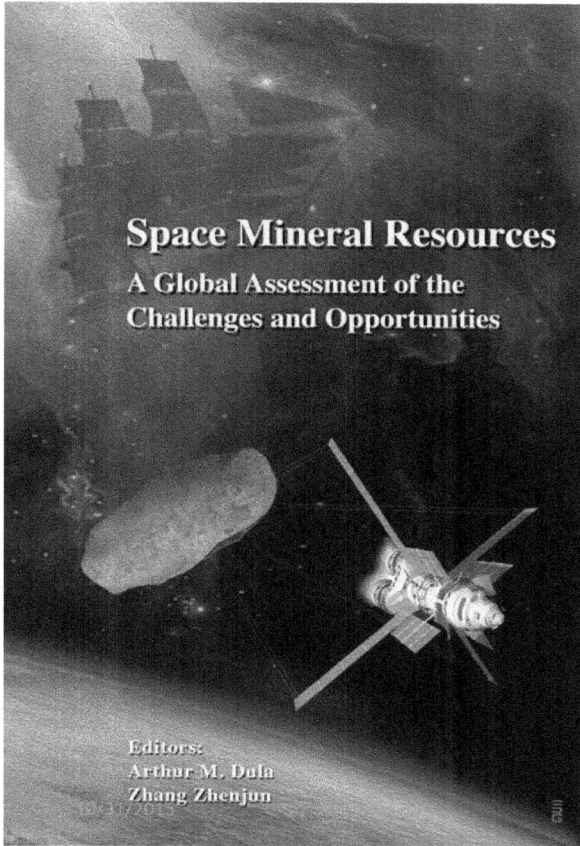

Space Mineral Resources
A Global Assessment of the Challenges and Opportunities

Editors:
Arthur M. Dula
Zhang Zhenjun

International Academy of Astronautics

The participants acknowledge that this study is only a beginning.

A second IAA study on this subject has already begun.

The participants hope that their work will be of value to those who follow.

13

The Case for Space Solar Power

ISBN : 978-0-9913370-0-2
Publication Date: January, 2014
Page Count: 488
Format: Hardcover - $49.95
eBook - $9.95

Space Elevators:
An Assessment of the Technological Feasibility and the Way Forward

ISBN : 978-2917761311
Publication Date: January, 2014
Page Count: 349
Format: Hardcover - $29.95
eBook - $9.95

Virginia Edition
SCIENCE DECK
Advancing Robert A. Heinlein's vision of humanity in space.
10/31/2015

Books for sale at the Heinlein Display
Ordered online at: www.virginiaedition.com

14

Essence of SMR

- Enhance the human condition on Earth
  - Provide Jobs
  - Re-invigorate education in science, technology, enginering, and mathematics [STEM]
  - Stimulate innovation
  - Provide a vision for return on investment [ROI] from space
  - Stimulate commercial investors for space activities
  - Provide avenues for commercial expansion into our solar system
- Initiate Commercial Movement into the Solar System
  - Provide a vision for movement off-planet
  - Provide profit motive for movement into space
  - Provide commercial products to national space exploration programs
  - Enable space exploration
  - Enable space colonization
  - Enable solar power satellites

10/31/2015   15

Cosmic Study Outline

To provide a logical, systematic and practical road map to promote and encourage near term evaluation, development and use of space mineral resources (SMR)

Executive Summary
1 **Introduction**
2 **Mining of Space Resources**
3 **Market Approach**
4 **SMR Roadmaps**
5 **Analysis of Systems**
6 **Modeling and Analysis**
7 **Policy and Legal**
8 **Finding, Conclusions and
    Recommendations**

10/31/2015

Executive Summary

"Don't undertake a project unless it is manifestly important and nearly impossible." Edwin Land, quoted in the Coral Reef Alliance letter, March 30, 2011. www.coral.org

Executive Summary:

- The exploitation of space mineral resources is becoming a commercial space endeavor for the benefit of humanity and profit
- The question on the table is not "how" to leverage space minerals resources; but, "how best" to leverage them
- Preliminary economic conclusions include (1) architectures based upon returning precious metals to terrestrial markets alone appears to be a non-starter, (2) the existence of in-space customers for propellants, consumables, structural materials, and shielding could make asteroid mining economically feasible, and (3) longer-term hybrid architectures with both terrestrial and in-space customers could become feasible as costs drop and market size increases.

10/31/2015   17

Major Conclusion & Finding

Major Conclusion: The process of mining water from asteroids, the Moon or Mars will ensure that the key elements are available at the spaceports of the future. Water will ensure that human exploration will expand beyond low Earth orbit with the profit motive driving the exploitation of resources.

Principle Finding: SMR ventures cannot wait for government programs to lower technological and programmatic risks. Commercial ventures must determine the optimum path for commercial success and aggressively lead the way beyond LEO. During the first half of the 21st century, space leadership will come from commercial enterprises and not depend upon government space programs.

10/31/2015   18

Cosmic Study Conclusions

The conclusions from this study fall into a few distinct categories:

- **Legal:**   The space elevator can be accomplished in todays arena!
- **Technology:** It can be accomplished with today's projection of where materials science and solar array efficiencies are headed.
- **Business:**   This mega-project will be successful for the investors
- **Culturally:** This project will drive a renaissance on the surface of the Earth with its solution to key problems, stimulation of travel throughout the solar system, and inexpensive and routine access to GEO and beyond.

10/31/2015   19

Today's Topics

- Introduction
- IAA Study
- Logistics – Develop Tools

- Logistics – Store Supplies
- Logistics – Examples
- Conclusions & Questions

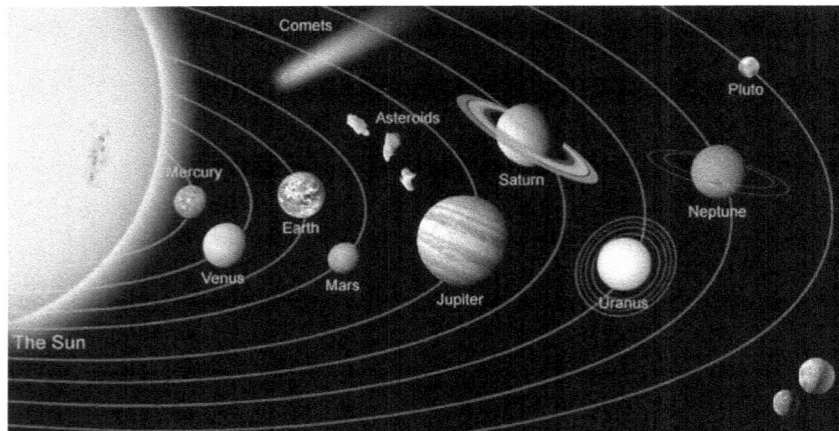

Image from 100 YSS

10/31/2015   20

Propellant Flow (quantity)

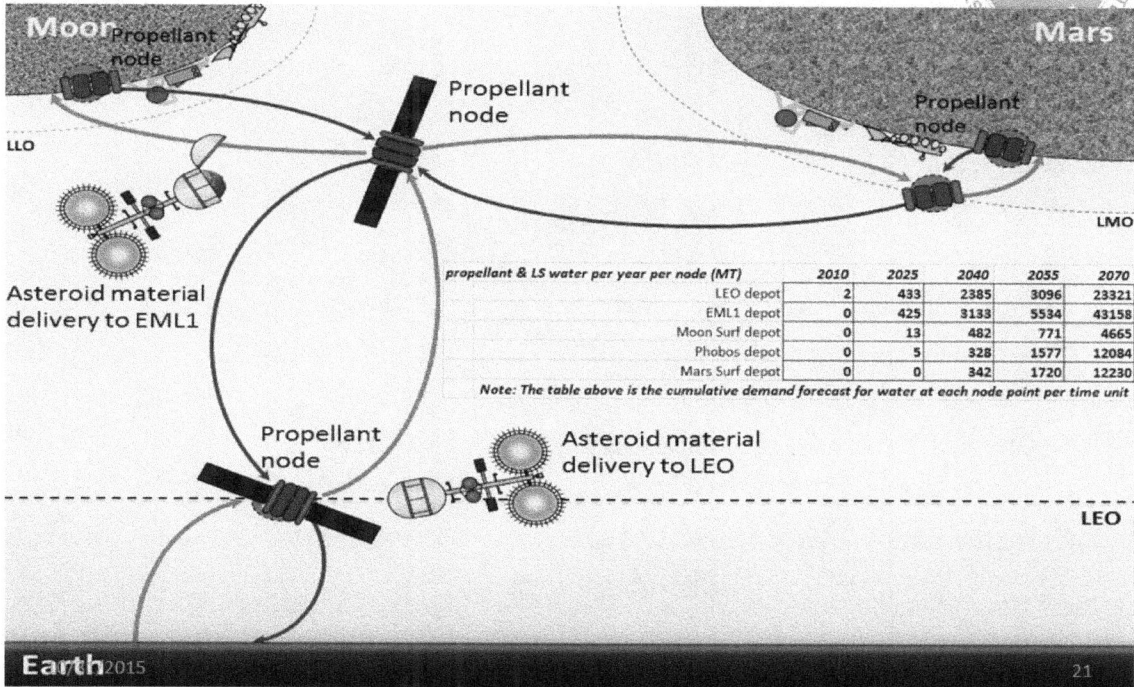

propellant & LS water per year per node (MT)	2010	2025	2040	2055	2070
LEO depot	2	433	2385	3096	23321
EML1 depot	0	425	3133	5534	43158
Moon Surf depot	0	13	482	771	4665
Phobos depot	0	5	328	1577	12084
Mars Surf depot	0	0	342	1720	12230

Note: The table above is the cumulative demand forecast for water at each node point per time unit

Earth Moon Infrastructure Velocity Requirements:

Delta-Vs in Earth's Neighborhood [Mankins, 2012].

Bummer – approximately 10,000 kilometers per second of energy required to get to LEO

10/31/2015

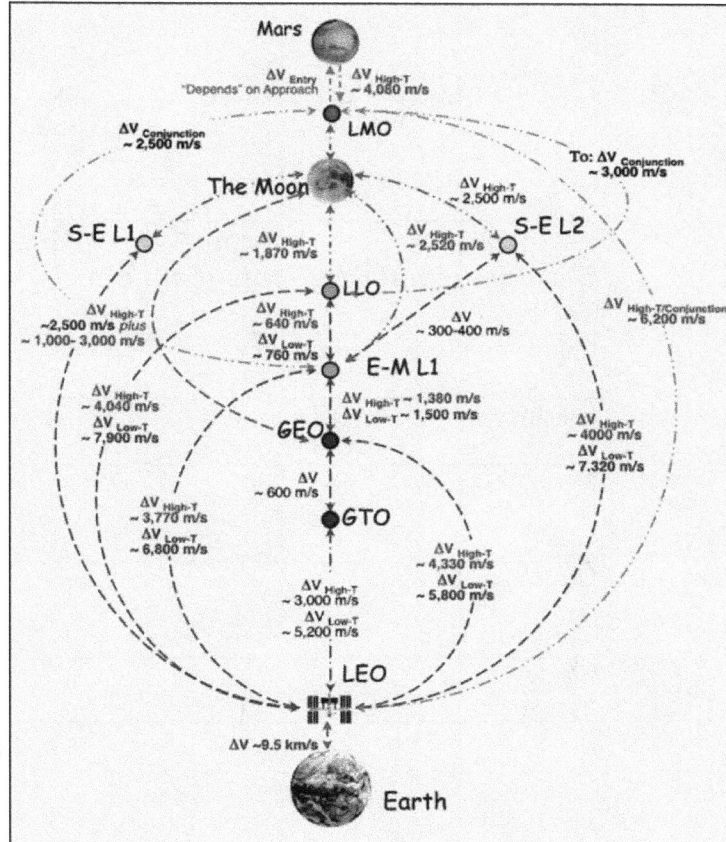

Billionaire Space Investors

rank	name	age	net worth	source	space investment
19	Jeff Bezos	49	$25.20	Amazon	Blue Origin
21	Sergey Brin	40	$22.80	Google	Google Lunar X Prize
20	Larry Page	40	$23.00	Google	Google Lunar X Prize, Planetary Resources
53	Paul Allen	60	$15.00	Microsoft	SpaceShipOne, SETI telescope array
138	Eric Schmidt	58	$8.20	Google	Planetary Resources
272	Sir Richard Branson	63	$4.60	Virgin Group	Virgin Galactic
527	Elon Musk	42	$2.70	PayPal, Tesla Motors	SpaceX
831	Guy Laliberte	53	$1.80	Cirque du Soleil	Visitor to ISS
922	K Ram Shriram	56	$1.65	Google	Planetary Resources
1031	Ross Perot, Jr.	54	$1.40	Oil & Gas	Planetary Resources
			$106.35	Total Net Worth	

Commercial Space Companies

Lunar Developmenmt	Asteroid Development	Mars Development	Space Tethers and Elevators
Golden Spike	Planetary Resources	SpaceX	Tethers Unlimited
Shackleton Energy, Co.	Deep Space Industries	Inspiration Mars	Liftport
Moon Express	Excalibur Exploration	Mars One	International Space Elevator Consortium
Excalibur Almaz			Japanese Space Elevator Association
Bigelow Aerospace			

Today's Topics

- Introduction
- IAA Study
- Logistics – Develop Tools

- Logistics – Store Supplies
- Logistics – Examples
- Conclusions & Questions

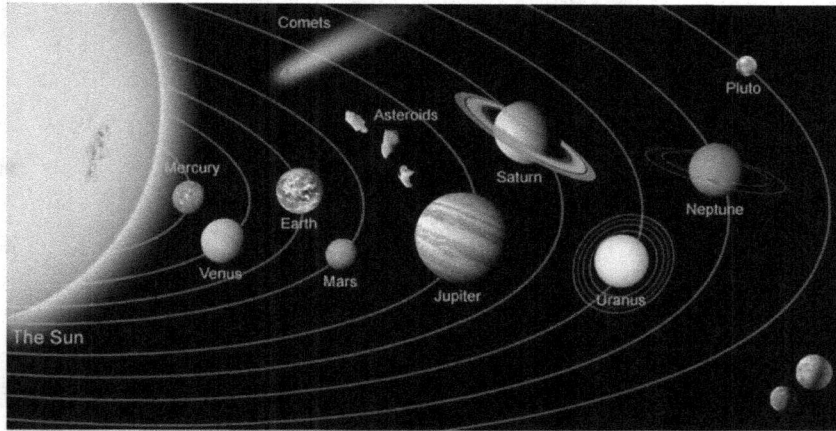

Image from 100 YSS

10/31/2015   25

Cruise Ships
Precursors to Interstellar

Name	Passengers	Crew	Total	Mass (tons)	Cost
Disney Wonder	2700	900	3600		
Carnival Legend	2124	930	3054	88,500	
Carnival Elation	2052	920	2972	70,367	
RC Oasis of the Seas	5400	2100	7500	225282	$ 1.2 billion US
RC Allure of the Seas	6296	2100	8396	225282	

7,000 people with 225,282 Metric Tons For seven days

Allure of the Seas from Wikipedia

10/31/2015

61

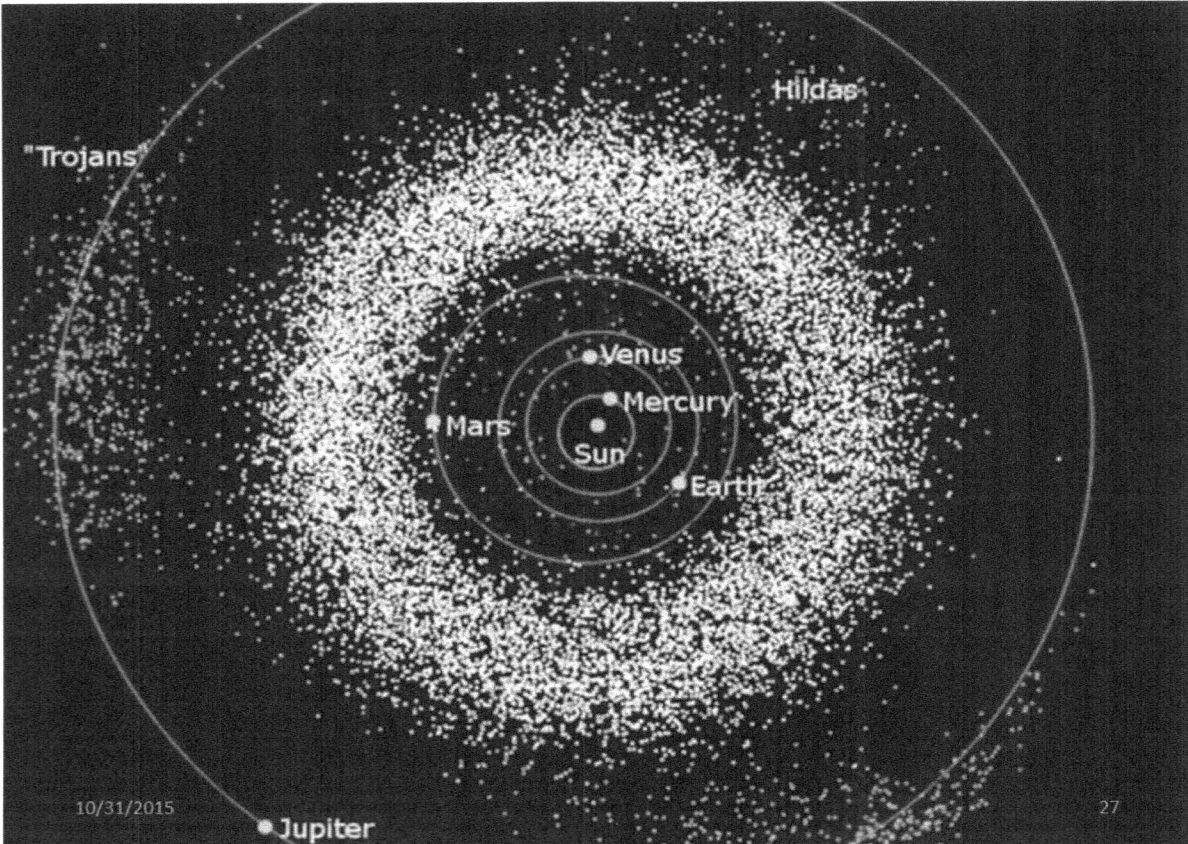

Asteroid 216 Kleopatra

An asteroid roughly the size of New Jersey (135 x 58 miles) located between Mars and Jupiter.

- Composed mostly of nickel (10%) and iron (88%)

- Kleopatra is not completely solid - its surface is about one meter of metal dust and loosely consolidated rubble, although its core may contain large (cubic miles) solid-metal nodes.

- 2003 world steel production was 854.1 million metric tons. At the world market price of $482/ton this is $320 billion.

- At this rate, 10% of Kleopatra would be worth over $200 trillion.

- Kleopatra alone has more material wealth than all of humanity has produced on Earth.

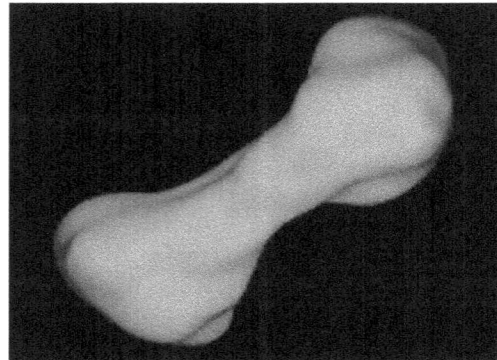

Capture and Move Resources

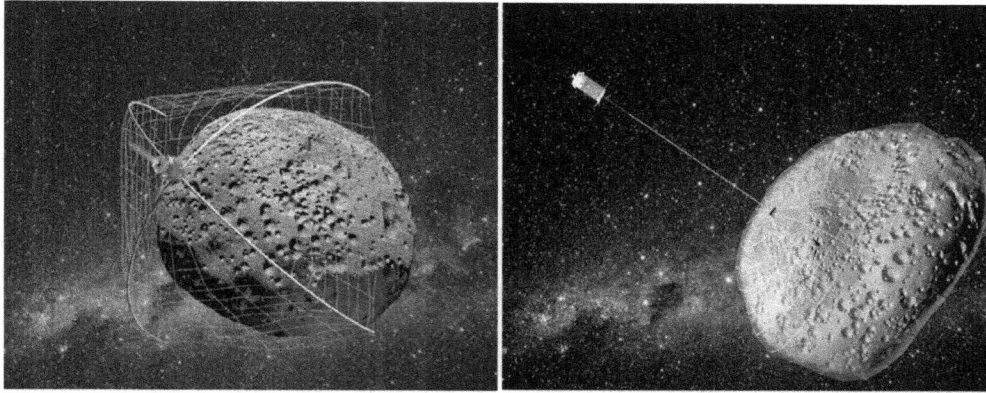

WRANGLER system - Asteroid Capture [Tethers Unlimited]

Lunar Resources

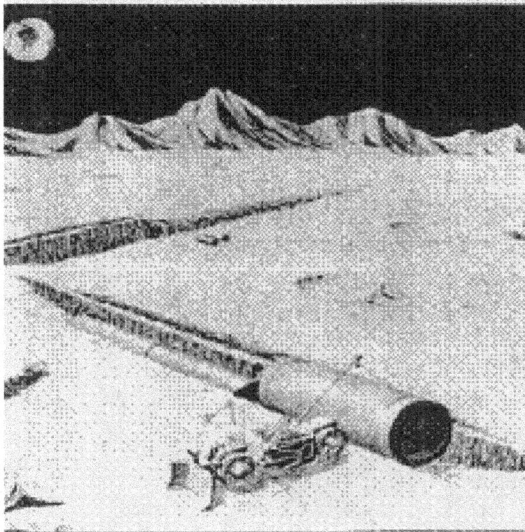

From Project Horizon Study, US Army, 1959]

- Resources are boundless on the Moon
- Low gravity enables movement by multiple methods
- One key development will be a Lunar Elevator for moving supplies to the EM L-1 Manufacturing arena
- A key resource is Lunar water for living and fuel.

Sell Water at Spaceports

	Mass kg	Payload [water] Mass kg	Price per kg	Price per metric ton
On–Pad	1,462,836			
In LEO	53,000	13,250	$7,547	$ 7.5 million
At GEO	21,000	5,300	$ 18,868	$ 18.9 million
At Earth-Moon EML-1	20,000	5,000	$ 20,000	$ 20.0 million
On Asteroid surface*	14,000	3,500	$ 28,571	$ 28.6 million
On Lunar Surface**	7,314	1,828	$ 54,705	$ 54.7 million
On Mars Surface***	13,200 (Insertion)	1,320 [surface]	$ 75,757	$ 75.8 million

"Water will be the Currency of Space!"

10/31/2015

31

Today's Topics

- Introduction
- IAA Study
- Logistics – Develop Tools
- Logistics – Store Supplies
- Logistics – Examples
- Conclusions & Questions

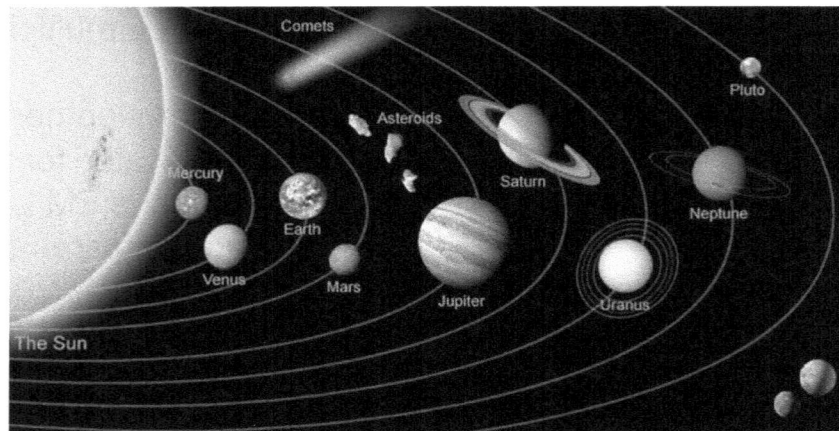

Image from 100 YSS

10/31/2015

32

Volatile Materials in Asteroids

- 10 to 50% of known large asteroids are likely hydrated CI-CM-like, possibly parent bodies of CI-CM chondrite meteorites

- CI-CM chondrites are typically 10-20% water by weight in the form of hydrated minerals with significant other volatile (e.g. CO_2) content that can be thermally extracted

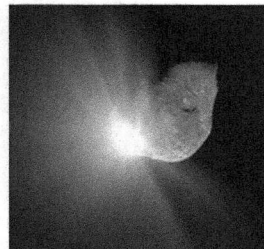

- CI-CM materials are friable and may be in rubble piles with regolith or in blocks on asteroids.

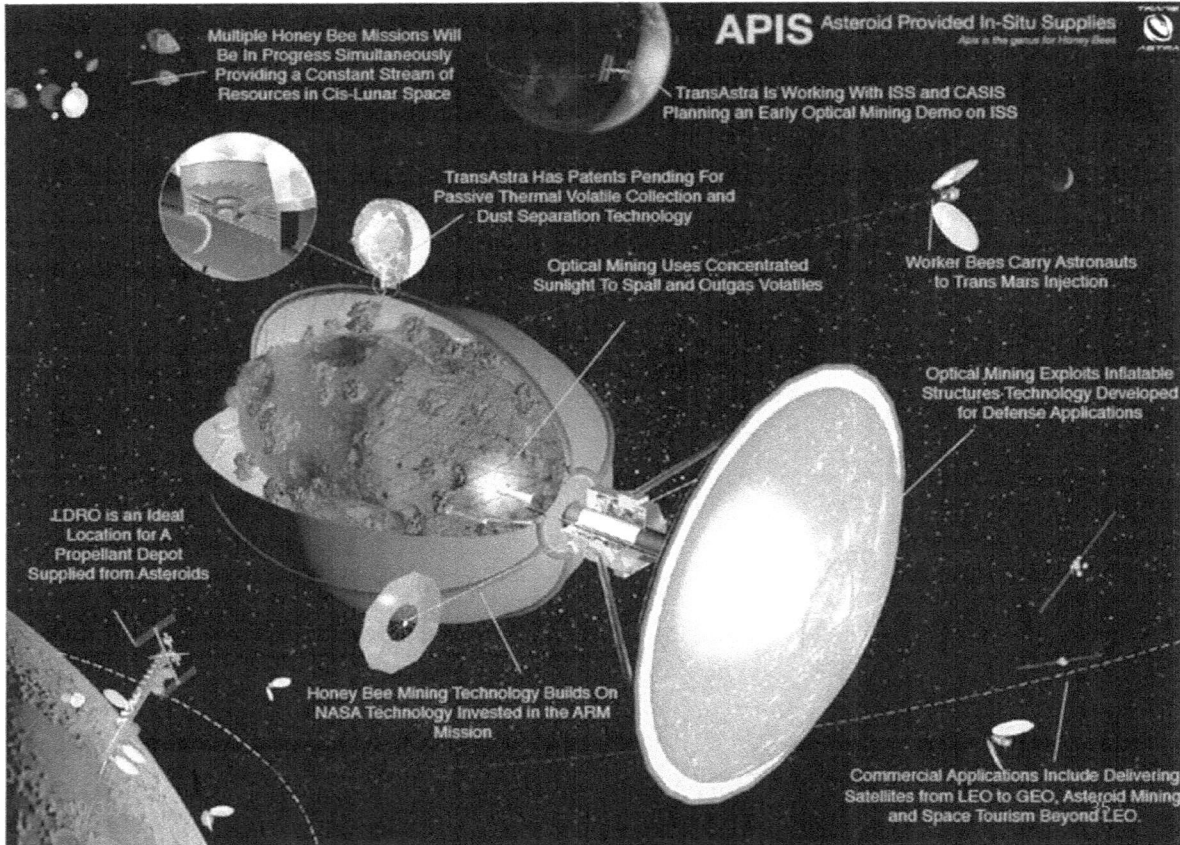

At Our Next Star: Need Resource Acquisition: A Vision

Images by chasedesignstudios.com

For Interstellar travel, mining and processing raw resources is essential:

- Develop Tools, Practice and Grow Skills in Earth-Moon Ecosystem
- Logistics support for Interstellar Flight will include processed SMRs
- Logistics re-supply along the way will require special designs
- Once arrival in target Solar System, will need resource replenishment

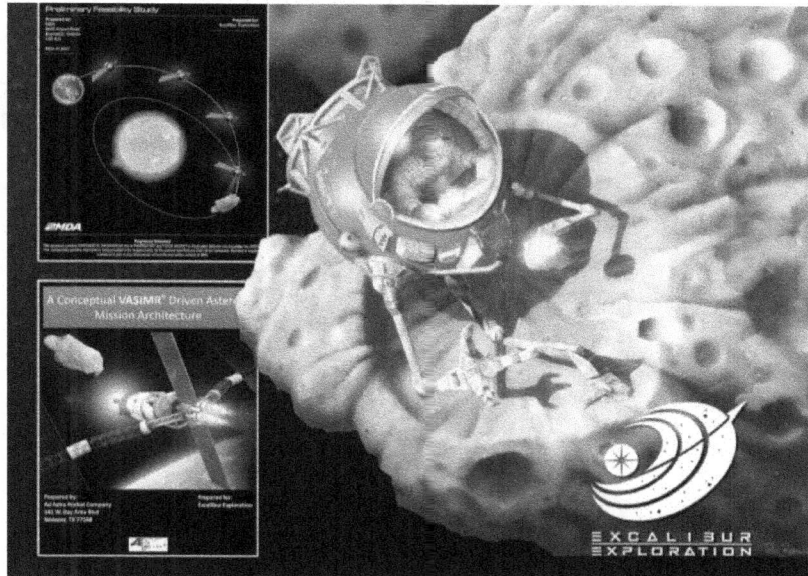

Today's Topics

- Introduction
- IAA Study
- Logistics – Develop Tools
- Commercial Ventures
- Logistics – Store Supplies
- Logistics – Examples
- Conclusions & Questions

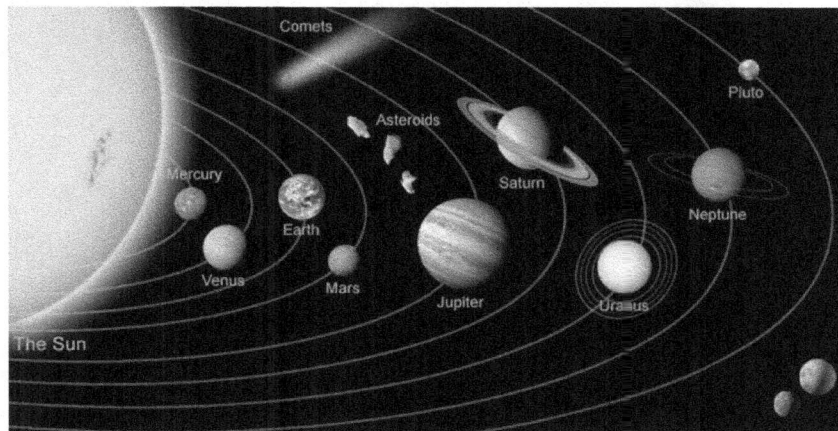

Image from 100 YSS

10/31/2015

38

Finding 1: Technological risk reduction and engineering design are within State of Art

The mining of asteroids and lunar regolith is within the current state of the technical art. The extrapolation of Earth-based mining seems to be a one-for-one trade with some significant alterations due to vacuum, low gravity and temperature extremes. Many proposed solutions have been suggested and tested [on Earth] leading to positive conclusions on this topic.

Known Near-Earth Asteroids
1980-Jan through 2012-Dec

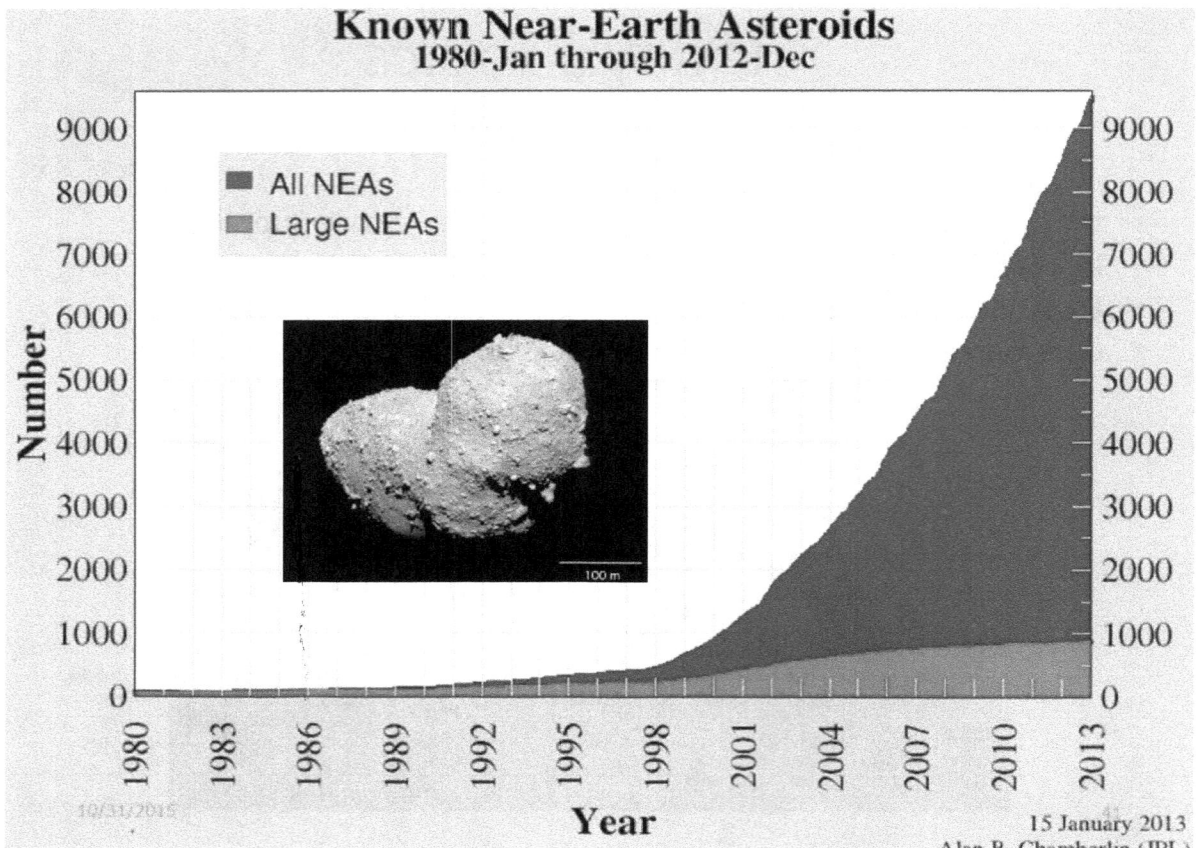

15 January 2013
Alan B. Chamberlin (JPL)

Abundances of ferrous and precious metals in asteroids

Metal	(1) Abundance in metal of average LL-chondrite asteroid	(2) Abundance in good" iron asteroid (90th percentile in Ir, Pt)	(3) Abundance in "best" iron asteroid (98th percentile in Ir, Pt)
Ferrous metals:			
Fe	63.7%	81-94%	82-94%
Co	1.57%	0.46-0.80%	0.43-0.75%
Ni	34.3%	5.6-18.0%	5.4-16.5%
Precious metals:			
Ge	1020 ppm	0.06-70 ppm	0.05-35 ppm
Re	1.1 ppm	1.1 ppm	2.4 ppm
Ru	22.2 ppm	20.7 ppm	45.9 ppm
Rh	4.2 ppm	3.9 ppm	8.6 ppm
Pd	17.5 ppm	2.6 ppm	1.2 ppm
Os	15.2 ppm	14.1 ppm	31.3 ppm
Ir	15.0 ppm	14.0 ppm	31.0 ppm
Pt	30.9 ppm	28.8 ppm	63.8 ppm
Au	4.4 ppm	0.16-0.70 ppm	0.06-0.6 ppm

Calculated from data given by Müller et al. 1971, Buchwald 1975, Malvin et al. 1984, Rasmussen et al. 1984, Hirata and Masuda 1992, and Morgan et al. 1992.

10/31/2015　　　　　　　　　　　　　　　　　　　　　　　42

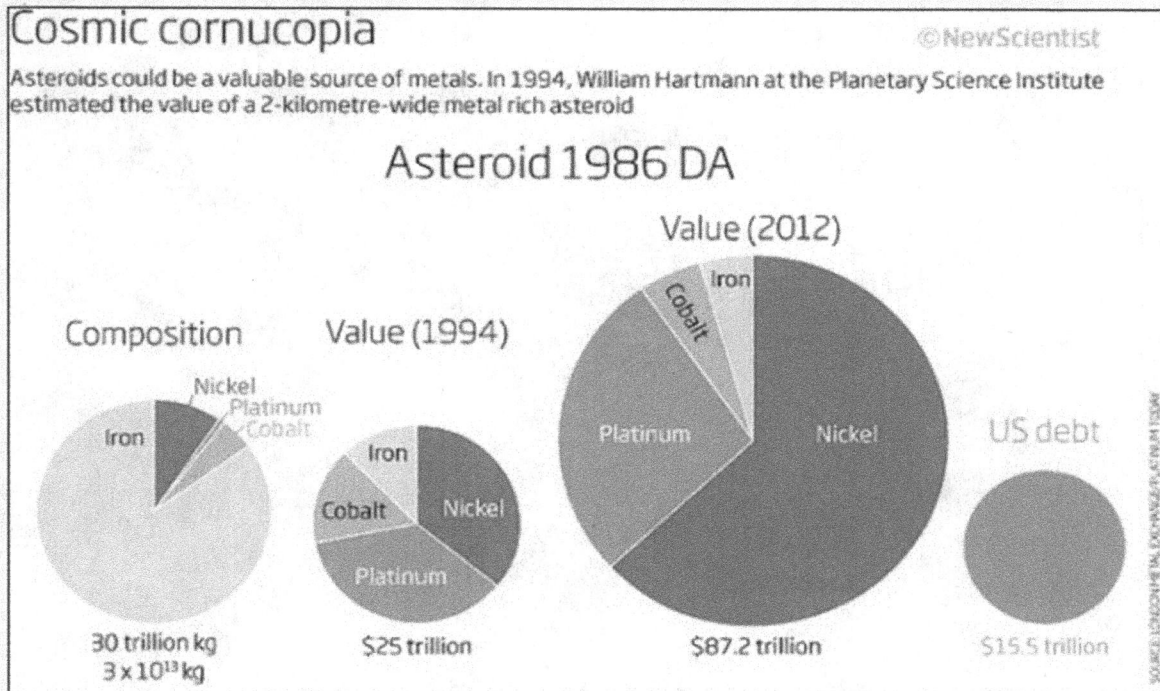

10/31/2015　　　　　　　　　　　　　　　　　　　　　　　43

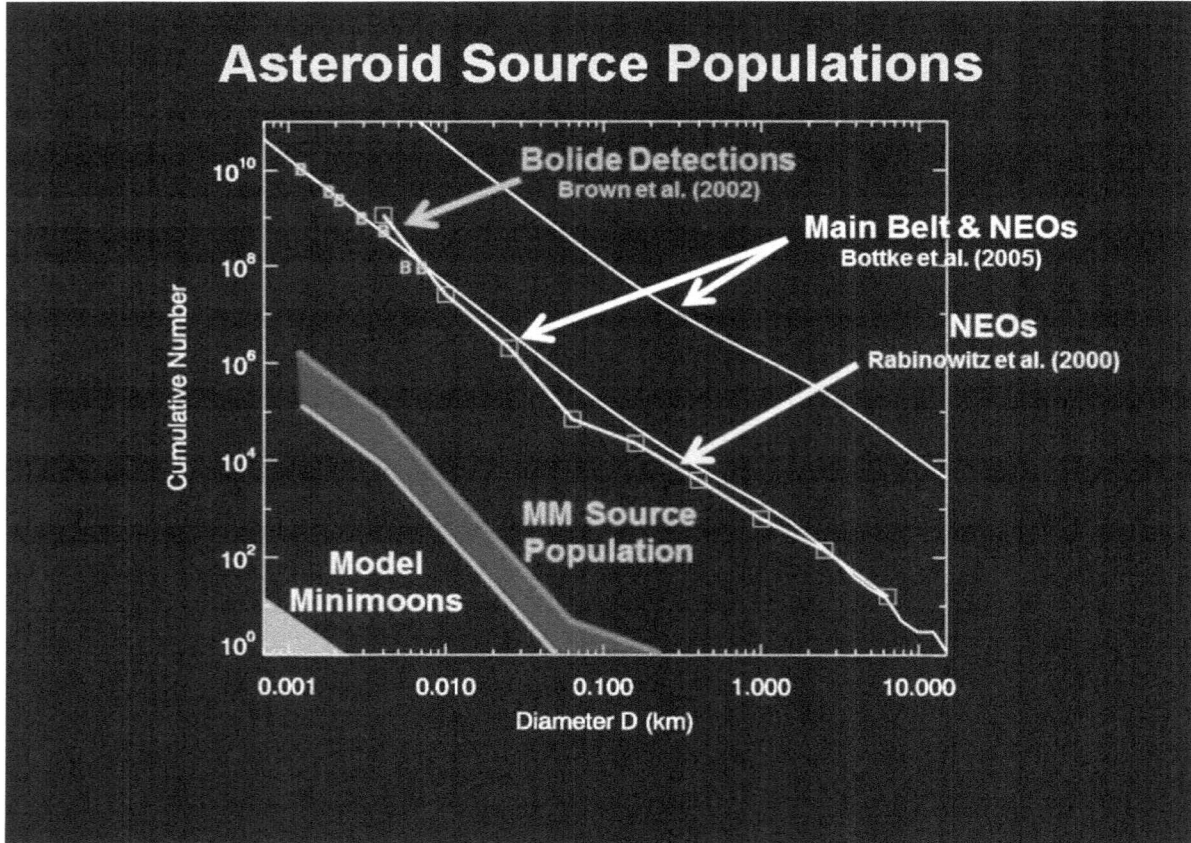

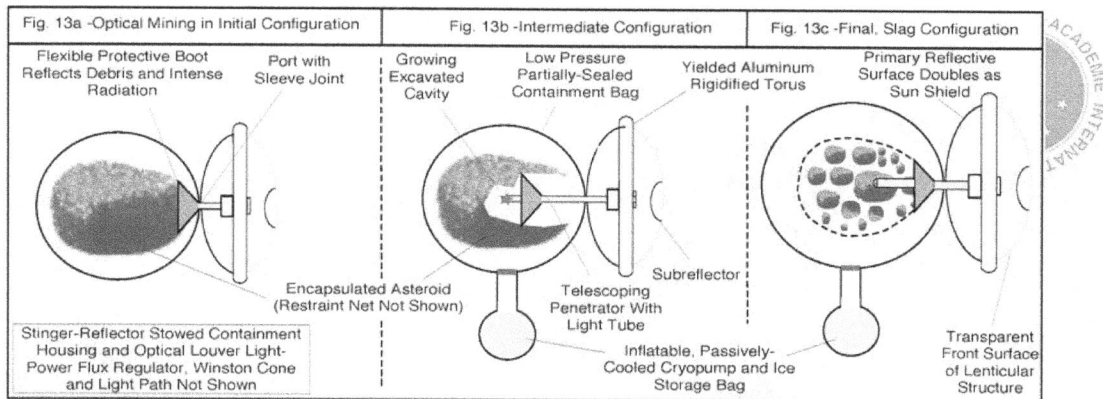

Fig. 13a -Optical Mining in Initial Configuration	Fig. 13b -Intermediate Configuration	Fig. 13c -Final, Slag Configuration

The Trans Astra Corporation has developed a new patent pending invention called "optical mining" – a way to excavate and process asteroids using sunlight. Space News called this invention "a possible game changer for space exploration." *The ability to tap mega-amounts of water from asteroids could be used directly as propellant in solar thermal rockets to provide inexpensive space transportation."* **The ground research into this new concept is funded by private capital and government contracts from NASA and others.**

Fig. 2 – Figure from a Paper in *Nature* Showing A Large Population of NEOs Far More Accessible Than the Moon (Binzel 2014)

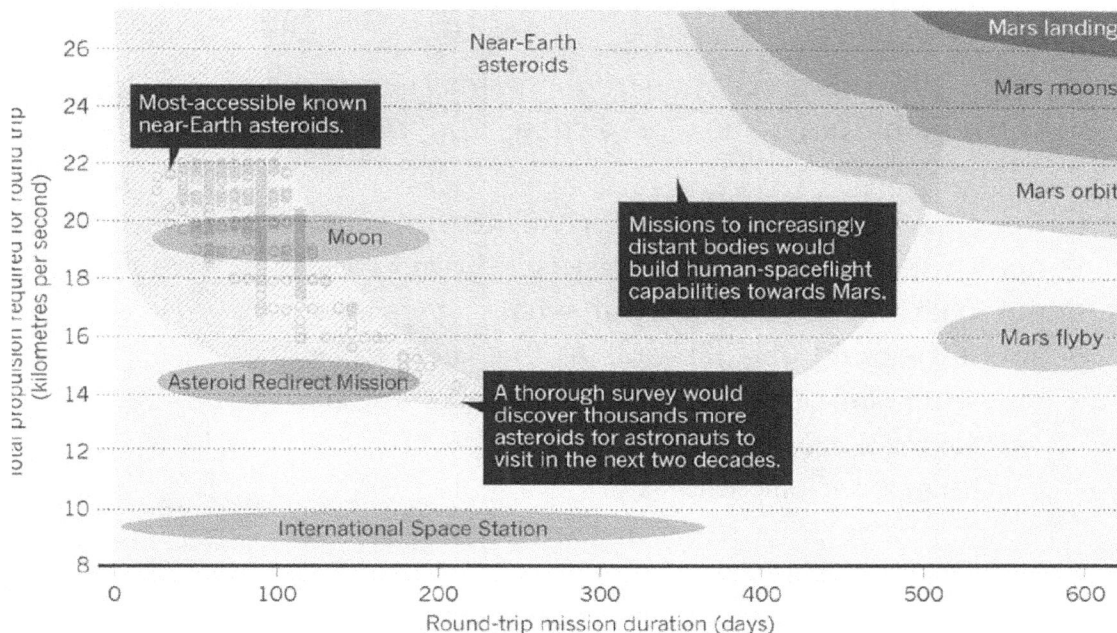

71

Why – Mining in Space?

- *When I awoke this morning, I looked around and saw nature in crises*
- If I expand this perception from my small community to the global population, predictions from the Club of Rome seem real.*
- Opening the resources of space will not only change our lives; it will change our destiny. The question is not what can I do about it; but, what can we all do about the multitude of problems that seem to be overwhelming our world. The answers seem simple:
  - Change the equation.
  - Change the assumptions.
  - Increase the resources and produce innovation, jobs and wealth along the way

*Donella H. Meadows, Dennis L. Meadows, Jörgen Randers, and William W. Behrens III, "Limits to Growth," 1972.

10/31/2015

Technological Approaches That Can Search for Life Close Up in Exotic Destinations

Seth Shostak, Ph.D.

SETI Institute

Why bother?

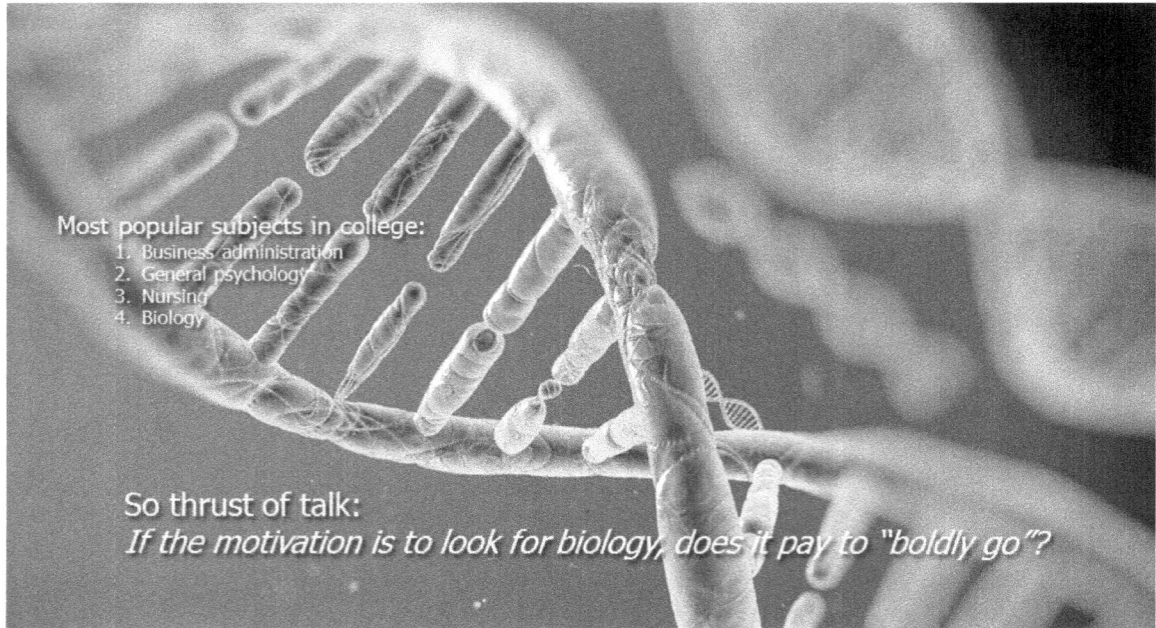

Most popular subjects in college:
1. Business administration
2. General psychology
3. Nursing
4. Biology

So thrust of talk:
If the motivation is to look for biology, does it pay to "boldly go"?

There are 10^{22} stars in the visible universe

(Milky Way: one trillion planets)

If 10% of stars have inhabited worlds, then nearest life is ~6 light-years away.

But finding that life ...?

There are two approaches:

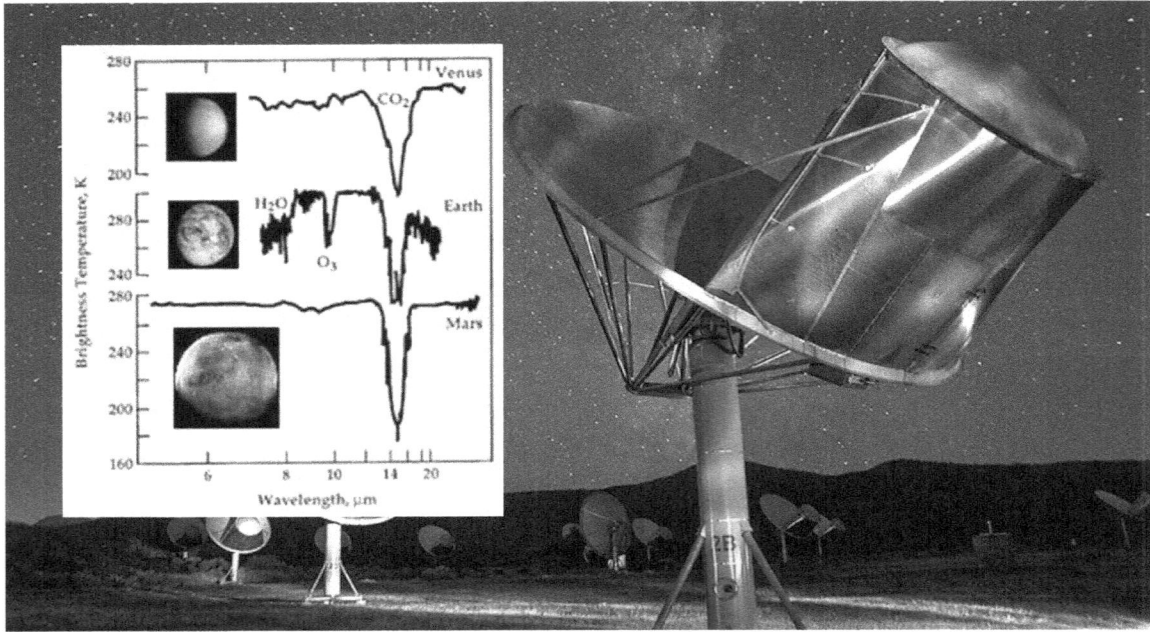

Hard way ...

Viking Lander, 1976

So if we send a spacecraft to a planet around Alpha Centauri (assuming there *is* one), would we be able to do better?

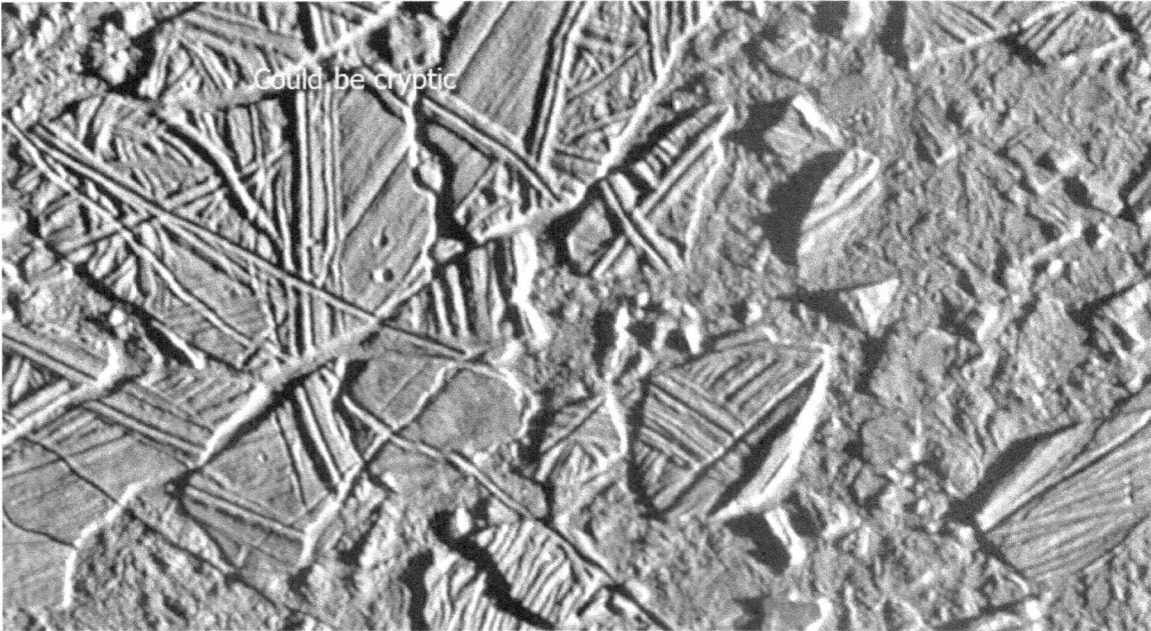

Could be cryptic

Getting up close may lead to inconclusive results.

So maybe forget the 100YSS, and instead embrace a
100 year telescope project?

Go for the "easy" way to find life, and forget the
unpredictable (and unmodifiable) "hard" way?

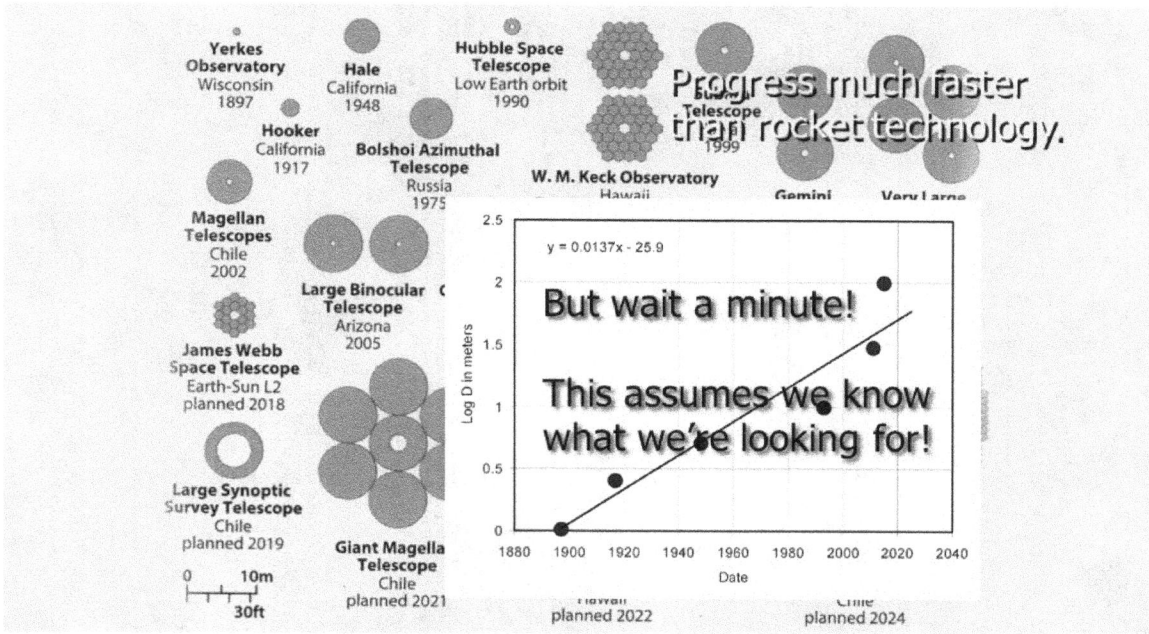

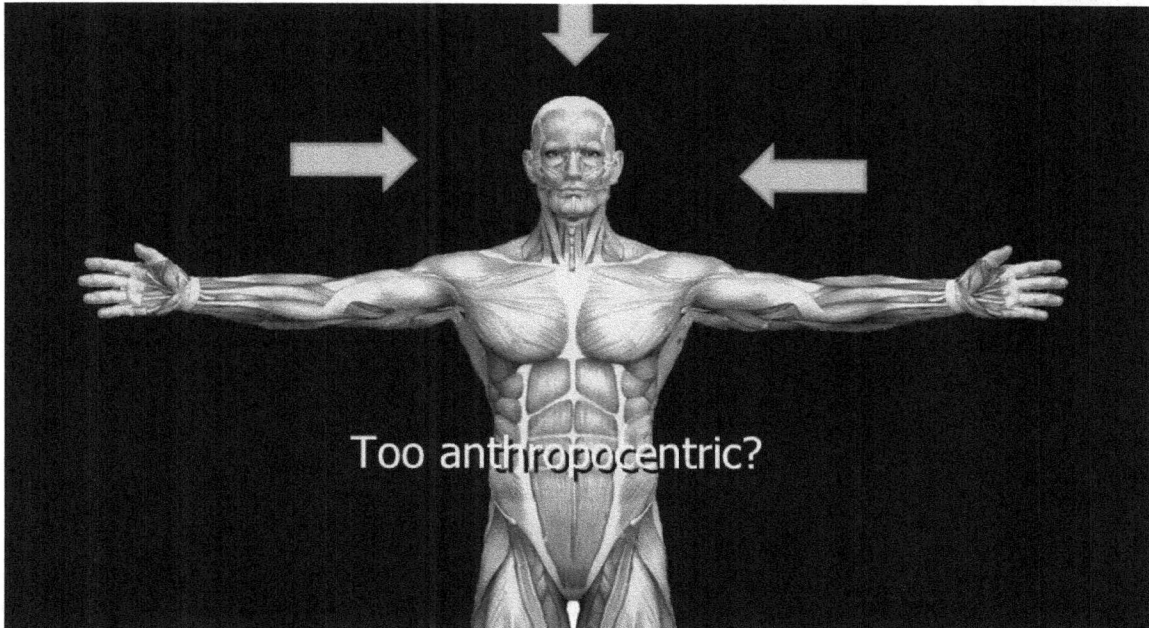

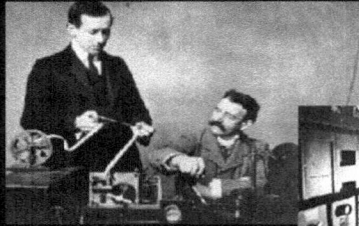

The real value will not be looking up close for stuff we might better find from far away.

But discovering the truly unknown.

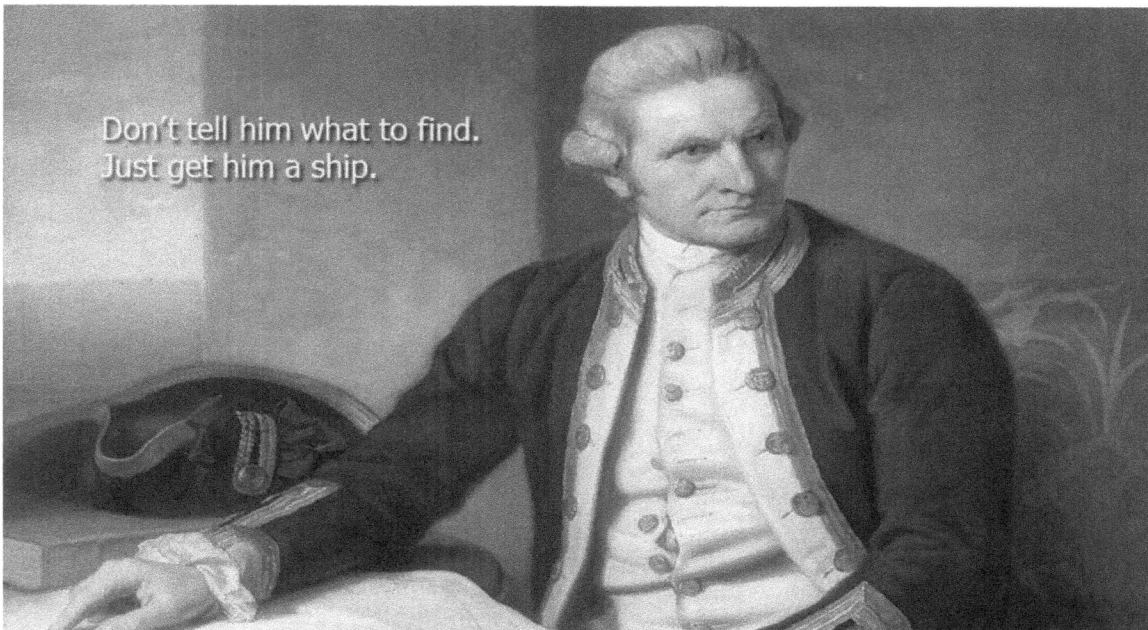

Don't tell him what to find.
Just get him a ship.

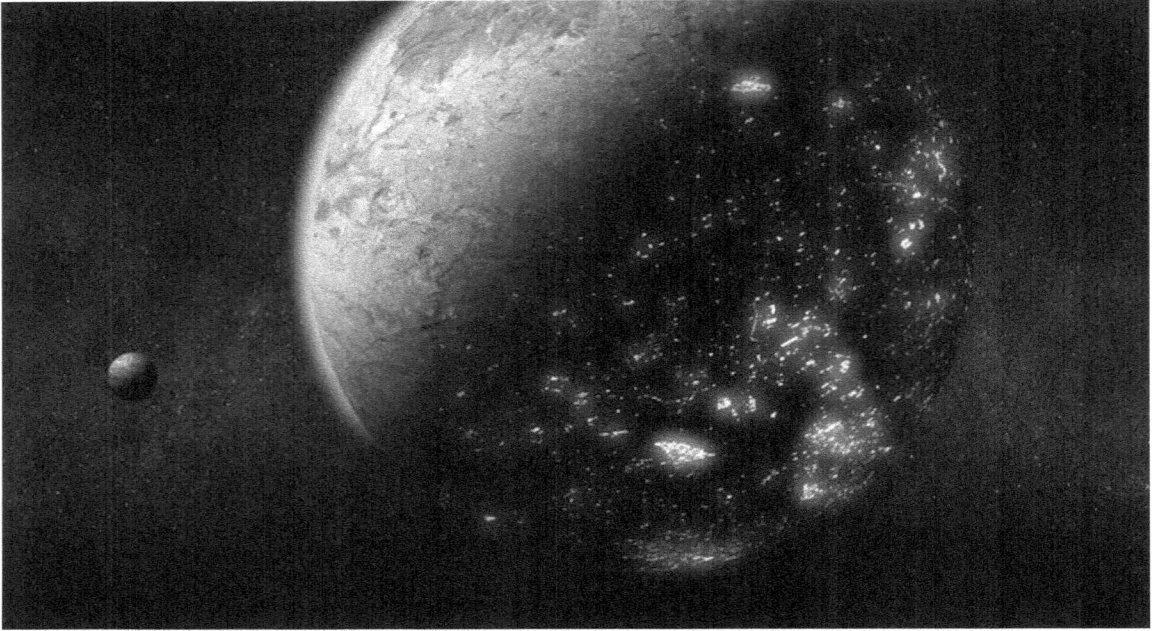

Get Real to Get Relativistic: Talk or Do?

Philip Lubin, Ph.D.

Physics Department, University of California Santa Barbara

lubin@deepsapce.ucsb.edu

Get Real to Get Relativistic
Talk or Do?

I **do not** want to be **talking** about what we might be doing 20 years from now.

I **want** to be **building** the first relativistic interstellar probes

I will show you how to begin on the real axis

It will NOT be easy

There are MANY difficult technical issues

It will NOT be cheap

But it is possible

It will profoundly affect humanities capabilities

Star Trek is good for movies
Not for reality
Get REAL

- Human spaceflight followed on robotic probes
- So should interstellar flight
- It you want to get a probe to the nearest stars in a lifetime
- → v/c > 0.05 (5% c – 88 yrs to α Centauri)
- Leave the propulsion system at home
- **Recent photonic advances radically change what we can do**
- DE → a real path to get there in less than 40 yr flight
- The only real way I know of to achieve beginnings of IC
- **"Beam me up"** was right – but in a different way
- IC (Interstellar Capability) will transform humanity
- But HICs (Human Interstellar Capability) comes later
- Read our **"A Roadmap to Interstellar Flight"** for details
- We need a roadmap before we build the road to the stars
- **The consequences of this technology are profoundly transformative**

What does NOT Work

- Chemical propellants – the norm
- Ion engines
- Solar sails
- Nuclear thermal engines
- Imaginary propulsion

DEEP-IN
Directed Energy Propulsion for Interstellar exploratioN
P. Lubin   Physics Dept – UC Santa Barbara
lubin@deepspace.ucsb.edu
www.deepspace.ucsb.edu

These are the Voyages of the StarChip WaferSize

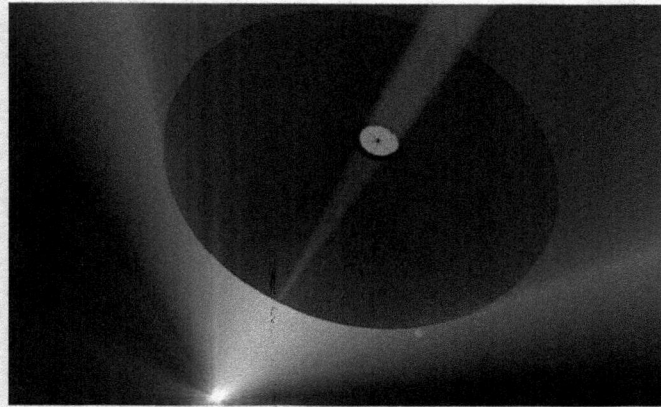

UCSB DEEPSPACE GROUP
Experimental Cosmology at UCSB

100YSS October 31, 2015

NIAC
NASA Innovative Advanced Concepts

- DEEP – Directed Energy Propulsion
- Phase I Study chosen June 2015
- **To explore directed energy for relativistic flight**
- **To explore wafer scale spacecraft as an option**
- **To develop roadmap to interstellar flight** – start small and build up
- To explore technological limitations
- To explore TRL limitations and roadmap to increase TRL
- To explore possible photon recycling option – passive increase in thrust
- To explore use of a single driver to launch an armada of SC
- To explore use inside solar system for rapid delivery
- Other uses of technology
  - Planetary defense
  - Asteroid and comet manipulation – including orbit and spin control
  - Space debris mitigation
  - Long range power beaming
  - SETI searches and beacons
  - Long range composition analysis
  - Storm mitigation and terraforming

85

Stars and ExoPlanets within 25 ly of Earth

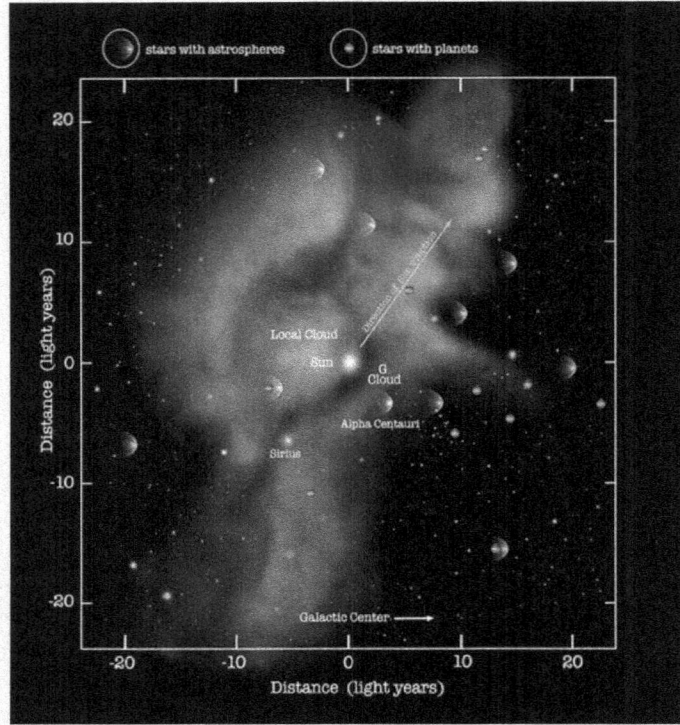

Human Accelerated Objects

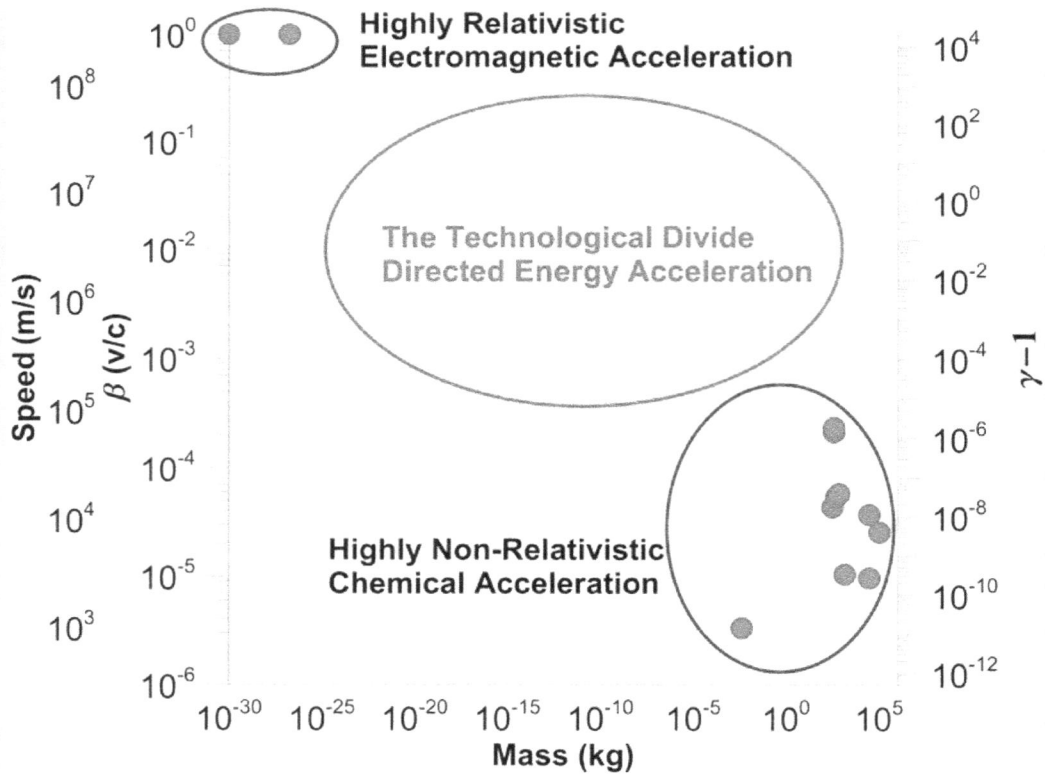

Photon Driven Spacecraft Propulsion

- Use **photon rail gun mode** of DE-STAR
- Propulsion – "**Don't leave home with it**" – No carried propellant → << mass
- Photon Force = P/c - absorption - P = power
- Photon Force = 2P/c - reflection
- 1 W ~ 6.6 nN (much more possible with photon recycling)
- 70 GW (approx power SLS) ~ 470 N ~ 100 lbs thrust – DE-STAR 4
- 1000 kg robotic craft -> a=F/m = 470/1000 ~ 0.47 m/s^2
- Time to travel distance d: t= $(2d/a)^{1/2}$ = $(mcd/P)^{1/2}$ = $(t_I \ t_E \ )^{1/2}$
- t_I =d/c – light travel time t_E =mc^2 /P = time to equal mass energy
- Ex: Approx 0.3, 1, 3, 9, 30 days to Mars for 1,10,100,1000,10000Kg
- **Speed at 1 AU ~ 1200km/s for 100 kg (> galactic esc vel)**
- **Maximum speed for equal sail and payload mass → v~m$^{-1/4}$**
- **Speed edge solar system ~ 2% ->3%c (cont illum) 100kg craft – DE-STAR 4**
- Problem – no brakes! – ion engine retro – ping pong second DE-STAR?
- **Allows first interstellar probes – communication all the way out**
- **Wafer scale spacecraft (~ 1g) can achieve 30% c   α-Centauri →20yrs**

Physics of Directed Energy Flight

http://www.deepspace.ucsb.edu/projects/directed-energy-interstellar-precursors
see "A Roadmap to Interstellar Flight" for details

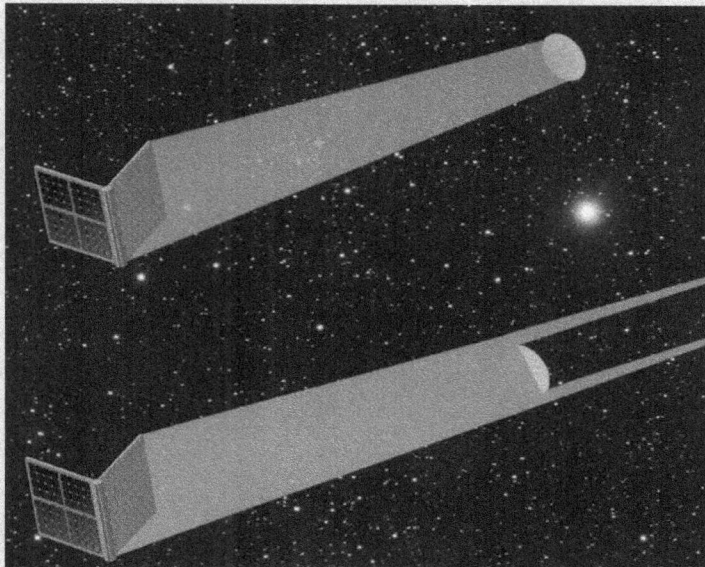

Microsoft is Ready to Go!

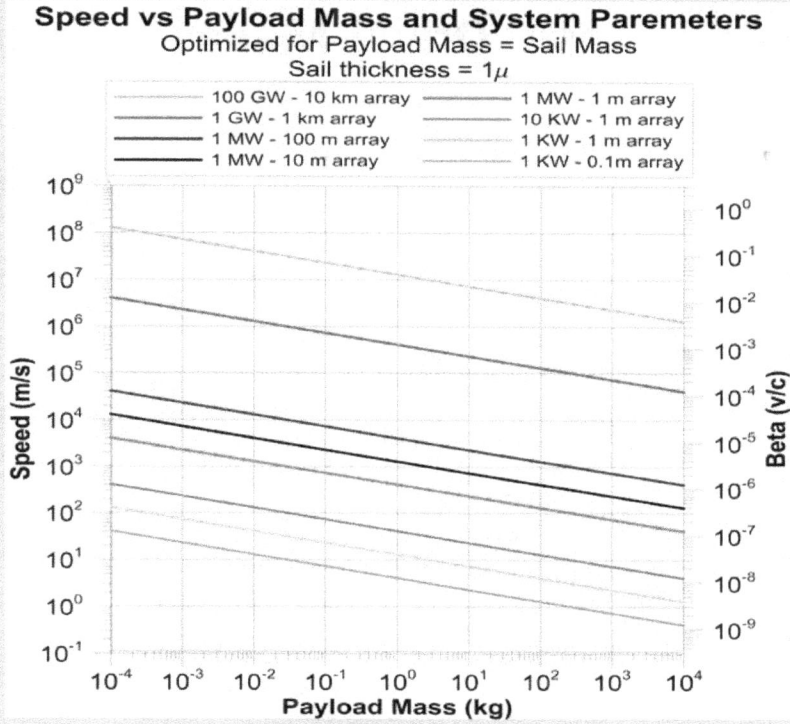

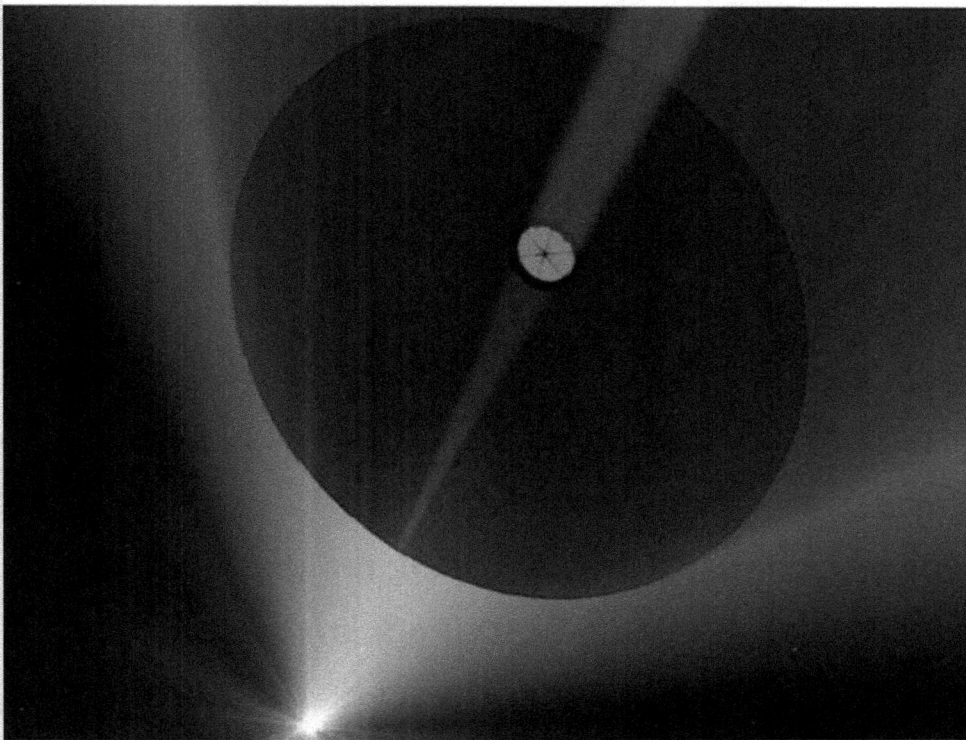

Phased Array Laser Driver Makes it Possible

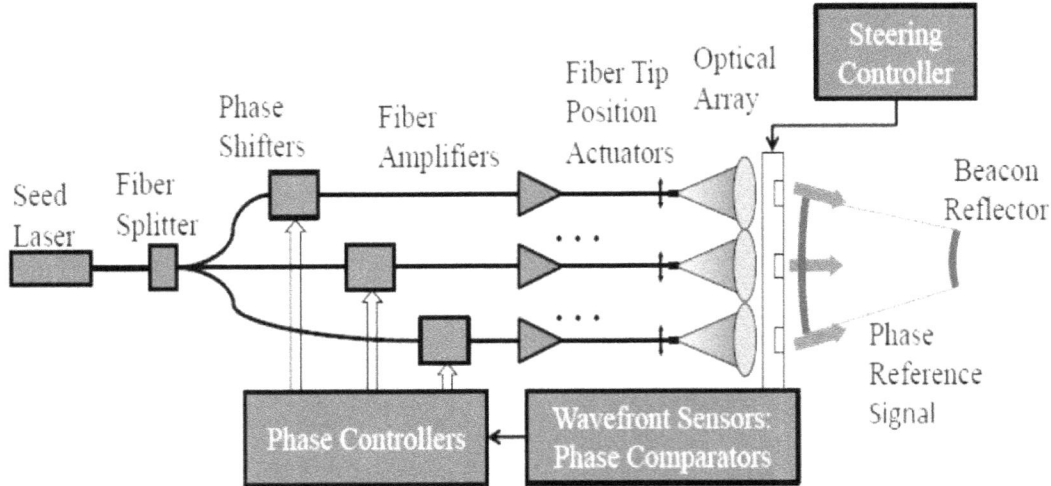

High Performance Yb Fiber Amplifiers

- Ytterbium (Yb) doped fibers remarkably efficient
- Pump at 976 nm – lase at 1064 nm
- 83-87% optical efficiency
- Wall plug efficiency limited by 976 nm pump
- 976nm laser diode pump eff (7/15) ~ 55% (InGaAs on GaAs)
- Wall plug total eff ~ 42% (this will rise - push pump to 70%)
- All solid state – long life in theory
- **Power mass density (kw/kg) rising rapidly ->1 soon**
- SBS (Stimulated Brillouin Scattering) limited
- Narrow band (KHz) -> long coherence – 10-100km – 100 w
- Wide band (10 GHz) -> shorter coherence – cm – 1-2 kw

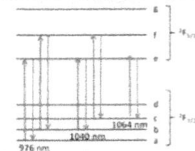

Efficiency of Fiber Amplifiers
Photon pump efficiency ~ 85% !

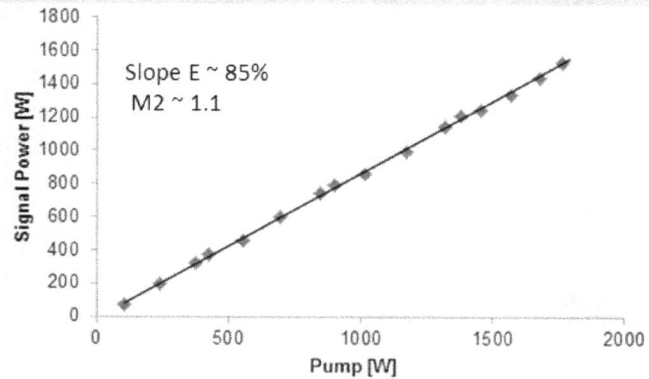

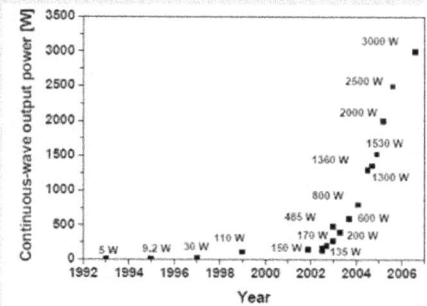

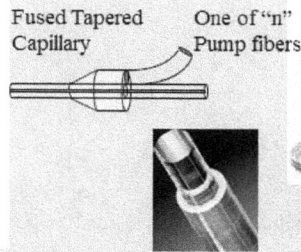

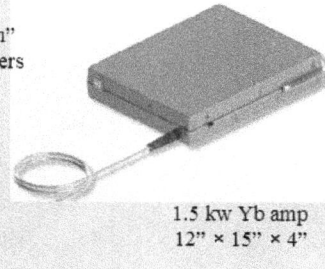

1.5 kw Yb amp
12" × 15" × 4"

Fiber laser "Moore Law" Scaling
20 month doubling

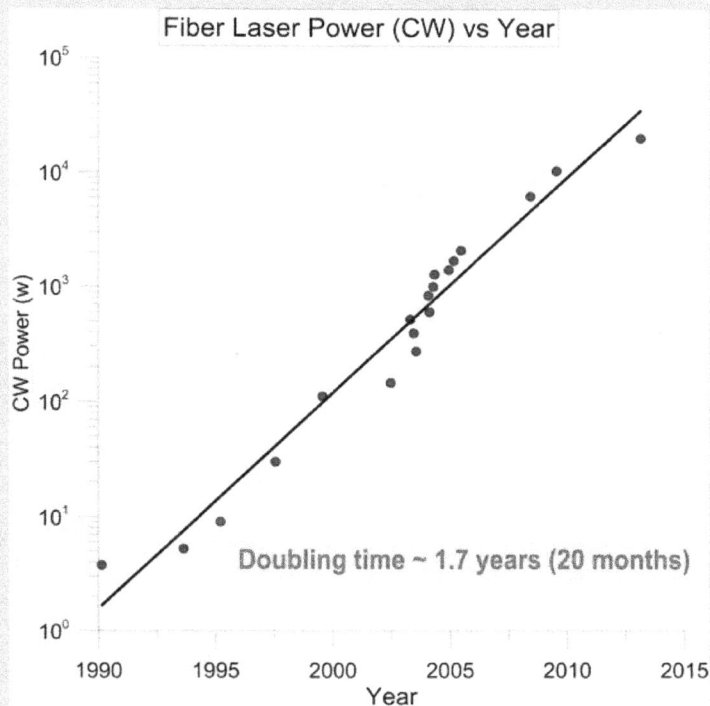

Doubling time ~ 1.7 years (20 months)

91

Low areal density Thin Film Optics
MOIRE – Membrane Optical Imager Real-time Exploitation
DARPA-Ball-LLNL
Achieved WFE RMS 30nm mandrel, 30nm film, 280 nm transfer, 60g/m^2

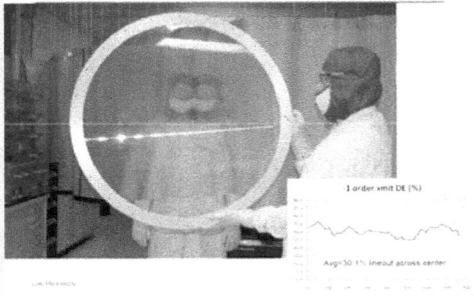

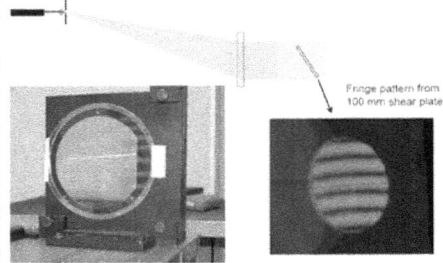

4 µm pitch   6 µm pitch   8 µm pitch

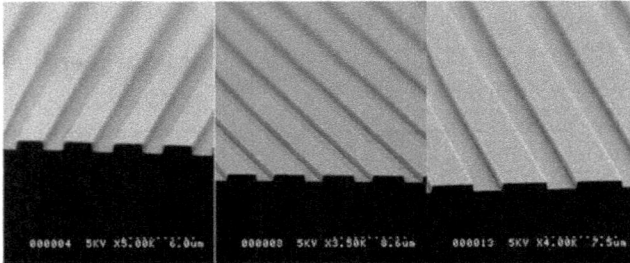

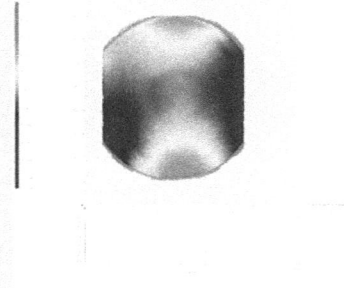

Summer 2015 lab testing

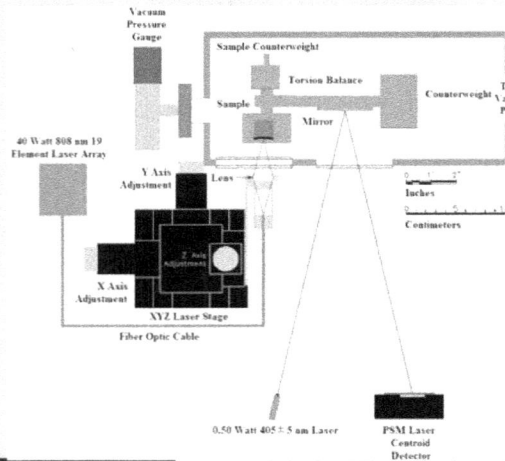

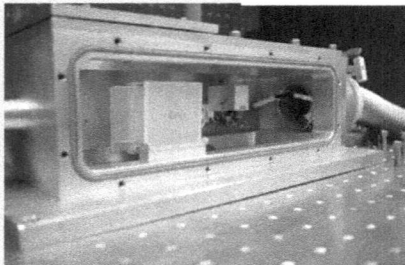

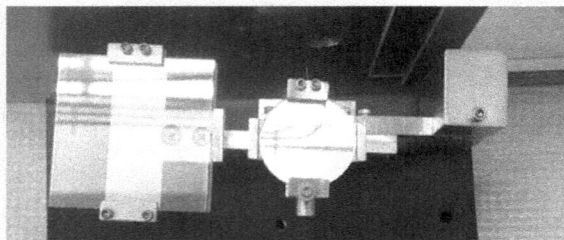

Efficiency (reflection case)

- r_w =Thrust per watt F=2P/c -> r_w =2/c
- r_m= Thrust per "mass flow = dm/dt"   dm/dt=P/c^2
- r_m = F/ P/c^2 = 2c (reflection) or c (absorption)
- **Inst ε = Mech P/ Drive P** = d/dt (mv^2 /2)= mv dv/dt /P =mav/P=Fv/P
- =2Pv/c/P=2v/c =**2β** - **eff improves with speed (non rel)**
- Average efficiency over acceleration time = β - faster is more eff!
- **Photon drives are the most efficient in terms of mass flow**
- **Photons drives are the least efficient in terms of power**
- **Non relativistic case (conv or ion)**
- **r_w =Thrust per watt = 2/v$_{rel}$**
- **r_m= Thrust per "mass flow" = v$_{rel}$**
- → **Similar conclusions for ion drives**
  - Ion drives efficient in terms of mass flow (compared to conv prop)
  - Ion drives not efficient in terms of power (compared to conv prop)
- → **Conclusion: Photon drive from a distance**

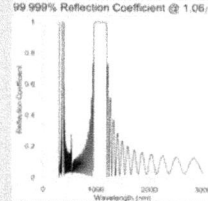

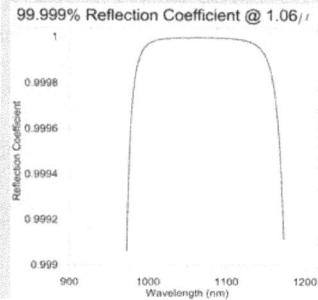

1000 sigma photon drive (no recycling) ~ 30 watts

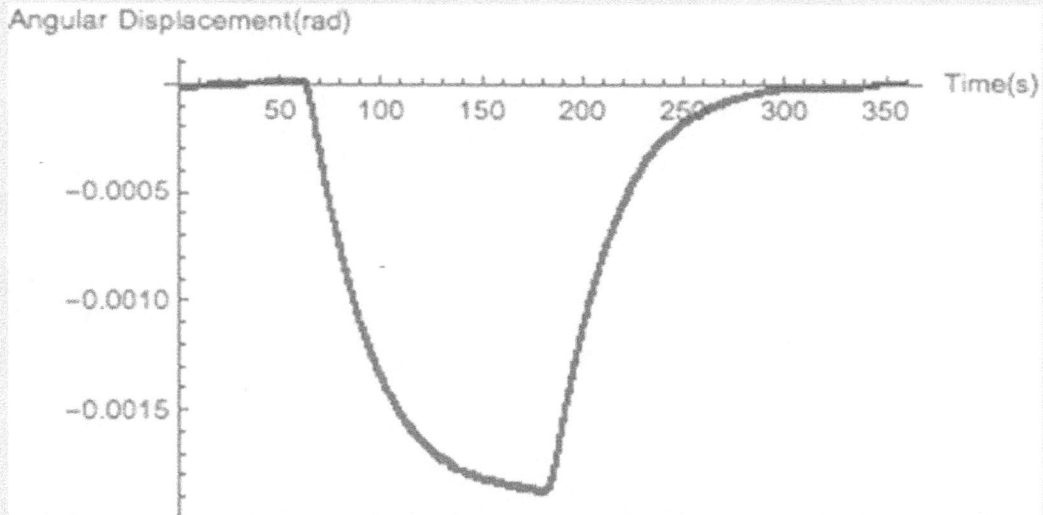

Photon Recycling option

- Use multiple bounces to increase thrust
- ONE photon CAN exert an arbitrary large amount of force
- Optical cavity mode
- Photons (reflect) ~ 6.6nN/w
  Ion engines ~ 40 µN/w ISP 3000 (JPL ARM) (higher ISP is WORSE)
- Conventional Propellant ~ 1 mN/w
- Conventional/Photons ~ 10^5 (IF no photon recycling)
- **IF we could get 10^5 bounces we would exceed conventional chemical propellants**
- **IF we could get to 6000 bounces we would exceed ion engines**
- 99.9999% (six 9's) reflectivity mirrors exist (LIGO)
- 99.999% (five 9') more common
- **Optical cavities with finese >10^5 exist today - BUT extremely difficult to build – lab scale so far**
- With ~ NO effort we have already achieved recycling ~ 10 (100-1000 seems possible)
- Bae Young has achieved 500 using active pumping recycling
- **Practical realization limited**
  - Precise alignment of mirrors
  - Diffraction and spillover (sidelobes)
  - Reflector losses
  - Surface roughness and dust on mirrors
  - Complex for long range space demo
  - Not as useful for relativistic craft – single bounce efficiency ~ 2β (ave efficiency over acceleration= β)
- Has long worked in lab BUT long range photon recycling difficult
- Example – 4 km LIGO arm has 99.9999% (~1 ppm) ref mirrors

~10 x photon recycling (30 watt in)

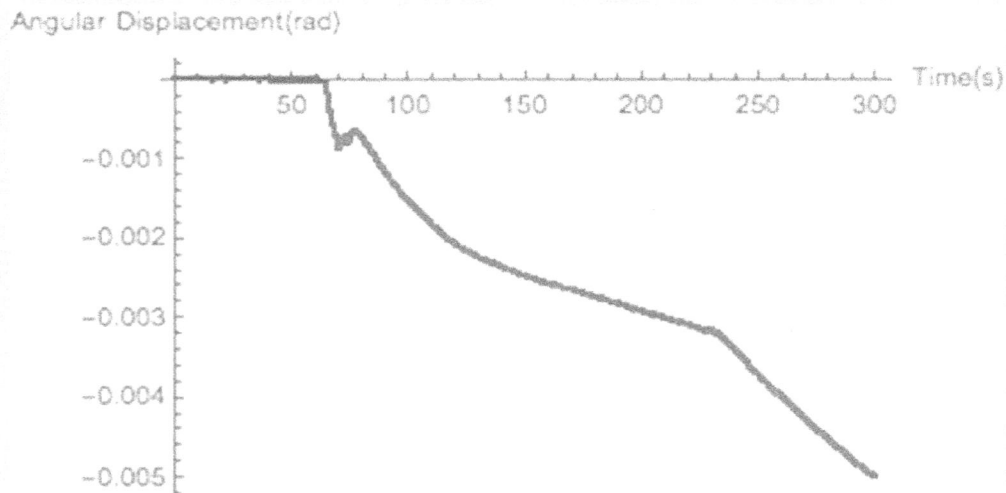

Example of modest multi pass laser propulsion
15m array (150kw) – 1 g payload – with orbit optimization
LEO to escape – no photon recycling

Full Scale WaferSat – 10 min acceleration to c/4
Laser on for 10 minutes then coast – no photon recycling
Alpha Centauri in 20 years
140 launches per day – 40,000 payload per year

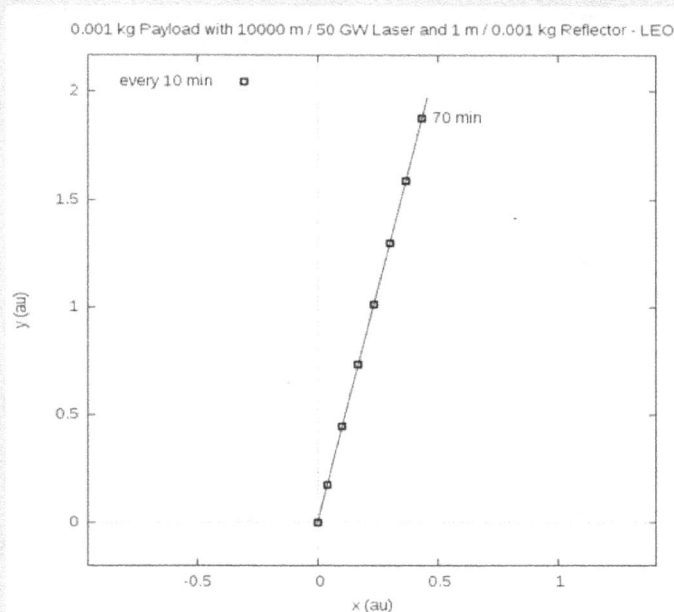

0.001 kg Payload with 10000 m / 50 GW Laser and 1 m / 0.001 kg Reflector - LEO

Interstellar Wafer Scale Satellite

Alpha Centauri in 20 years

We can make this happen

- •DE-STAR 4 Photon driver
- •No photon recycling assumed
- •1 meter laser "sail"
- • All dielectric sail
- •10cm wafer scale spacecraft
- •26% c in 10 min acceleration
- •20,000 g acceleration!
- •Launch 140 missions per day
- •>40,000 launches per year
- •1 watt laser comm burst mode
- •10 cm wafer only laser comm
- •1 AU – Mars in ½ hour
- •Pass Voyager 120 AU in 3 days
- •Grav solar lens 500 AU – 6 day
- •1000 AU in 12 days
- •→ 1 Kg (CubeSat mass) @ 5% c

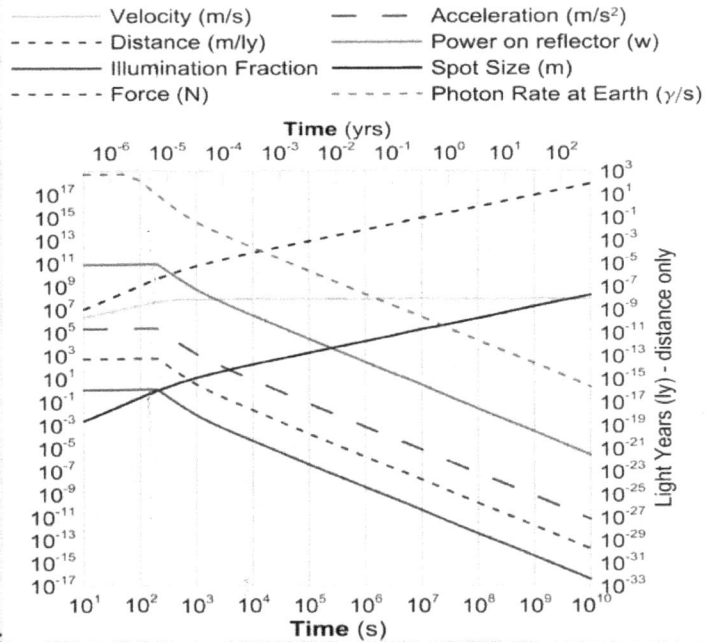

Could use SC Reflector with Laser Comm →higher data rate

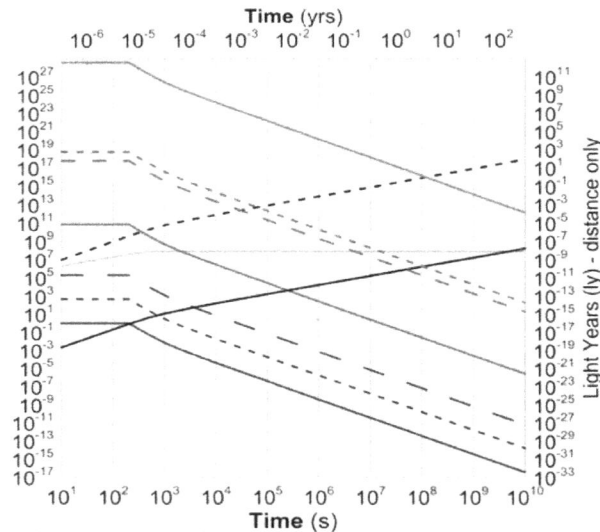

96

Long range interstellar communications

- Optical link calculation
- DE-STAR 4 -> 3×10^{29} γ/s $\vartheta \sim \lambda/D \sim 10^{-10}$ rad
- Kepler exoplanets ~ 1000 ly (10^{19} m) away
- -> 10^{11} γ/m^2 - s with spot size ~ 10^9 m (~ diam Sun)
- **Brighter than a magnitude 0 star (10^{10} γ/m^2 - s)**
- Link between two DE-STAR 4 units at 1000 ly?
- Get ~ 10^{19} γ/s received at each end!
- → ultra high speed link over much of the galaxy
- BUT live streaming is severely delayed
- What do we mean by communicating?

Implication of this technology on SETI

We can detect comparable civilizations across the entire horizon

SDI spin off – Search for Directed Intelligence

SMDI search finished with positive results

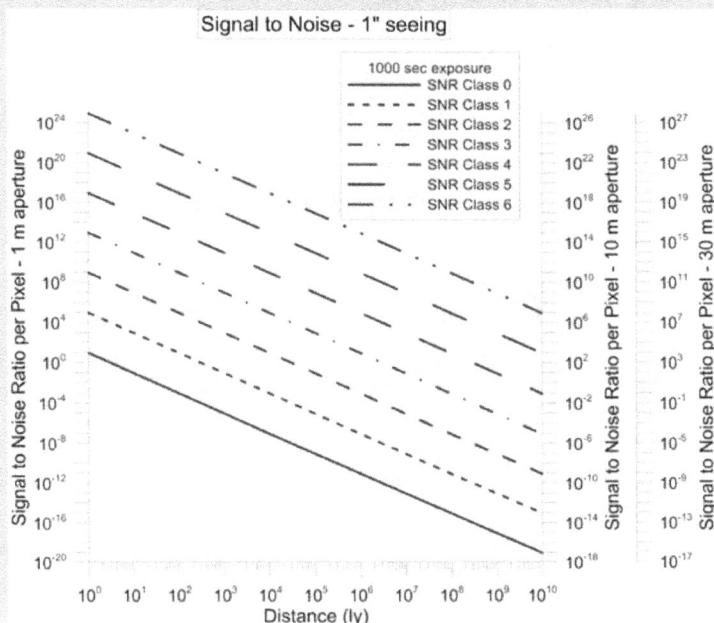

Practical Roadmap and Milestones
Walk before you run – **modular scaleable emphasis**

- **DE-STAR -1** -> CubeSat 10 cm aperture – photon recycling tests
- **DE-STAR 0** -> test unit – possibly ISS space debris defense
- **DE-STAR 0** on STARLITE – stand on planetary defense – see our papers
- **DE-STAR 1** -> space debris mitigation - wafer LEO to GEO and LEO to escape
- **DE-STAR 2** – ground testing – photon launch assist
    - LEO to escape, Moon , Mars with WaferSat
    - ISS defense and space debris clearing
    - Close asteroid deflection
- **DE-STAR 3** – Serious photon drive
    - Begin standoff planetary defense
- **DE-STAR 4** – full relativistic drive for WaferSat
    - Fast interplanetary photon drive for large payloads
    - Long range power projection to distant SC
    - Ultra high speed interplanetary communications

Historical Trends in PV – Comparison
4 decades -> 3 orders magnitude price drop
Swanson's Law

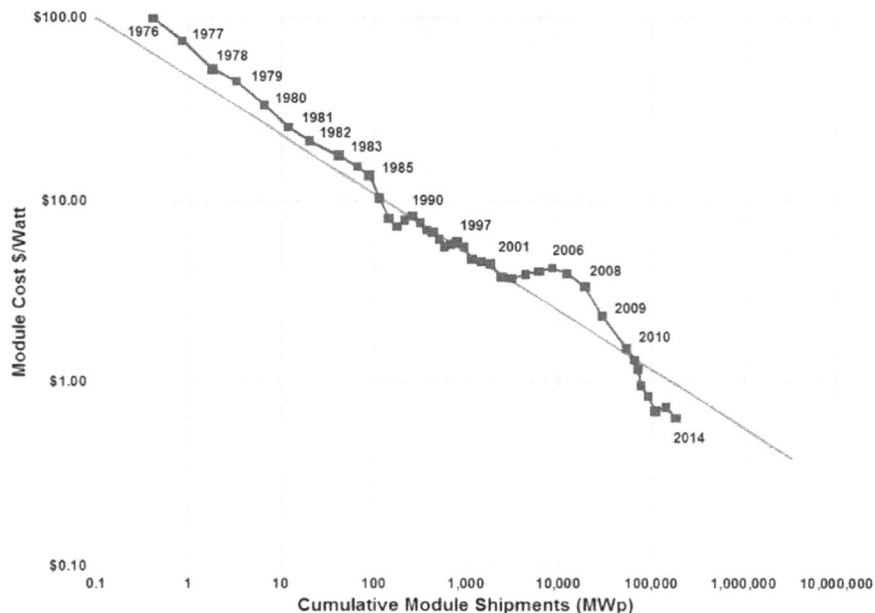

Technology Development, Mass Production and Costs
6 decades -> 11 orders of magnitude price drop!

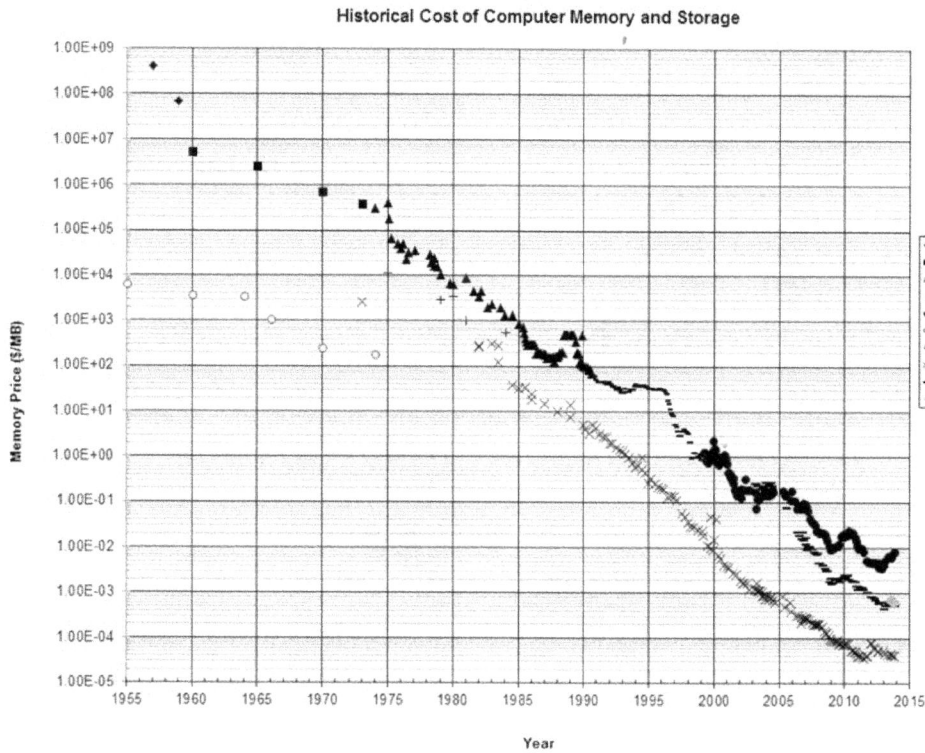

Historical Cost of Computer Memory and Storage

Mass Production vs Time
115 years (1900-2015 incl CPU and GPU) -> 17 orders of magnitude price drop!

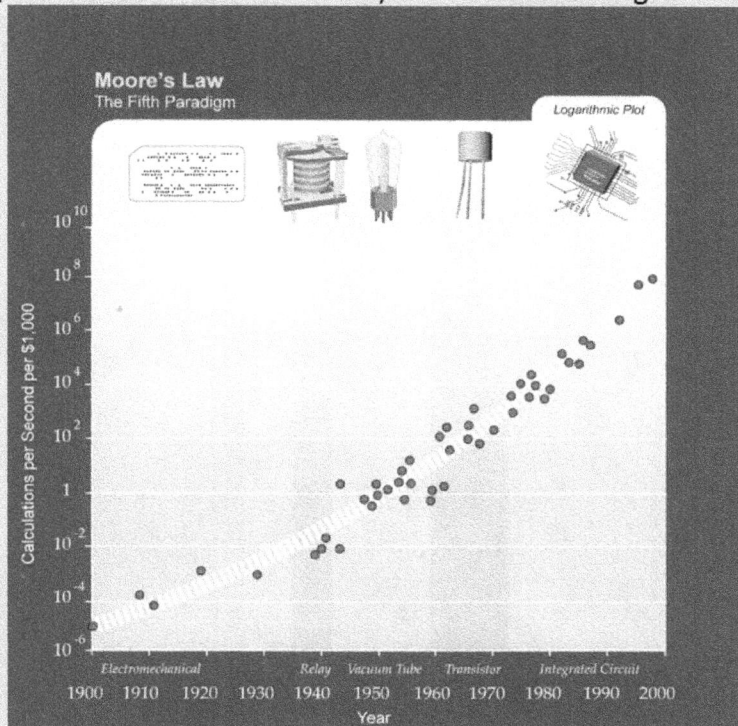

Conclusions

- **Recent photonics advances allow science fiction → science**
- **There is no known reason why we cannot do this**
- There is a roadmap to relativistic flight
- **The system is completely scalable and modular – allows any size**
- Allows for economy of scale
- **All modules are the same → mass production → radical cost reduction**
- Allows for technological co-development
- We can begin small and evolve to interstellar capability
- MANY other applications
- Enables many mission scenarios allowing cost amortization
- Is it hard? YES – Is it impossible? - NO
- This is not for the faint of heart!
- It is time to begin this inevitable journey – together
- → See the many papers on our website
- www.deepspace.ucsb.edu

• Thank you

100 YEAR STARSHIP™

The State of the Universe

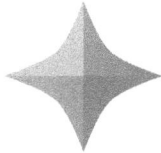

Finding Earth 2.0

Jill Tarter, Ph.D.
SETI Institute

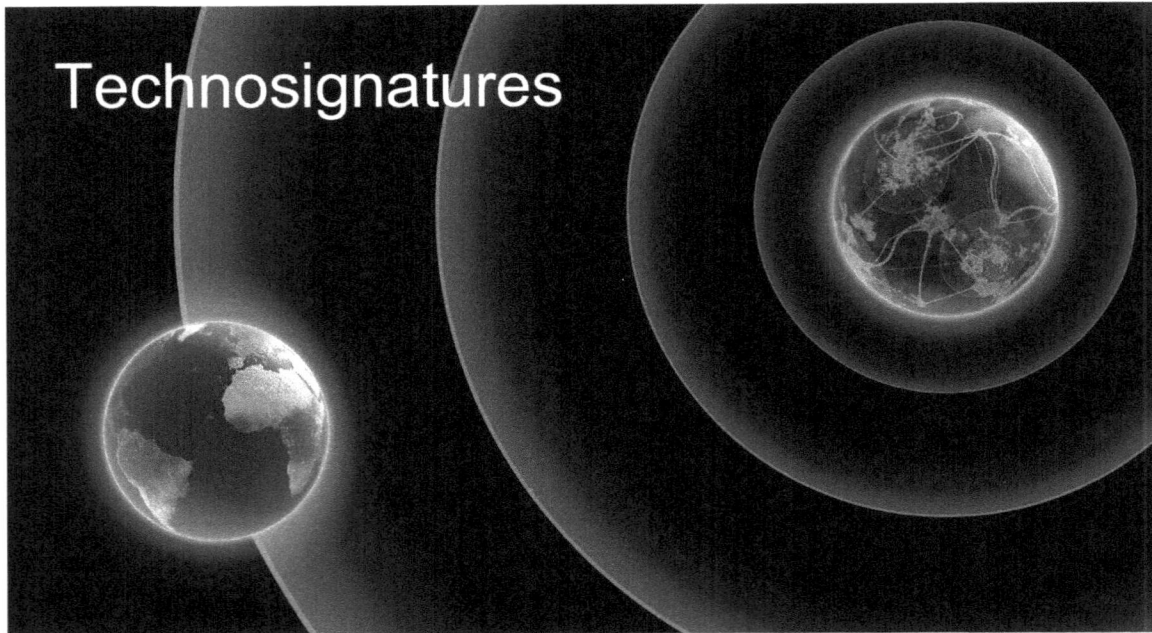

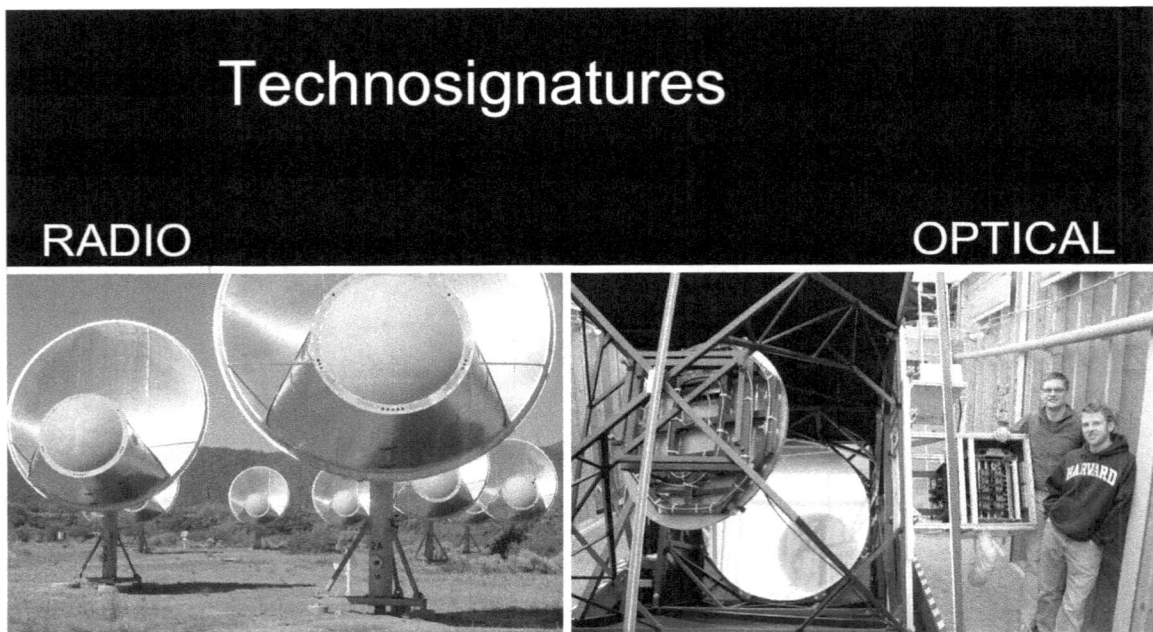

The Cosmic Haystack Is Huge

Nine Dimensional
- 3 – space
- 1 – time
- 2 – polarizations
- 1 – frequency
- 1 – modulation scheme
- 1 – sensitivity

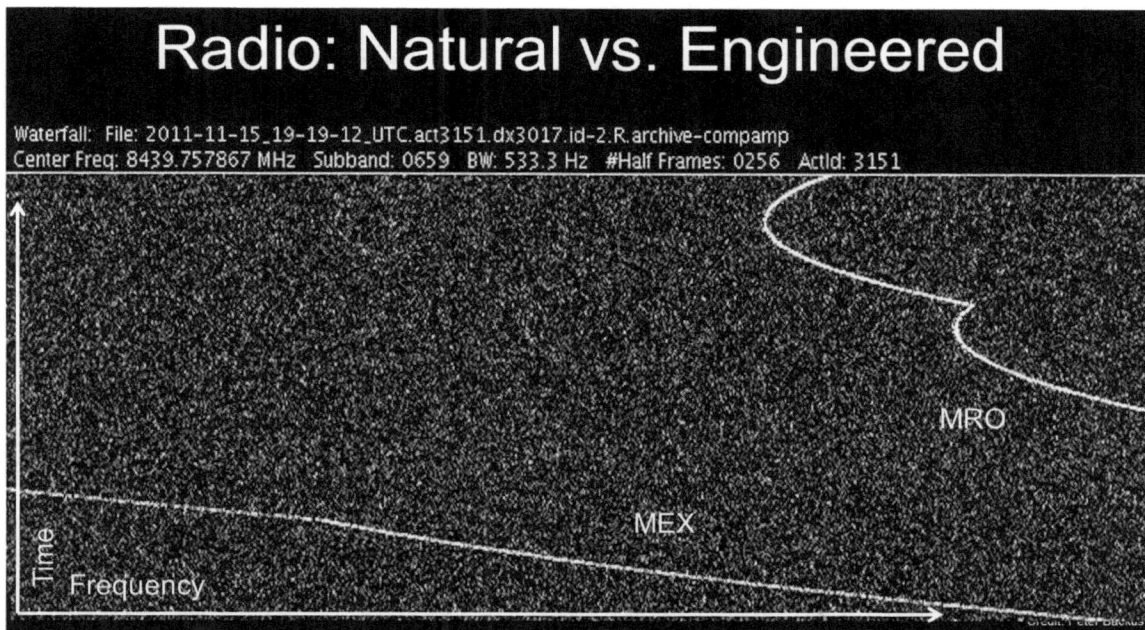

9

Radio: Natural vs. Engineered

Waterfall: File: 2011-11-15_19-19-12_UTC.act3151.dx3017.id-2.R.archive-compamp
Center Freq: 8439.757867 MHz Subband: 0659 BW: 533.3 Hz #Half Frames: 0256 ActId: 3151

MRO

MEX

Time

Frequency

Project Phoenix: 1994 – 2004
1000 stars x 1700 MHz = 1.7 x 10⁶ Star-MHz

Mopra

Parkes

Woodbury

140 Ft.

Lovell

Arecibo

2.2 x 10⁵ Star-MHz

2.9 x 10⁵ Star-MHz

4.8 x 10⁵ Star-MHz

NSS: PCs + accelerators

NASA-derived
TSS: full custom

The ATA-42

6.1 m Offset Gregorian Antenna - LNSD

SonATA (SETI on the ATA) Since 2011

- Use 3 Beamformers to target 3 systems simultaneously with 100 million channel detectors
  - <u>All</u> Kepler candidates
  - Exoplanets from *exoplanets.org* $\delta > -30°$
- Over the entire terrestrial MW window 1-10 GHz

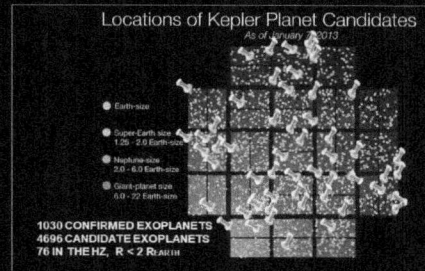

Locations of Kepler Planet Candidates
As of January 2013

- Earth-size
- Super-Earth size 1.25 - 2.0 Earth-size
- Neptune-size 2.0 - 6.0 Earth-size
- Giant-planet size 6.0 - 22 Earth-size

1030 CONFIRMED EXOPLANETS
4696 CANDIDATE EXOPLANETS
76 IN THE HZ, R < 2 REARTH

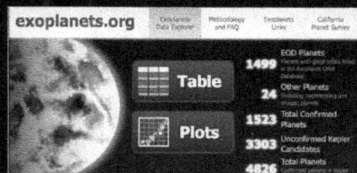

1499		EOD Planets
24		Other Planets
1523		Total Confirmed Planets
3303		Unconfirmed Kepler Candidates
4826		Total Planets

4.5×10^7 Star-MHz

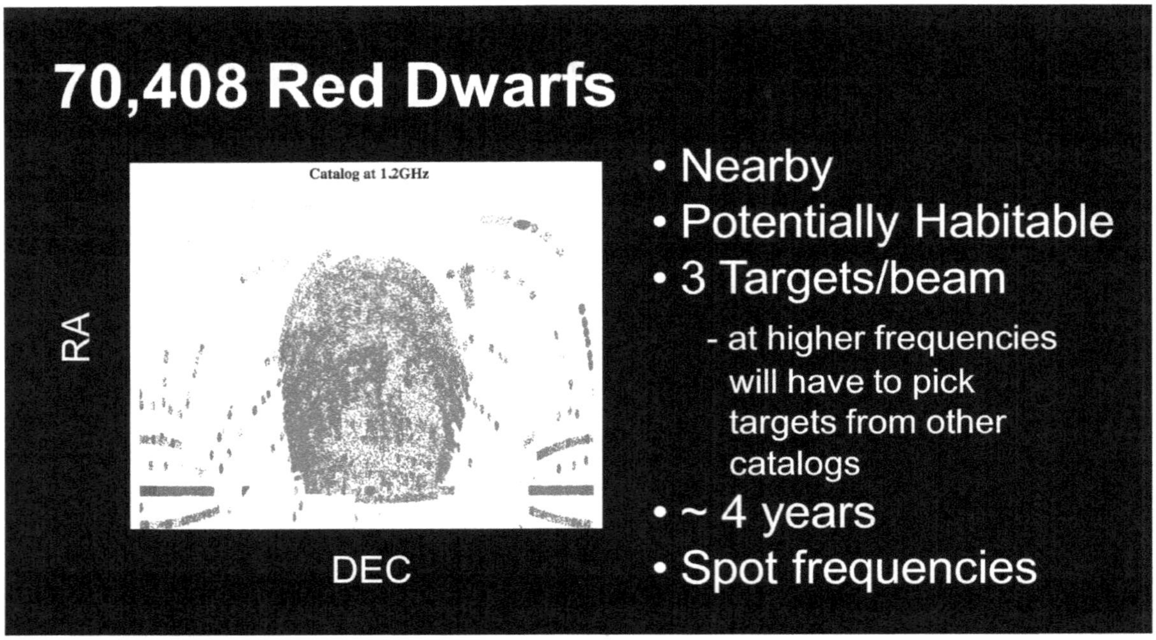
70,408 Red Dwarfs
Catalog at 1.2GHz
RA
DEC
• Nearby
• Potentially Habitable
• 3 Targets/beam
- at higher frequencies will have to pick targets from other catalogs
• ~ 4 years
• Spot frequencies

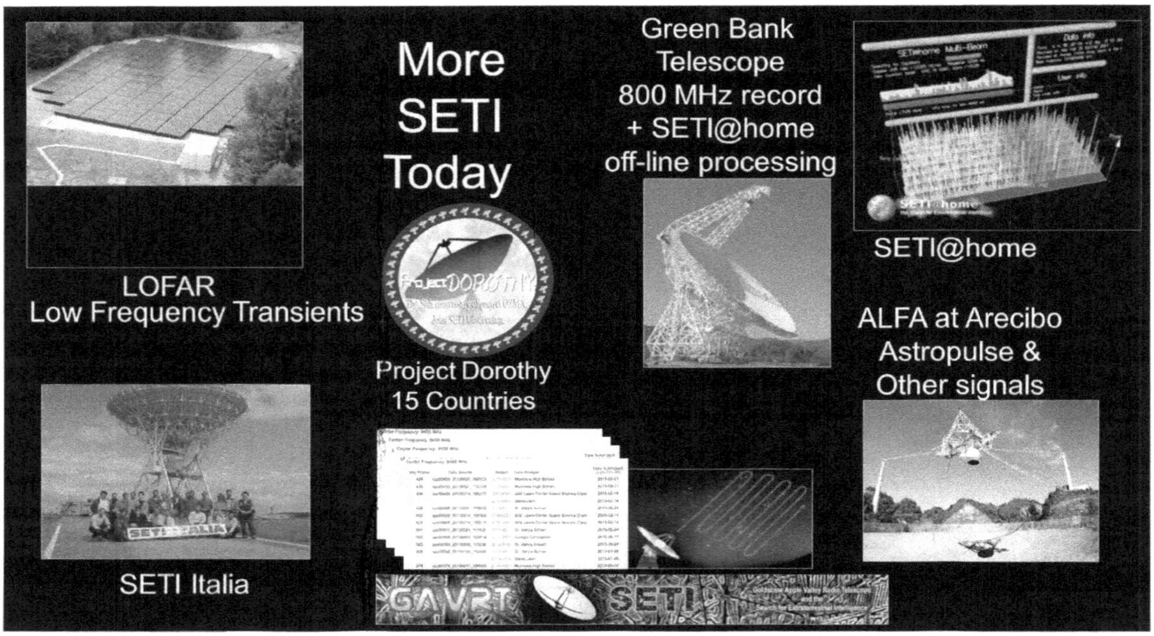
More SETI Today
LOFAR
Low Frequency Transients
Project Dorothy
15 Countries
Green Bank Telescope
800 MHz record
+ SETI@home
off-line processing
SETI@home
ALFA at Arecibo
Astropulse &
Other signals
SETI Italia
GAVRT
SETI

10 γ' s/m²/ns

Harvard OSETI Sky Survey of Northern Sky

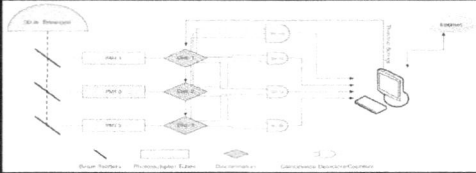
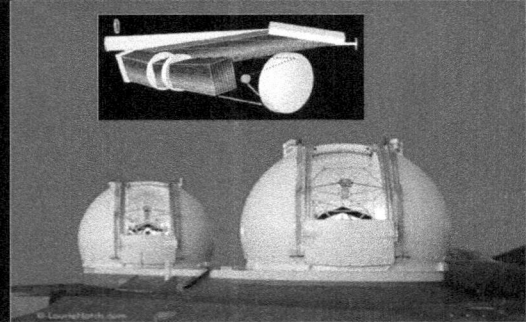

Berkeley Optical SETI

100 γ's/m²/ns

Leuschner
Observatory

HIRES at
KECK
Observatory
1000 KOIs
10 γ's /hr in laser line

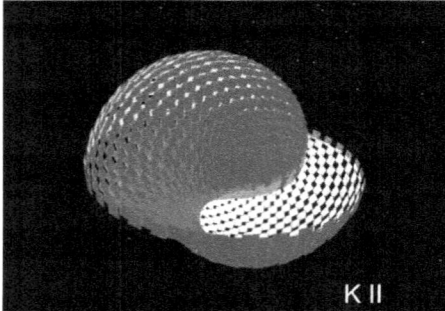
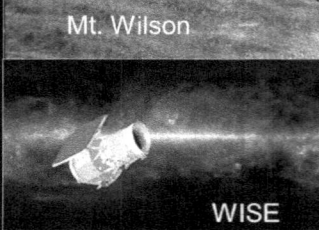

K II

Mt. Wilson

WISE

K III

The Future

JWST and then ???   WFIRST-AFTA
w. starshade

IMAGING / BIOSIGNATURES   LUVOIR

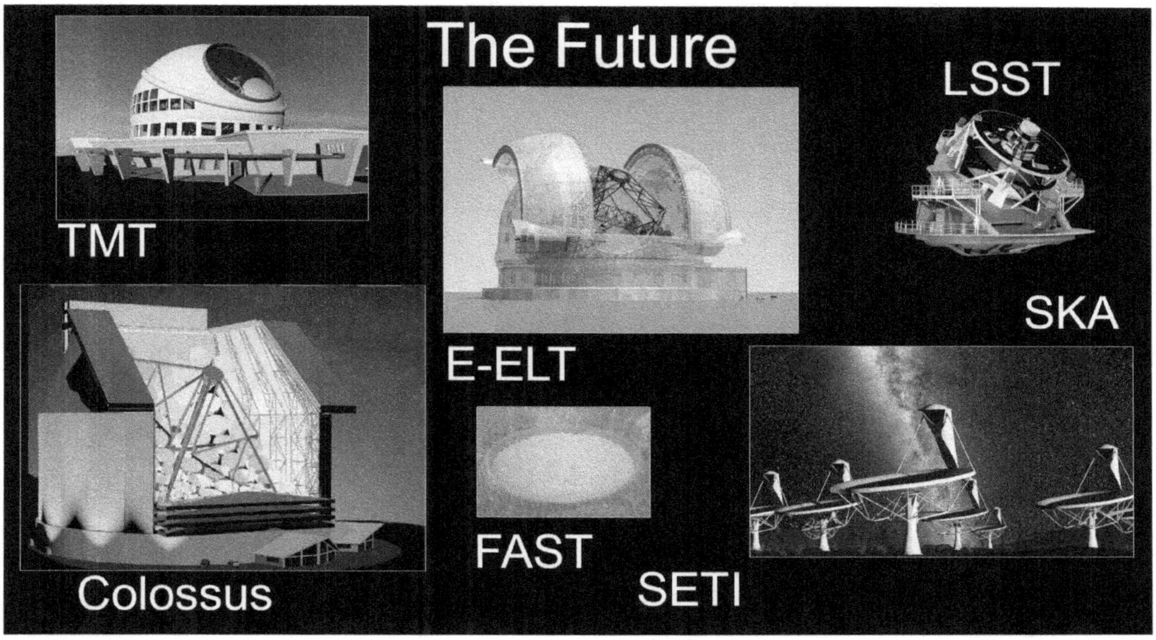

The Future

LSST

TMT

SKA

E-ELT

Colossus

FAST

SETI

The More Immediate Future

Breakthrough Listen Initiative

Breakthrough Message Initiative

$100M over 10 years
Yuri Milner

On a finite world, a cosmic perspective isn't a luxury; it is a necessity.

Caleb Scharf (2014)

Earthrise/ William Anders/ Apollo 8

A Planet for Goldilocks

Natalie Batalha, Ph.D.

NASA Ames Research Center

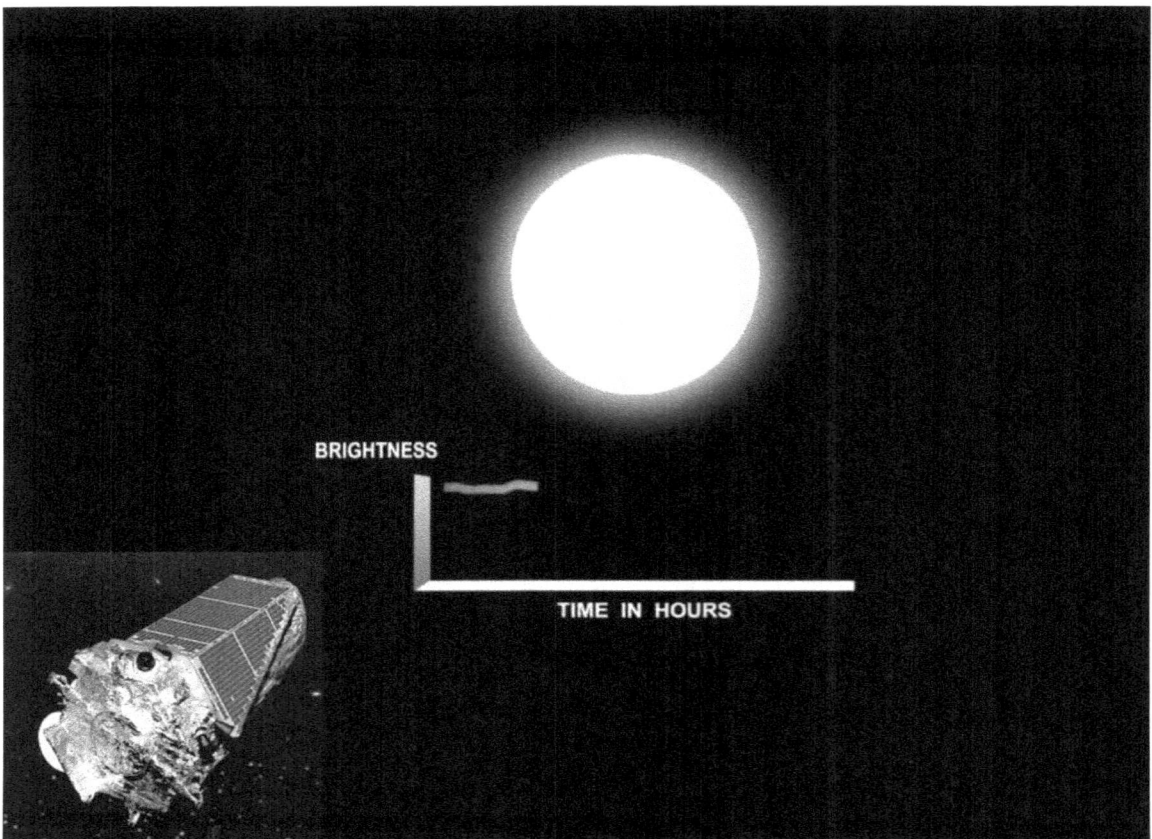

BRIGHTNESS

TIME IN HOURS

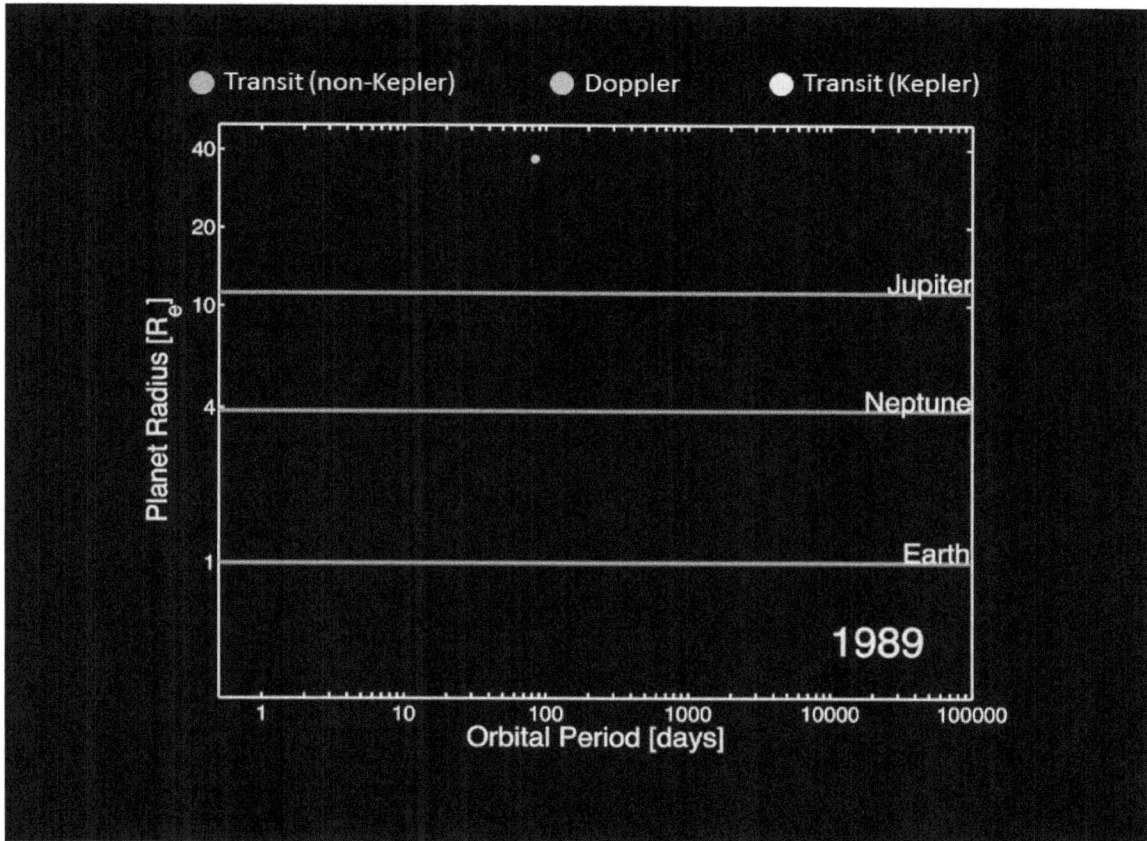

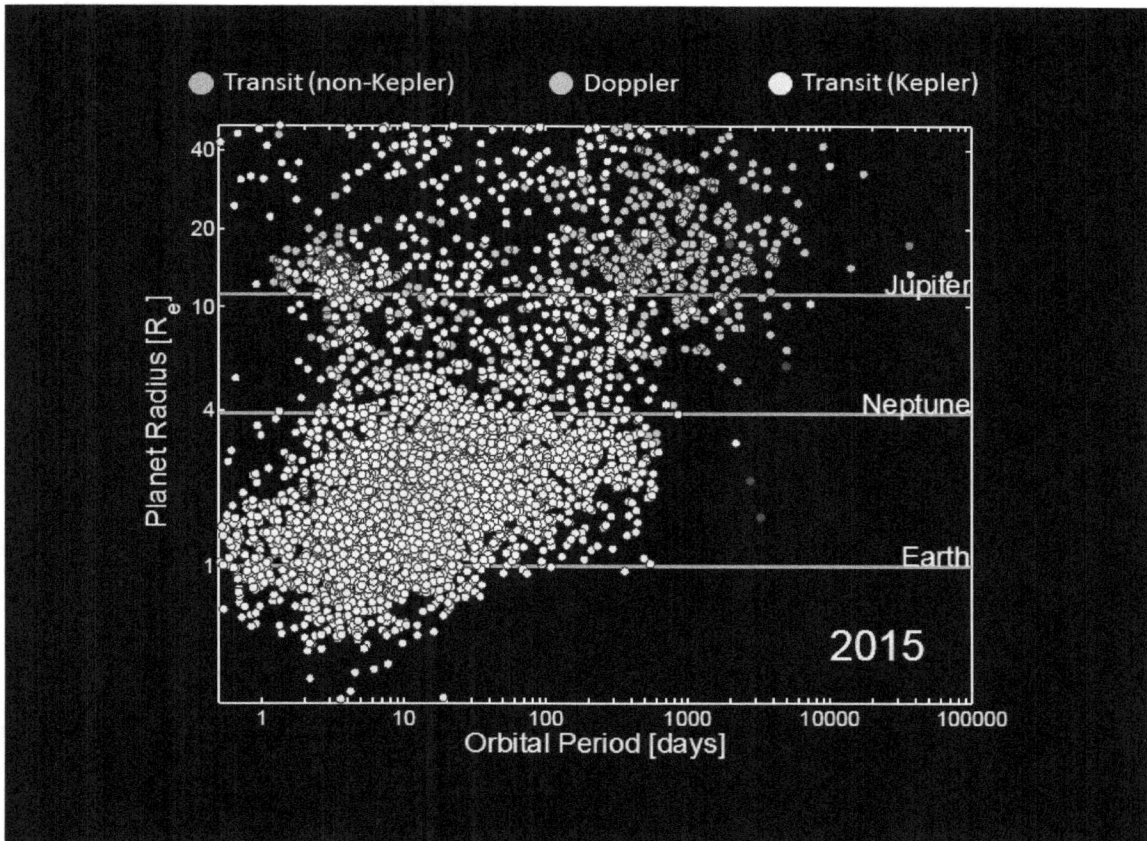

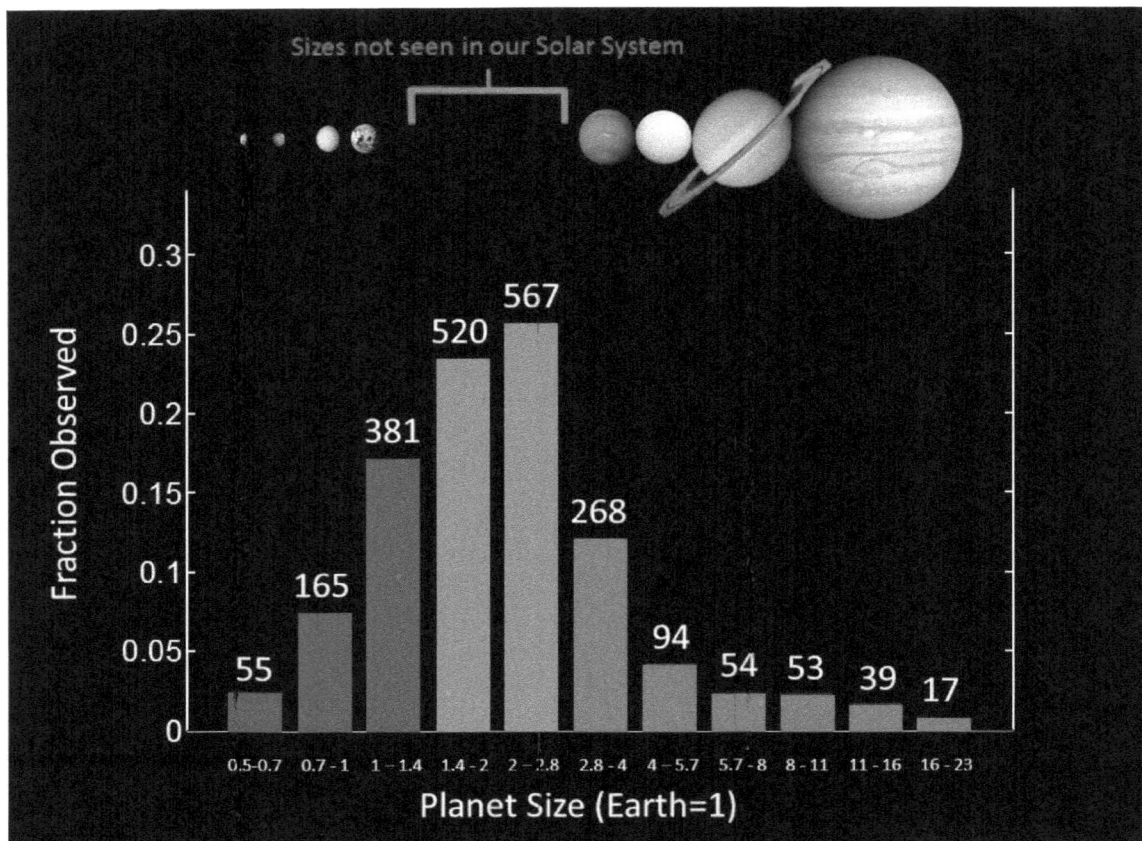

Kepler-10c: Mega-Earth?

Radius = 2.35 R_{earth}

Mass = 17 x M_{earth}

Density = 7 gcc

Kepler-138d: Beer-belly-Earth?
Radius = 1.7 R_{earth}
Mass = 1 x M_{earth}
Density = 1.3 gcc

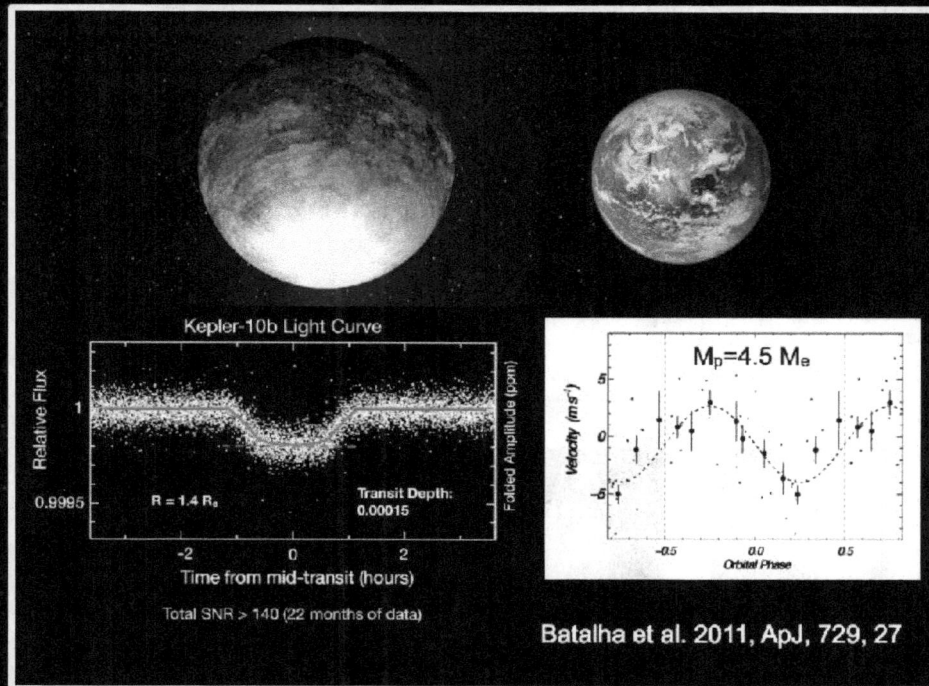

Batalha et al. 2011, ApJ, 729, 27

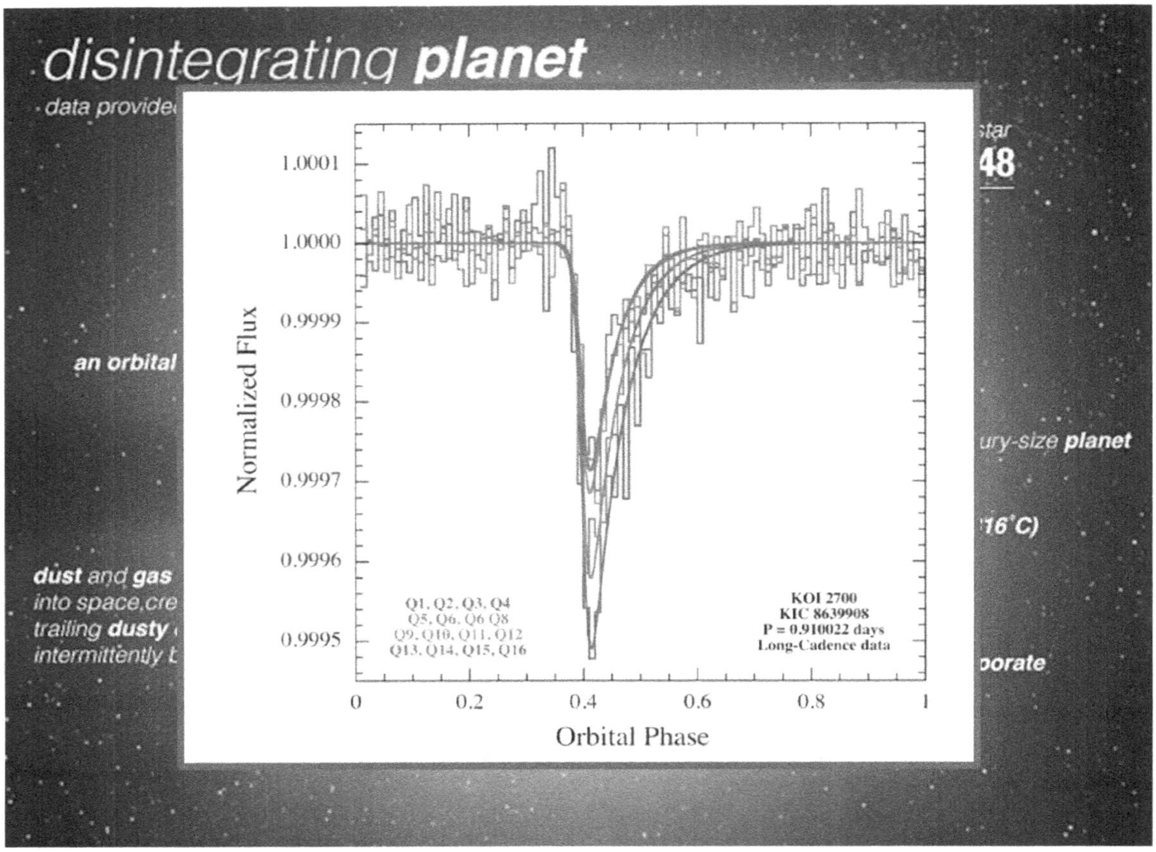
disintegrating planet
data provide
star
48
an orbital
ury-size planet
16°C)
dust and gas
into space,cre
trailing dusty
intermittently
oorate
Normalized Flux
1.0001
1.0000
0.9999
0.9998
0.9997
0.9996
0.9995
Orbital Phase
0
0.2
0.4
0.6
0.8
1
Q1, Q2, Q3, Q4
Q5, Q6, Q6 Q8
Q9, Q10, Q11, Q12
Q13, Q14, Q15, Q16
KOI 2700
KIC 8639908
P = 0.910022 days
Long-Cadence data

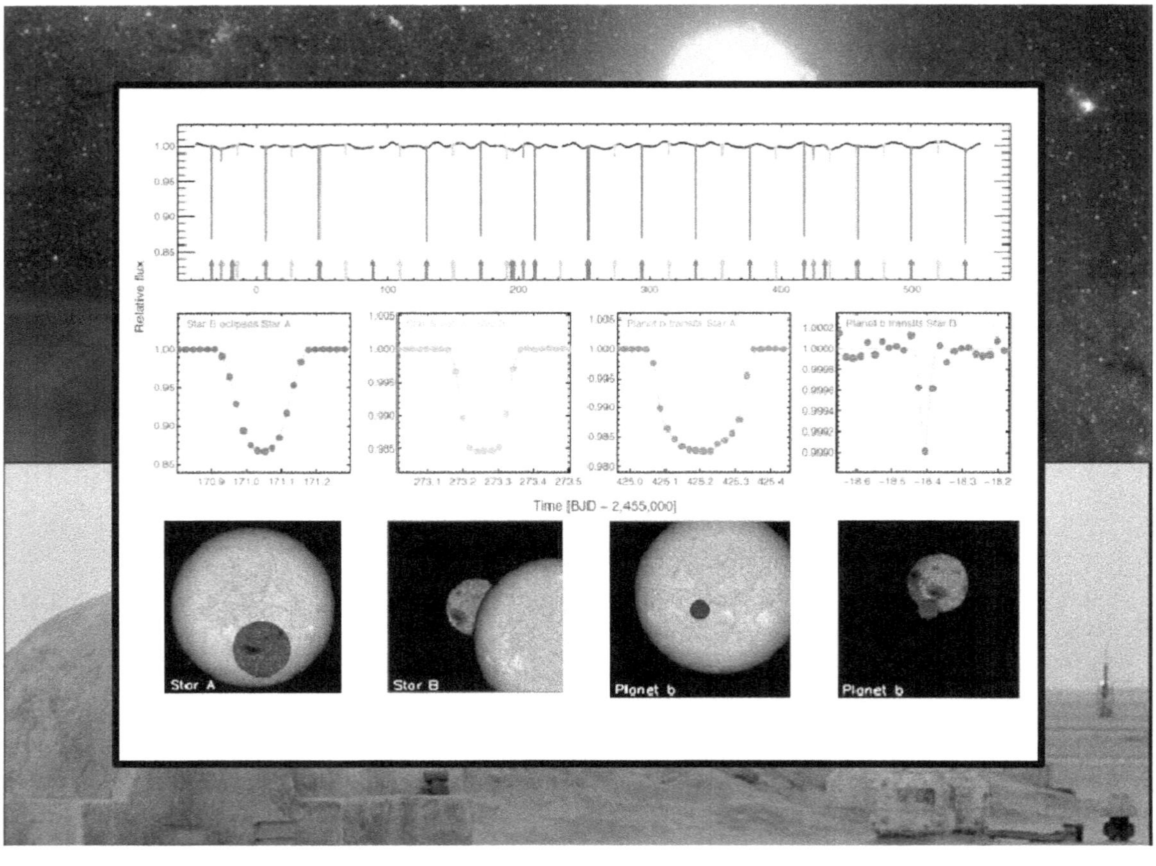
Relative flux
Star B eclipses Star A
Planet b transits Star A
Planet b transits Star B
Time [BJD − 2,455,000]
Star A
Star B
Planet b
Planet b

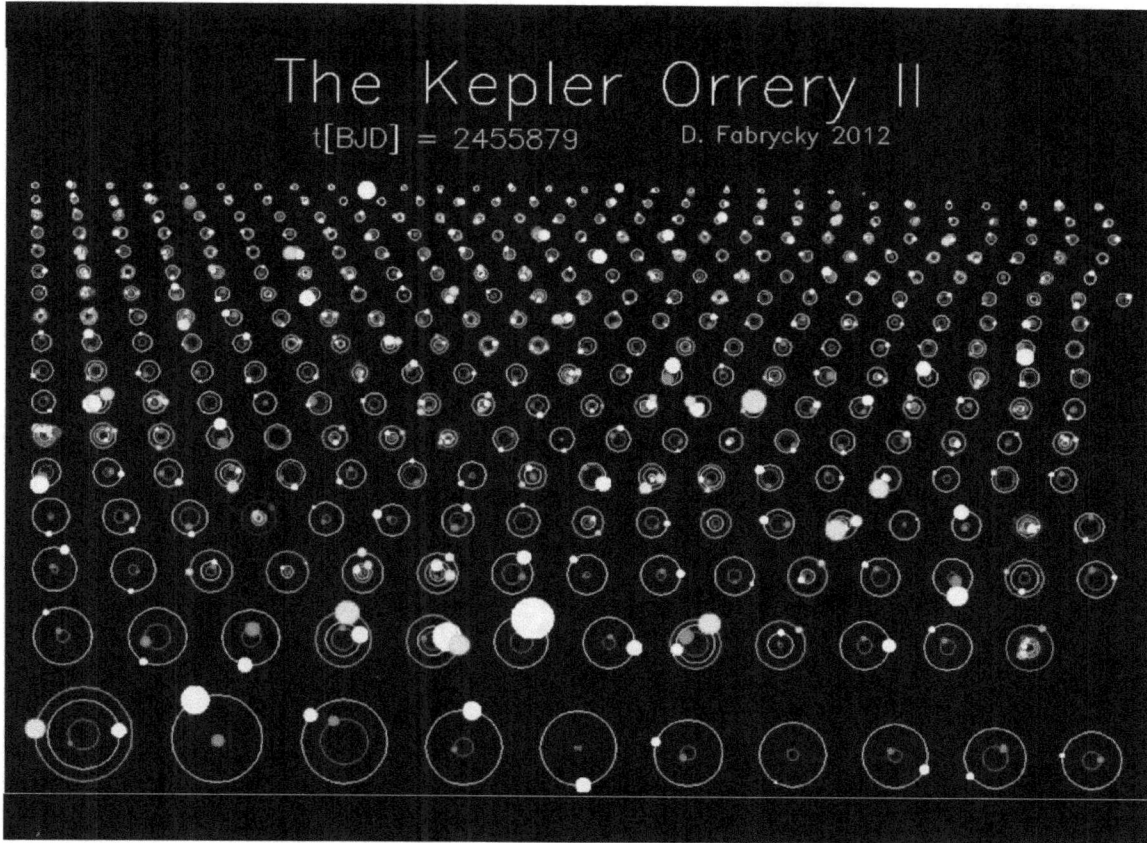

The Kepler Orrery II
t[BJD] = 2455879   D. Fabrycky 2012

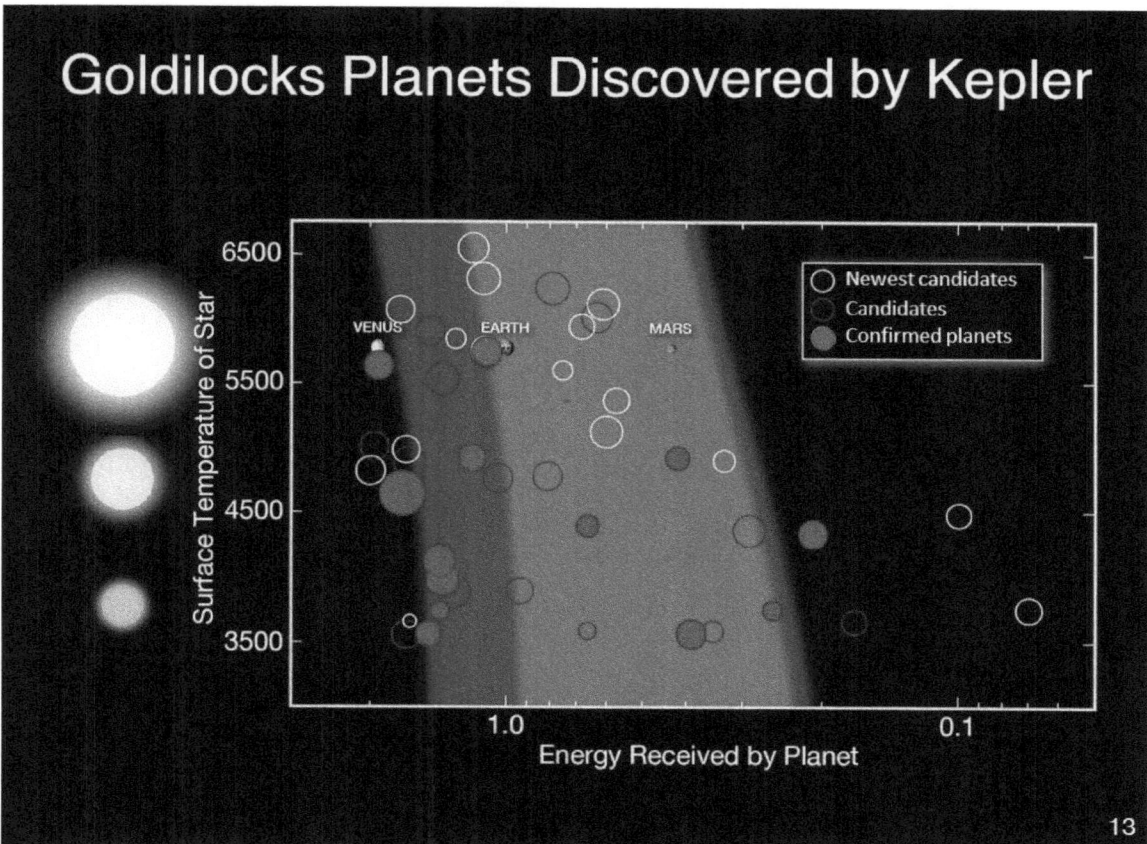

Goldilocks Planets Discovered by Kepler

KEPLER-452b:
EXOPLANET MOST LIKE EARTH
10% LARGER THAN SUN
ORBITS IN HABITABLE ZONE OF G2-TYPE STAR
365 DAYS IN A YEAR
DAYS IN A YEAR 385
1.6 TIMES THE SIZE OF EARTH

EXPERIENCE THE GRAVITY OF
HD 40307g
A SUPER EARTH
RELAX ON
KEPLER 16b
THE LAND OF TWO SUNS
WHERE YOUR SHADOW ALWAYS HAS COMPANY
Kepler-186f
WHERE THE GRASS IS ALWAYS REDDER ON THE OTHER SIDE
Exoplanet Travel Bureau
15

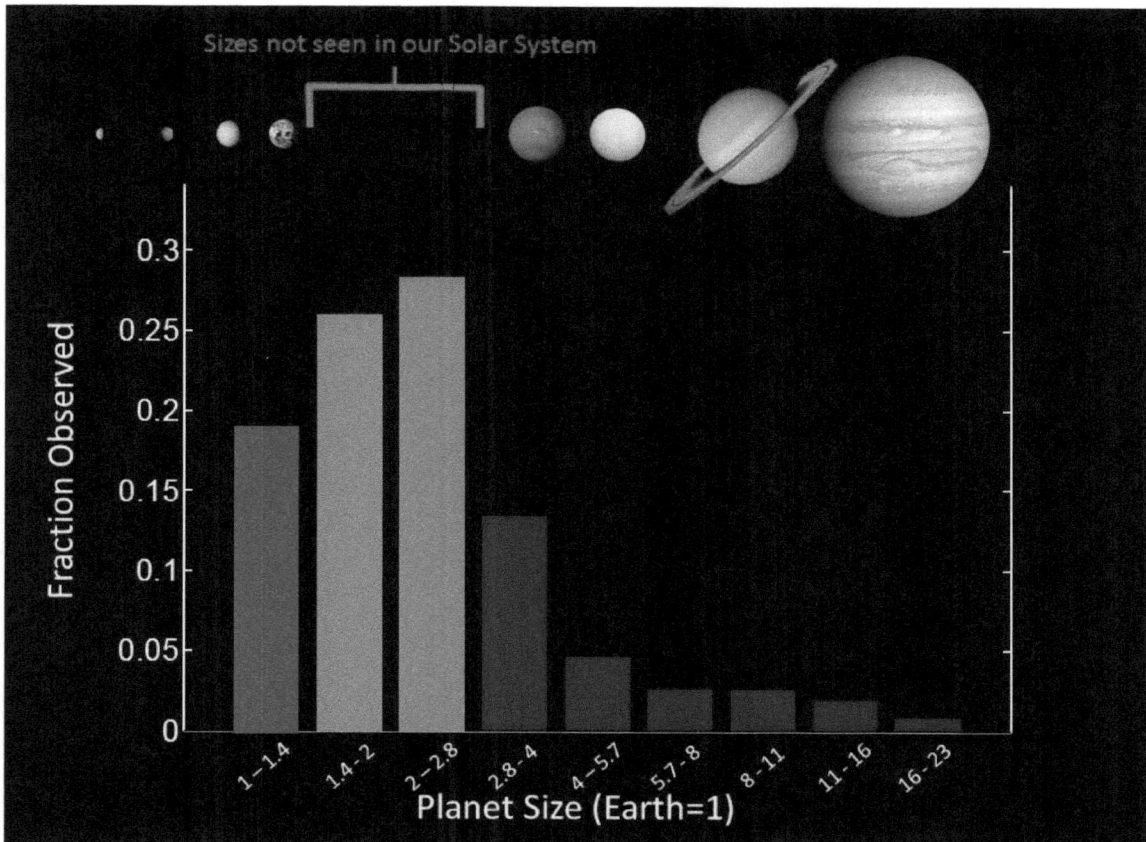

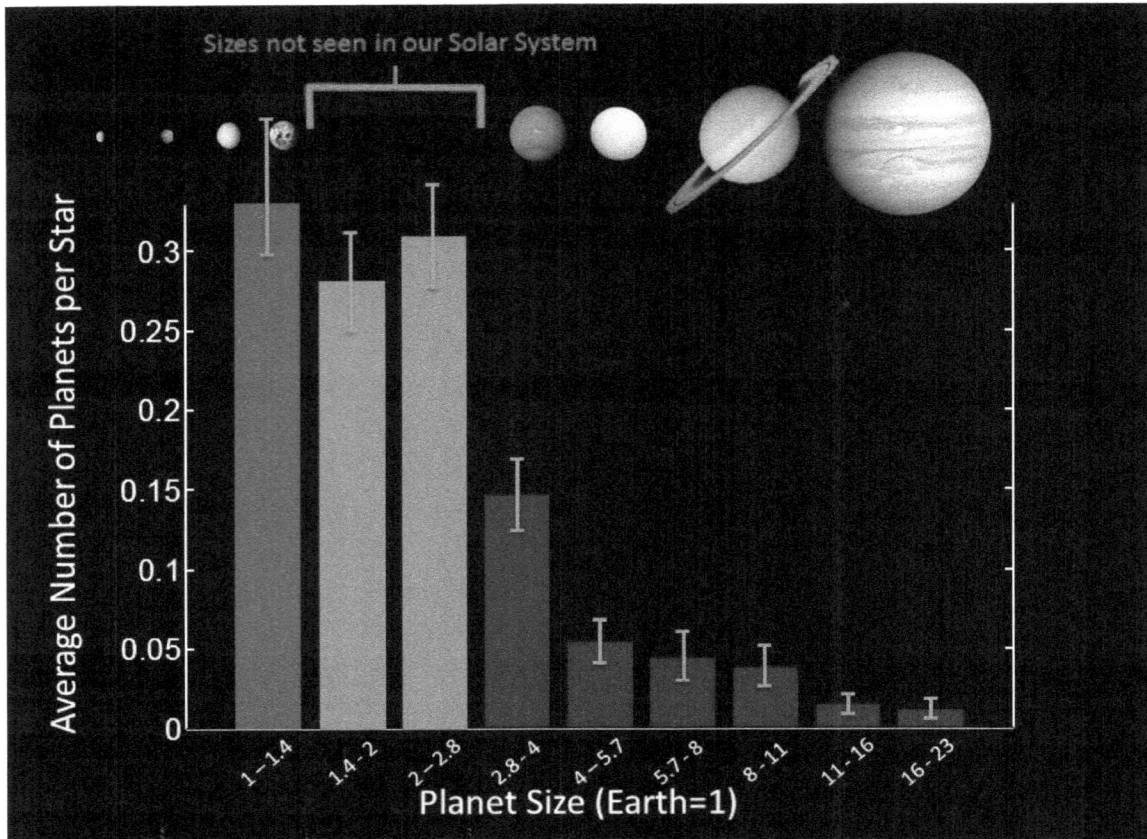

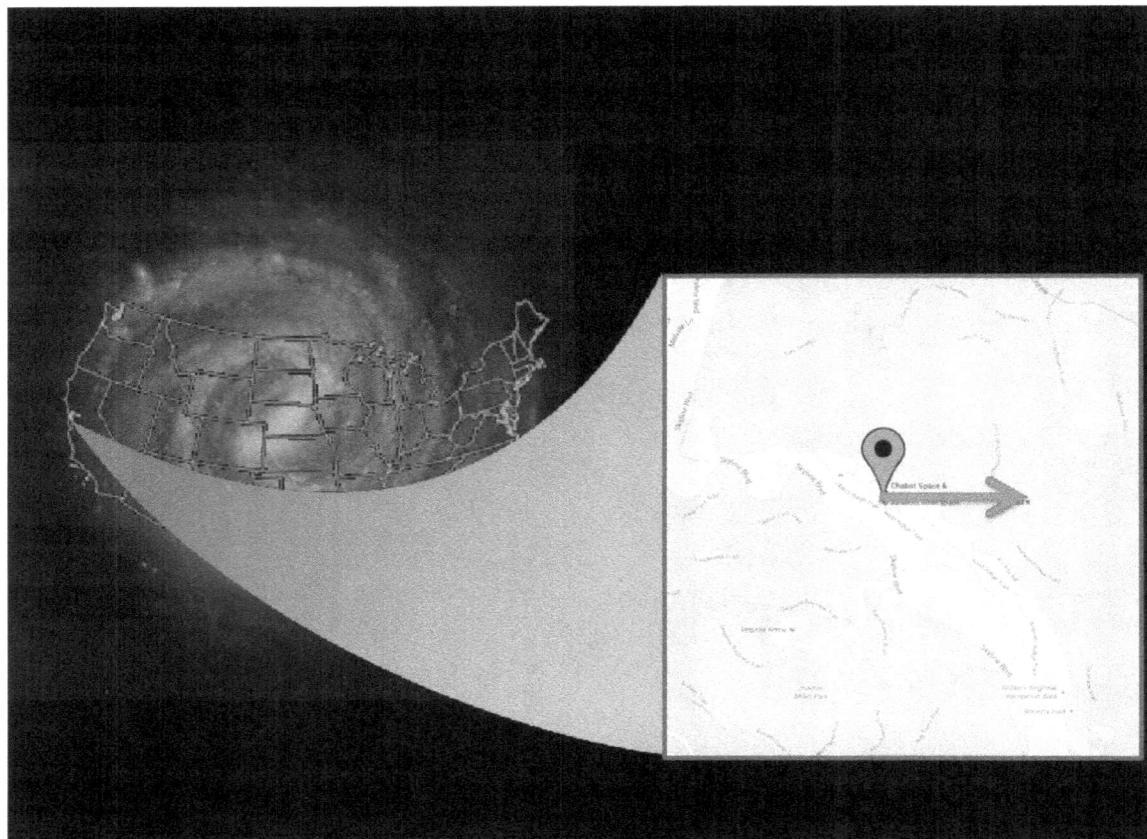

122

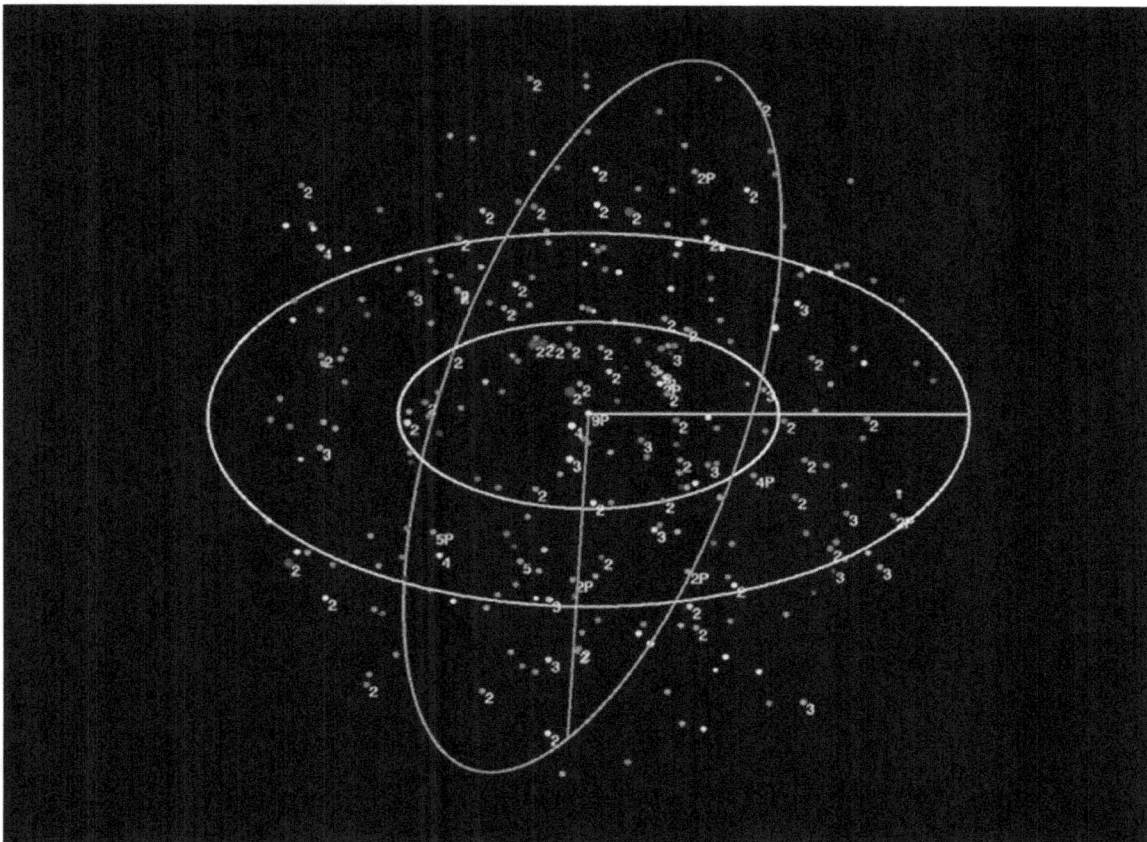

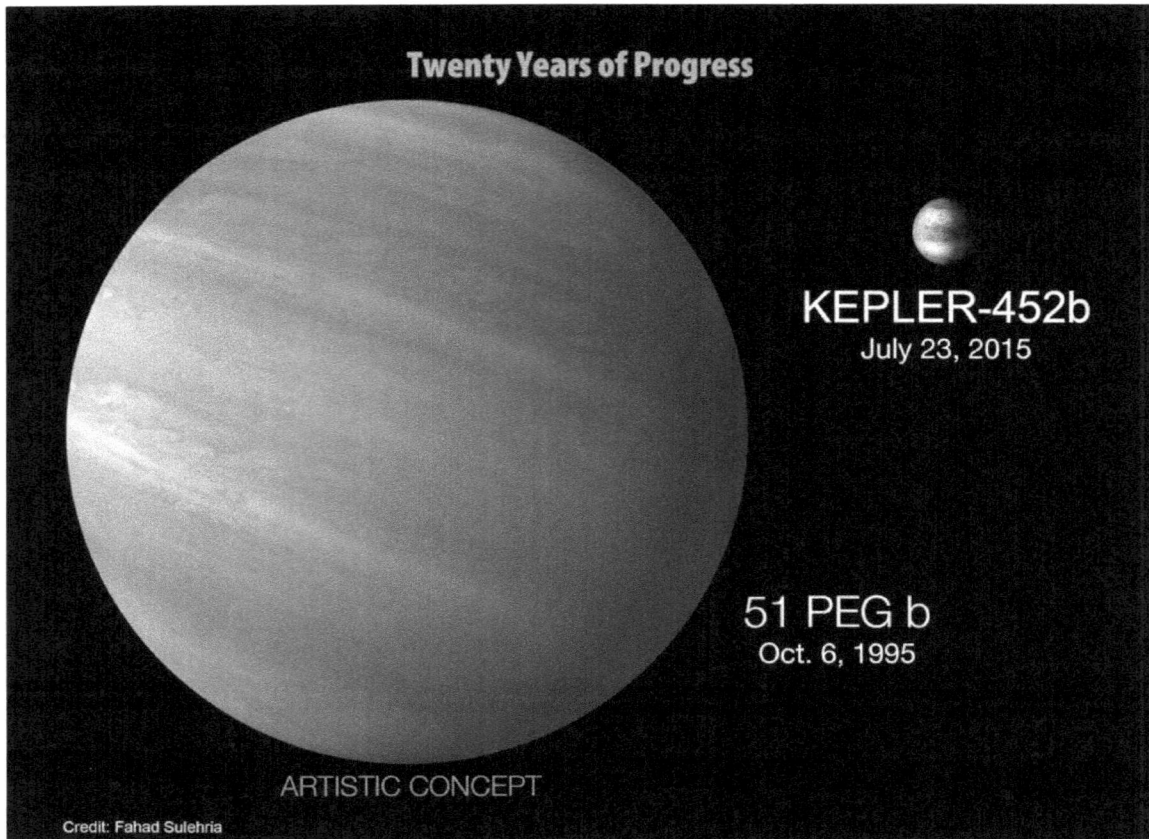

Twenty Years of Progress

KEPLER-452b
July 23, 2015

51 PEG b
Oct. 6, 1995

ARTISTIC CONCEPT

Credit: Fahad Sulehria

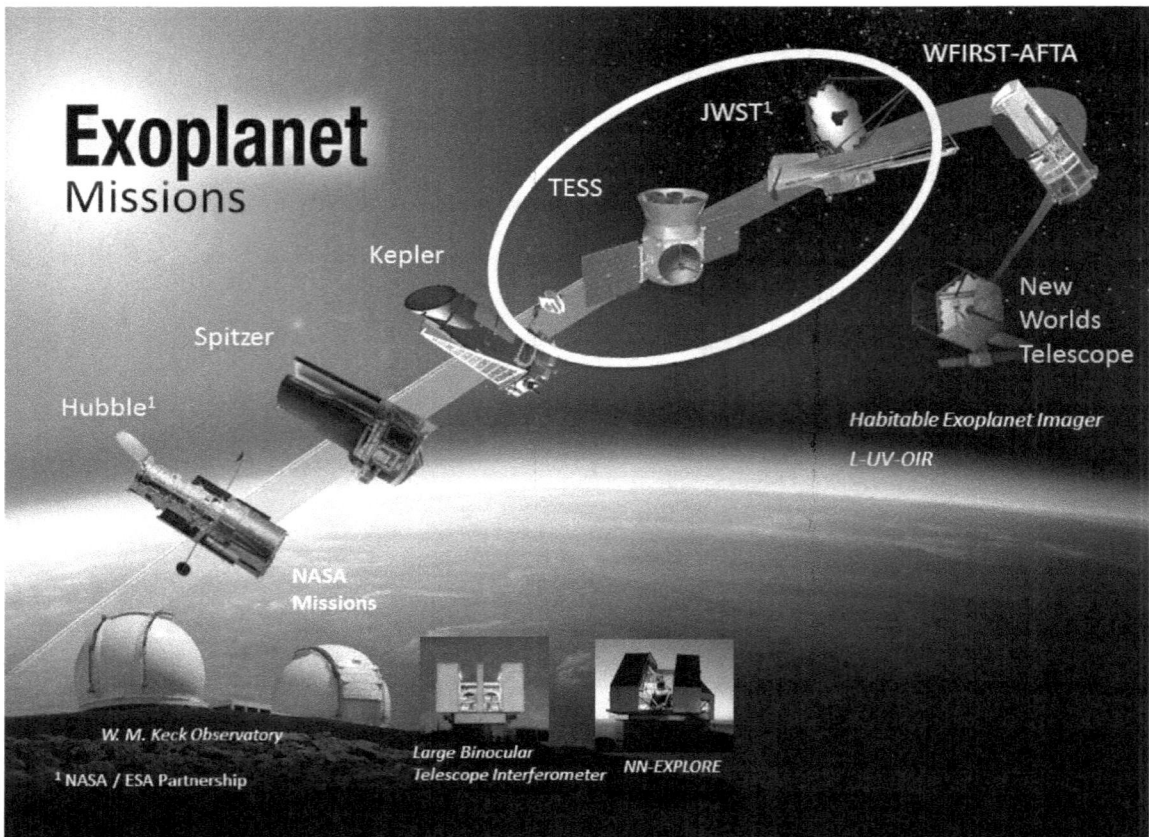

Exoplanet
Missions

WFIRST-AFTA

JWST[1]

TESS

Kepler

Spitzer

Hubble[1]

New
Worlds
Telescope

Habitable Exoplanet Imager

L-UV-OIR

NASA
Missions

W. M. Keck Observatory

Large Binocular
Telescope Interferometer   NN-EXPLORE

[1] NASA / ESA Partnership

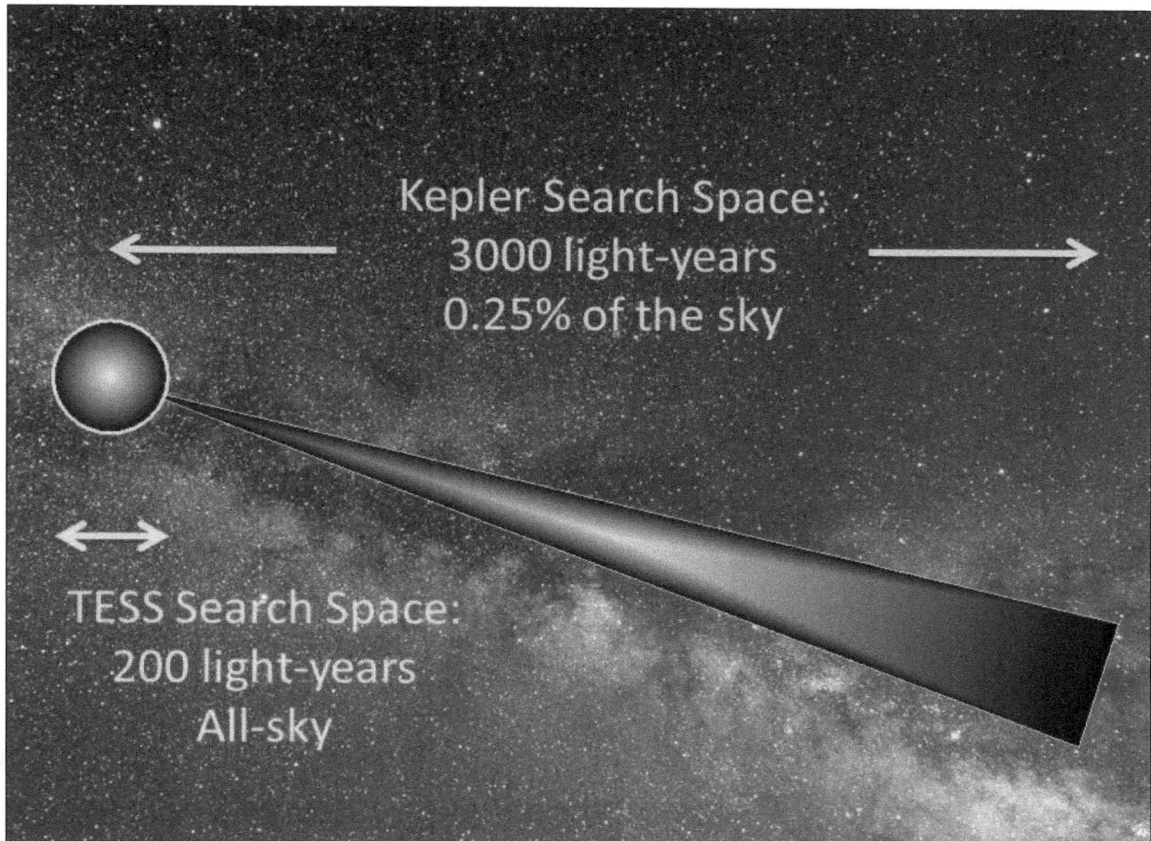

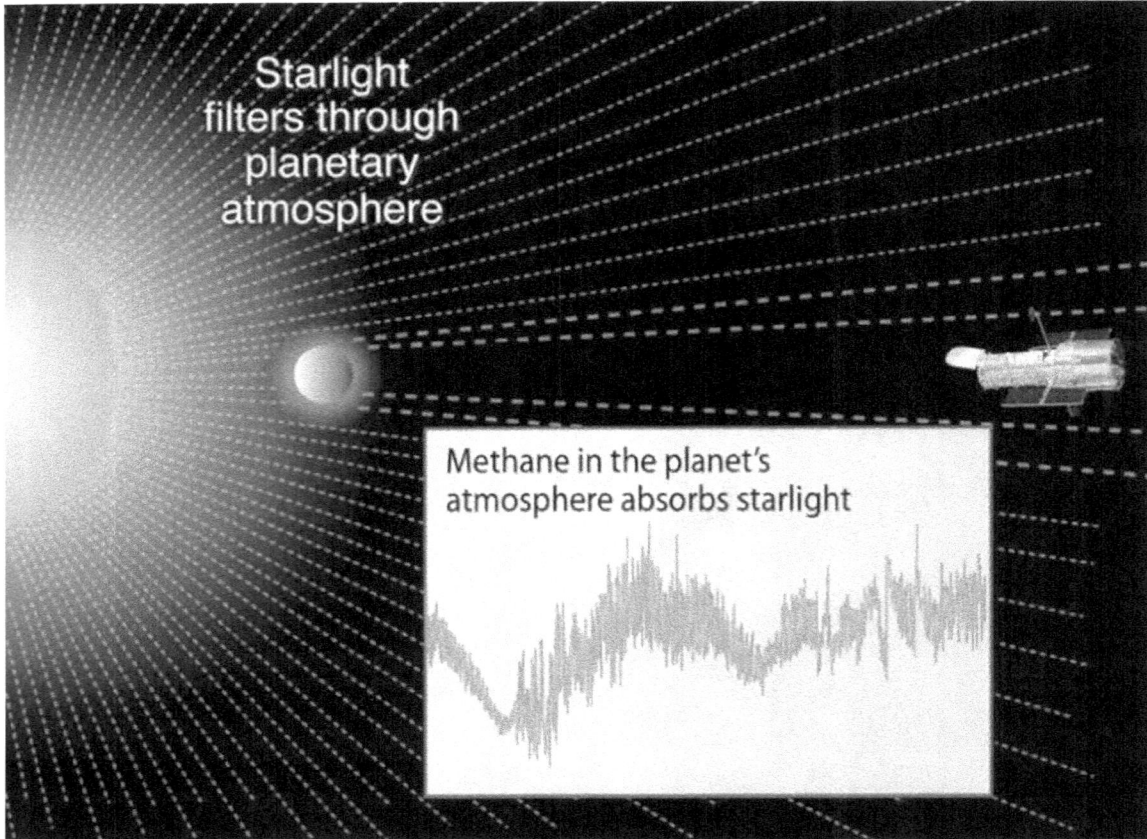

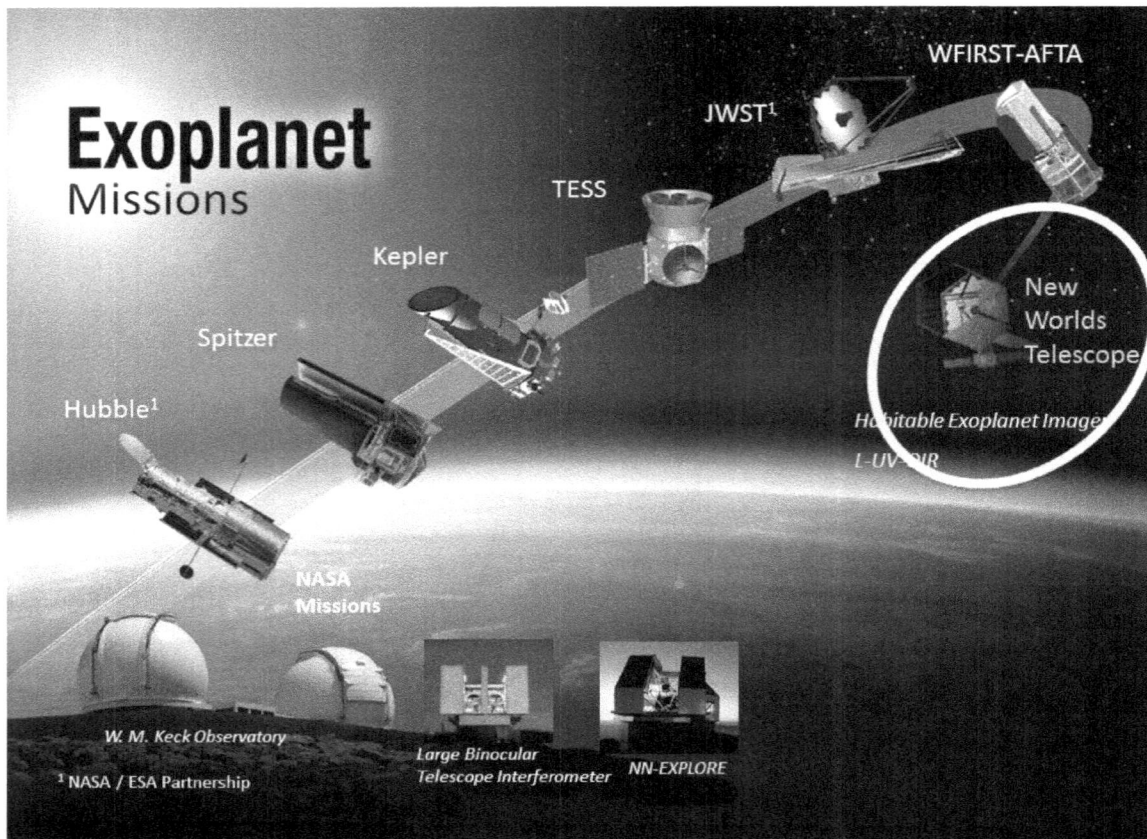

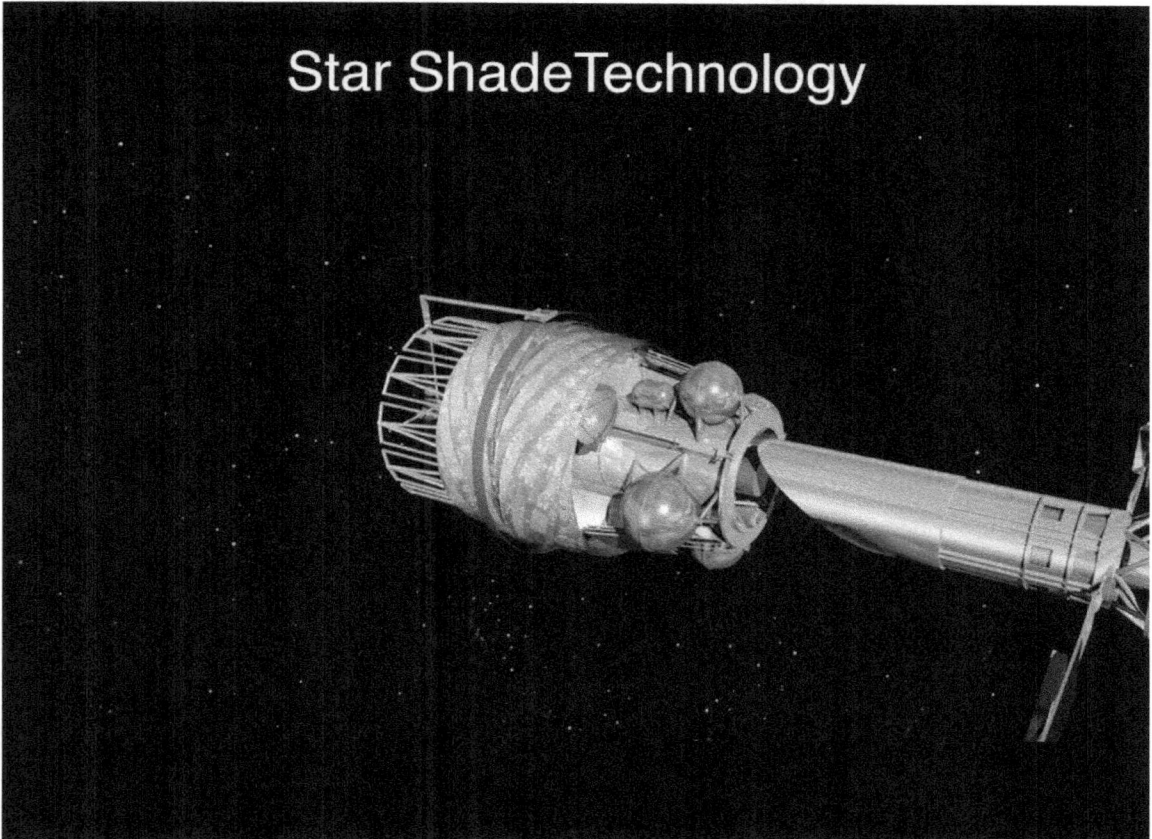

Star ShadeTechnology

Ground-based Coronagraph Discoveries

HR 8799 bcde

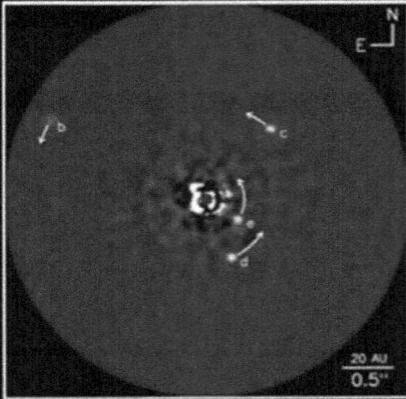

51 Eri b

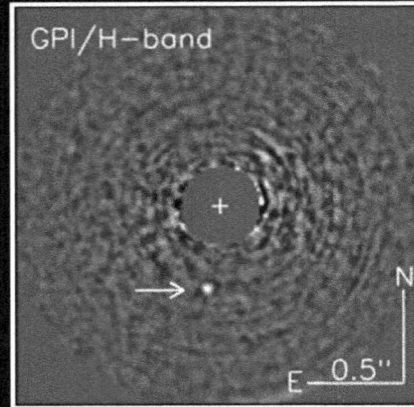

Solar System from 30 light-years away

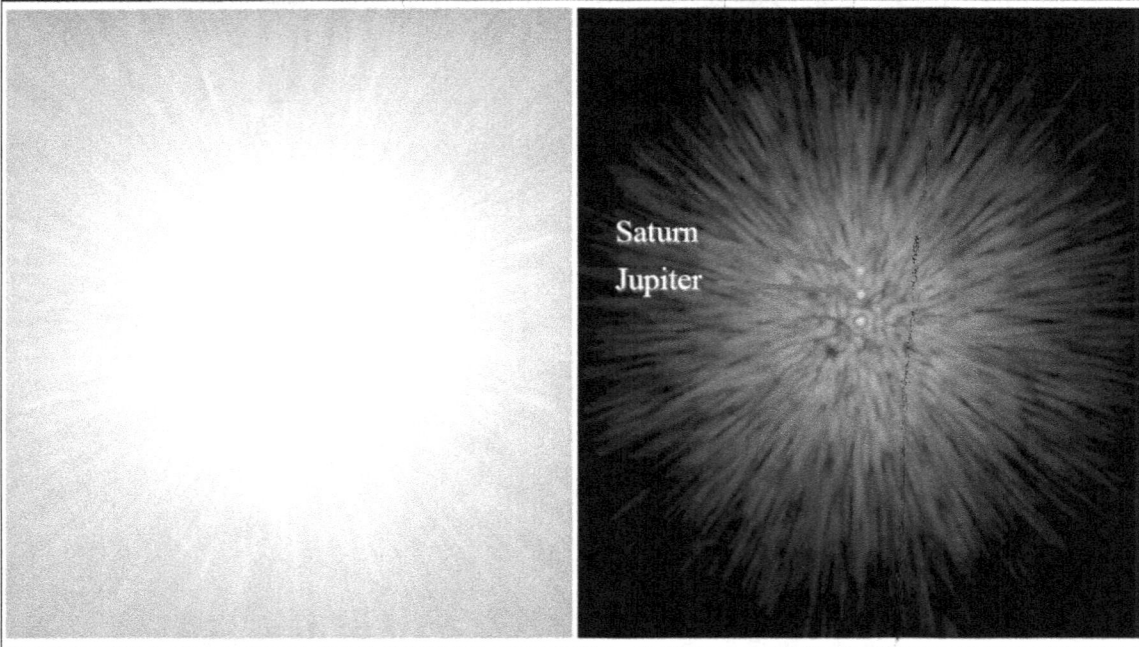

Large Space Telescope Required

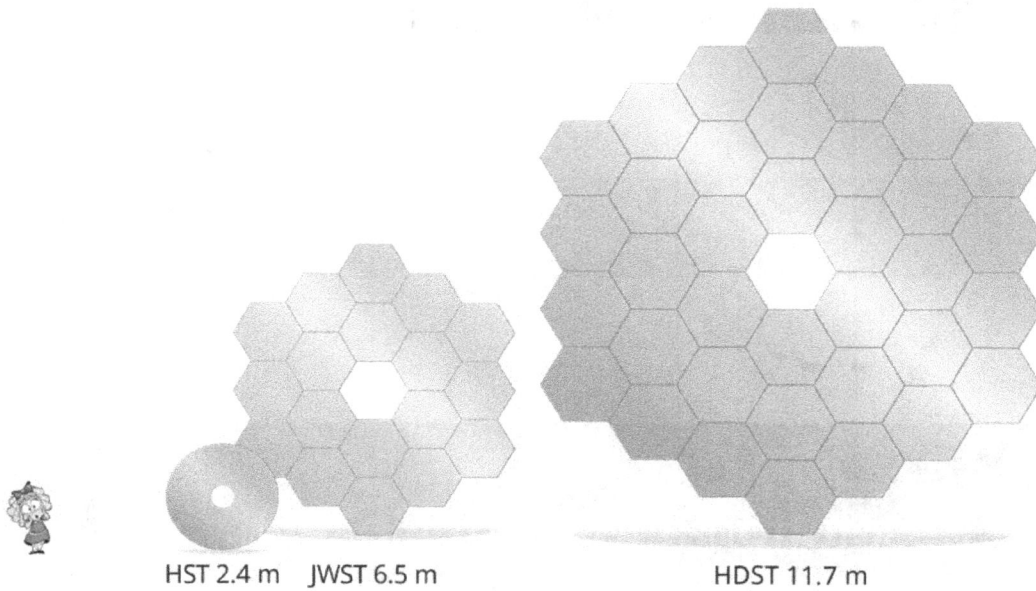

HST 2.4 m   JWST 6.5 m   HDST 11.7 m

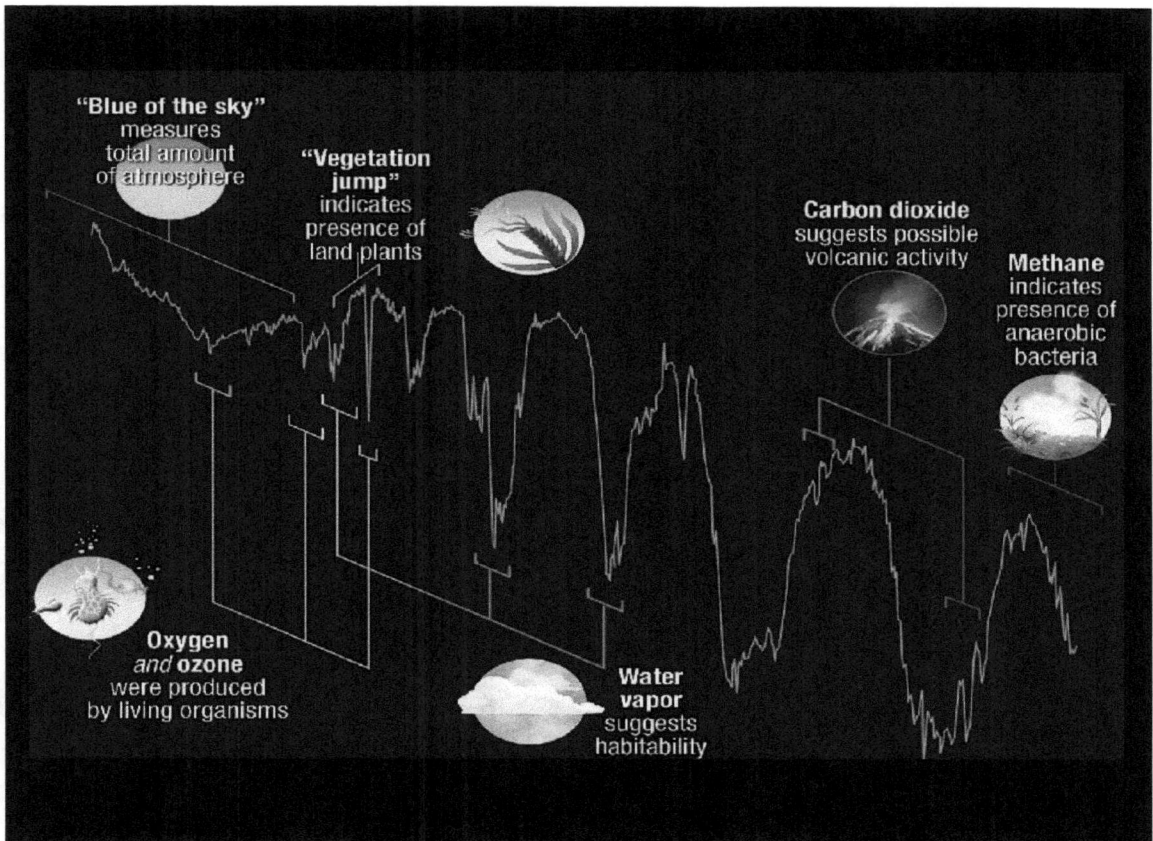

"Blue of the sky" measures total amount of atmosphere

"Vegetation jump" indicates presence of land plants

Carbon dioxide suggests possible volcanic activity

Methane indicates presence of anaerobic bacteria

Oxygen *and* ozone were produced by living organisms

Water vapor suggests habitability

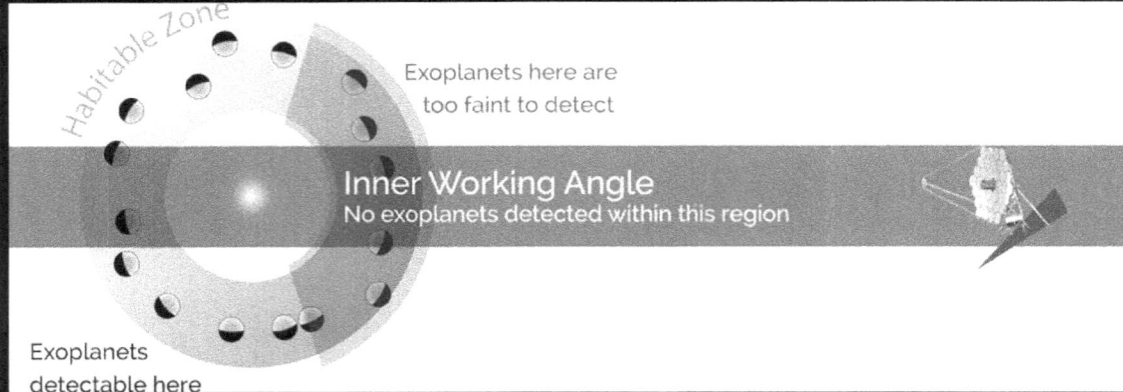

Why a Large Telescope?

Habitable Zone

Exoplanets here are too faint to detect

Inner Working Angle
No exoplanets detected within this region

Exoplanets detectable here

Resolution and Light Collecting Area

37

K2
Kepler's Second Mission

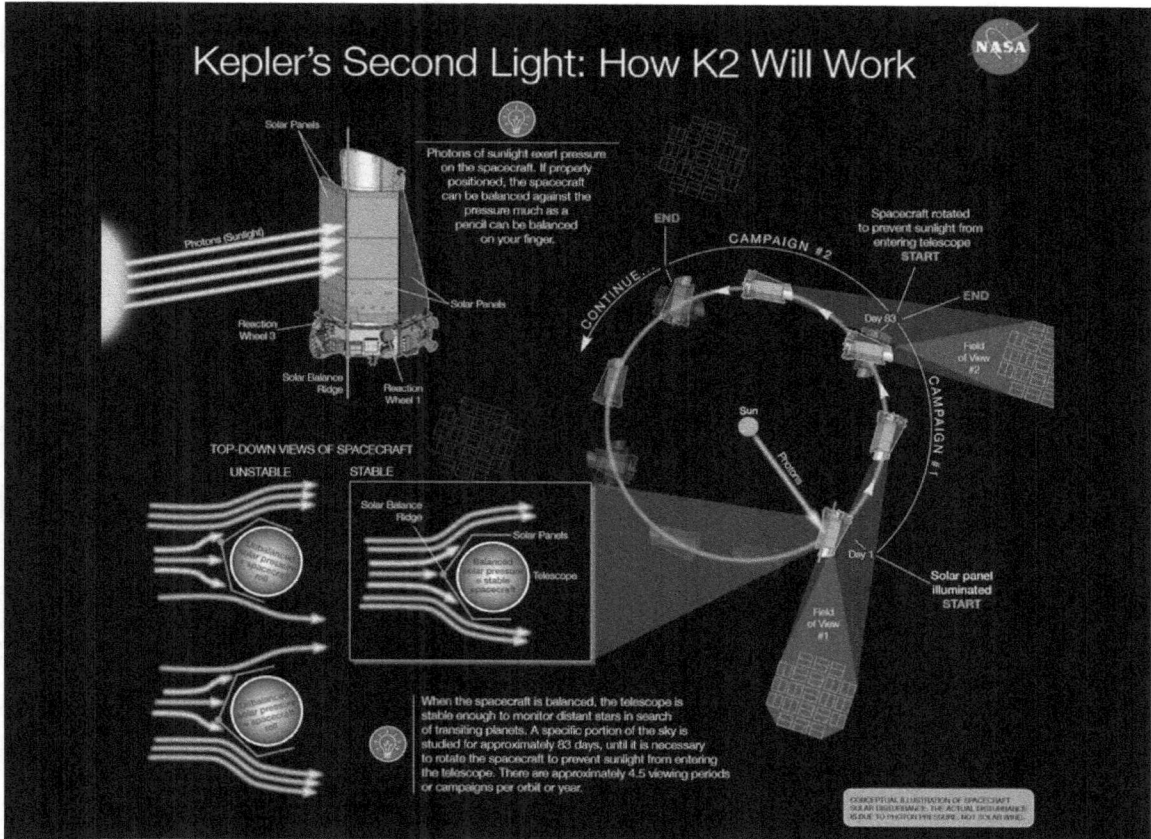

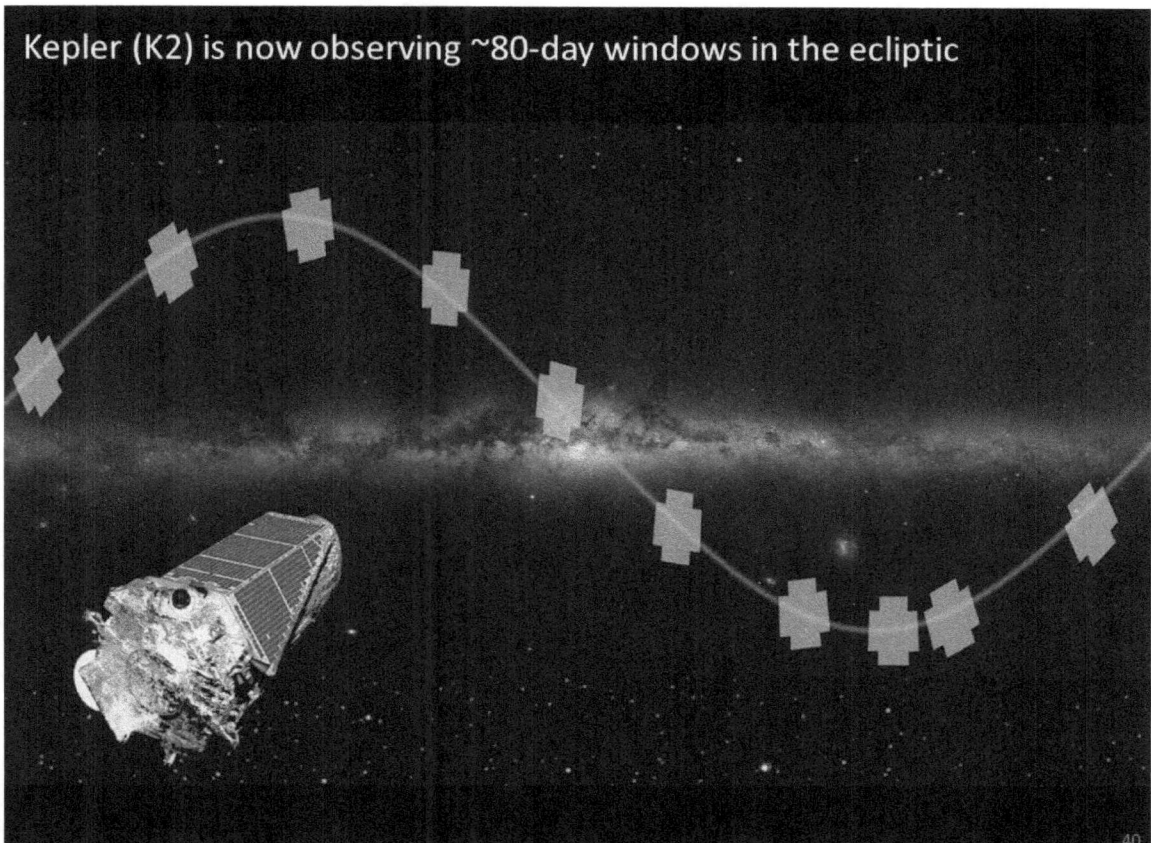

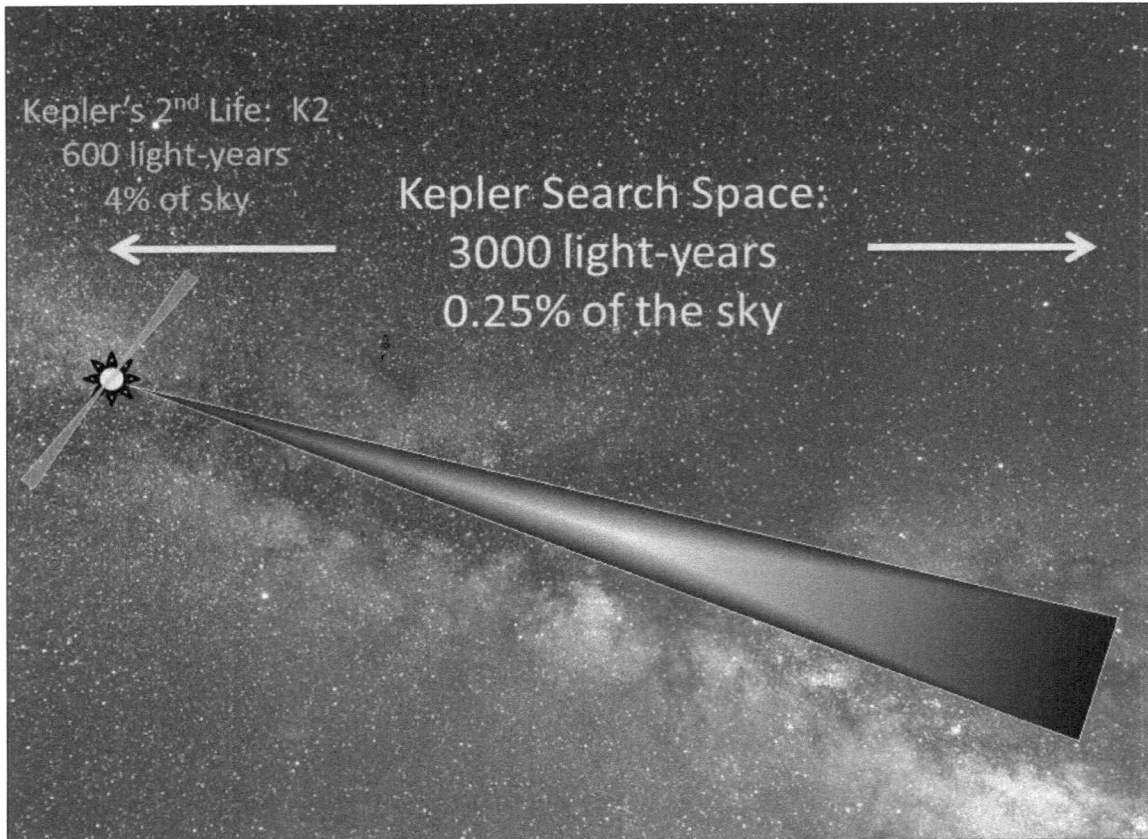

Kepler's 2nd Life: K2
600 light-years
4% of sky

Kepler Search Space:
3000 light-years
0.25% of the sky

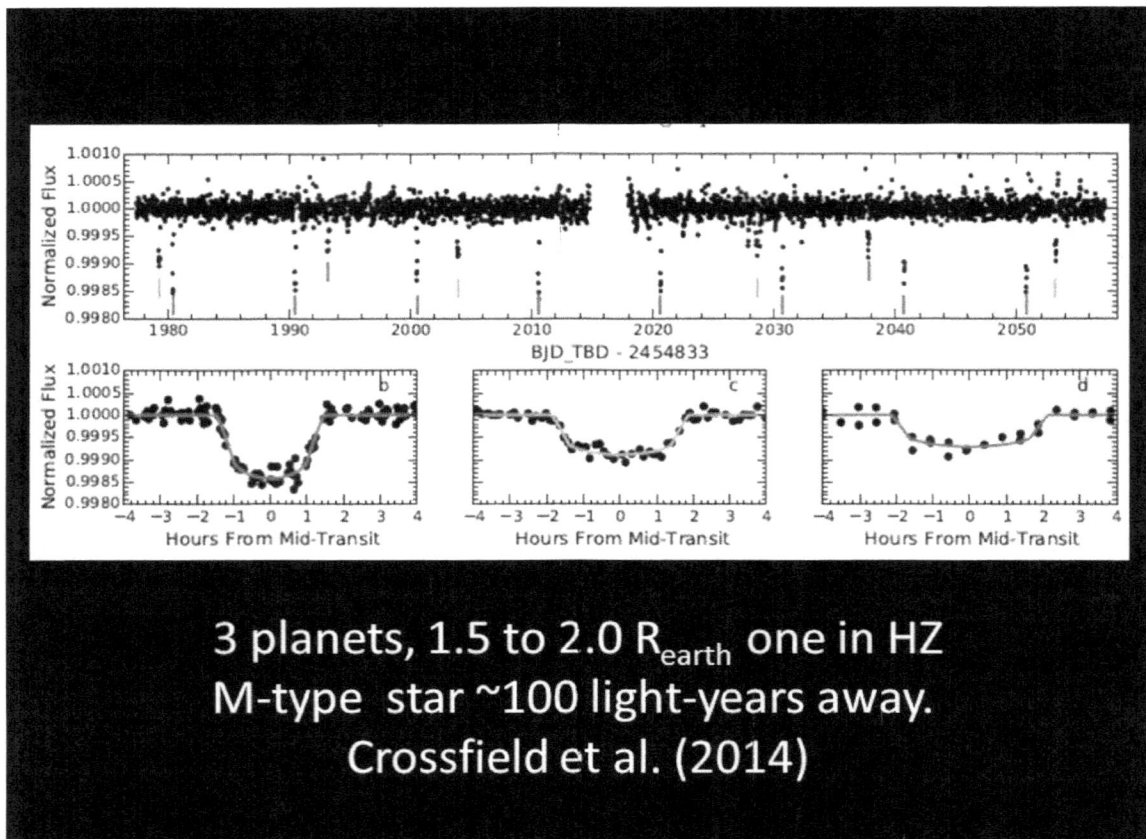

3 planets, 1.5 to 2.0 R_{earth} one in HZ
M-type star ~100 light-years away.
Crossfield et al. (2014)

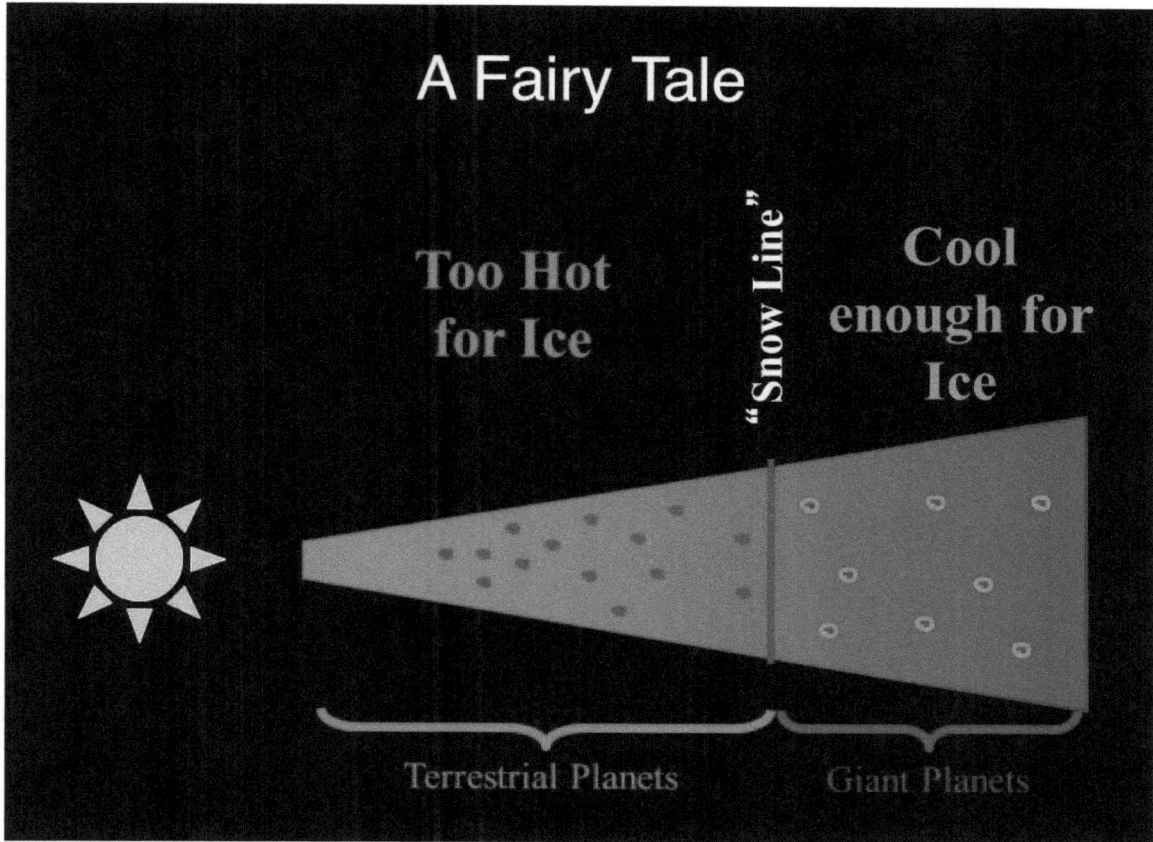

Remotely Detectable Biosignatures

Niki Parenteau, Nancy Kiang, Bob Blankenship, Esther Sanroma,
Enric Palle, Tori Hoehler, Bev Pierson, Vikki Meadows

NASA Astrobiology Institute, VPL

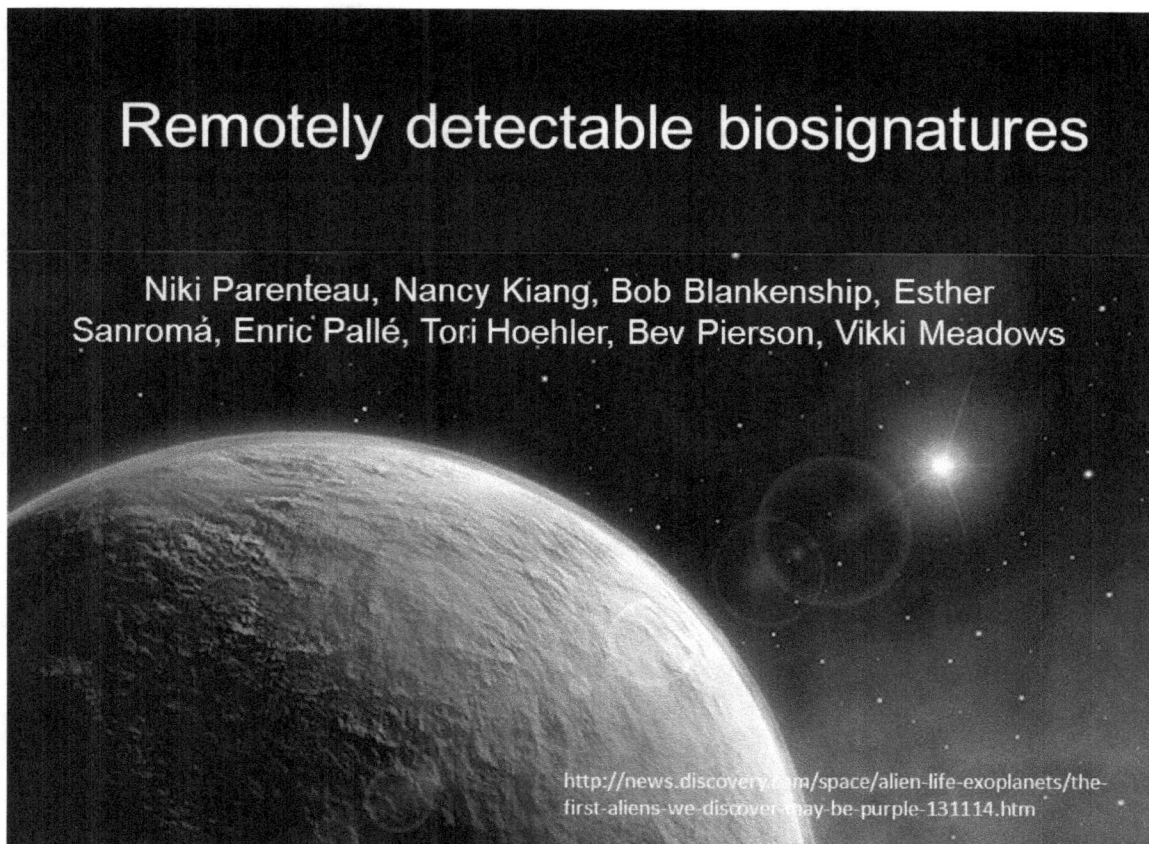

Linking the microscope to the telescope

Biologists
Niki Parenteau, *SETI & NASA Ames*
Nancy Kiang, *GISS*
Bob Blankenship, *WUSTL*
Tori Hoehler, *NASA Ames*
Bev Pierson, *UPS*

Astronomers
Esther Sanromá, *IAC*
Enric Pallé, *IAC*
Vikki Meadows, *UW*

Outline

- Examples of biosignatures
- Problems
  - False positives
  - Modern Earth
- New experimental work
- Detectability
  - disk averaged spectra
- Future work
  - multiple lines of evidence

Life's Global Impact

To be detectable, non-technologically capable life must modify its environment on a global scale
 - atmosphere, surface, temporal behavior

Exoplanet biosignatures

- Chemical disequilibrium in atmosphere (Lovelock, 1965)
- Biogenic gases and their byproducts (Pilcher, 2003; Domagal-Goldman et al., 2011)
  - e.g., O_2/O_3, N_2O, CH_3Cl
- Planetary surface features
  - Red edge of vegetation (Segura et al., 2005; Seager et al., 2005; Des Marais et al., 2008)
  - Polarization due to chiral biomolecules (Wolstencroft and Breon, 2005; Sparks et al., 2009)
  - Non-photosynthetic pigments
    - Bacteriorhodopsin (DasSarma, 2007)
    - Carotenoids (Schweiterman et al., 2015)
    - Misc. pigments (Hegde et al., 2015)

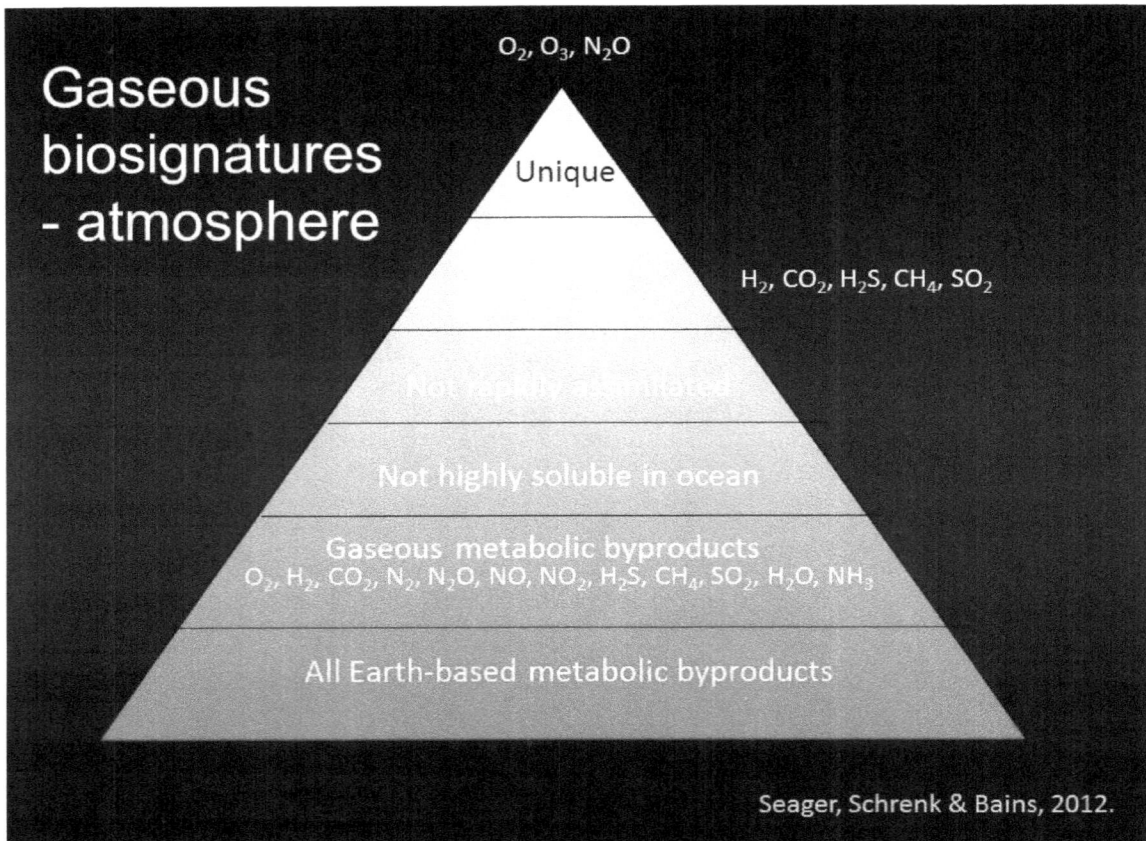

Gaseous biosignatures - atmosphere

O_2, O_3, N_2O

Unique

$H_2, CO_2, H_2S, CH_4, SO_2$

Not rapidly assimilated

Not highly soluble in ocean

Gaseous metabolic byproducts
$O_2, H_2, CO_2, N_2, N_2O, NO, NO_2, H_2S, CH_4, SO_2, H_2O, NH_3$

All Earth-based metabolic byproducts

Seager, Schrenk & Bains, 2012.

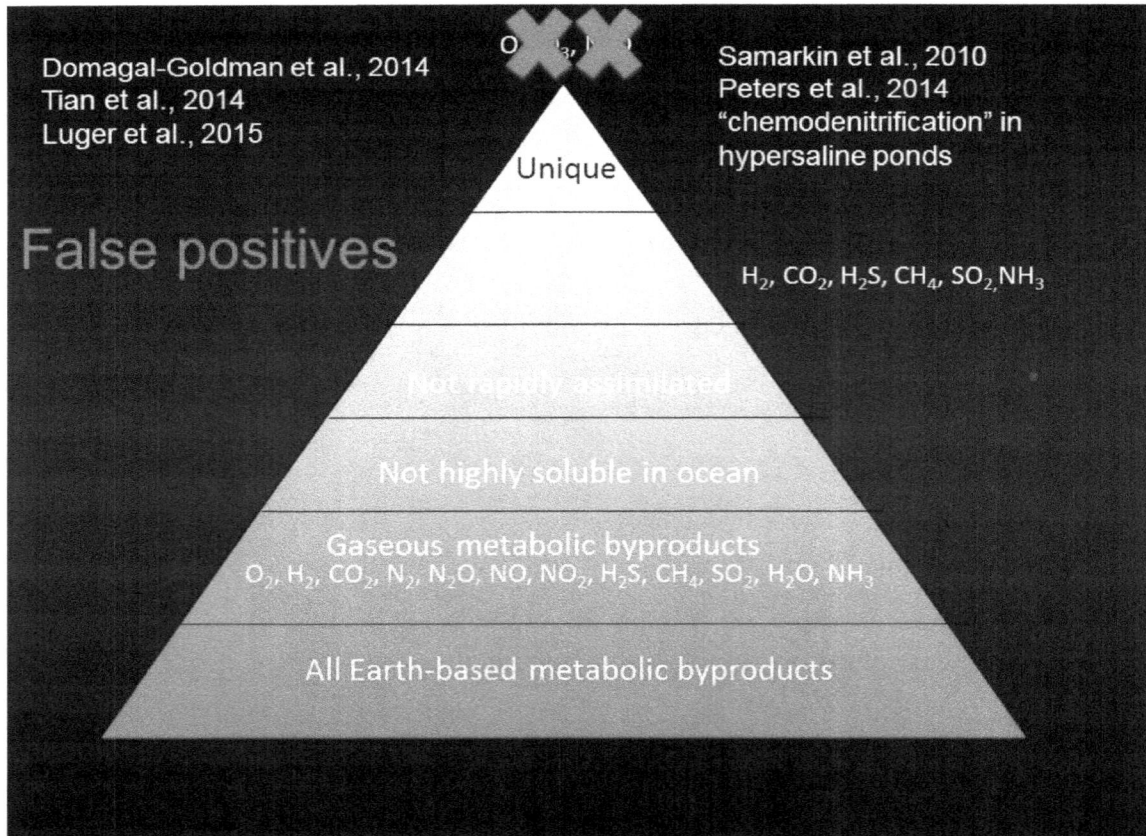

Domagal-Goldman et al., 2014
Tian et al., 2014
Luger et al., 2015

Samarkin et al., 2010
Peters et al., 2014
"chemodenitrification" in hypersaline ponds

False positives

Unique

$H_2, CO_2, H_2S, CH_4, SO_2, NH_3$

Not rapidly assimilated

Not highly soluble in ocean

Gaseous metabolic byproducts
$O_2, H_2, CO_2, N_2, N_2O, NO, NO_2, H_2S, CH_4, SO_2, H_2O, NH_3$

All Earth-based metabolic byproducts

138

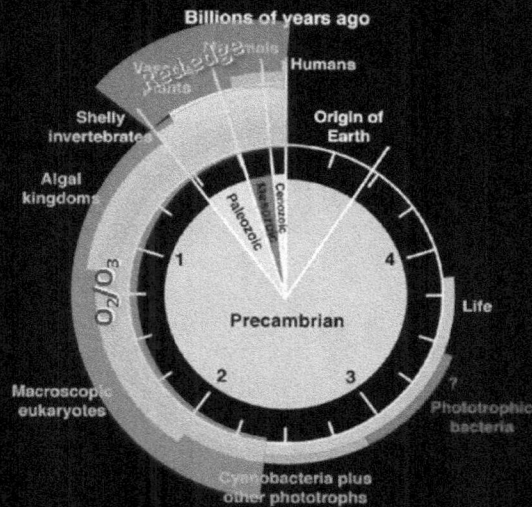

Strongest known planetary-scale biosignatures
- atmospheric oxygen
- vegetation red edge

Des Marais (2000)
Science 289:1703-1705

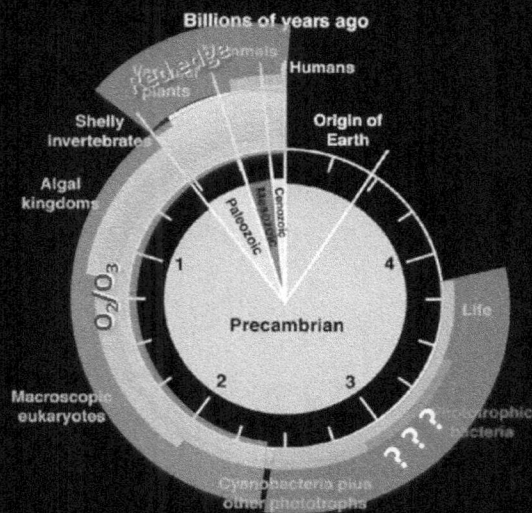

Remotely detectable biosignatures of anoxygenic phototrophs and cyanobacteria pre-GOE are poorly characterized

- Early Earth as an exoplanet

- Earth has looked different through time

- Work is needed to identify biosignatures for the first ~40% of Earth's history

Des Marais (2000)
Science 289:1703-1705

Photosynthetic pigments of anoxygenic phototrophs are well suited to absorb the NIR radiation of M dwarf stars

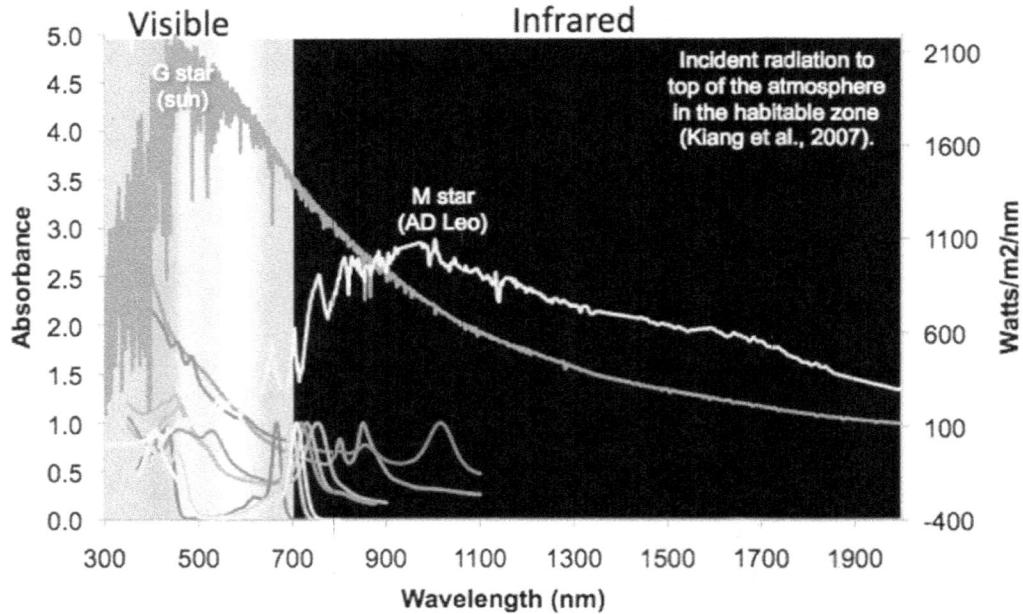

Chlorophyll pigment absorbs in the red, reflects in the near-infrared, which we can't see.

Also reflect green light, which is why plants are green.

Vegetation red edge

Gates et al., 1965; Knipling, 1970

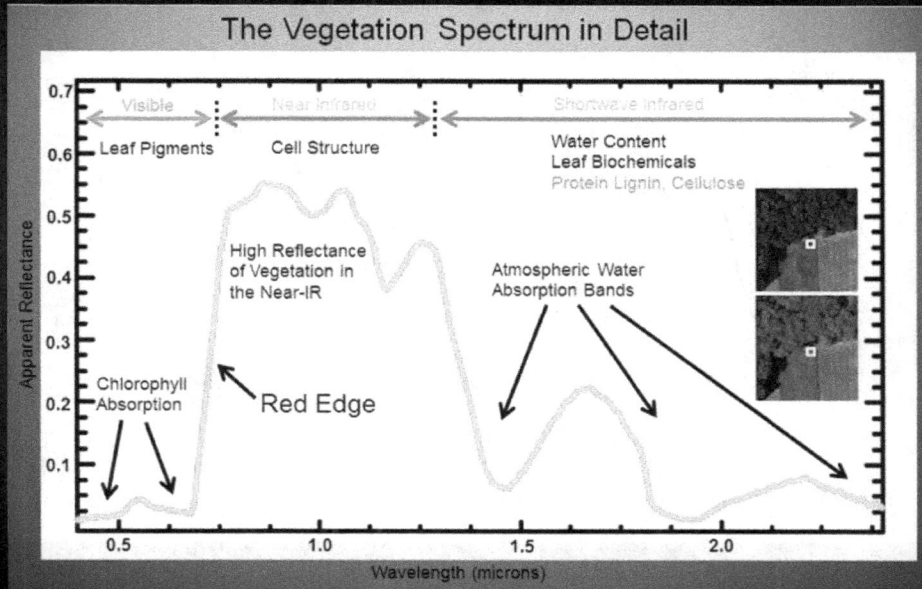

The Vegetation Spectrum in Detail

http://www.markelowitz.com/Hyperspectral.html

Reflectance spectra
pure cultures of anoxygenic phototrophs

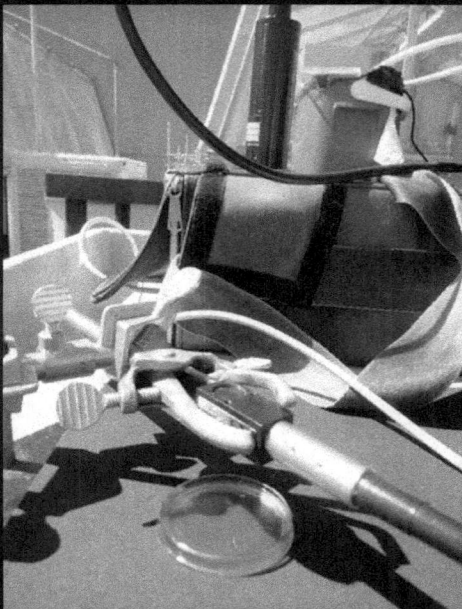

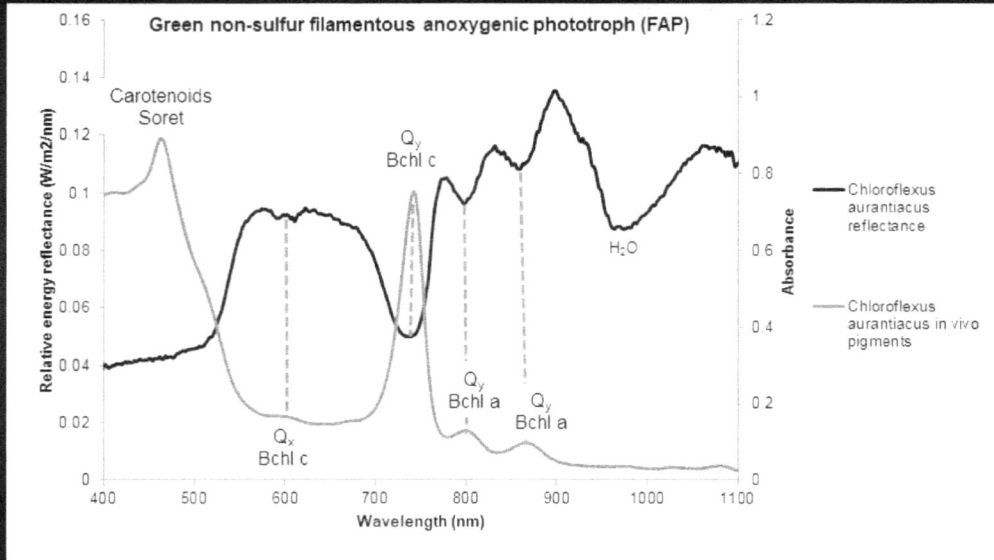

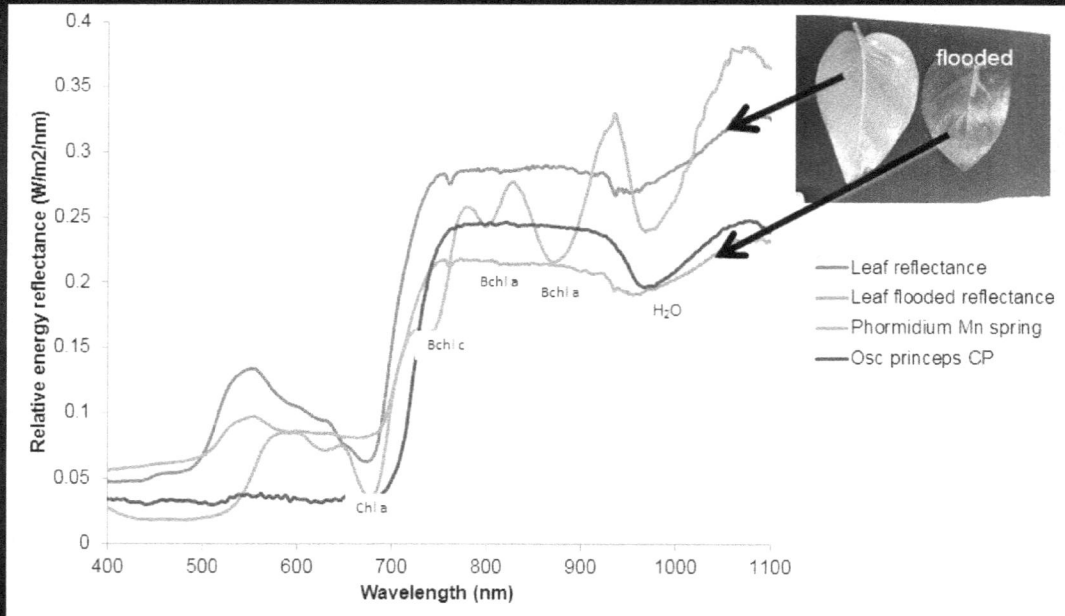

Environmental samples
Microbial mats

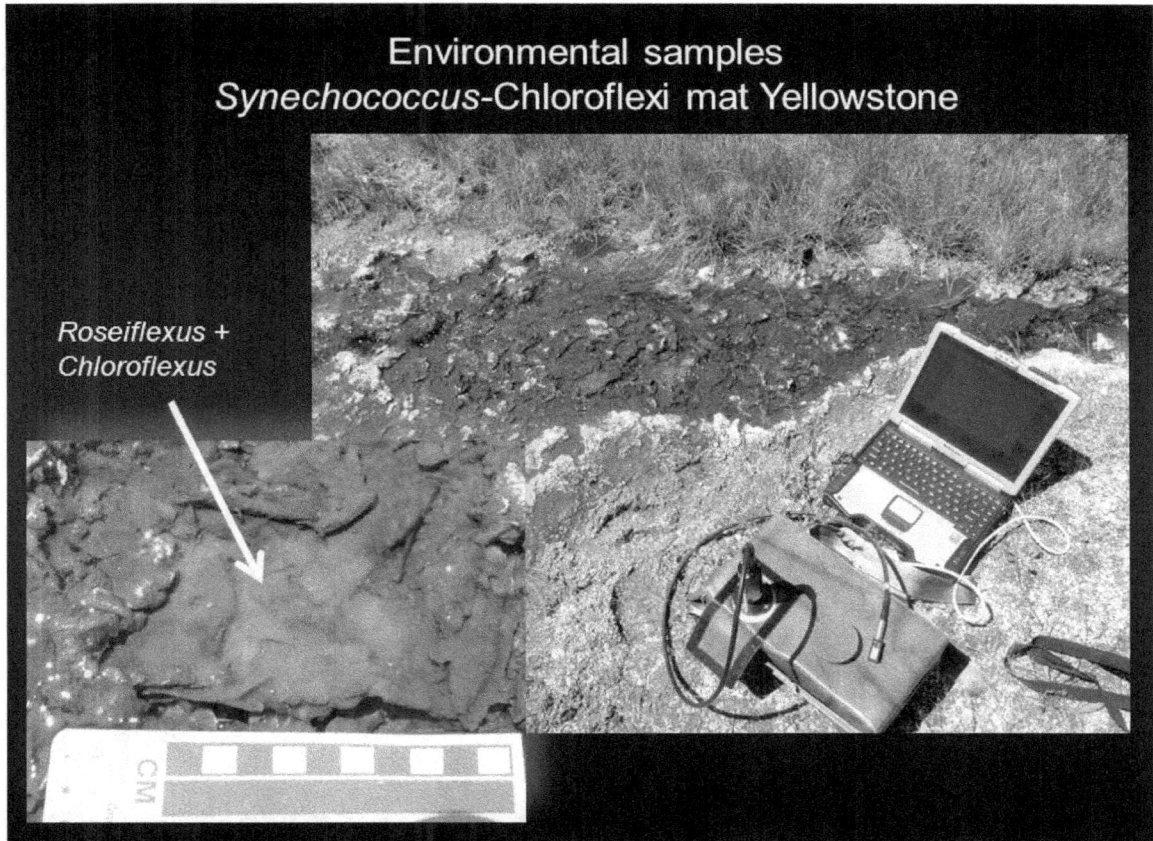

Environmental samples
Synechococcus-Chloroflexi mat Yellowstone

*Roseiflexus +
Chloroflexus*

Laminated microbial mat
Phormidium-Chloroflexi mat Yellowstone

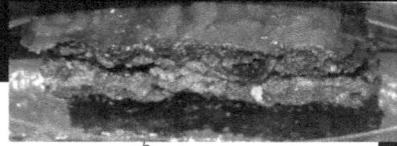

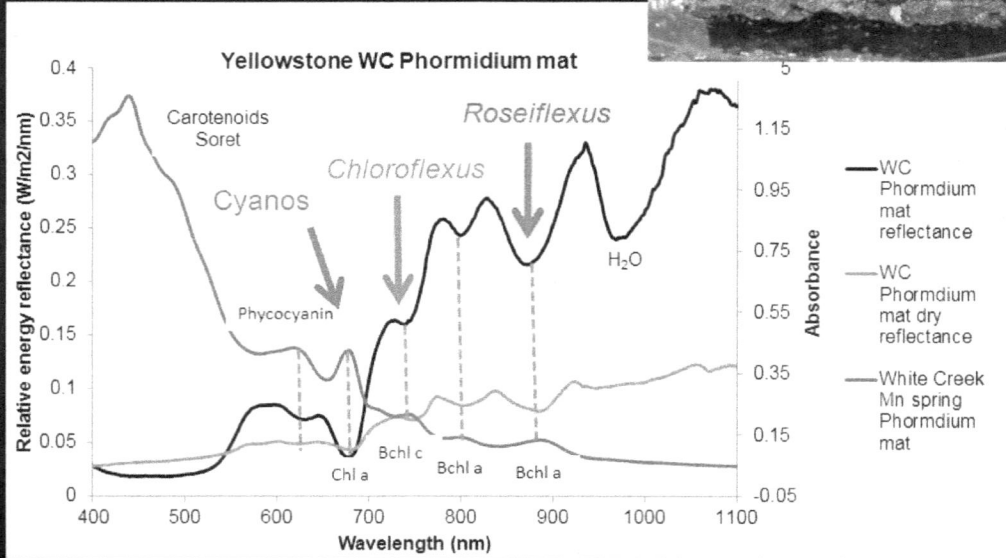

Microbial life on M dwarf exoplanets - Purple planet?

144

Detectability

Characterizing the purple Earth: Modelling the globally-integrated spectral variability of the Archean Earth

E. Sanromá[1,2], E. Pallé[1,2], M. N. Parenteau[3,4], N. Y. Kiang[5], A. M. Gutiérrez-Navarro[6], R. López[1,2] and P. Montañés-Rodríguez[1,2]

Instituto de Astrofísica de Canarias (IAC), Vía Láctea s/n 38200, La Laguna, Spain

mesr@iac.es

• radiative transfer modeling
• disk averaged spectra

http://news.discovery.com/space/alien-life-exoplanets/the-first-aliens-we-discover-may-be-purple-131114.htm

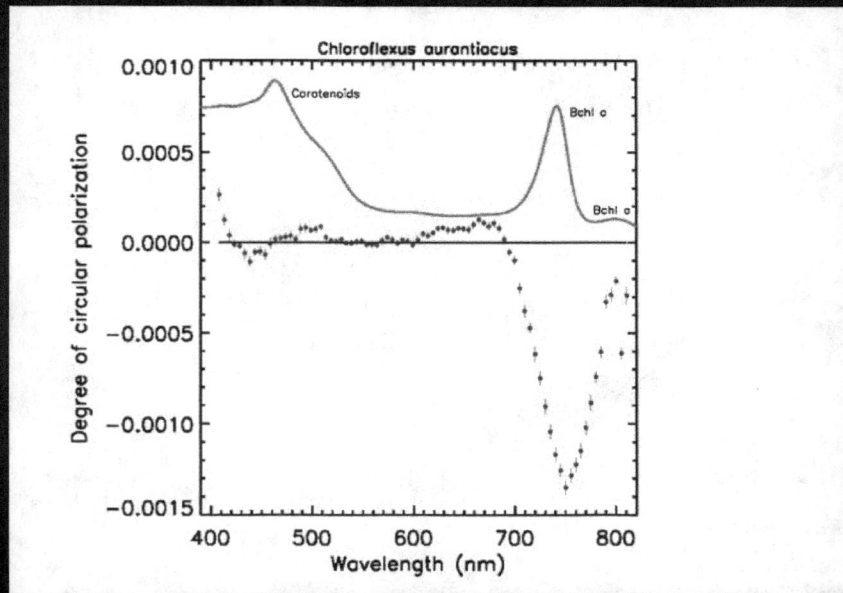

Multiple lines of evidence
Polarization due to chiral biomolecules

Summary

- New biosignatures
  - primitive phototrophs
  - community signatures
- Understanding rules for modeling
- False positives? Multiple lines of evidence...
  - e.g., reflectance + spectropolarimetry + gas?
- Detectability
  - Disk averaged spectra: realistic % areal coverage

Acknowledgments

- NAI VPL
- Yellowstone National Park
  - Christie Hendrix
  - Stacey Gunther
- Wilbur Hot Springs
- Nancy Kiang
- Bob Blankenship
- Vikki Meadows
- Esther Sanromá
- Enric Pallé
- Bill Sparks
- Tori Hoehler
- Bev Pierson

vpl

100 YEAR STARSHIP™

What's Just Right for Life?

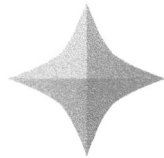

What's Right for Life: Can Life Exist Without Water?

Morgan L. Cable, Ph.D., Jonathan Lunine,
Jack Beauchamp, and Christophe Sotin

NASA Jet Purpolsion Laboratory

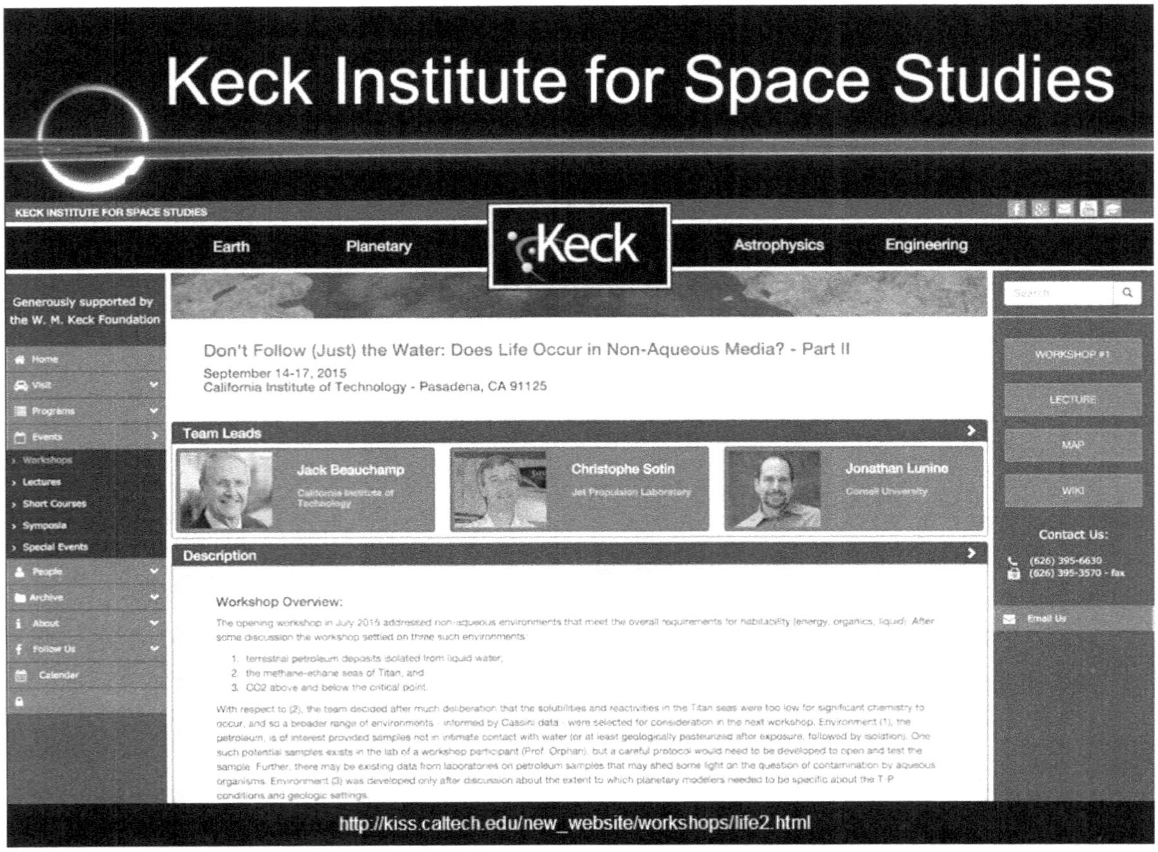
Keck Institute for Space Studies
KECK INSTITUTE FOR SPACE STUDIES
Earth
Planetary
Keck
Astrophysics
Engineering
Generously supported by the W. M. Keck Foundation
Home
Visit
Programs
Events
Workshops
Lectures
Short Courses
Symposia
Special Events
People
Archive
About
Follow Us
Calendar
Don't Follow (Just) the Water: Does Life Occur in Non-Aqueous Media? - Part II
September 14-17, 2015
California Institute of Technology - Pasadena, CA 91125
Team Leads
Jack Beauchamp
California Institute of Technology
Christophe Sotin
Jet Propulsion Laboratory
Jonathan Lunine
Cornell University
Description
Workshop Overview:
The opening workshop in July 2015 addressed non-aqueous environments that meet the overall requirements for habitability (energy, organics, liquid). After some discussion the workshop settled on three such environments:
1. terrestrial petroleum deposits isolated from liquid water;
2. the methane-ethane seas of Titan, and
3. CO2 above and below the critical point.
With respect to (2), the team decided after much deliberation that the solubilities and reactivities in the Titan seas were too low for significant chemistry to occur, and so a broader range of environments - informed by Cassini data - were selected for consideration in the next workshop. Environment (1), the petroleum, is of interest provided samples not in intimate contact with water (or at least geologically pasteurized after exposure, followed by isolation). One such potential samples exists in the lab of a workshop participant (Prof. Orphan), but a careful protocol would need to be developed to open and test the sample. Further, there may be existing data from laboratories on petroleum samples that may shed some light on the question of contamination by aqueous organisms. Environment (3) was developed only after discussion about the extent to which planetary modelers needed to be specific about the T-P conditions and geologic settings.
WORKSHOP #1
LECTURE
MAP
WIKI
Contact Us:
(626) 395-6630
(626) 395-3570 - fax
Email Us
http://kiss.caltech.edu/new_website/workshops/life2.html

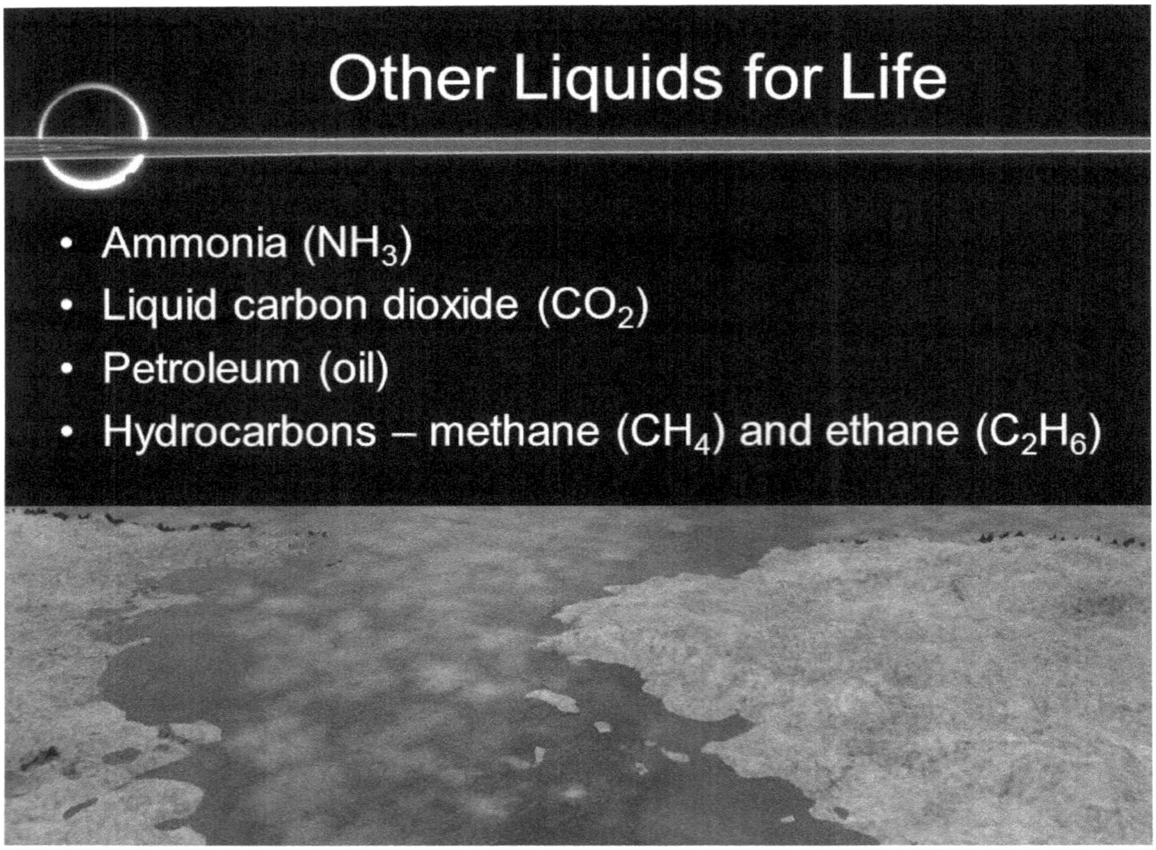
Other Liquids for Life
• Ammonia (NH_3)
• Liquid carbon dioxide (CO_2)
• Petroleum (oil)
• Hydrocarbons – methane (CH_4) and ethane (C_2H_6)

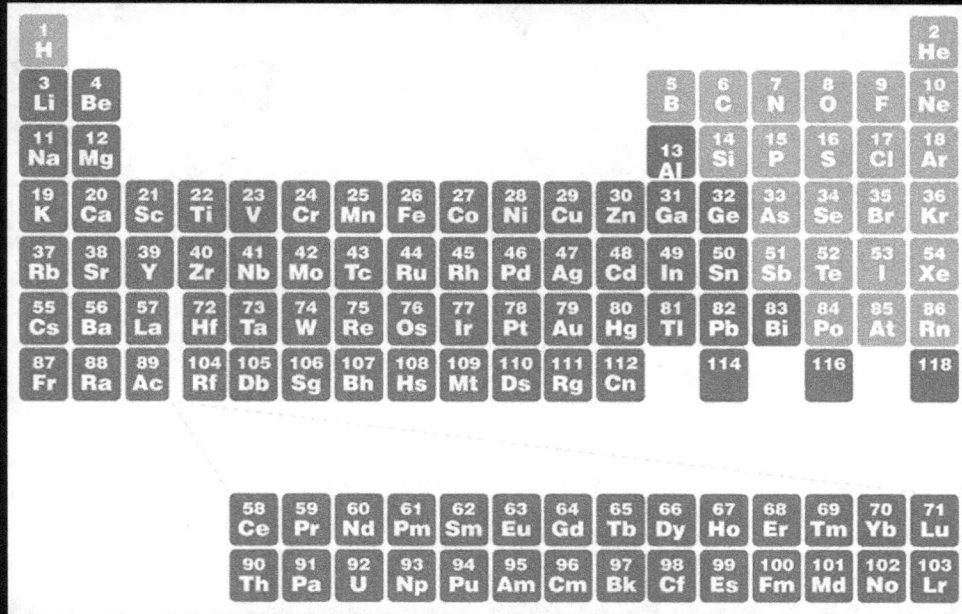

Focus on Carbon-Based Life

Image credit: NASA/CXC/M.Weiss)

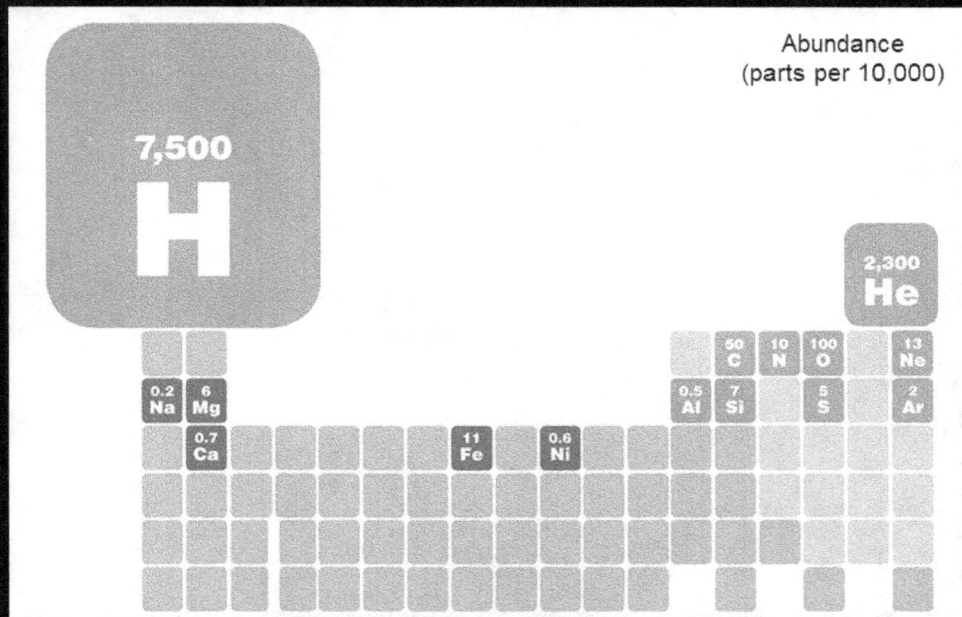

Focus on Carbon-Based Life

Image credit: NASA/CXC/M.Weiss)

Other Liquids for Life

- Ammonia (NH_3)
- Liquid carbon dioxide (CO_2)
- Petroleum (oil)
- Hydrocarbons – methane (CH_4) and ethane (C_2H_6)

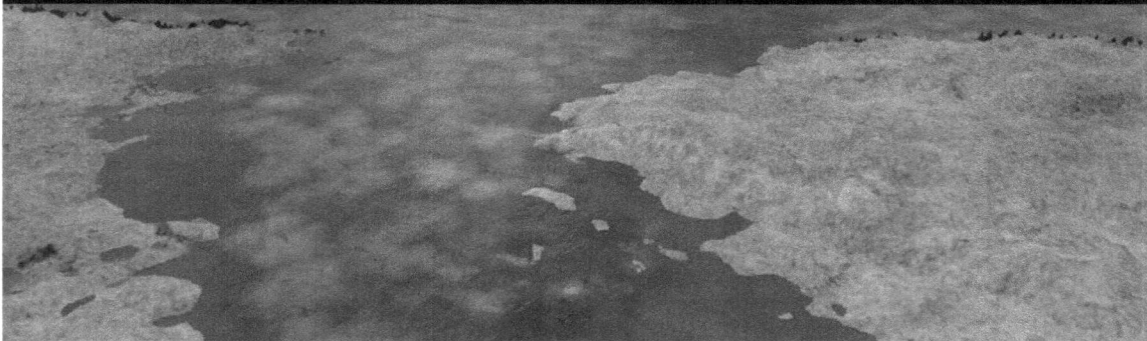

Ammonia

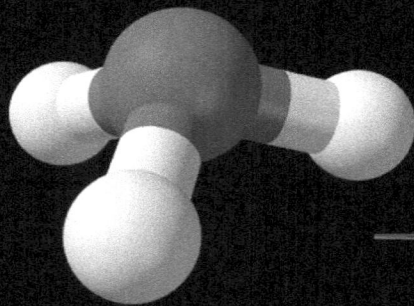

NH_3

Image credit: NASA-JPL/Caltech

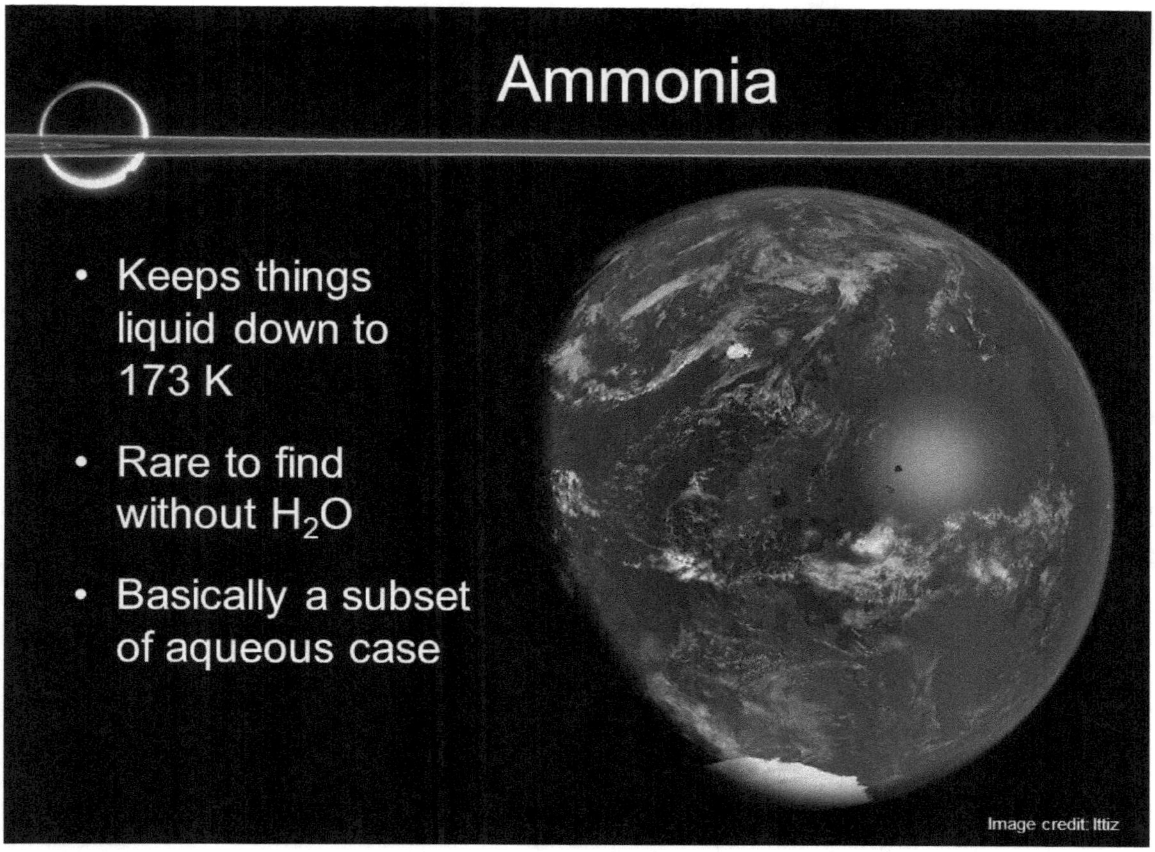

Ammonia

- Keeps things liquid down to 173 K

- Rare to find without H_2O

- Basically a subset of aqueous case

Image credit: Ittiz

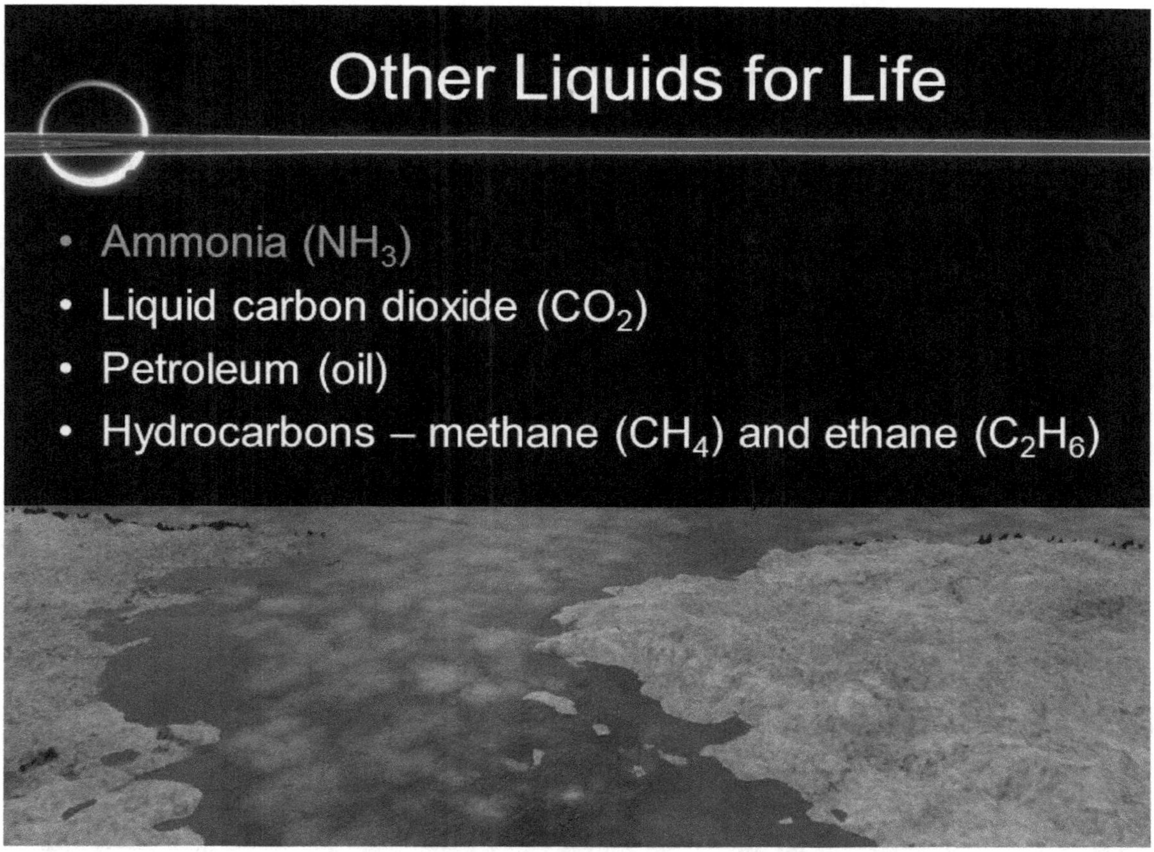

Other Liquids for Life

- Ammonia (NH_3)
- Liquid carbon dioxide (CO_2)
- Petroleum (oil)
- Hydrocarbons – methane (CH_4) and ethane (C_2H_6)

Morgan L. Cable, Jonathan Lunine, Jack Beauchamp, and Christophe Sotin

Other Liquids for Life

- Ammonia (NH_3)
- Liquid carbon dioxide (CO_2)
- Petroleum (oil)
- Hydrocarbons – methane (CH_4) and ethane (C_2H_6)

Carbon Dioxide

- Worlds to consider:
  - Venus
  - Earth (subsurface/ tectonic plates)
  - Mars (subsurface, ice caps, clathrates)
  - Niche environments (CO_2 bubbles, pockets)
  - Exoplanets with CO_2 oceans

CO_2

Carbon Dioxide

- Common in planetary atmospheres
- Comes in two liquid flavors:
  - Liquid CO_2
  - Supercritical CO_2

Image credit: NASA, ESA, & G. Bacon/STScI, STScI-PRC14-06

Supercritical Carbon Dioxide

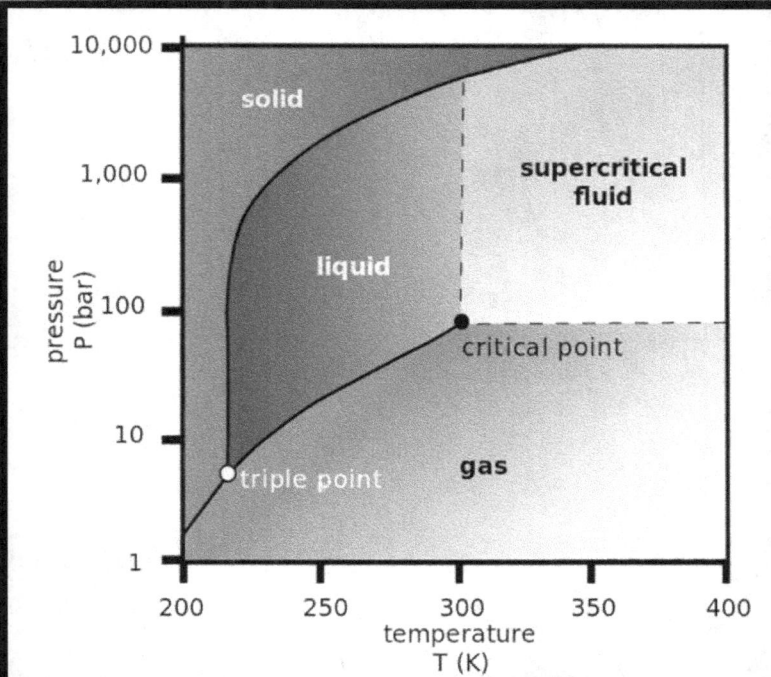

Image credit: Finney and Jacobs

Supercritical Carbon Dioxide

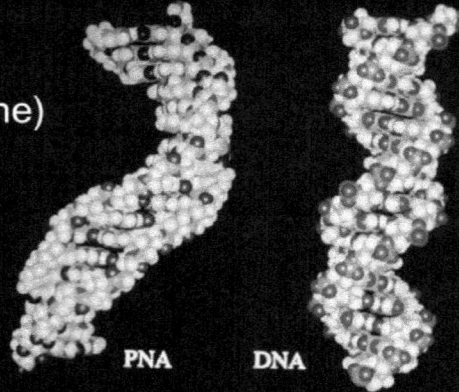

PNA   DNA

- Different from water
  - Nucleobases (purine, pyrimidine) aren't soluble
- But, many unique properties
  - Acetylated sugars show increased solubility in supercritical CO_2
  - Peptide nucleic acids (PNA) can act like DNA and serve as the information backbone in supercritical CO_2
  - Also polyethers are very soluble

Avoiding Venus

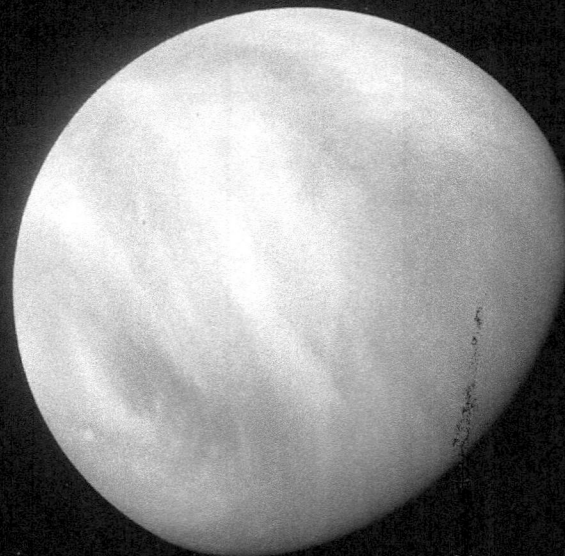

Venus

- Challenging to prevent runaway Greenhouse effect
- Adding dissolved N_2, H_2 into the ocean might help stabilize
  - Allows atmosphere to lose heat to space better
  - H_2 atmospheres are stable for Earth-sized planets

Supercritical Carbon Dioxide

- ## More theoretical work is needed

  - Radiative convective calculations

  - 3 dimensional global circulation models (GCMs)

  - Include N_2, H_2

  - Trace species

  - Effect of CO_2 clouds

- ## Lab work is ongoing

  - Peptide backbones as a type of information carrying molecule

HD 189733b (First exoplanet w/CO_2)

Important Issues to Consider

- Contact with/influx of prebiotic molecules as building blocks

- Solubilities of organics, catalytic species

- Unique reactivity of prebiotic molecules in CO_2 environments

- Formation of reverse micelles in supercritical CO_2

- Formation of CO_2 'bubbles' in water

- Contact with minerals and/or catalytic surfaces in geological environments

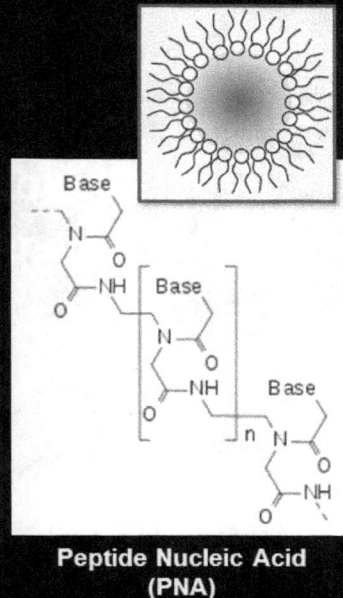

Base

Base

Base

**Peptide Nucleic Acid
(PNA)**

157

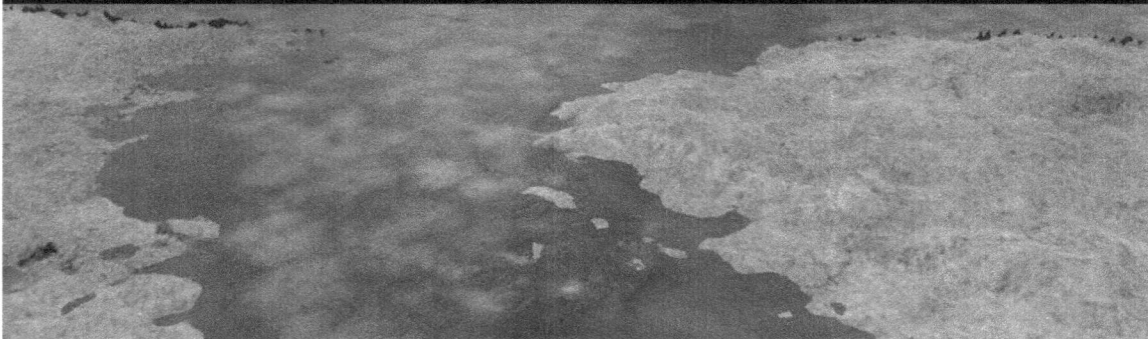

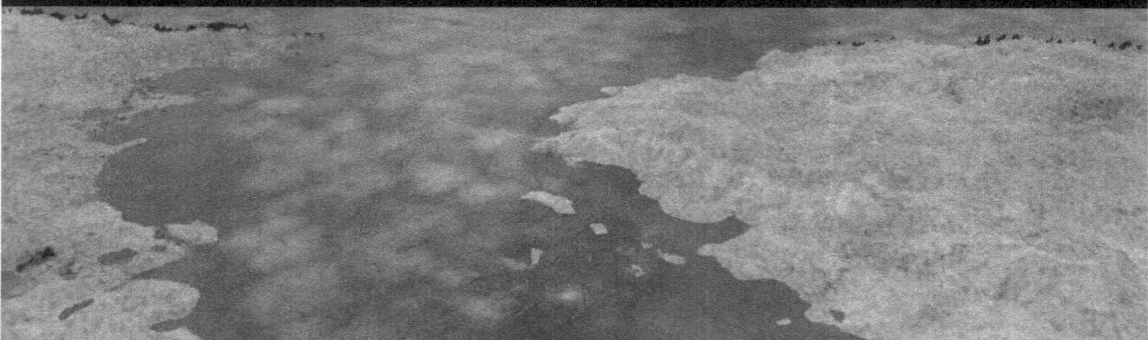

Petroleum

Dodecane

Napthalene

Pentane

Petroleum

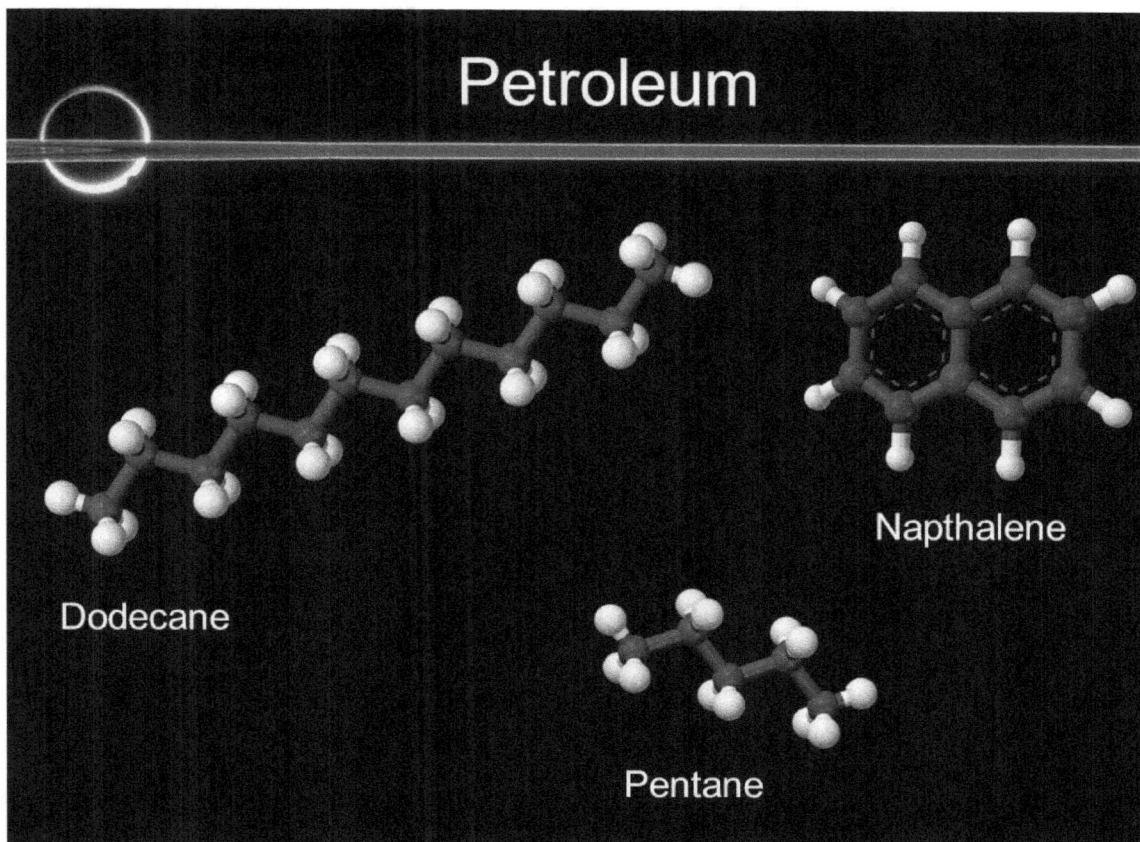

- What could form the compounds of life in petroleum?
  - Isoprenoids
  - Long-chain fatty acids
  - Greasy amino acids (polyisoleucine, polyphenylalanine)
- Sample complexity is enormous
- Difficult to analyze
- Working to identify an environment where there is water-free petroleum
- Life as we know it might use petroleum at interfaces with water

Petroleum

- What could form the compounds of life in petroleum?
  - Isoprenoids
  - Long-chain fatty acids
  - Greasy amino acids (polyisoleucine, polyphenylalanine)
- Sample complexity is enormous
- Difficult to analyze
- Working to identify an environment where there is water-free petroleum
- Life as we know it might use petroleum at interfaces with water

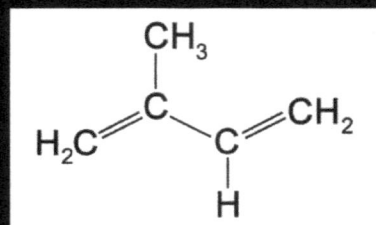

Isoprene

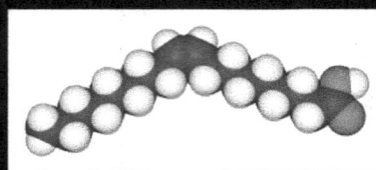

Oleic Acid (C18)

Petroleum

- There could be life in petroleum, but finding it will be challenging

- Look for petroleum deposits on Earth where contact with water has been broken for some amount of time

- Redox and metabolic experiments to identify:

  - Petroleum-associated water-based life as we know it

  - Hydrocarbon-based life as we don't know it

- Baseline abiotic processes in lab, then look at a biological process and see what the deviation is

Other Liquids for Life

- Ammonia (NH_3)
- Liquid carbon dioxide (CO_2)
- Petroleum (oil)
- Hydrocarbons – methane (CH_4) and ethane (C_2H_6)

Other Liquids for Life

- Ammonia (NH_3)
- Liquid carbon dioxide (CO_2)
- Petroleum (oil)
- Hydrocarbons – methane (CH_4) and ethane (C_2H_6)

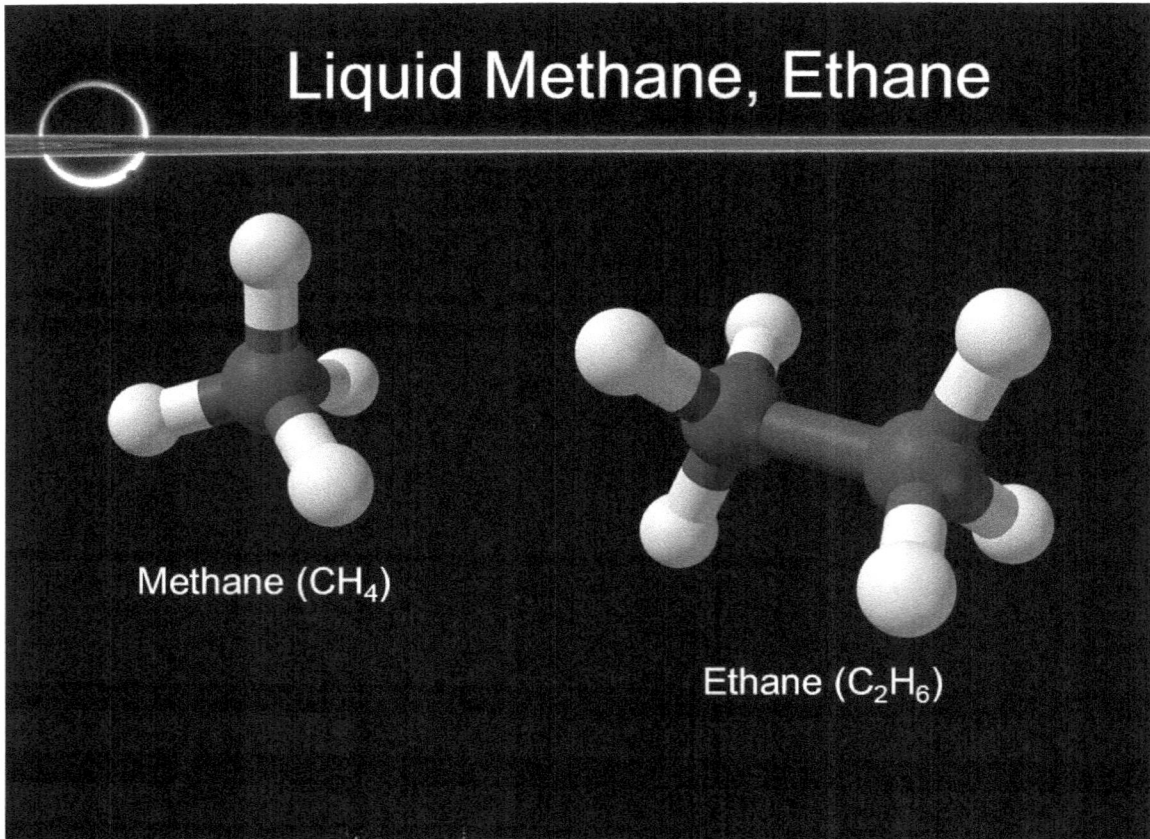

Liquid Methane, Ethane

Methane (CH₄)

Ethane (C₂H₆)

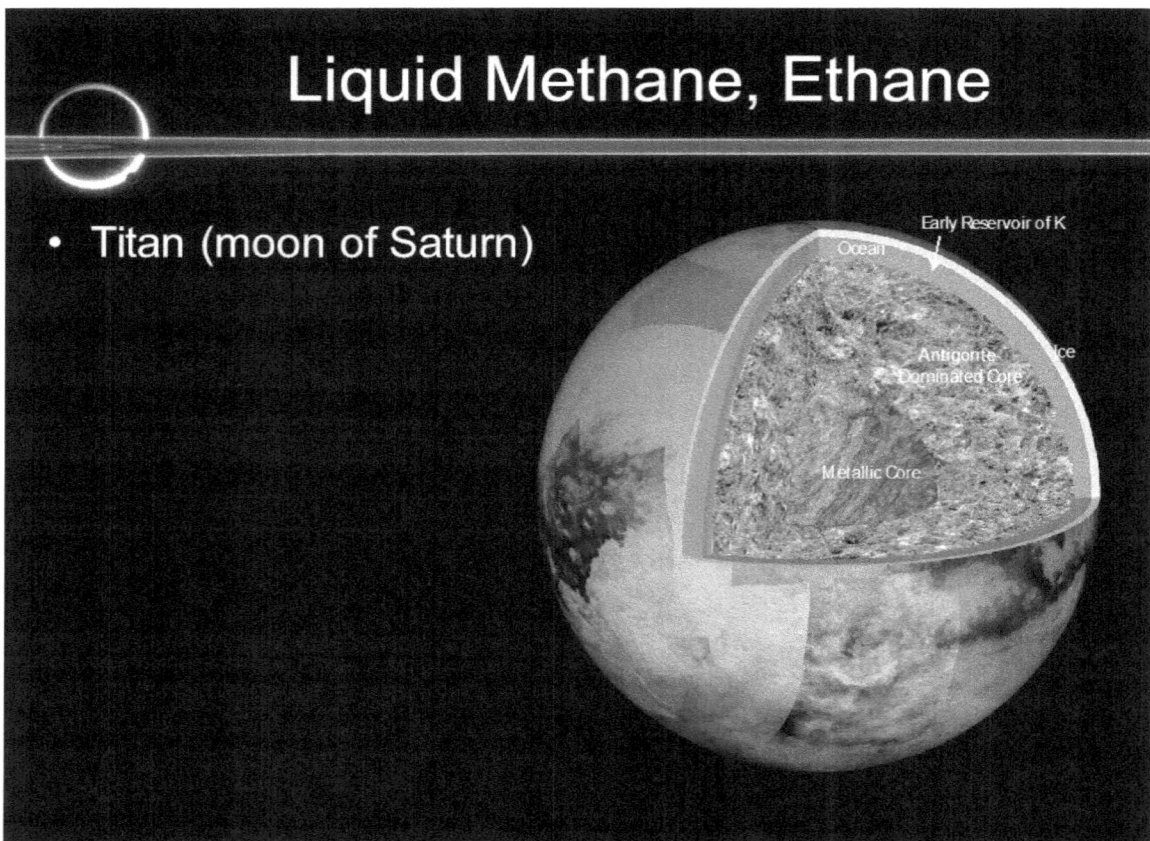

Liquid Methane, Ethane

- Titan (moon of Saturn)

Artist's depiction of Titan lake (Ron Miller)

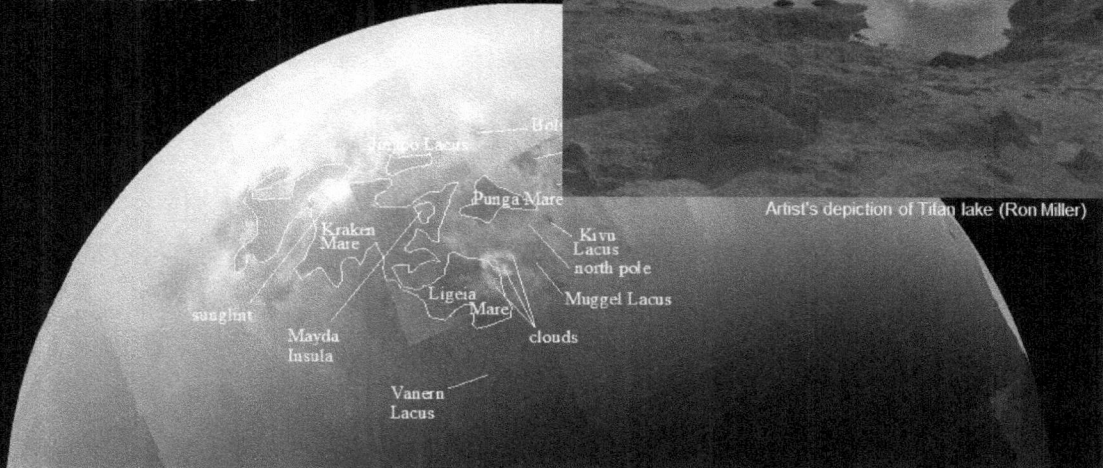

Liquid Methane, Ethane

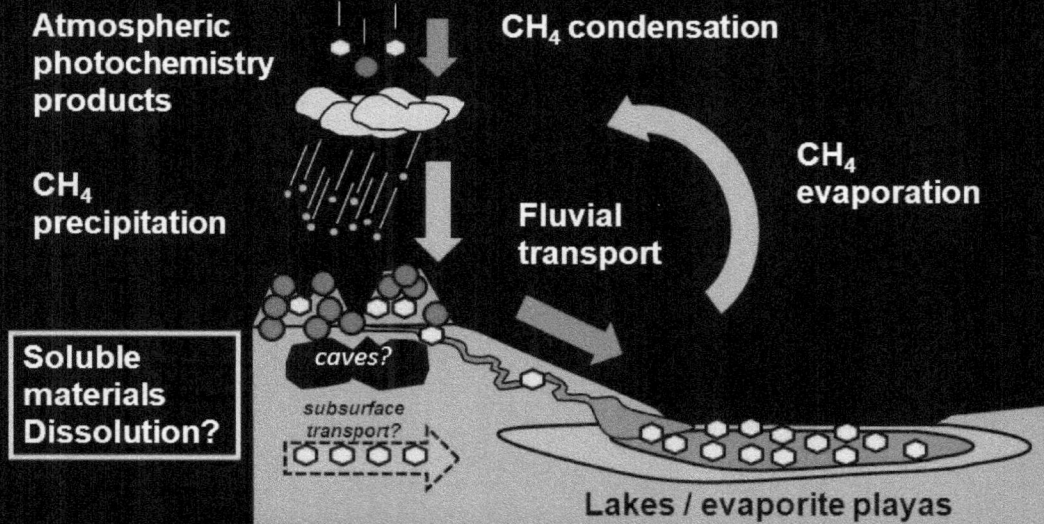

- Titan (moon of Saturn)
- Lakes made of methane and ethane

Liquid Methane, Ethane

Malaska et al., Workshop on the Habitability of Icy Worlds (2014), Abstract 4020

Liquid Methane, Ethane

• Solubility of most molecules in liquid methane and/or ethane is poor

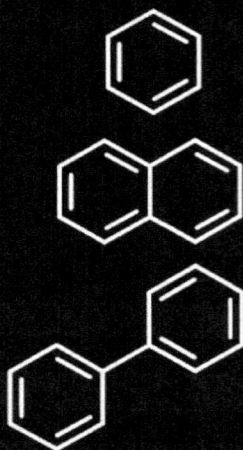

Compound	Solubility in Liquid Ethane at 94 K (mg/L)
Benzene	18.5 ± 1.9
Napthalene	0.159 ± 0.003
Biphenyl	0.039 ± 0.006

Sucrose in water (25 °C): 3,750,000 mg/L
Stearic acid (~olive oil): 3 mg/L

Malaska et al. Icarus 2014.

Looking to the Shorelines

Solid-Phase Chemistry

- Carbamation

$$R-NH_2 \ + \ O=C=O \ \longrightarrow \ R-\overset{H}{N}-C\overset{O}{\underset{OH}{}}$$

Carbamic acid

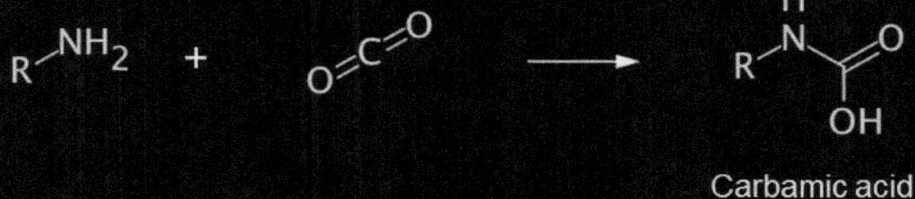

- Imine, ether formation
  - Polyimine could form a catalytic site

- Azide and alkyne polymerization
  - HCN polymerization

Looking for Organic/Ice Interfaces

RADAR – effective dielectric constant is low ε = 1.6 – 2.0 [1] **Consistent with organics with voids**	H₂O ice $\varepsilon = 3.1$
	Hydrocarbon material $\varepsilon = 2.0 - 2.4$
	Avg. Titan surface $\varepsilon = 1.6 - 2.0$

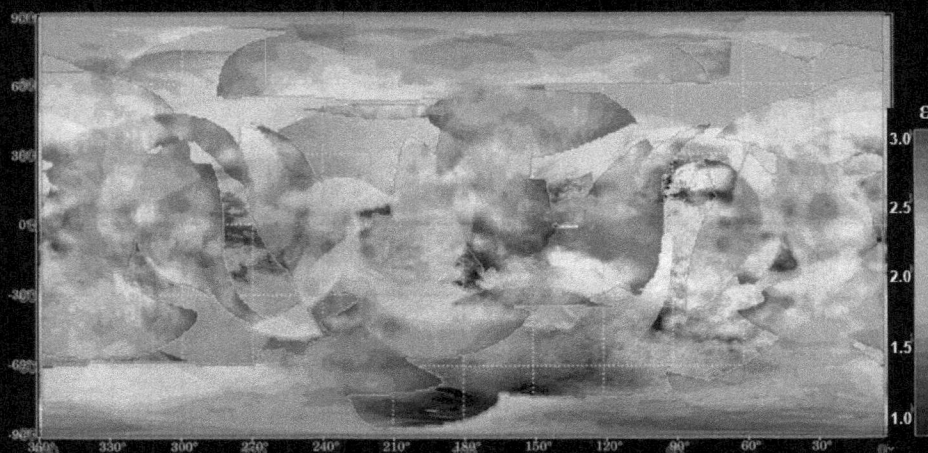

[1] Janssen et al., Icarus 200 (2009) 222-239.

Low Temperature Considerations

- **Weaker forces dominate**
  - Noncovalent interactions
- **We need to rethink our requirements for life**

Earth	Titan	Interaction/Bond Strength
Ionic Bond	Covalent Bond	Permanent or semi-permanent
Covalent Bond	Hydrogen bonds, π-bonds	Can be made/broken on timescales relevant for life (ATP, proteins, etc.)
Hydrogen bonds, π-bonds	van der Waals forces	Loose associations that help hold together secondary structures

Prebiotic versus Protobiotic

- Instead of looking for life itself, target the mechanisms that lead to prebiotic chemistry (aka, Protobiotic chemistry)
  - Geologic patterning
  - Chemical gradients
  - On Earth, this led to more advanced chemistry, which eventually led to life

- How would we see this?
  - Look for differences in orientation and association of chemical subunits
  - Domination of one type of material on Titan (might change with seasons)

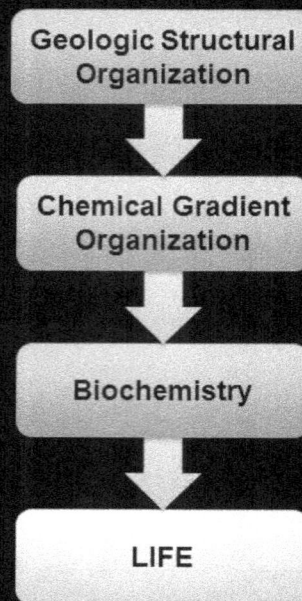

Geologic Structural Organization

↓

Chemical Gradient Organization

↓

Biochemistry

↓

LIFE

Overall Study Findings

- Mars' charter is explicitly, "Follow the Water"
- Our conclusions have found that you may find interesting things apart from water
- Seek out the interfaces!
- Weak interactions may be the key

NASA shouldn't just be looking for areas with liquid water

Happy Halloween from Cassini!

Image credit: NASA-JPL/Caltech

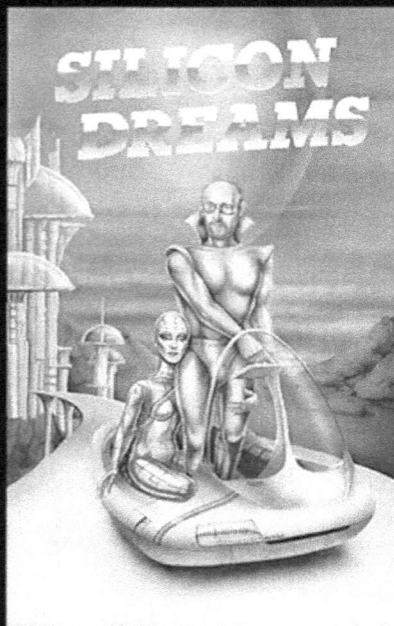

Backup

Why Not Silicon-Based Life?

- Silicon is not as abundant as carbon

- Si-Si bonds are ~50% weaker than C-C bonds

- CO_2 is much more volatile and soluble than SiO_2 (i.e., available to do chemistry)

- Rare to find long strings of silicon bonded together

Focus on Carbon-Based Life

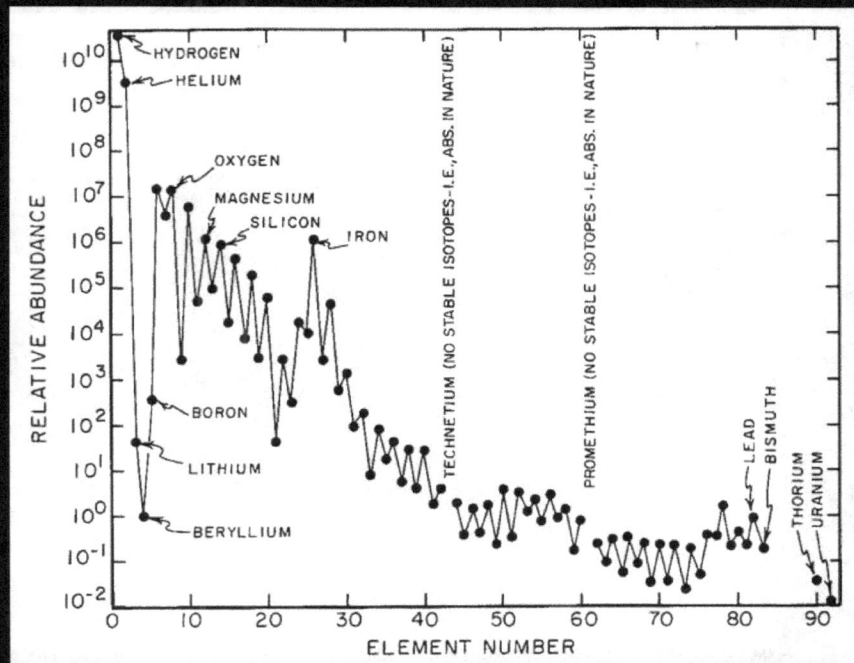

Methane/Ethane Seas Have Transient Features

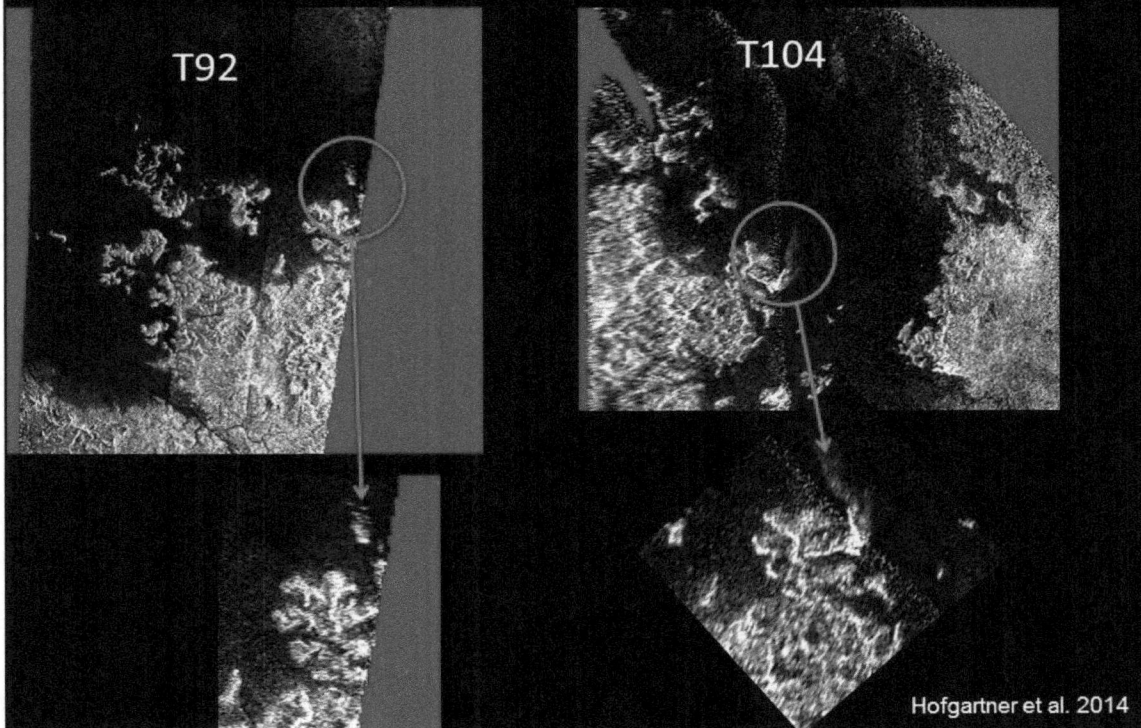

T92

T104

Hofgartner et al. 2014

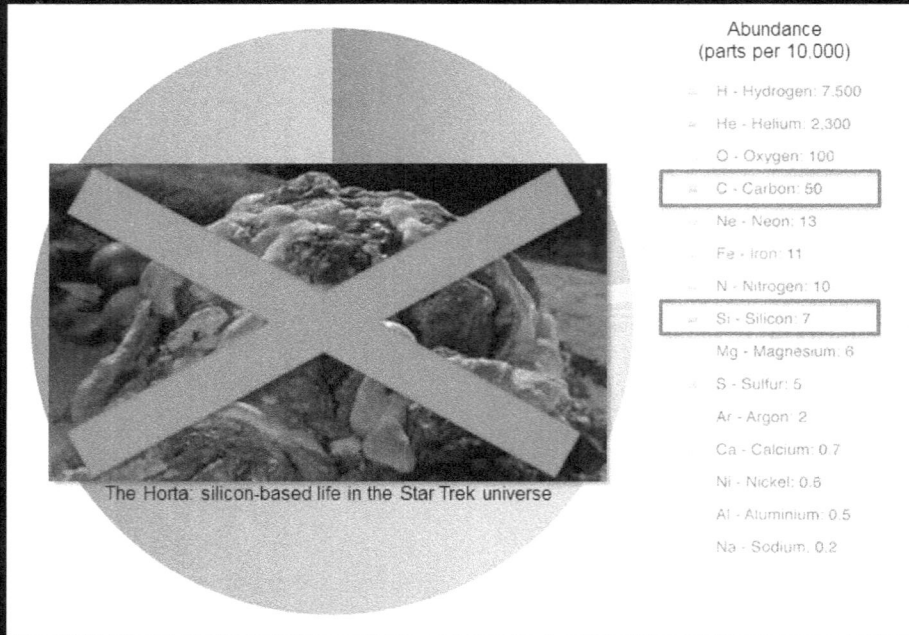

An Earthling's Guide to the Habitable Zone

Christopher J. Burke

Kepler Project, SETI Institute

christopher.j.burke@nasa.gov

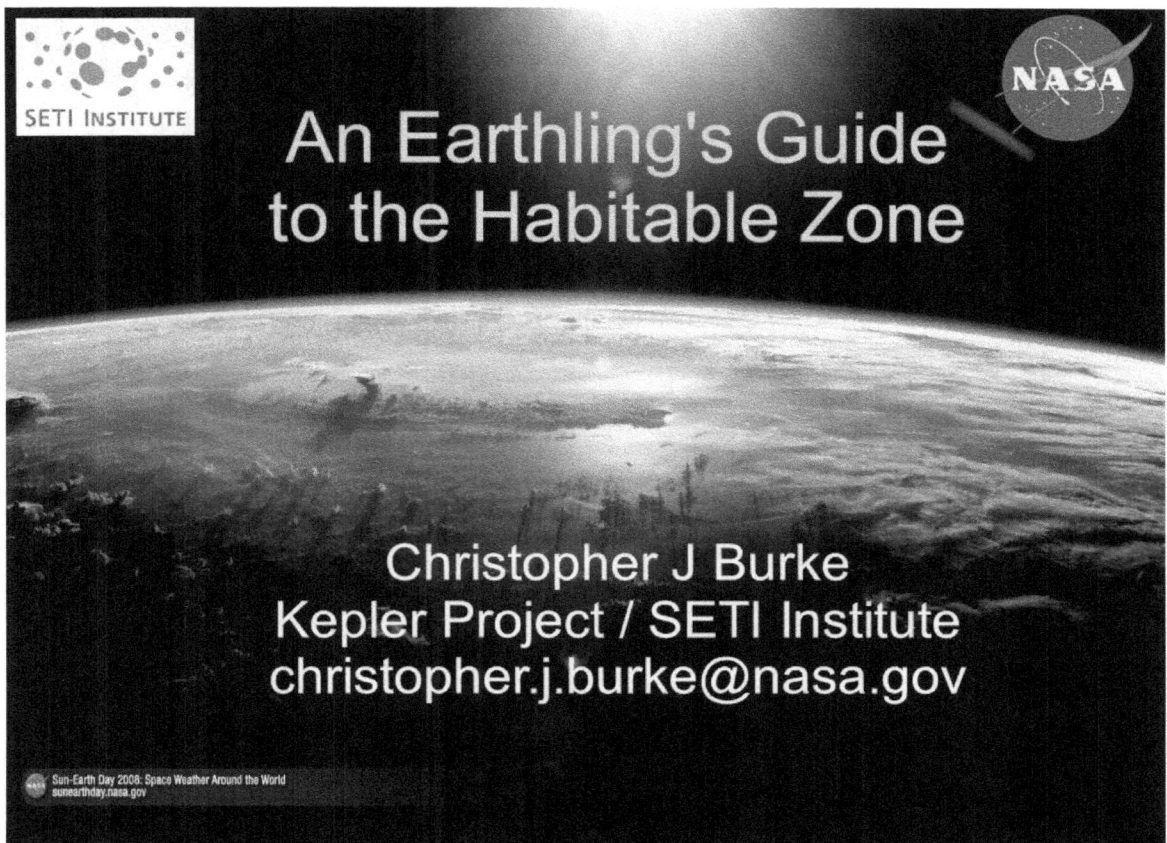

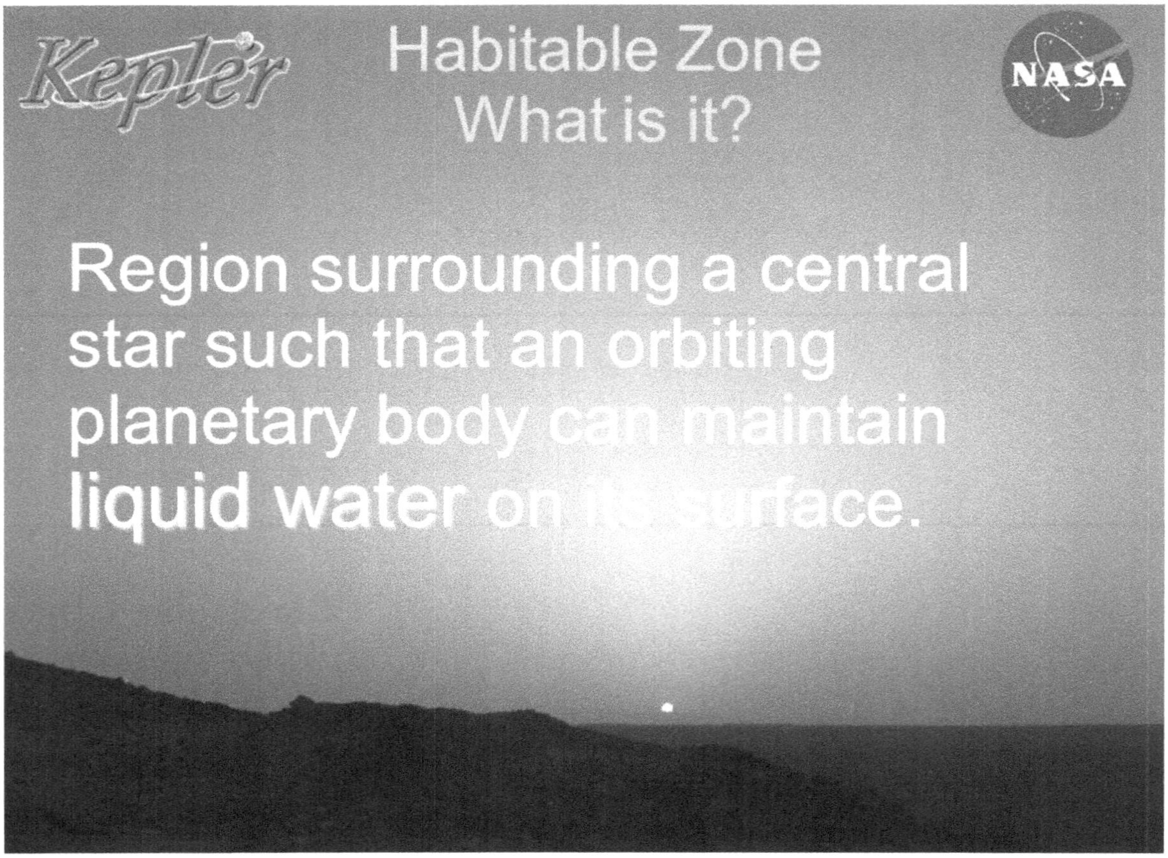

Kepler | Habitable Zone
What is it?

NASA

Region surrounding a central star such that an orbiting planetary body can maintain liquid water on its surface.

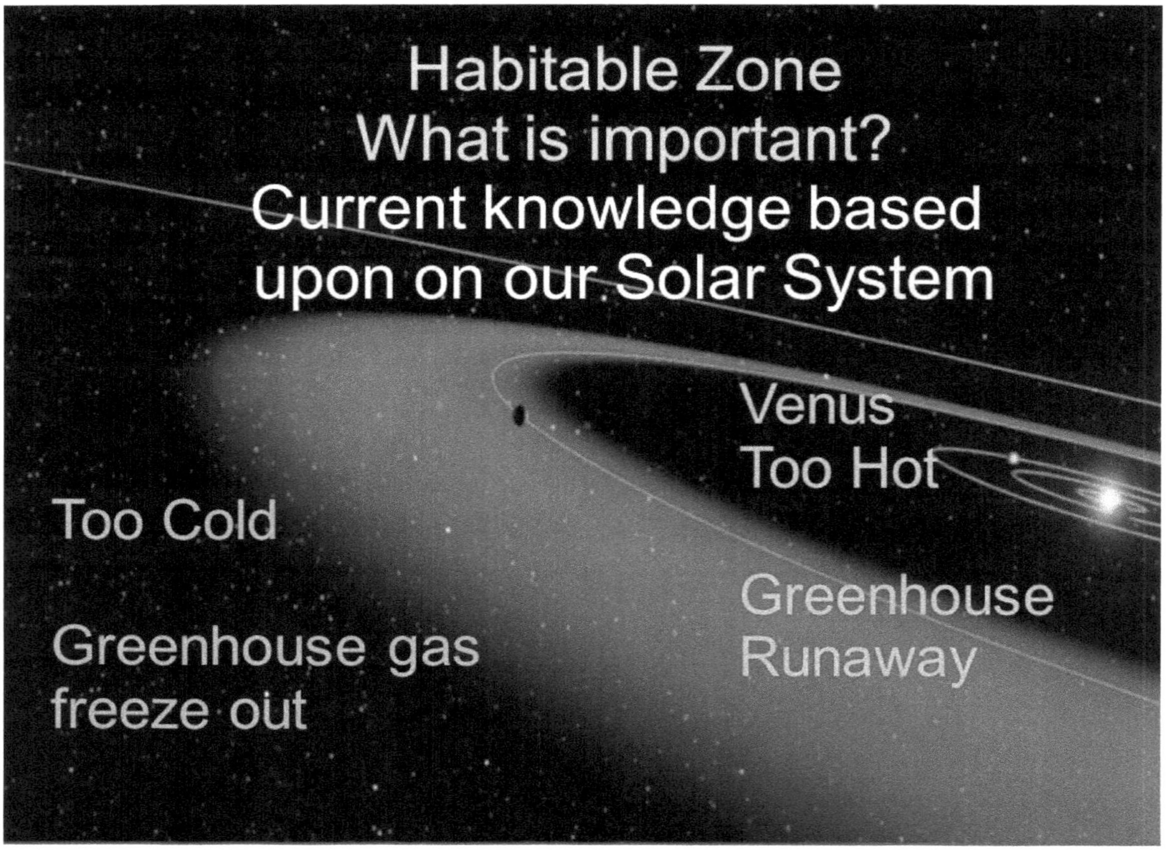

Habitable Zone
What is important?
Current knowledge based upon on our Solar System

Too Cold

Greenhouse gas freeze out

Venus
Too Hot

Greenhouse
Runaway

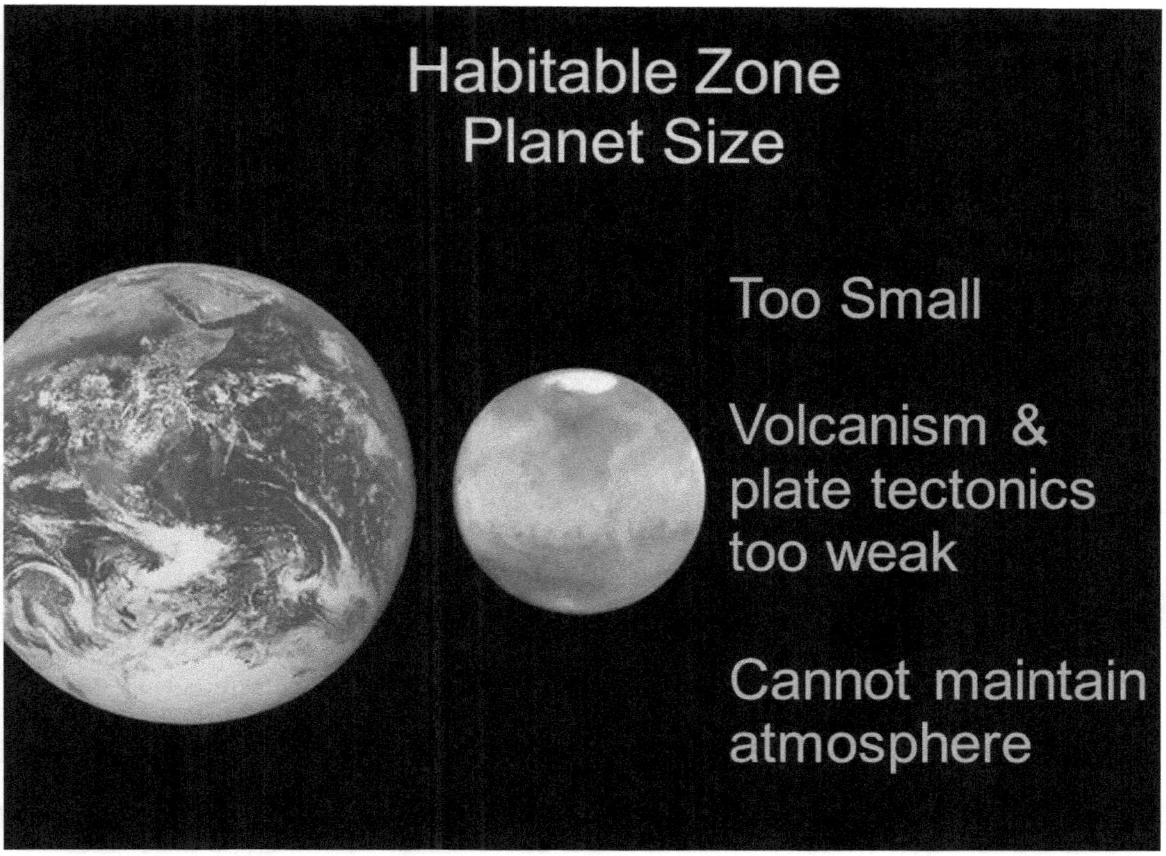

Habitable Zone
Planet Size

Too Small

Volcanism &
plate tectonics
too weak

Cannot maintain
atmosphere

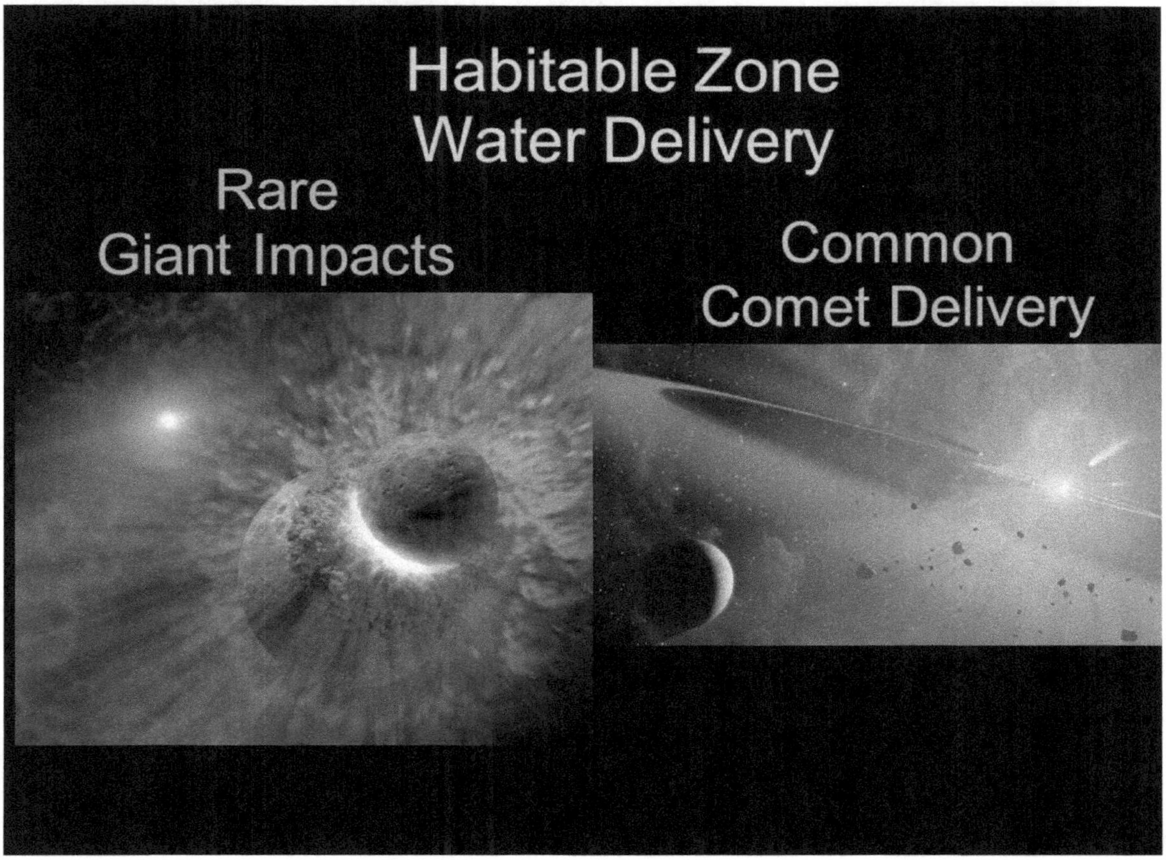

Habitable Zone
Water Delivery

Rare
Giant Impacts

Common
Comet Delivery

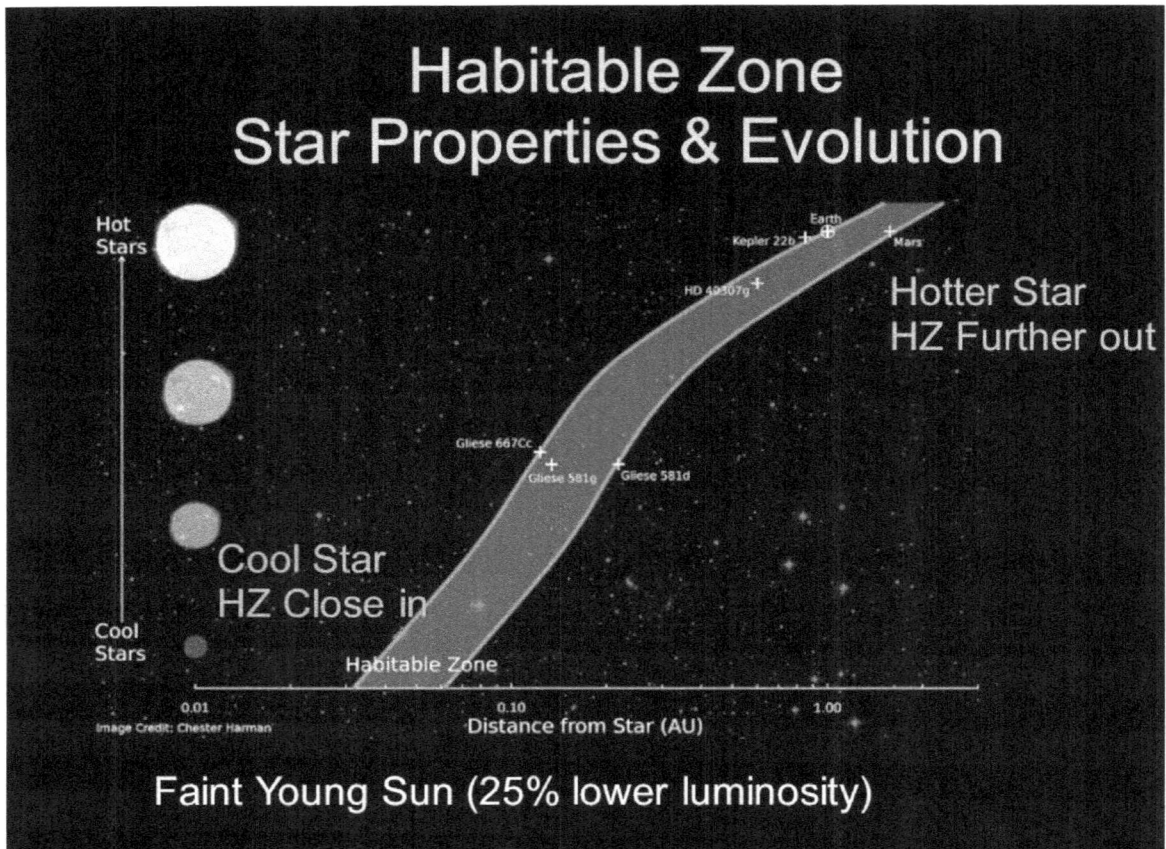

Habitable Zone
Star Properties & Evolution

Hot Stars

Cool Stars

Hotter Star
HZ Further out

Cool Star
HZ Close in

Habitable Zone

Image Credit: Chester Harman

Distance from Star (AU)

0.01   0.10   1.00

Kepler 22b   Earth   Mars

HD 40307g

Gliese 667Cc

Gliese 581g   Gliese 581d

Faint Young Sun (25% lower luminosity)

Habitable Zone
Potential For Large Diversity

Chaotic planetary dynamics
Planet rotation rate
Cloud formation
Orbital eccentricity
Tidal heating energy habitable moons
Moon necessary?
Haze formation
Ice, ocean, land, and biomass albedo feedback
Magnetic field for atmosphere protection
H_2 greenhouse
Extreme UV radiation
Dry planet HZ extension
Radiogenic Heating

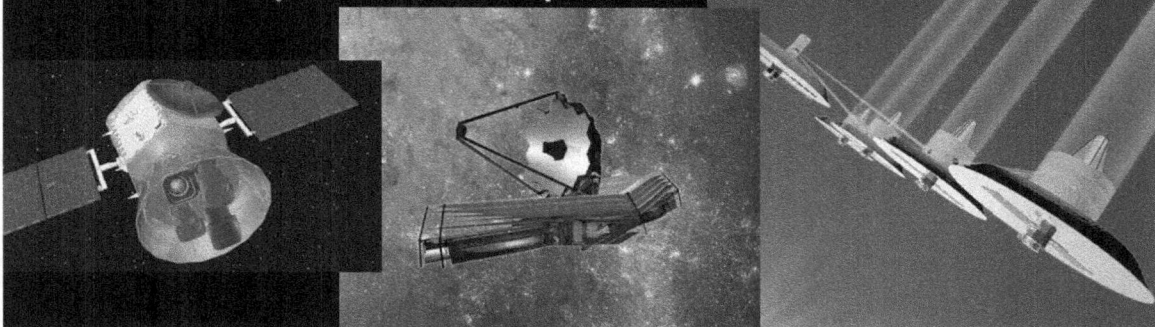

Habitable Zone
For Earth 2.0?

What are the most important processes impacting habitability?

Is the Solar System relevant?

Concrete answers will require observations of extrasolar planet atmospheres

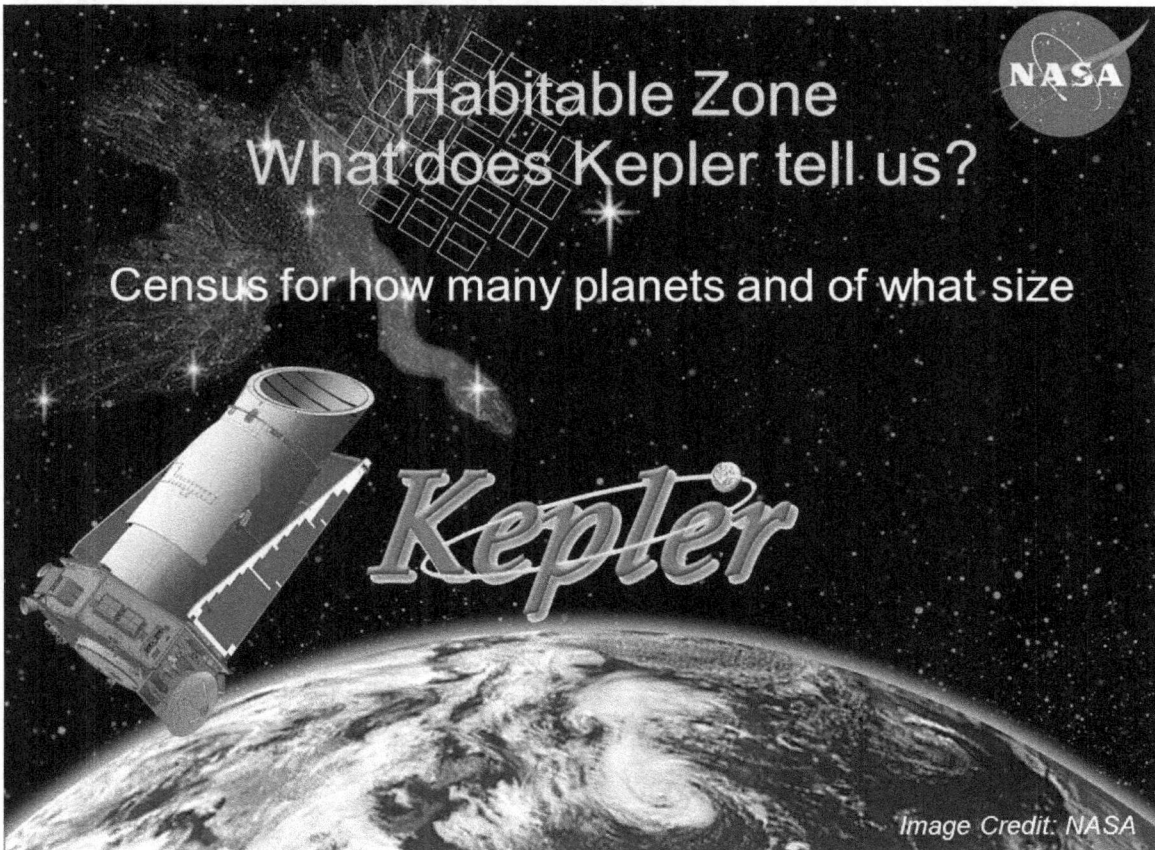

Habitable Zone
What does Kepler tell us?

Census for how many planets and of what size

Image Credit: NASA

1 Yr Terrestrial Planet Rate From Kepler

Number of planets per star with a planet having an orbital period and radius within 20% of Earth's

Best Value 10%
Lower limit 1%
Upper limit 200% (or 2)

Burke et al. (2015)

Habitable Zone
What does Kepler not tell us?

Kepler only measures planet radii, planet orbital periods, and numbers of planets

Habitable Zone
What does Kepler not tell us?

?

Kepler only measures
planet radii,
planet orbital periods,
and numbers of planets

Summary

Understanding habitability of an Earth 2.0 is still in its
infancy.

Kepler finds terrestrial planets with 1 year orbits are
not exceedingly rare (>1% ; 10% typical).

The next steps for finding and characterizing
extrasolar planet atmospheres are coming to fruition

100 YEAR STARSHIP™

Who's Got Next?

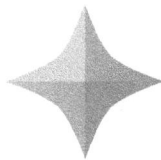

Dawn of the African Space Era

Mmboneni Muofhe
Deputy Director Genral, Department of Sicence and Technology, South Africa

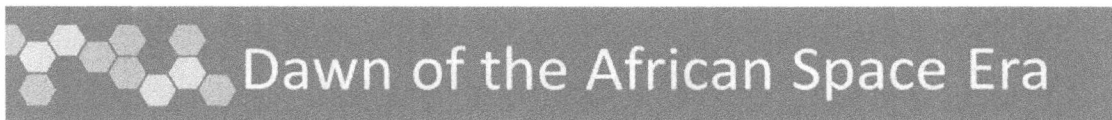

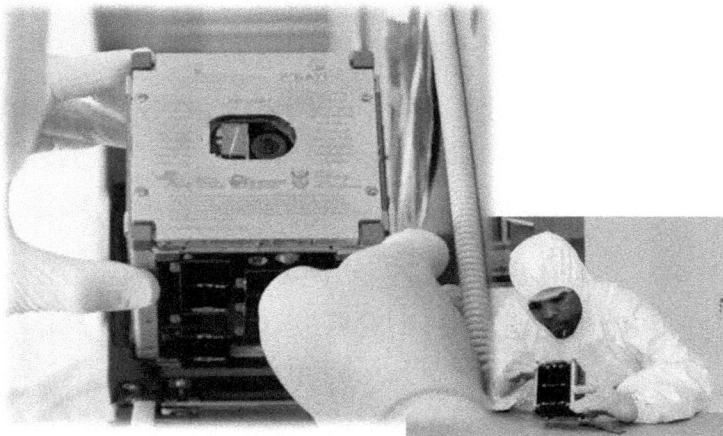

MMBONENI MUOFHE

Countries with Space Programmmes

- Nigeria
- South Africa
- Egypt
- Morocco
- Algeria

Towards a continental Agency

- African Space Policy
- African Space Strategy
- Establishing a PAN African Space University
  - South Africa to host
  - Finalising the model
  - Identified possible host institution
  - Negotiating host agreements
  - Planning a needs assessment

Global Challenges

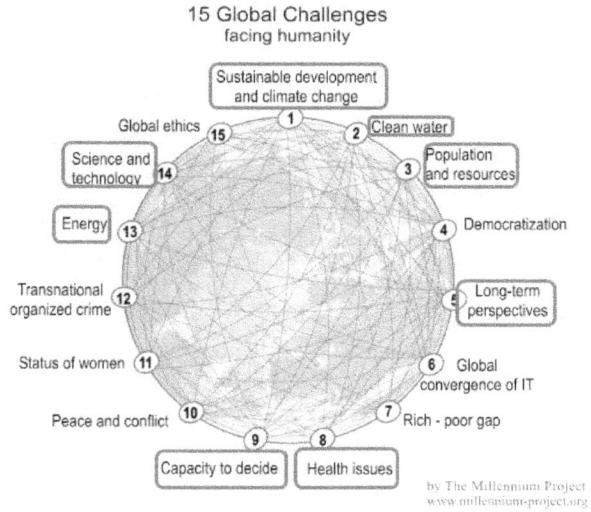

15 Global Challenges
facing humanity

by The Millennium Project
www.millennium-project.org

Building a knowledge economy

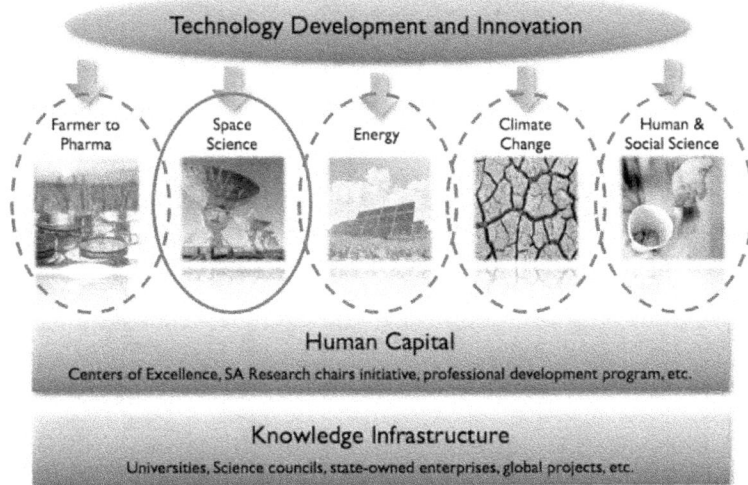

DST 10-year Innovation Plan

Tshepiso – What has been achieved

50 Master's graduates

15 Engineers-in-training

> 15 000 learners reached over past 5 years

17 peer reviewed journal papers

60 conference papers

Innovation – Africa Space Innovation Centre (ASIC)

SARChI Research Chair

Hi-tech job creation (8 positions in ASIC)

Equipment and world class infrastructure > R10 million

Multi-National Institute

South African Programmes

Earth Observation

Image acquisition
Image distribution
Value-added
services
Value-added
products

Space Science

Geo-space
observations
Space physics
Space weather
EM technologies

Space Operations

Launch support
In-orbit testing
Emergency
support
Carrier monitoring
Hosting

Space Engineering

Satellite
development
Industry
development
Technology
development

- BUIDING ON A SUCCESFUL CUBESAT PROGRAMME

Student demographics

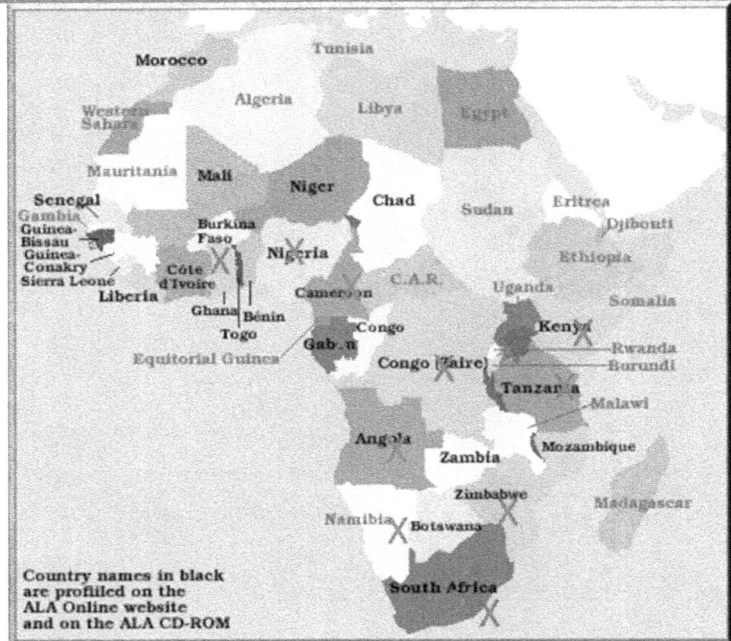

Country names in black
are profiled on the
ALA Online website
and on the ALA CD-ROM

Current missions: ZACUBE-2

- S-Band Transmitter
- ADCS System
- VHF/UHF Communication System
- Vessel tracking payload
- **Technology demonstrator for Operation Phakisa**
- 5 MP Camera Payload
- Possible IR camera
- Engineering model 2015

Making STEM cool

South Africa, 2013

Zambia, 2012

Satellite Development Facilities
FSATI, CPUT

Soldering Area
PCB polulation

Satellite Production Lab
FM Model Area + Clean Room

Clean Room
Class 100 000

Microwave and RF Labs
Additional computer and lecture rooms

Satellite Development Lab
12 Workspaces + Store room

Post Graduate Facilities
48 research spaces for Masters and Doctoral students

Post Graduate Facilities
48 research spaces for Masters and Doctoral students

The earth – our future

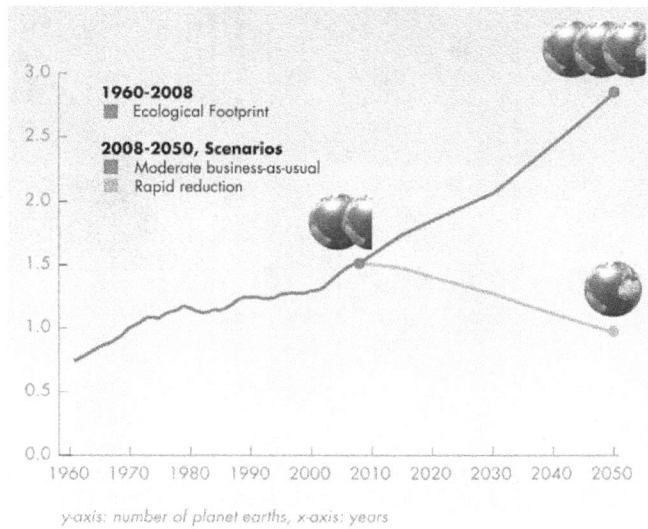

1960-2008
Ecological Footprint

2008-2050, Scenarios
Moderate business-as-usual
Rapid reduction

y-axis: number of planet earths, x-axis: years

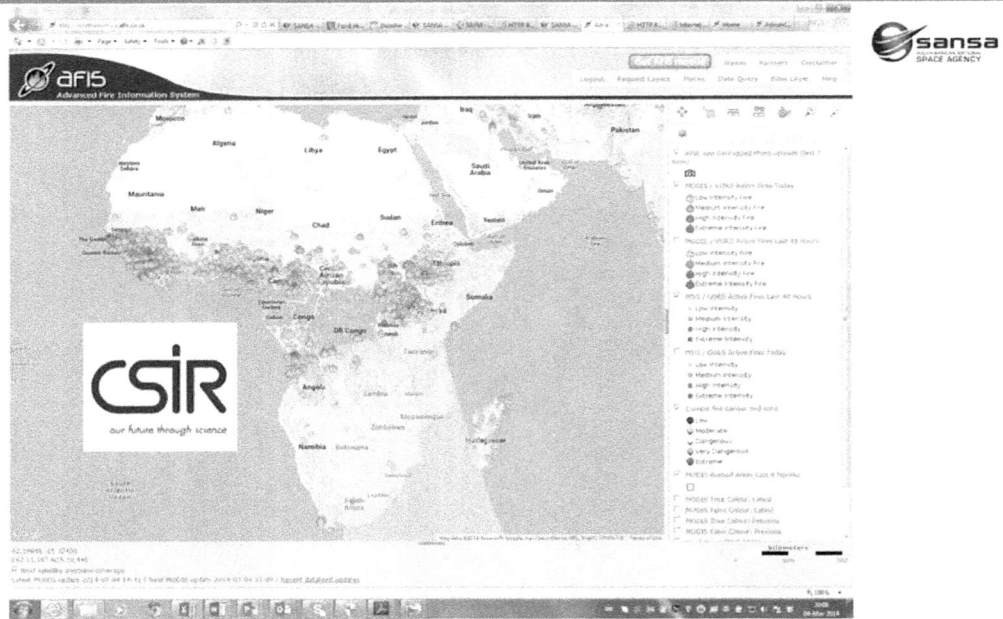

Fire Warning Services

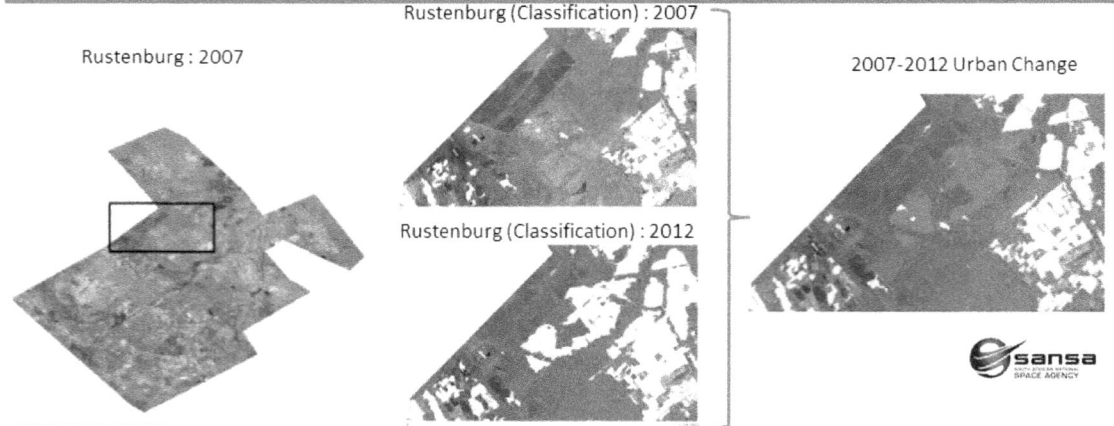

Example: Urbanisation Monitoring & Planning

Rustenburg : 2007

Rustenburg (Classification) : 2007

Rustenburg (Classification) : 2012

2007-2012 Urban Change

Problem

- Global urban population of 47% (2000), projected to increase to 67% (2050)
- 90% of growth is in developing countries
- Africa has the fasted urbanisation rate @ 3.5%
- Current African urban population of about 33% forecasted to increase to 50% by 2030

Consequences

- Inadequate amenities
- Strain on natural resources
- Social strife and strikes
- Poverty and crime
- Pollution

Space Solution

- Intelligence for urban planning
- Development of SMART cities
- Identification of problem areas

Research Platform: Geo-space laboratory

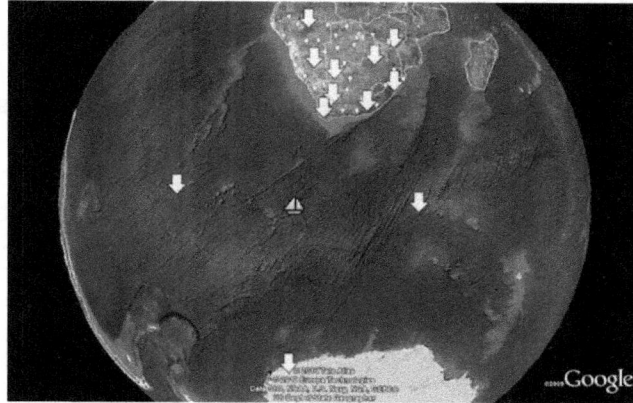

Geographic Advantage

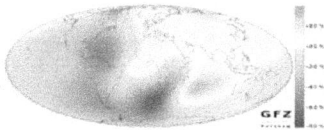

Wide Observational Network

Global Data Distribution & Access – example:

- ISES
- INTERMAGNET
- SuperDARN

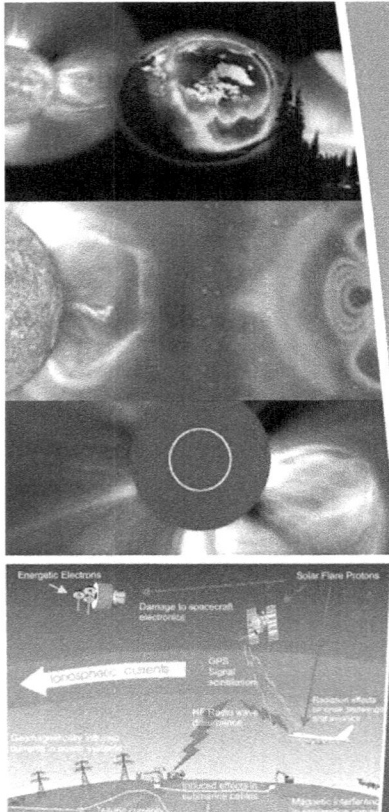

SPACE WEATHER: Regional Warning Centre for Africa

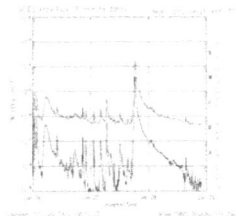

- Monitoring the Earth-Space system, predictions and forecasts
- Distributing data and creating new knowledge on the system

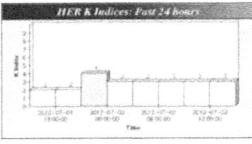

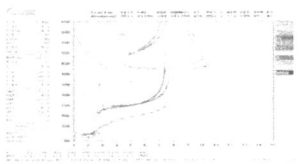

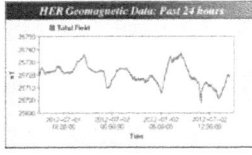

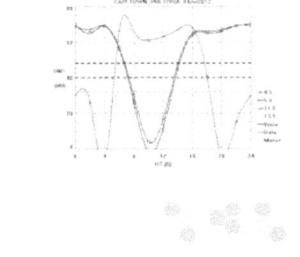

Example: Service Delivery Monitoring & Planning

Housing Development: 2012

Housing Image: 2013

Automated Housing Extraction: 2013

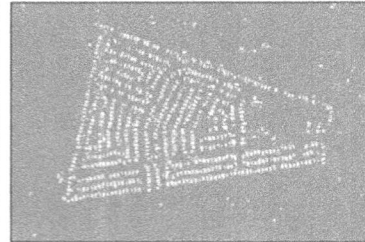

Potential Usage

- ❖ Independent report on progress made on infrastructure development
- ❖ Information to support population estimates
- ❖ Information to support infrastructure planning
- ❖ Information to support policies on environmental, natural resources and disaster management.

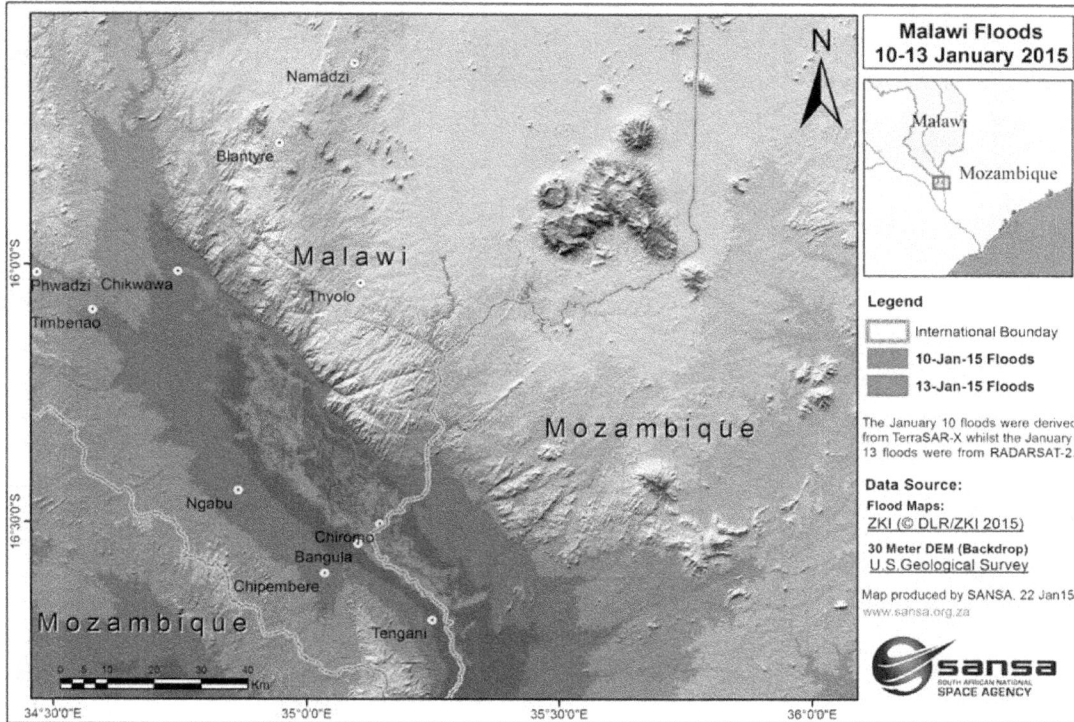

Census Listing Capturing system

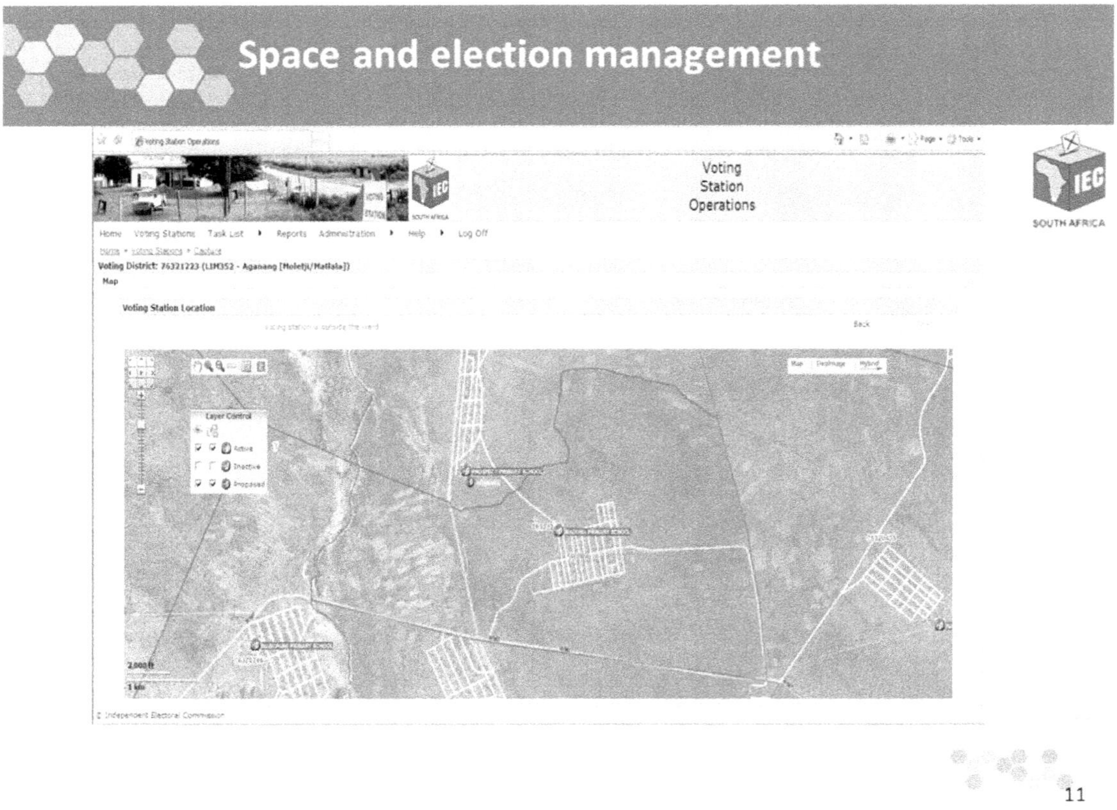
Space and election management
Voting
Station
Operations
IEC
SOUTH AFRICA
Home Voting Stations Task List Reports Administration Help Log Off
Voting District: 76321223 (LIM352 - Aganang [Moletji/Matlala])
Map
Voting Station Location
Back
Layer Control
Active
Inactive
Proposed
Map
Hybrid
2,000 ft
1 km
© Independent Electoral Commission
11

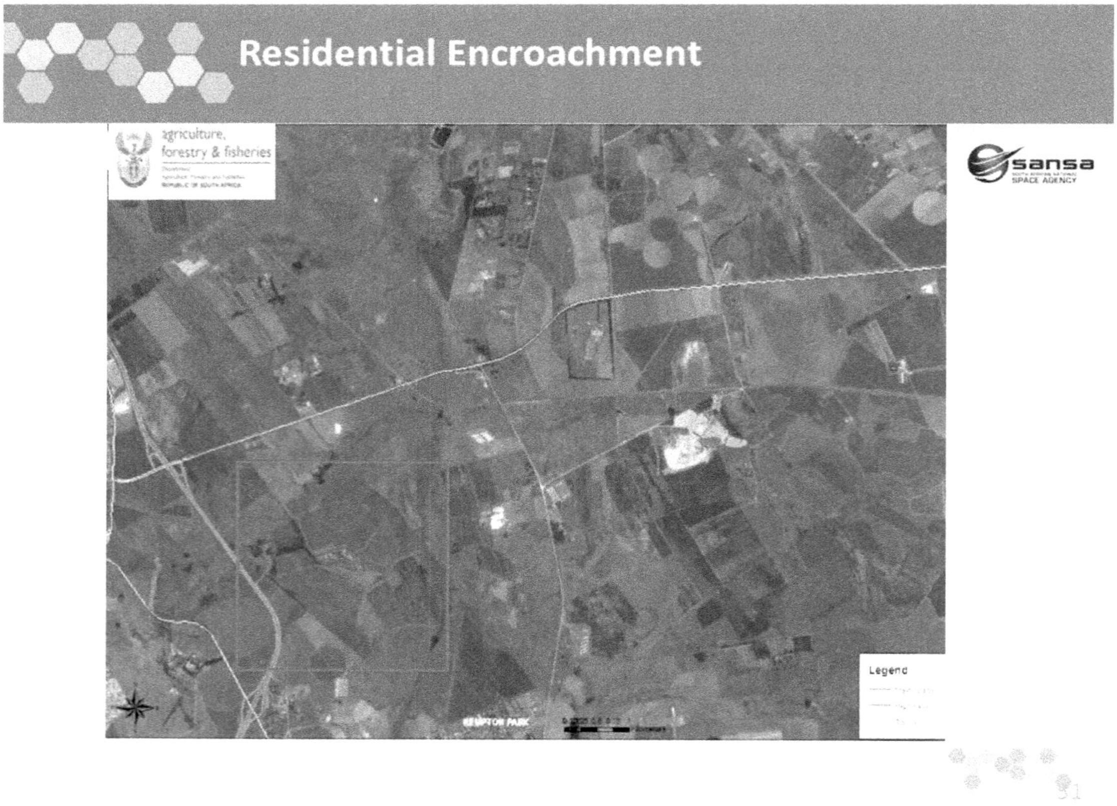
Residential Encroachment
agriculture,
forestry & fisheries
sansa
SPACE AGENCY
Legend
KEMPTON PARK

Mining Encroachment

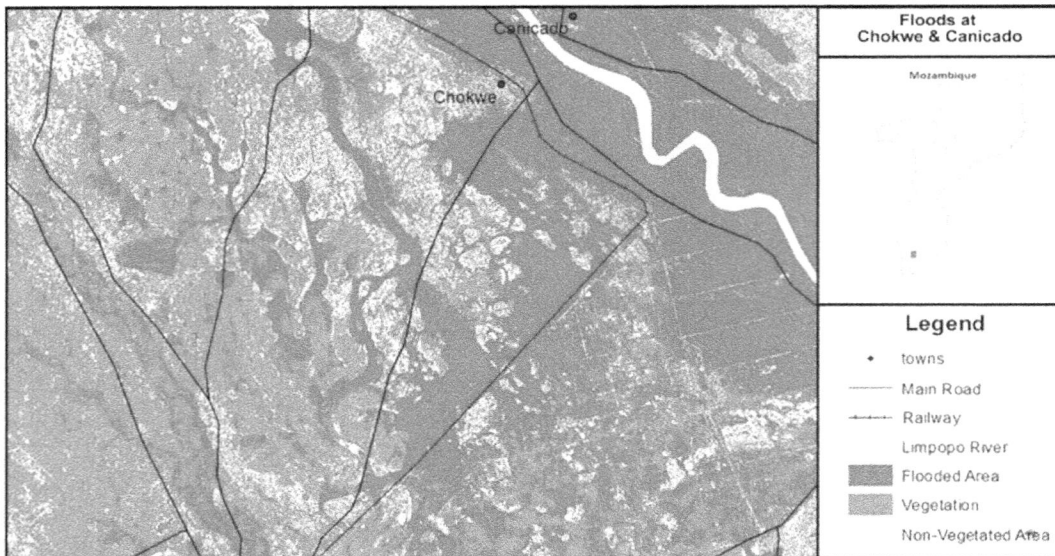

Post-flood Management & Assessment

The Wall

A unique mini-workshop and networking opportunity to "throw your ideas against The Wall and see what sticks!"

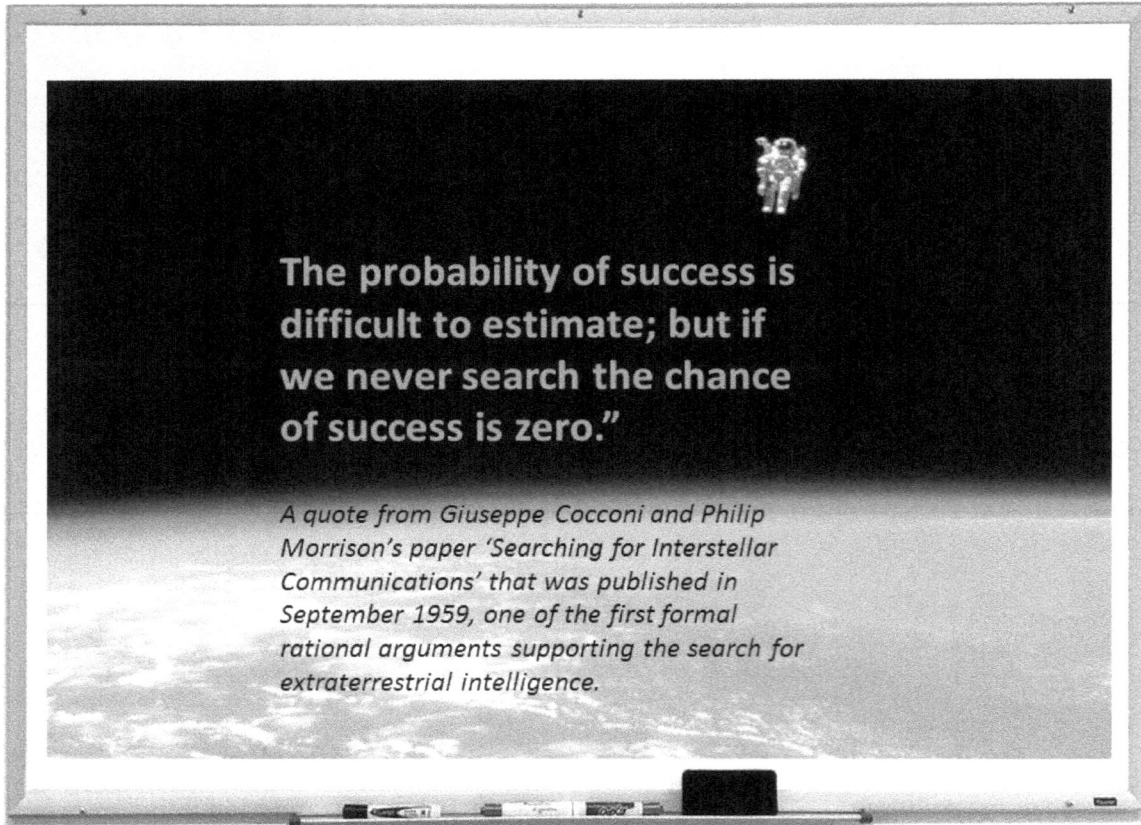

The probability of success is difficult to estimate; but if we never search the chance of success is zero."

A quote from Giuseppe Cocconi and Philip Morrison's paper 'Searching for Interstellar Communications' that was published in September 1959, one of the first formal rational arguments supporting the search for extraterrestrial intelligence.

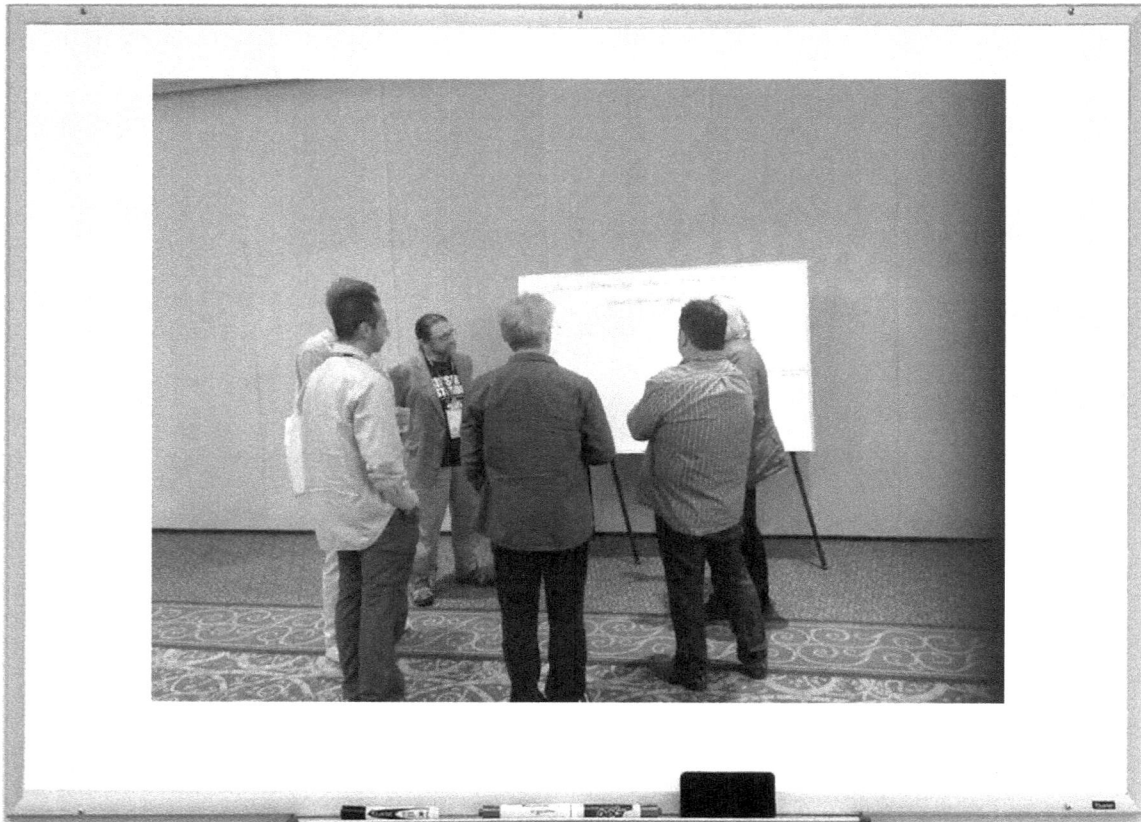

QUESTIONS

How should be raised in space?

How much time should be spent on cultural activities?

How do we distribute work in a new settlement?

Do we need to elect a supreme leader?

Should there be space currency?

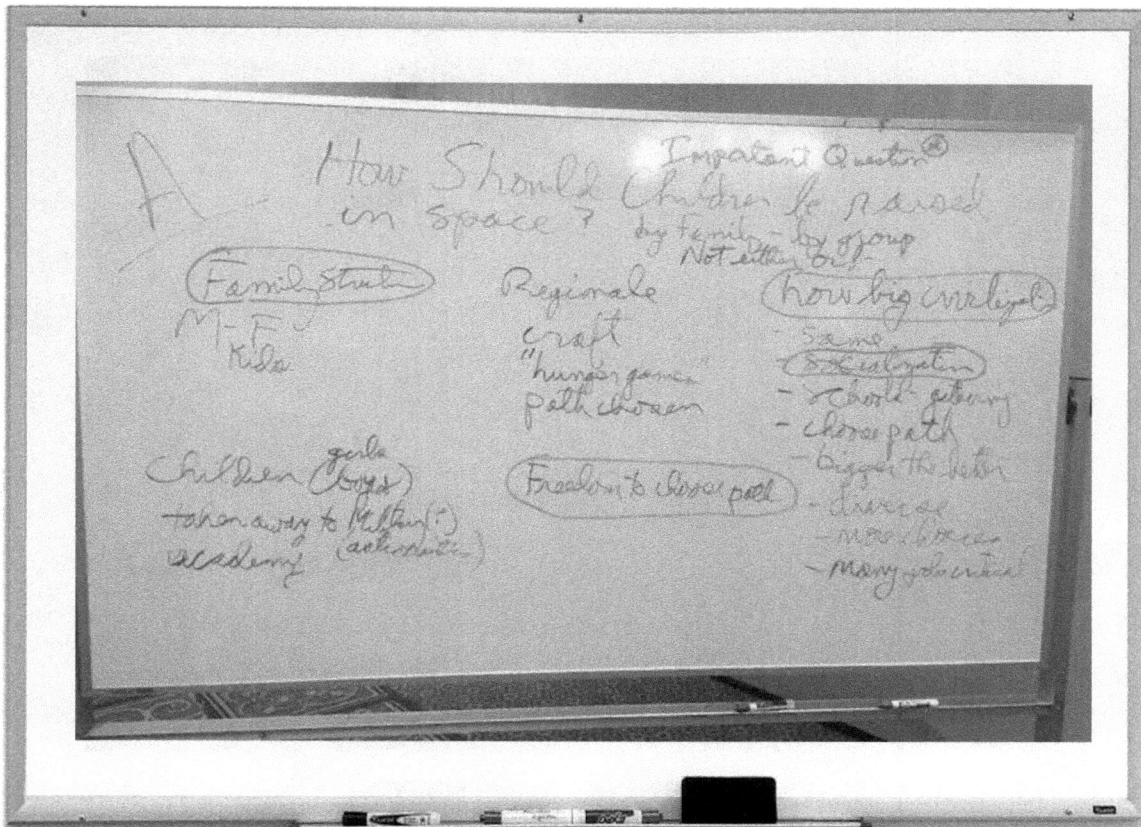

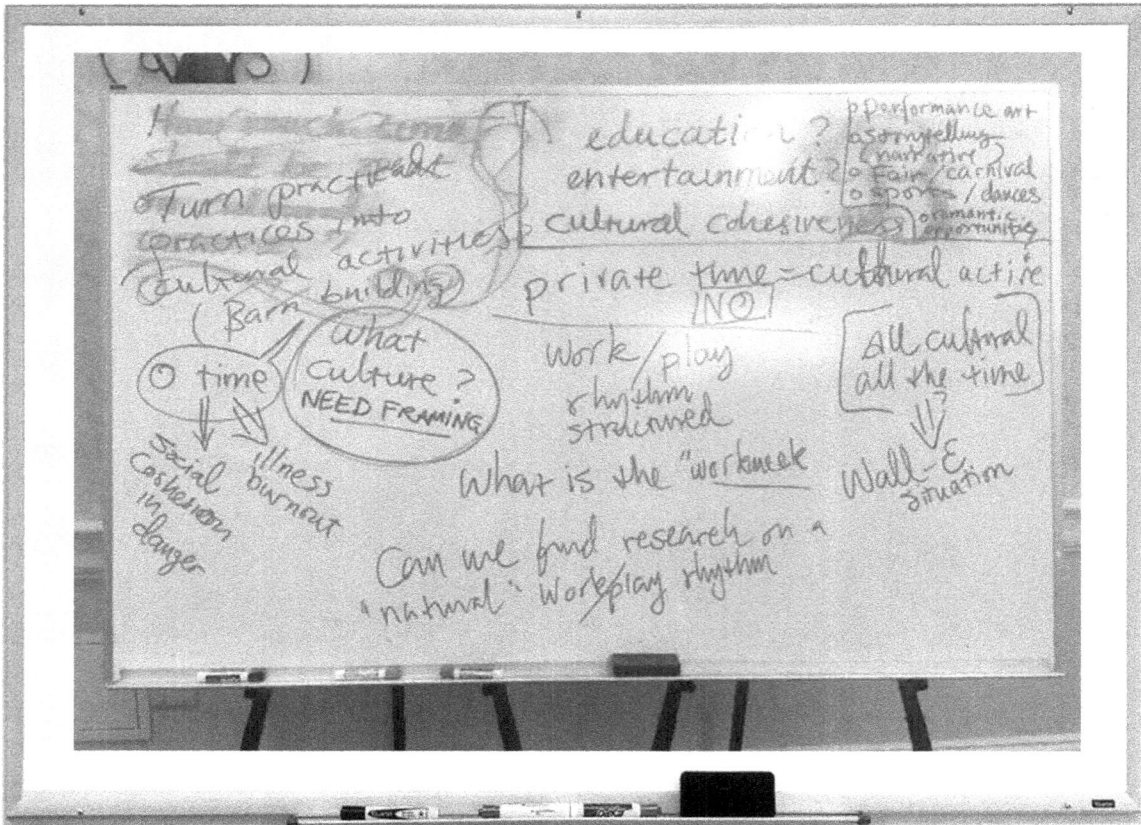

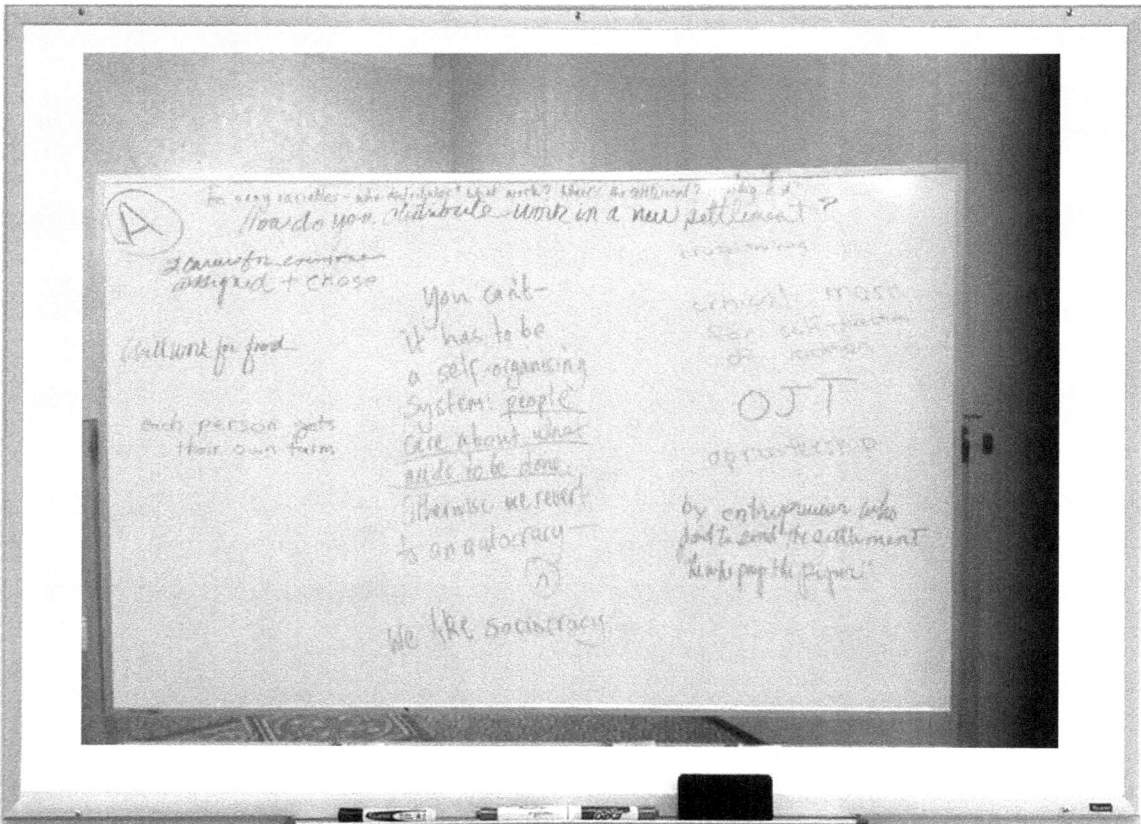

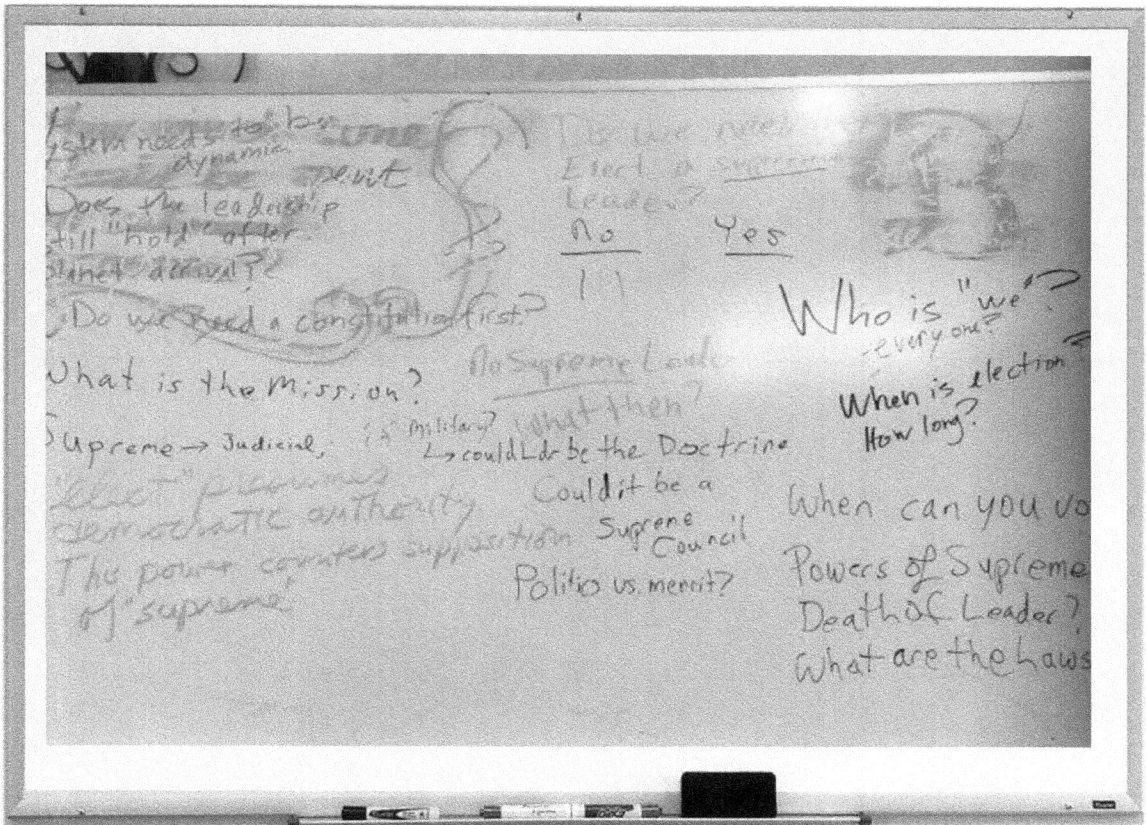

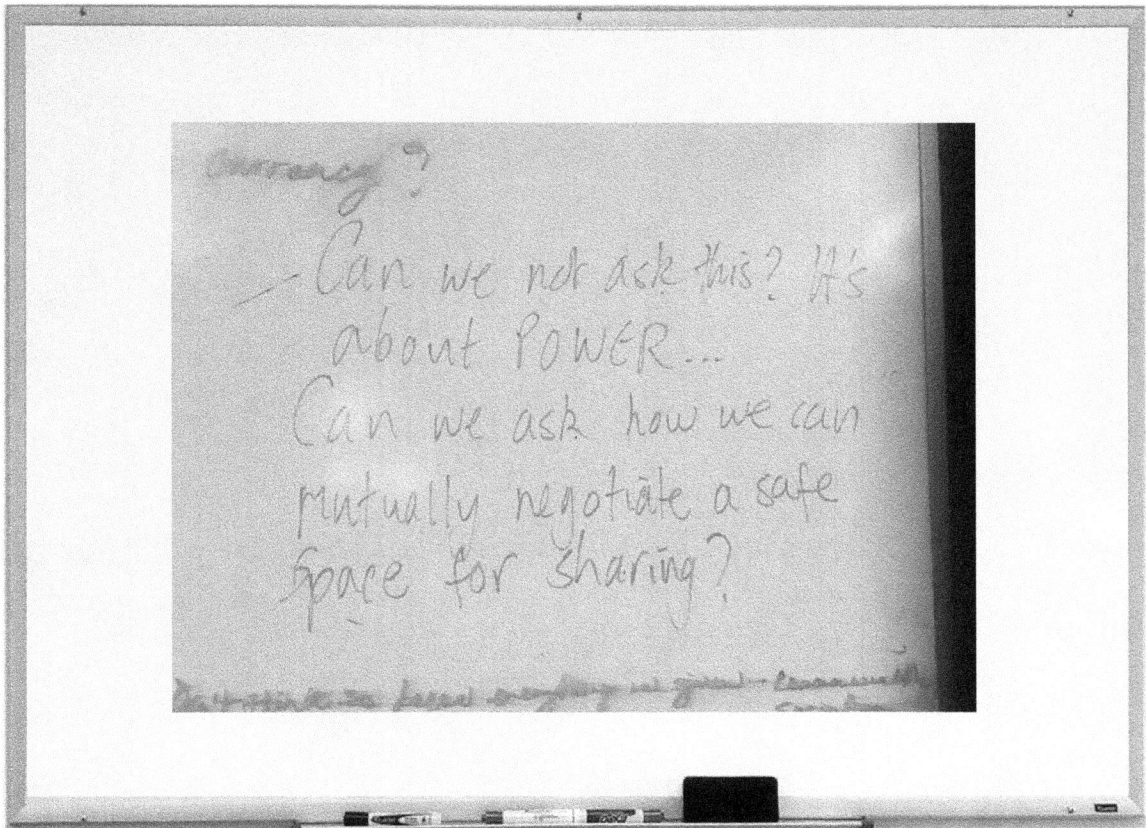

Technical
Tracks

Pamela Reilly Contag, PhD

Overall Technical Track Chair, 2105 100YSS Public Symposium

CEO, Cygnet Biofuels

The 100YSS 2015 Public Symposium had an all star cast of speakers carrying forward the theme of interstellar travel. From energy sources to communications, health care to education and designing for life in space and the civilization we may create. The result was four days of morning plenaries, presentations, evening events that generated deep conversations about where, how and when will we enter space with the intent to colonize. At the same time, participants never forgot the charge to forge technology that will be used beneficially here on earth.

Two things were embedded in every conversation. First, our biggest challenge is the vastness of space. The emptiness that is not really empty but contains a medium that we have little skill in analyzing, dark matter and dark energy. Secondly, our venture into space requires the kind of energy source that is currently theoretical even hypothetical. To cover the vast distances or even hover in Earth's inner space requires a source of energy, whether antimatter, dark matter, quark nuggets, moonlight or starlight, that we are currently not able to harvest.

We have a technology gap in space and here on Earth that will require mass partnership, participation and dedication in order to close. The timeline we have set for ourselves is short and the gap is large and will require us to leap frog over the incremental and strive for the breakthrough science and technology. If we want to survive as a civilization in space, our breakthrough must also be in our civilization here on Earth.

I want to thank all of the speakers, participants and track chairs for making each year, a colossal event.

Overall Track Chair Biography

Pamela Reilly Contag, PhD

Assistant Professor, Biomedical Engineering Department, Rutgers University

Pamela R. Contag, Ph.D., founded and is Managing Partner of the Starting Line Group, LLC a virtual ecosystem for the commercialization of advanced technologies. She is also currently the CEO of Cygnet Inc. Cygnet develops technology platforms for the research and development of advanced materials, biologics and industrial enzymes.

Dr. Contag founded Xenogen Corporation in 1995 and took Xenogen public in 2004. She served as CEO, President and Founder at Xenogen from 1995 to 2006, and concurrently, the CEO of Xenogen Biosciences from 2000-2006 when Xenogen merged with CaliperLS. In 2000, Xenogen Corporation was listed as one of the "Top 25 Young Businesses" by Fortune Small Business and in both 2001 and 2003 received the R&D 100 award for achievements in Physics. In 2004, Xenogen was named in one of the top 100 fastest growing companies by the San Francisco Times and received the Frost and Sullivan Technology Innovations awards. Dr. Contag was named one of the "Top 25 Women in Small Business" by Fortune magazine. She was also awarded the Northstar Award from Springboard Enterprises.

In 2005, Dr. Contag founded Cobalt Technologies, Inc., a venture backed company that produces biobutanol from renewable feedstock. She was the Chairman and CEO of Cobalt Biofuels from 2005-2008. In 2008 Cobalt was named one of the top 20 Cleantech Companies and in 2009 one of the top 100 Cleantech Companies. In 2007, Dr. Contag co-founded ConcentRx, Inc. a biotechnology company developing a unique cancer therapy developed by three Researchers from Stanford University. Dr. Contag founded Cygnet BioFuels in 2009. Cygnet BioFuels, her second biofuels company, is a company focused on the utilization of novel organisms for feedstock and biofuel production. In 2011 Dr. Contag was awarded "Cleantech Innovator of the Year" award for Cygnet technology.

Dr. Contag has held board positions, public, private and not-for-profit sectors. Dr. Contag was a Director of Xenogen Corporation (Nasdaq) (1995-2005) and a Delcath (Nasdaq) Board Member (2008-2011). In the private sector she was CEO and Chairman of Cobalt Technologies (2005-2008), Cygnet Biofuels (2009-present), Director at ConcentRx (2007-present) She also joined in 2009 the DOE Biomass technology Advisory Committee and two nonprofit boards, Springboard Enterprises, an accelerator of women entrepreneurs and the Molecular Sciences Institute as executive chairman and in 2011 merged that entity into MSI/VTT, and remains a Director. Dr. Contag also consults in biotechnology for academics and industry, including consulting Professorship at Stanford School of Medicine in the Department of Pediatrics (1999-present), the Dean's Advisory Board of the Johns Hopkins Bloomberg School of Public Health (1999-2005). In 2010 Dr. Contag joined the Merrick Engineering Consultancy specializing in the energy field and in 2011 Dr. Contag was named to the Start-up America Foundation National Board.

With more than 25 years of microbiology research experience, Dr. Contag is widely published in the field of Microbiology and Optical imaging and has over 35 patents in Biotechnology. Dr. Contag received her Ph.D. in Microbiology at the University of Minnesota Medical School in 1989 studying Microbial Physiology and Genetics (for Alternative Fuels) and completed her Postdoctoral Training at Stanford University School of Medicine in 1993 specializing in "Host/Pathogen Interactions".

BECOMING AN INTERSTELLAR CIVILIZATION
Chaired by Kathleen Toerpe PhD
How will the process of looking for and then definitively finding "another Earth" impact the social, cultural, economic, educational, religious, legal, political and ethical aspects of life here on Earth? How and in response to what do we create the belief systems that guide us? Who will we be and what will define our societies, morality, ethics, cultures, laws, economies, relationships and identities? Does biodiversity of life on Earth become more or less important or valued? What types of introspection or outward ambition are prompted? What are the implications for education—locally, nationally and globally? Do we purposefully broadcast "We are here" or become more conscious of our "radio wave spillage" into space? Does the U.N. get more or less money? Does the military become more important and funded more? Is now the time to think about becoming "Earthlings"? What treaties need to be in place? Will more investment in space-based tech become the trend? Presenters were asked to present on how civilizations and its mechanisms will influence or be influenced by finding an Earth 2.0 in the next 5, 10, 15 and 25 years.

DATA, COMMUNICATIONS, AND INFORMATION TECHNOLOGY
Chaired by Ron Cole
Sending and receiving information by interstellar travelers or robotic vehicles requires development of new methods to traverse the vast emptiness between stars. Additionally, in the absence of routine and timely communication with Earth, a probe or traveler must be self-sufficient in gathering, generating, compiling, storing, analyzing and retrieving data while ensuring these systems are operational over the lifetime of the mission and beyond. For this year's symposium, we had four excellent presentations that address the concerns of this technical panel. Following are gist of those presentations and email addresses of the authors for your follow-up for more details.

DESIGNING FOR INTERSTELLAR
Chaired by Karl Aspelund, PhD
Presenters in Designing for Interstellar are asked to consider parameters for designing probes, and vehicles or habitats for robotic or human crews –Earth, Earth orbit and deep space based—that will actively accelerate finding an analogue Earth and which may be implemented within the next 5, 10, 15 and 25 years. What aspects of design will be impacted by the various methodologies instituted to discover a planet outside of our solar system capable of supporting terran life? What are the design parameters that should be met to optimize the chances and rapidity in which such a planet may be identified?

Designs for probes and crewed vehicles must address the unique characteristics and extreme conditions of isolated research bases, deep space and interstellar space. The equipment, structures, tools, materials, cleaning and maintenance processes—the accoutrements of life and work— surround and create an operating environment or habitat. Such an environment protects, nourishes and facilitates daily activities. For living things, the environment must support the myriad physical needs. For higher order creatures, physical, mental and emotional requirements must be met as well.

Understanding, optimizing and manufacturing design for sustainability are critical for success—with a living crew or robotic probes.

LIFE IN SPACE: HEALTH, ASTROBIOLOGY, EARTH BIOLOGY, AND BIOENGINEERING
Chaired by Terry Mulligan, MD
How will the myriad fields making up the life sciences impact and be impacted by finding an indisputable Earth 2.0? Presenters are asked to consider the following areas for discussion from the perspective of experiments, projects and work that may reasonably be started/achieved within the next 15 years.

PROPULSION AND ENERGY
Chaired by Hakeem Oluseyi, PhD
A major aspect of discovering details of exoplanets is to get closer to them, to take samples and test actual physical properties beyond our solar system. Profound breakthroughs in the generation, storage and control of energy for propulsion, as well as communications and data gathering instruments are required to get to the interstellar medium in 10-20 years, much less to reach another star. Such breakthroughs are accompanied by robust leaps in theory and technology paradigms, and also incremental advances in engineering technology deployable in the next 5, 10 and 15 years.

Presenters in Propulsion and Energy are asked to present research and supportable ideas on how to address the design and deployment of instruments, probes and vehicles that will accelerate travelling beyond our solar system and closer to exoplanets within the next 25 years.

INTERSTELLAR SPACE, STARS, AND DESTINATIONS

Chaired by Margaret Turnbull, PhD

Finding Earth 2.0 is rooted first and foremost in astronomy, planetology and astrophysics. Whether using ground based observation, earth orbiting technologies or deep space probes within or just outside of our solar system, understanding and advancing knowledge, instruments and theories of the history, formation, composition and evolution of the universe, galaxies, stars and planets is fundamental to finding earth analogues.

Presenters are asked to provide research, outline novel concepts, propose instrument designs and methods, review data and define capabilities, knowledge and mission parameters key to furthering the understanding: the composition of exosolar systems; the identification of exoplanets in the "goldilocks zone"; planets that are rocky; exoplanet atmosphere composition, size; as well as defining the interstellar medium and aspects of maps, navigation and guidance.

In addition, as our gaze is drawn many light years away, focusing on closer objectives as stepping-stones to deep space will be essential. Beyond Mars, what missions should be designed to eventuate successful travel to another star? How should potential destinations be evaluated?

POSTER SESSION

Chaired by Timothy Meehan, PhD

Great ideas arise through unique individual observations, from people of all ages and educational backgrounds. Students are especially encouraged to submit to this session.

The Poster Sessions are an opportunity to present snapshots of early concepts and experiments. Presentation in the poster format allows in-depth discussion in a small group setting. Presenters are welcome to present on any of the topics from the other technical tracks as well as other topics germane to the theme Finding Earth 2.0. Suitability is at the discretion of the Track Chair.

100 YEAR STARSHIP™

Becoming an Interstellar Civilization

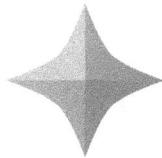

Chaired by Kathleen Toerpe, PhD

Deputy CEO for Program and Special Projects, Astrosociology Research Institute

Track Description

How will the process of looking for and then definitively finding "another Earth" impact the social, cultural, economic, educational, religious, legal, political and ethical aspects of life here on Earth? How and in response to what do we create the belief systems that guide us? Who will we be and what will define our societies, morality, ethics, cultures, laws, economies, relationships and identities? Does biodiversity of life on Earth become more or less important or valued? What types of introspection or outward ambition are prompted? What are the implications for education—locally, nationally and globally? Do we purposefully broadcast "We are here" or become more conscious of our "radio wave spillage" into space? Does the U.N. get more or less money? Does the military become more important and funded more? Is now the time to think about becoming "Earthlings"? What treaties need to be in place? Will more investment in space-based tech become the trend? Presenters were asked to present on how civilizations and its mechanisms will influence or be influenced by finding an Earth 2.0 in the next 5, 10, 15 and 25 years.

Track Summary

Session B in the 2015 *Becoming an Interstellar Civilization* track cast a strong gaze at the past, while simultaneously focusing on the future. The session was thematically bookended by two distinct analyses of what a roadmap for interstellar travel could look like. Abigail Sherriff's "A Roadmap to Interstellar Travel: Societal Challenges" (mentored by faculty adviser, Chris Welch) pinpointed the human social and cultural landmarks along the way and raised a core dilemma regarding humanity's interstellar mission: Is it really "mission complete" if we sacrifice our humanity along the way? In other words, how much can we remold our current social, religious, cultural, and ethical beliefs and practices before it is no longer humans like us reaching the stars? Antoine Faddoul's "Roadmap to the Stars" completed the mission overview by analyzing the structural, environmental, and technological hurdles that must be overcome before humanity ever lifts off. Using NASA's 1989 Integrated Space Plan to highlight how far we've come, Faddoul's projected roadmap posits a self-supporting moon base, permanent lunar colonies, and crewed missions to Mars all as indisputable prerequisites to eventual interstellar exploration.

With the mission defined, the session then looked at the use of an unlikely historical analogue as a travel guide. Karl Aspelund and Joelle Rollo-Koster's "The More Things Change . . . A 13th Century Model for Space-based Communities" examined how the hierarchical and village-centered feudal societies of medieval North Atlantic countries could guide the design and creation of future interstellar Worldship communities. With both societies defined by harsh environments and rigid hierarchies (aristocratic/religious in the former and scientific/militaristic in the latter), Aspelund and Rollo-Koster predicted that a generational Worldship society would result in a closed-loop "monoculture," marked by a lack of social mobility, the eventual dilution of capitalism, and the rise of technological authoritarian power structures.

215

In "Tomorrowland: Recreating the Cultural Ambition for Interstellar Travel," Jaym Gates brought the session back to the present by analyzing how the gaming industry has created a host of virtual analogues for space exploration, with space-themed games an increasingly large niche among new gaming titles. The widespread popularity of these games has especially engaged youth from lower socio-economic communities who are often underserved by current space outreach efforts.

In the session's last presentation, "Memory Issues for 100YSS Missions: Lifelogging, Social Media, and Virtual Time Capsules," JoAnn Oravec introduced a new word into space vocabulary: spacenemonics – the care and managing of memories on future long-term space missions. Interstellar travel, especially in generational or suspended animation ships, will require the transmission not only of immense amounts of technical data among the crew and residents, but also the curating and sharing of personal memories necessary to self-identity and those communal memories necessary to social cohesion and functioning. The process of data logging individual or communal memories or creating data time capsules also has important implications for treating memory-related diseases in the here and now.

Overall, Session B offered analyses that tied the past to the future, while offering benchmarks to which present researchers and explorers can aspire – a fitting conclusion to the *Becoming an Interstellar Civilization* track.

Track Chair Biography
Kathleen Toerpe, PhD

Deputy CEO for Program and Special Projects, Astrosociology Research Institute

Kathleen Toerpe, PhD, is a social and cultural historian who researches the human dimension of outer space through an emerging field called "astrosociology." She is the Deputy CEO for Programs and Special Projects and a Senior Research Scientist with the Astrosociology Research Institute (ARI). She is also an Executive Editor for the new peer-reviewed Journal of Astrosociology. Kathleen heads up 100 Year Starship's Education Special Interest Group and volunteers as a NASA/JPL Solar System Ambassador. She provides space outreach consulting, resources, and training through Stellar Outreach, LLC, and teaches courses in the social sciences and in critical and creative thinking at Northeast Wisconsin Technical College. Chairing the Interstellar Education track at the 2014 Symposium has been an honor. She can be found online at kathleentoerpe.com, at Twitter at @ktoerpe, or reached at ktoerpe@100yss.org.

A Framework for the Design of Ethical Robotic Probes

David Burke

Galois, Inc.

Abstract

One of the challenges inherent in sending a robotic probe to an exoplanet is that of interacting effectively with any potential life forms it encounters. In such a situation, how important is it that the probe, as an ambassador from Earth, acts morally and ethically? How do we even begin to define what those terms mean in this context?

We've just started to wrestle with similar challenges here on Earth. In recent years, there has been increasing interest in the problem of how to program robots to behave in a moral and ethical fashion when interacting with humanity (Asimov's "Three Laws of Robotics" being an early and well-known example). Various approaches have been proposed, ranging from reducing moral behavior to a set of irreproachable rules for the robot to enforce, to providing robots with a "moral calculus" module that computes the greatest good for the greatest number, to instilling robots with virtuous qualities (courage, truthfulness, kindness, etc.)

Whereas all of these approaches have merit, robot morality is still a complex, unsolved problem. And yet, building ethical robots on Earth is a simpler problem to tackle than the scenario above, where we can't easily make assumptions about the exoplanet's life forms: we can't presume they have human-like emotions or rational faculties; we will know very little about their evolutionary history, and we have no idea how they will conceptualize or perceive our probe.

In this paper, we discuss how people might constructively begin to think about this challenge, how to merge the various ethical perspectives mentioned above, and to suggest some next research steps towards an meta-ethical framework that would enable effective moral interactions between the probe and any denizens of an exoplanet it encounters.

Keywords
robotics, ethics, robot probe, thermodynamics, prime directive

217

1. Introduction

Given the challenges of sending humans on interstellar journeys, first contact with some life form on an exoplanet is overwhelmingly likely to be through a robotic probe. And without some as-yet undiscovered method of faster-than-light communication, the probe is going to be on its own, functioning autonomously without meaningful human guidance. Hence, the probe is a de facto ambassador for humanity, and the stakes for its behavior are very high. In such a situation, can we ensure that the probe will act morally and ethically?

It is easy to construct fanciful situations such as the probe arriving on an exoplanet that contains intelligent life, and the probe is under pressure to take sides on in some sort of political dispute. Such scenarios are very common in science fiction: as an example, consider the television show 'Star Trek' and the so-called 'Prime Directive', which states that humanity must not interfere with the natural development of evolution and society on other worlds. Fictional starship captain Jean-Luc Picard once stated the reasoning behind such thinking: "Whenever mankind interferes with a less developed civilization, no matter how well-intentioned that interference may be, the results are invariably disastrous." [1] (on the other hand, that interference makes for compelling drama, which is why the Prime Directive was violated so frequently.)

In recent years, there has been increasing interest in the problem of how to program robots to behave in a moral and ethical fashion when interacting with humanity, to say nothing of interactions with extraterrestrial life. Isaac Asimov's "Three Laws of Robotics" is an early and well-known example of exploring the consequences of robot ethics programming (and like 'Star Trek's Prime Directive, it was ostensible violations of these three laws that generated the plots for Asimov's stories). Various approaches have been proposed, ranging from reducing moral behavior to a set of irreproachable rules for the robot to enforce, to providing robots with a "moral calculus" module that computes the greatest good for the greatest number, to instilling robots with virtuous qualities (courage, truthfulness, kindness, etc.)

Whereas all of these approaches have merit, robot morality is still a complex, unsolved problem. And yet, building ethical robots on Earth is a simpler problem to tackle than the scenario above, where we can't easily make assumptions about the exoplanet's life forms: we can't presume they have human-like emotions or rational faculties; we will know very little about their evolutionary history, and we have no idea how they will conceptualize or perceive our probe. In this paper, we discuss how people might constructively begin to think about this challenge, how to merge the various ethical perspectives mentioned above, and to suggest some next research steps towards an meta-ethical framework that would enable effective moral interactions between the probe and any denizens of an exoplanet it encounters.

2. The Challenge in Historical and Contemporary Context

Before specifically tackling the challenges of designing and specifying ethical behavior for robots, it is worth leveraging the rich tradition of philosophical thought about ethics as a guide to our computational instantiations.

Very roughly, we can sort ethical thinkers into three different schools of thought about how morality fundamentally works:

1. Deontological: These are those philosophical and religious thinkers who gravitate towards the idea that there is (or should be) a set of rules or natural laws that describe our ethical obligations towards one another. Kant is an exemplar of this tradition, in which people's actions are judged by their intention to conform with natural law.
2. Consequentialism: In this school of thought, morality is essentially a matter of cost & benefit, the "greatest good for the greatest number". Moral acts are judged by their consequences on society; Jeremy Bentham was an early exponent of this philosophical worldview.
3. Virtue Ethics: In this school of thought, there are cardinal virtues to which all humans should aspire to: humility, courage, prudence, and so forth. If we can mold people's characters so that they adopt these virtues, then moral behavior will naturally result. This was Aristotle's view, for instance.

The very fact that there continue to be (at the very least…) three schools of thought on the subject is evidence that we don't really know the answer. It is very easy to design scenarios which illustrate the unintended consequences of a slavish adherence to any of these worldviews. For instance, for any rule that might be put forth as a categorical statement, say, "Don't kill", it isn't hard to come up with an exception scenario.

The potential weaknesses that lurk in the consequentialist approach were explored by the philosopher Philippa Foot in her "Trolley Problems" – a famous set of ethical thought experiments. [2] In these scenarios, you are asked to take actions that sacrifice innocent lives to save others, and these demonstrate that the "greatest good for the greatest number" is more complicated than you might initially think. All of the various schools of though suffer from these kinds of difficulties, and the way forward is even less clear once you start imagining that you need to generate the kind of formal behavioral specification that would be required if you were to direct a machine's actions. As a result, the field of 'machine ethics' is an active and contentious one. [3,4,5]

This issue isn't just an abstract, philosophical one, because the future is coming at us very quickly. DARPA (Defense Advanced Research Projects Administration) funded the roboticist Ronald Arkin to look at how to design architectures for battlefield robots. In his final report, Arkin concluded that there is reason to believe that robots can be more ethical in combat than humans, since they don't suffer from fatigue, anxiety and other emotions that could lead to irrational actions in battle.

The popular press has also shown great interest in the subject, with headlines like this appearing in the last couple of years:

- "What if your self-driving car decides that one death is better than two – and that one is you?"
- "Killer Robots with Automatic Rifles Could Be on the Battlefield in 5 Years"

The smart money is on the bet that semi-autonomous and autonomous robots will be increasingly ubiquitous in the near future, and that we will be relying on them to make sensible decisions in domains that have a serious impact on human welfare. And yet, there is no consensus as to the underlying framework to support robot morality. How do we avoid the potential negative consequences of this future? And if we suggest a way forward, is it applicable to the autonomous exoplanet ambassador scenario?

In the rest of this paper, we will unveil a framework to support thinking about morality as a general principle, abstracted away from any particular human concerns. To do this, we will begin by looking at ethics as an abstraction of energy flows. Looking more carefully at the phenomenon of energy flows, we will find that universe is full of persistent, far-from-equilibrium gradient reduction mechanisms, life in particular. And this insight will lead us to a preliminary formulation for a proposed Prime Directive for robotic probes. This new formulation can serve as a meta-ethical framework and a foundation for further investigation.

3. Ethics as a Resource Design Problem

In an effort to bring some clarity to the topic of human morality, we're first going to step back and take a naturalistic approach: imagine that we are extraterrestrial beings trying to scientifically understand the concept of human morality: what problem, exactly, is ethics trying to solve in the first place? In this way we divorce the issue from the usual human-centered emotional associations with Right and Wrong, Good and Bad, Heaven and Hell, and so forth.

In our view, human morality is the question of how human beings live in groups and manage resources so as to mediate the natural tension between the individual and the group. That is, which resources does the individual keep, and which resources are shared with the group (and the details of that sharing). Anthropologists and biologists describe human beings as prosocial animals – it is unnatural for us to live alone for long periods of times, and we have evolved to live in groups. But that doesn't mean it is easy, and morality is the challenge of maintaining that balance through the allocation of resources.

Early in human evolution, the resources in question were what we call 'direct' resources – material resources such as food, shelter, clothing, and access to mates. Other resources, such as land, and social status, can be thought of as indirect: if you control land, you control the ability to dispense material resources, and in the case of social status, you can use your prestige or role in a group to influence other people to give you resources (for example, commoners paying tribute to a king simply because of his role).

It is fascinating to consider how resources have become increasingly indirect and abstract as societies have evolved. We now live in a staggeringly complex social world, filled with abstractions such as financial derivatives, or elaborate rituals such as professional sports. In a modern culture, these resources matter a great deal – witness how some of the most wealthy and prestigious members of our society have roles such as hedge fund managers or professional athletes. Also, it can be argued that in a society in which basic material resources such as food and shelter have been satisfied, abstract resources such as social status, fame, and prestige become more important.

These supposedly abstract resources can generate very material consequences, however: for instance, think of the amount of time and energy spent fighting for the future direction of the Hugo Awards during 2015 as a result of the "Sad/Rabid Puppies" campaigns. Clearly, the battles weren't about the value of the material trophy itself – they were about the argument as to who is deserving of the prestige that comes with winning an award that is considered to confer high status (and the associated financial consequences of incremental book sales, invitations to speak at events, and so forth). When one side of such a controversy is accusing the other side of unethical behavior, it is because what is at stake is the prospect of future resources. Again, morality is fundamentally about the question of resource allocation, where those resources matter to somebody or some group.

4. Resources, Flows, and Thermodynamics

It has been often said that "Energy is the Ultimate Resource". So if we're going to talk about morality as being essentially a matter of resource allocation among organisms (humans in particular), then it is worth thinking about the energy, and its potential applicability to the problem at hand.

The science of thermodynamics can be thought of as the science of energy flows. Classical thermodynamics is often considered to be a well-mined, even dull subject that centers on heat engines, steam tables, and the like. It is also sometimes thought of as a depressing subject, due to the 2nd law of thermodynamics, which supposedly predicts an inevitability to our universe dying a heat death after it has run out of usable energy. However, the science of thermodynamics is now undergoing a renaissance, manifested in the sub-field of "nonequilibrium" thermodynamics or "open system" thermodynamics, which includes the study of self-organized systems that evolve and gain in complexity, seemingly (but not really) in violation of the 2nd law.

One of the key lessons from this new research is that flows, whether these flows comprise energy, matter, or information, are the key to understanding how systems evolve. Over time, systems (both natural and man-made) tend to acquire better configurations to provide more access for the currents that flow through them. This insight has been captured with the following characterization, which is known as the Constructal Law:

"For a finite-size system to persist in time (to live), it must evolve in such a way that it provides easier access to the imposed currents that flow through it." [9]

This design principle illuminates why we see similarities in nature between river deltas (flows of water) and lungs (flows of oxygen), or for the similarities in shape between human-designed systems such as airport terminals (flows of people) and heat sinks (flows of heat).

A complementary way of looking at the nature of flows is to say:

"Nature abhors a gradient" [10]

This aphorism explains why hurricanes and tornadoes form - to reduce heat gradients in the atmosphere. Their complex structure (their spiral swirls, in this case) allow them to dissipate energy faster than if they were simple, amorphous structures.

An explosion can be thought of as a violent gradient reduction (for example, when a match triggers the dissipation of the potential energy inherent in a canister of gasoline), and so that raises the question – if gradient reduction is the name of the game, why don't we see more explosions in nature? The answer has to do with the dimension of time – gradient reduction is not just a phenomenon that needs to take place at one moment in time, and the results of an explosion can render a system incapable of doing any future gradient reduction. For example, if you induce your automobile to explode, you certainly reduce the fuel gradient that existed in the gas tank. But now your automobile is no longer capable of reducing future fuel gradients.

This line of thinking is very suggestive when it comes to understanding the phenomenon of life – all living creatures are continually reducing gradients (turning the energy from food into a combination of muscle action and the waste products), and it has been proposed that life arose in the first place because living organisms are more effective at gradient reduction over time than any non-living entities. Through reproduction, we ensure that even when our bodies finally succumb to entropy and die, we have created replacement gradient reducers by bearing children. Life as energetic flow that persists through time has the name autopoesis. [11] If you're looking to define meaning of life, you could do much worse than to claim that the meaning of life is simply gradient reduction: Life on Earth persists because it tends to effectively reduce the gradient between the hot sun and its cold surroundings.

5. Gradients and Far-from Equilibrium Conditions

As mentioned previously, the most exciting development in the science of thermodynamics is the research being done on "open" thermodynamic systems. In these systems, sufficient energy continually comes into the system so that the system can support complex, self-emergent behavior. These open systems are also known as "far-from-equilibrium" systems. One of the insights from this work is the realization that gradient reduction and far-from-equilibrium conditions are duals of each other – neither phenomenon is causal with respect to the other. Both natural and designed energy flows can induce gradients that persist at far-from equilibrium conditions, and these conditions can induce further gradients for reduction.

The concept of ecological succession is a good example of this. In ecological succession, we have a cyclical process with a duration on the order of a hundred years. Consider a meadow that has just experienced a fire that destroyed virtually all the existing plant life. If we observe this meadow over the next 100 years, we will observe various phases. In the first phase, taking place over the first five years or so, the meadow is settled by 'pioneer species' – hardy grasses, mostly. After these have been established, we begin to see the appearance of intermediate species – a collection of grasses and shrubs with the addition of some species of fast-growing trees such as pines. As the decades pass, a 'climax community' arises as the meadow is populated by mature trees such as oaks. This climax community persists until another major fire signals the beginning of a new cycle.

We can characterize the process of ecological succession by observing the complexity of the system and the energy flows it contains. As ecological succession occurs, we observe:
- The system moves farther from thermodynamic equilibrium
- The system contains greater amounts of free energy (i.e., potential energy)
- System components have longer energy residence times
- The system contains a hierarchy of energy flows, creating a more complex overall structure.

Similar phenomena can be observed if you take a look at a modern technological society. Structures such as dams and flying aircraft are operating at far-from-equilibrium (for instance, equilibrium for the dam/water system is for the dam to fall apart, and water on each side to reach the same level; for the airplane, it is violating the ever-present gravity gradient). Energy flows and gradient reduction can even contribute insights to the study of potential life on other planets. As an illustration of this, there is the fascinating story of how James Lovelock was consulting to NASA in the 1960's during a time when robotic missions to Mars were being planned in order to look for evidence of life. Lovelock insisted that life on Mars was exceedingly unlikely: the fact that the Martian atmosphere existed in chemical equilibrium, unlike the Earth's atmosphere, which is in a constant far-from-equilibrium state, was a strong piece of evidence against the hypothesis that life exists on Mars.

The key takeaway here is that complex energy flows and the associated gradient reductions are inextricably linked to far-from-equilibrium conditions. This is as true of human-made artifacts as it is to natural ones such as the ecological state of a meadow, or the physiological state of an organism. Energy flows, persistent gradient reduction and complex structures all go together such that where you have one, the other two are always present.

6. Conclusion: A Prime Directive for Robotic Probes

Now we'd like to put the pieces together, and make explicit the implied argument that gradient reduction and its associated concepts can provide guidance in the design of ethical robotic probes. Retracing the steps, our argument thus far goes like this:
1. Ethics is about the allocation of resources
2. Resources are, fundamentally, energy flows – ones that matter to an organism
3. Both natural and human-made energy glows can induce gradients that persist at far-from-equilibrium conditions
4. Persistent gradient-reducing energy flows give rise to complex structures that tend to induce those flows.

Therefore, to act ethically in presence of persistent energy flows is to respect the gradients that matter.

Morality is simply respecting existing gradients. A sophisticated intelligence (of the kind we hope a robotic probe will be armed with) in this context means the ability to detect diverse kinds of gradients: radar, pressure, ultrasound, temperature, radiation, capacitance, humidity, infrared, and so forth. The more gradients the probe can sense, the more aware it will be of which ones are likely to matter: those that are persistent, and whose interruption lead to a violation of the integrity of other flows.

In terms of morality, a simple corollary to the principle we have just distilled becomes our "first draft" of a proposed Prime Directive:

Prime Directive for Robotic Probes: Do not interfere with the maintenance or growth of persistent energy flows on any world you visit.

As we become more skilled at the awareness of and the management of the energy flows on our own world, we will no doubt refine this draft Prime Directive into something more nuanced, and equip our robotic ambassadors to the stars with the wisdom necessary to make mutually beneficial first contacts with the other inhabitants of this universe.

References

1. "Prime Directive" retrieved from Wikipedia.org

2. Edmonds, David, *Would You Kill the Fat Man? The Trolley Problem and What Your Answer Tells Us about Right and Wrong*, Princeton University Press, 2014

3. Anderson, Michael & Anderson, Susan Leigh, editors, *Machine Ethics*, Cambridge University Press, 2011

4. Lin, Patrick & Abney, Keith, editors, *Robot Ethics: The Ethical and Social Implications of Robotics*, MIT Press, 2011

5. Wallach, Wendell, *Moral Machines: Teaching Robots Right from Wrong*, Oxford University Press, 2010

6. Arkin, Ronald, *Governing Lethal Behavior in Autonomous Robots*, Chapman and Hall/CRC Press, 2009

7. "What if your self-driving car decides one death is better than two – and that one is you" appeared in the online *Washington Post*, October 28th, 2015

8. "Killer Robots With Automatic Rifles Could Be on the Battlefield in 5 Years" appeared in *WIRED* online, October 18th, 2013.

9. Bejan, Adrian & Lorente, Slyvie, "Design with Constructal Theory, Wiley, 2008

10. Schneider, Eric & Sagan, Dorion, *Into the Cool: Energy, Flow, Thermodynamics, and Life*, University of Chicago Press, 2006

11. Maturana, Humberto & Varela, Francisco, "Autopoiesis and Cognition: The Realization of the Living", D. Reidel Publishing Company, 1980

12. Lovelock, James. *Gaia: A New Look at Life on Earth*, Oxford University Press, 2000

Interstellar Embryos through Interstellar Teens: Negotiating Human Development Through Education In An Off-World Context

Janet de Vigne, MEd

PhD student – Education and Social Justice

University of Lancaster UK, Bailrigg, Lancaster LA1 4YW

janet_devigne@yahoo.com

Abstract

Bringing up Terra-based children is fraught with difficulty; guiding children through development in an extra-terrestrial context may prove even more so. Neuroscience is increasingly feeding into education in the 21st century, so what are the neurobiological game changers today, and where might education be headed in the next thirty years? This paper presents recent findings that impact on child development and classroom practice, examining issues such as managing the teenage brain and its attraction to risk; potential neurological losses should we choose not to include, for example, linguistic diversity or music; and the development of computer or child-driven education systems. Along the way we will discuss issues of culture and society and their merits and demerits: what do we choose to keep? What do we choose to lose, in terms of Terran culture and mores (Shakespeare, Milton, the Bible)? How do we encourage the best of being human in an extra-terrestrial context? How do we maintain human rights and guarantee liberty? What definition of liberty will we deploy, particularly in the case of children?

Keywords

education in space, off-world education, children's rights, technology and education, neuroscience and education, child development, brain development

1. Introduction

The idea of sending humans off world is gathering pace as Mars One recruits and trains volunteers to carry the race into space with no chance of return. Where there are humans, there must therefore be reproduction, be it natural or mechanically aided, or there would be no point. The survival of the species must depend on its capacity to reproduce and educate the new generation to carry us forward. This paper will argue that ensuring a healthy, successful community of humans will depend on the ways in which children are educated, focusing specifically on the period of adolescence, popularly conceived as problematic. If we wish to avoid already imagined scenarios of tyranny, such as the chaos of The Lord of the Flies (1954) [1] or Total Recall (1990) [2], considering the possible roots of the exercise of control might be wise. I suggest that an equitable society supported by the principles of social justice and democracy would be the one to aim for, as shared responsibility for the success (meaning non-implosion and destruction) of a small, off-world community must mean the practice of quasi-utopian ideals of conflict resolution, listening to each other, valuing difference, and practising tolerance. It would be wise to capitalise on the strengths of the community, as the risks arising from disaffection would endanger it.

2. The Context

Twenty-first century Western society is really not very nice to our teenagers [3]. Managing this crucial period of human development is a trial for all concerned. Middle-class parents might choose to ignore it by deploying the "keep them busy" tactic, hoping that extracurricular activities will ease their child into adulthood by distracting them from any possibly inconvenient behaviours. Other parents (from any and all backgrounds) might view aspects of the teenager with increasing bewilderment–what is happening? A "normal," polite, respectful, and chatty child turns overnight into a monosyllabic, raging monster, full of resentment and irrational disaffection. This period is universally recognised and caricatured in numerous comedic performances, in the UK (and now on YouTube™) successfully by British comedian Harry Enfield in his portrayal of "Kevin the Teenager"–a spotty, strangely dressed creature, constantly screaming at his gentle, long suffering, and increasingly perplexed parents, desperate to 'fight' where there is no conflict, and convinced of the injustice being meted out to him with no real evidence for the same [4]. Off-world irrationality—behaviour and thinking—is greatly to be feared. It doesn't bode well on earth either, but there appears to be space for teens to work it out without compromising the existence of the entire human race. (Adult irrationality is of course another matter.)

Attempting to control irrational behaviour usually does not work; the consequences of any attempted control might well be worse than the initial problem. Managing it may be a better option, by which I mean providing ways in which issues can be played out with little or no risk to the social and physical environment. Rites of passage—practices perpetuated to this day by older societies than ours—have a role here, as does heightened play: the value of Shakespeare in emotional growth and Theory of Mind (effective team building) development (El Sistema)? More of this anon.

Why, then, is the period of adolescence so dangerous? What is really happening in terms of biological development, and how might it be possible to manage this in a positive way? Only considering management, and thereby conceptualising the teen years as a problem, may not be the best way forward. If instead we viewed this time as a positive phase, where teenagers might perform a valuable role in the building of a new society, we might find a more helpful and acceptable way of negotiating the turmoil.

2.1   The Neurobiology of Adolescence

Many people use the surge in neurobiological activity of the teenage brain to explain—and write off—the more unacceptable facets of teen behaviour. The real story is of course more complex, and it should be possible to see every supposedly risky aspect as having a positive side. Teenagers love risk; how could this be built upon to benefit the community? Similarly, the desire to negotiate a social world, to compete, the vulnerability to peer pressure, and the unpredictability of endocrinological impulses including sex hormones—perhaps the most notorious area of development at this stage—all these might in fact be regarded as positives. Some consider that this period is an evolutionary necessity, occurring in order for the young adults to be encouraged to leave the community and go off to forge their own destiny without the confines of parental influence. Other mammalian behaviour seems to substantiate this, because of course ours is not the only species to go through adolescence. Understanding our neurobiology holds the key.

Advances in medical technology mean that it is now possible to observe the brain without physically interfering with it. As a consequence, it has been possible for a while to measure the development of the brain, and blame teenage behaviour on this in a rather determinist manner. We know now that the brain develops from back to front, and that the mechanisms of the frontal lobes act as the "brakes": the consideration of consequences of any given action. With an undeveloped "consideration" function, a teenager is unlikely to be able to make an accurate assessment of the consequences of his or her actions. Extrapolating from this, it is easy to fear the unpredictability of an irrational act that would imperil an entire off-world community, and in our time, this is seen as the most dangerous aspect of teenage development. Is it really as simple as this, though? Strengthening the cognitive-control mechanism occurs throughout adolescence, and there is a maturational gap between this and the remodeling of the brain's socio-emotional reward system. What is now better understood is the relationship between peer related stimuli sensitising the brain's reward system to the value of risk—this is (in part) a social process [5]. The picture becomes more complex as we realise the dynamic developmental relationship between behavioural practice and neurobiological development. (Susan Greenfield noted something similar in the development of the hippocampus in London taxi drivers. "Experiences, then, are reflected in the strength and extension of brain connections." [6]) Recent research by Casey and Caudle (2013) shows that teenagers perform better in terms of making decisions than adults and children, unless they are being subjected to peer pressure. In other words, when in the company, exclusively, of their peers, teenage risk taking becomes dangerous [7]. Albert et al's research demonstrates the attraction of risk among peer groups: "It's no fun doing it on your own." [8] There is a hugely increased propensity for risk taking when in the company of other adolescents. It seems that they are not able to negotiate "socio-emotional and choice related incentive cues" [9] and although high-risk behaviour shows high levels of testosterone (boys) and estradiol (in girls), levels of estradiol are lower than those of testosterone [10].

The potential here for good and ill is clear; I suggest that it would be possible to use this function to the benefit of the community by mixing groups—say of pilots—but allowing teens the responsibility of making decisions. The problem that then arises is the old issue of when and where a human should be considered mature—I suppose that my argument here is that such old definitions need to be suspended if and where the function of an adolescent might outstrip that of an older adult. There are obvious philosophical implications. It might be, however, that the dynamic relationship between cause and effect could have a very positive effect on development.

2.1.1   Where the hormones, there moan I . . .

It seems, therefore, that "specific hormones impact specific neural systems to influence specific behavioural tendencies" [11]. No surprises here, but what is interesting is that the level of the hormone may not be related to the biological signal that it gives. In adolescence there is a "neuro-behavioural nudge" towards social bonding. The oxytocin and vasopressin system, shown to be central to this, and development of social cognitive processes such as the development of trust, changes at puberty. Even if these changes appear to be small initially, over time they may have a greater impact. To explore this in greater depth, even though there is not much human research in this area, there appears to be a reflexive relationship between the social environment and hormonal levels. Research by Peper and Dahl (2013) shows a relationship between the mediating effect of the environment on behaviour and on levels and production of testosterone [12]. These effects can begin positive or negative trajectories that will manifest in later life—effective management of these to the benefit of the individual concerned then seems crucial:

Interactions between socio-affective processing and cognitive control 'heat up' considerably   during puberty in the ventral striatum. If these interactions are positive, good things will   ensure; if they are negative, and even if they begin in a small way, they may result in   depression and substance abuse. These effects will not be 'one offs' they will 'cascade' across adolescence. [13]

Of further interest is the evidence that appears to show a relationship between socio-emotional hormonal activity and processing of guilt and shame—these seem to take place in the same area within the anterior temporal lobe [14]. As behavioural drivers, the toxic potential of guilt and shame is well documented, particularly in the growing field of psychosomatic trauma management [15].

2.2.2   Teenagers and mental health

The importance of the mental health of teenagers cannot be underestimated. It is acknowledged by health care professionals in the UK, that we do not care for our adolescents in a good way. Bearing in mind the positive and negative trajectories indicated by the reflexive relationships between environment and hormone production, it should be noted that one in five teens are likely to have mental health problems, and of these, fifty percent will continue into adulthood with the same issues. To a degree and with readily accessible help available to adults who can afford it, earth-based problems stand a chance of effective management, hopefully leading to resolution. Is this a statistic that could be so readily borne in space? I suggest not—and the research would indicate that it is possible to minimise the risks if we put our minds to it. Effective negotiation of this crucial developmental phase might include: watching for signs of depression (these may manifest themselves as irritability), managing peer interaction (balancing time and needs to mitigate the possible negative effects of negative peer pressure), being open about the neuro-behavioural nudge to the wider social world, and making informed and relaxed discussion spaces for the topic of sex. Safe spaces for effective anger management will be a big part of this, as emotions will be running high. We will need highly skilled emotional intelligence-related practices to negotiate a successful way through feelings, irrationality, and risky behaviour.

Decisions will need to be made, then, about what to include in a possible learning environment for teens. Topics and curricula that promote social cohesion while building trust and team work (Why not mention love and friendship—essential in a small community?) and encouraging the practice of conflict resolution and forgiveness, as well as personal strategies for releasing negative emotions are perfectly possible. The risk at the moment is that in the rush to ensure competence in the wide variety of scientific functions that will be necessary for survival, we will forget the human skills inherent in emotional intelligence. Knowledge has been advancing in these areas also, particularly in organisational development and conflict resolution. It is my contention that the period of adolescence will throw these into sharp relief, and it will be the responsibility of everyone in the community to help the teens in through and out the other side safely.

3.   Strategies and a Curriculum for Emotional Intelligence

One of the most effective media for helping children work through emotion successfully is heightened play. This has been demonstrated in the work of Cicely Berry, director and voice coach with the Royal Shakespeare Company, resident in the UK. In the DVD Where Words Prevail, Berry works with a group of street children, mostly teenagers, in Rio de Janeiro. Using the language of Shakespeare, heightened language that makes it safe to articulate very strong feelings, students clearly and powerfully express the reality of their lived experience, achieving catharsis [16] or, if you prefer, an emotional detox. This type of practice could be very valuable in a small community—a safe space in which to articulate issues that could cause disruption, or a place to simply let off steam. What then shall we choose to take with us in terms of culture and cultural practice? Which great global authors would help to manage heightened emotion in this context? (Decisions about language will also need to be made, as at the moment we seem to be assuming that everything will be in English).

Music, too, is a powerful force for team work and social cohesion, shown perhaps most notably in the work of Jose Antonio Abreu and the "El Sistema" movement:

In my view, the root cause of all our social problems is exclusion. We must fight to ensure   that a larger number of people – everyone, if that's possible - have access to this wonderful   world – the world of music, the world of the orchestra, the world of song, the world of art. [17]

Abreu is very clear that the El Sistema method of teaching children from all backgrounds to play popularly named classical music (in the Western tradition) is the way forward for social cohesion and the practice of equality. Many other benefits are associated with music—again, it acts as a mechanism of emotional catharsis while teaching team work, listening, discipline, attentiveness, and fluency in the language of music, arguably as good for the brain as fluency in another language. But might this be a misplaced attempt to include an imperialist imposition in a small, off-world community? It would appear not. A practice that brings the community together to communicate in a different way from the exchange of information, including their emotions, might be an excellent way to encourage positive brain development in the teen. Might there be room on a spacecraft for the instruments? Why not? Many claims are made for El Sistema, [18] among them reduction in crime levels, social cohesion and better learning. Accessibility is open; there are no pre-existing entrance requirements. Although some criticise the system for its discipline, the movement is pretty much universally accepted as a very good thing, and has now been instituted in Scotland where it has revolutionised several socio-economically deprived areas. Given that the root cause of social unrest is the resentment coming from exclusion, Abreu seems to have achieved his aim in focussing communities on the positive advantages of music. Over 310,000 children have so far benefitted from El Sistema.

4. Conclusion

Paying attention to human development to a degree not currently practised on earth will be essential in an off-world context if the species is to survive. Psychology and neuro-biology has much to offer in terms of understanding the most difficult period in the trajectory of human development, but as yet the relationship of emotions and hormones has not been made explicit, and it is clear that the socio-affective influences the socio- cognitive, and vice versa, to the detriment or benefit of all concerned. To help teenagers achieve their best potential will benefit not only them, as individuals, but also, clearly, work to the benefit of the entire group—it is not only possible to enable this, but it should be considered as part of the role of every person present. Would this were the case on earth today! Teenagers and all interstellar humans will need every skill we have and are developing—it is a concern that with a focus on scientific and engineering competence, these skills will not be considered as teachable. It is possible to learn skills and capitalise on talent, but the thorny issues of a successfully negotiated adolescence and an emotionally intelligent emergent adult should not be left to chance. To do so would be to place the species at risk, if we really are relying on interstellar embryos to carry humanity into the stars.

References

1. Golding, W. "The Lord of the Flies," London: Faber and Faber Limited, 1954.

2. Verhoeven, P., Director. Total Recall. [Film]. USA: Carolco Pictures, 1990.

3. St. Louis, Connie. "Life as a Teenager," London: BBC Radio 4 http://www.bbc.co.uk/radio4/science/lifeasa-teenager.shtml, 2015.

4. Enfield, Harry. "Kevin the Teenager," Tiger Aspect for BBC Comedy, London, 1997.

5. Albert, D., Chein, J., and Steinberg, L. "The teenage brain: Peer influences on adolescent decision making," Current Directions in Psychological Sciences, vol. 22, no. 2, pp. 114-120, 2013.

6. Greenfield, S. "Mind, brain and consciousness," The British Journal of Psychiatry, vol. 181, no. 2, pp. 91-93, 2002.

7. Casey, B.J., Caudle, K. "The teenage brain: Self control," Current Directions in Psychological Science, vol. 22, no. 2, pp. 82-87, 2013.

8. Albert, D., Chein, J., and Steinberg, L. "The teenage brain: Peer influences on adolescent decision making," Current Directions in Psychological Sciences, vol. 22, no. 2, pp. 114-120, 2013.

9. Albert, D., Chein, J., and Steinberg, L. "The teenage brain: Peer influences on adolescent decision making," Current Directions in Psychological Sciences, vol. 22, no. 2, pp. 114-120, 2013.

10. Peper, J.S., Dahl, R.E. "The teenage brain: Surging hormones—brain-behaviour interactions during puberty," Current Directions in Psychological Science, vol. 22, no. 2, pp. 134-139, 2013.

11. Peper, J.S., Dahl, R.E. "The teenage brain: Surging hormones—brain-behaviour interactions during puberty," Current Directions in Psychological Science, vol. 22, no. 2, pp. 134-139, 2013.

12. Peper, J.S., Dahl, R.E. "The teenage brain: Surging hormones—brain-behaviour interactions during puberty," Current Directions in Psychological Science, vol. 22, no. 2, pp. 134-139, 2013.

13. Crone, E.A., Dahl, R.E. "Understanding adolescence as a period of social-affective engagement and goal flexibility," Nature Reviews Neuroscience, vol. 13, pp. 636-650, 2012.

14. Zahn, R., Moll, J., Krueger, F., Huey, E.D., Garrido, G., and Grafman, J. "Social concepts are represented in the superior anterior temporal cortex," Proceedings of the National Academy of Sciences, vol. 104, p. 6430–6435, 2007.

15. van der Kolk, B. "The Body Keeps the Score," London: Allen Lane (Penguin), 2014.

16. Berry, C. Where Words Prevail [DVD], London: Sojourner Media, 2005.

17. El Sistema - trailer for the upcoming film. [Film]. Venezuela: UmbrellaClassical https://www.youtube.com/watch?v=276oR_tEmbs, 2008.

18. Majno, M. "From the model of El Sistema in Venezuela to current applications," Annals Of The New York Academy Of Sciences, vol. 1252, "The Neurosciences and Music IV: Learning and Memory," pp. 56-64, 2012.

Tomorrowland:
Recreating the Cultural Ambition for Interstellar Travel
or Dudes, Let's Go to Mars

Jaym Gates

Editor, War Stories; Communications Specialist

Abstract

At no time since the Moon Mission have we had such great national and international interest in leaving Earth's surface. NASA's scientists are rockstars, private corporations are building business plans for the asteroid belt, and billionaires in Hollywood and Silicon Valley are pushing for the stars with their vast resources. Even more positively, this interest is not caused by political tension, but by a rising awareness and curiosity.

How do we capitalize on the interest to spur our interstellar dreams? How can pop culture and the ever-unreliable media become our biggest allies in capturing the hearts and minds necessary to fuel humanity's greatest pilgrimage?

Keywords

Pop culture, civilization

1. Introduction

> "This guy comes up to me and goes, 'I went into science because of your work on *Deadspace*. It made me fall in love with the stars. Now I'm working on how we can get to other planets.'"
> - Writer Jeremy Bernstein (*Deadspace, Leverage*)

Science fiction has always looked to the stars, imagining what is out there. Science has been right beside it, the two worlds often hand-in-hand, influencing and informing each other. Now, as geeky pop culture explodes into the mainstream, we have an unprecedented opportunity to turn that deep well of enthusiasm to cultivating a culture that not only looks at the stars, but also looks at them critically . . . and with intent.

Authors, game designers, and movie directors are building a gateway to the sciences, and few sciences are more thrilling, more mythological than the interstellar, giving creators a wide slate of inspiration. Kids who grew up on *Star Wars, Star Trek, Battlestar Galactica*, the *Halo* and *Mass Effect* franchises, and even Firefly are now cap-italizing on those early inspirations.

At no time since the Moon Mission have we had such great cultural interest in leaving Earth's surface. NASA scientists are overnight internet and media darlings, millions apply for a long-shot chance at a one-way ticket to Mars, space-based movies and games net millions in profit, and Hollywood and Silicon Valley luminaries devote private resources and time to off-world initiatives. Science is sexy. STEM education is growing.

How do we capitalize on such a fickle market? Hollywood is notorious for throwing science under the bus at the first sign of conflict with creativity. Pop culture changes faster than we can keep up with it. Reporters find far more page views in "Listicles" and "what went wrong" clickbait than in serious looks at the advancement of science. Technology and art both have issues with equality, diversity, and access.

And yet, even with all of that, we're seeing nearly daily breakthroughs in multiple fields, many of them from teenagers, non-Western countries, and unexpected backgrounds. *Interstellar, Avatar*, and *Gravity* were astounding successes, while upcoming TV show *The Expanse* is al-ready one of the most anticipated titles of the year. There are over 256 results in the "space games" category on leading game re-tailer, Steam.

> **Interstellar**
> * Based on scientific theory
> * Written to encourage human spaceflight
> * Researched at NASA and SpaceX
> * Artistic development led to new visual understanding of black holes
> * *Interstellar* explores the relationships among 'science and faith and science and the humanities.'" - David Brooks/ New York Times [3]
> * $675 million gross [4]

But still, the cultural image of the drive for space is divorced from the gritty, boots-on-the-ground reality: big, remote, and in-accessible. It is the realm of billionaires and Ph.Ds., ivory towers and sterile labs. Space is still a dream, not a "plan," to all but a few.

It's a dream of a future as remote and inaccessible as the White House for the people who will likely make the difference in whether or not we leave the earth's atmosphere in the next one hundred years.

Last year, I went to a theater to see Interstellar for a second time. It was in my hometown, a small, rural, gold-country town with low education, high poverty, and high drug use, which relies on logging, mining, and ranching for employment.

Five teenage boys sat a few rows in front of us, laughing, ribbing each other, mocking previews, and generally being teenage boys, well into the beginning of the movie. But as the movie continued, I stopped watching it, and watched them.

Their voices became quieter. Their comments fewer. They stopped squirming. Finally, they were all sitting forward, riveted, utterly still and silent. This continued as the credits rolled, a few of them surreptitiously wiping their faces.

Suddenly, one of the boys exclaimed, "Dudes, let's go to Mars!" and the spell was broken. Loud, excited, thrilled, an entirely new world opened to them. "I'll ask my mom if she'll take us!" Silly stuff, overlying a deep and profound realization: anyone can change the world. Anyone can go to space. Even rural farm boys and rebellious girls.

And that's what we need. That's what the pop culture fascination with space, science, and science fiction offers: a chance to throw the doors wide open and create an environment that welcomes everyone with an idea, a drive, a passion. But more than that, it offers the opportunity for the majority of the world to start dreaming about what their part in our future might be.

Art and science and social development working hand-in-hand, projects like the Atlantic Council's "The Art of Future War" and the 100 Year Starship's Canopus Award, the collaborations between the Science Fiction and Fantasy Writers of America and school science programs, these are the building blocks of this partnership between science and art.

2. The Next Generation

Star Wars and Star Trek created a foundation for a golden age of geekdom. They began to bring a fringe hobby to the mainstream by showing a glimmer of what might be possible: big dreams, big ideas, big space, and big future. It's hard to find someone today who doesn't know about *Star Trek* or *Star Wars*, even if they're not part of the geek community. The success of those franchises built a multi-generation foundation of people who dreamed about other worlds. They, in turn, raised their children to believe that it was possible to leave the earth. And yet, funding for schools are being slashed constantly, teachers and schools are stretched to the breaking point. Adding another goal or duty to their load is unthinkable.

> ### Gravity
> - Near-universal critical acclaim
> - Film explored key mental concepts of space exploration
> - Created to illustrate isolation, survival, planning
> - "Cuarón shows things that cannot be but, miraculously, are, in the fearful, beautiful reality of the space world above our world." - Richard Corliss/*Time Magazine* [5]
> - $723 million gross [6]

So where do these opportunities come from?

Northwest Indian College, on the Lummi Reservation in the state of Washington, recently landed in the news after developing a tiny rocket club. Lacking funding and resources, they made do with what they had, going on to challenge Ivy League school science programs.

Inner city school programs are empowering African-American and Hispanic girls, teaching them to code, to build drones, and to value science as a career. Teenage girls and boys from disadvantaged backgrounds are making breakthroughs on medicine, energy, and technology. On instructor stated: "The joke was funny because this was just a tiny, two-year college, with no engineering program. Getting into space was the last thing on the minds of these students; they were just trying to escape poverty. Next thing they knew, NASA was calling them up" [7].

These kids will be working with what we leave them, but they aren't waiting for us. Science and art both have a tendency to value the ivory tower of learning, to keep the wealth of knowledge in select halls, but we are losing a valuable resource. " 'It comes down to sometimes, "Oh, do you have a paperclip, I need to put a paperclip in here to make sure this is secure," she said. 'And so, honestly, it's just whatever you have that works, you need to use it.' And it did work" [8].

Involving teenagers—particularly ones without resources, at high risk of prison and poverty—in innovation offers solutions to many problems. They are used to making do with almost nothing, to finding cheap ways to make their lives survivable. Giving them the tools and safety to develop the world they will inherit, and a cause to work toward, may give us a valuable, completely untapped resource of priceless energy and intelligence.

There are plenty of ways to engage kids and teenagers in developing the future. Design challenges, internships, and free educational resources accessible to everyone. Open a narrative to kids, give them the tools to tell their own stories and make their own dreams. Teach them that science doesn't have to be expensive, that the best innovation comes from garages and dining room tables (Iron Man 3 and Big Hero 6 were excellent examples of this). Teach them to collaborate and communicate, skills that are fading from our school system.

This can't just be Ivy League schools and wealthy neighborhoods. Toughness and resourcefulness are essential to off-world endeavors. The ability to use what is at hand, to survive with almost nothing—those are all very necessary skills. They can't do it on their own though. They need resources, and they need inspiration.

Empower the kids to find their own stories, and we'll all benefit.

3. The Empire Strikes Back

Geek media is big money right now. *The Walking Dead, Big Bang Theory, Game of Thrones, Supergirl, Daredevil, Jessica Jones, The Flash, Arrow, Agents of S.H.I.E.L.D, Agent Carter, American Horror Story, iZombie, Mr. Robot, Castle, Galavant, Once Upon a Time, The Vampire Diaries,* and *The Expanse* are just a sample of the TV shows making headlines in the last year, pulling millions of viewers, many of whom do not identify as geeks.

In the theater, *Star Wars: Episode VII, Avengers, Mad Max, Jurassic World, Inside Out, Ant-Man, The Martian, Hunger Games,*

> ### Halo
> - 46 million copies since 2001
> - 5.8 billion hours of gameplay
> - *Halo 3*: $170 million in first-day sales
> - *Halo 4*: $200 million in first-day sales
> - The *Halo 4* soundtrack debuted at 50 on the US Billboard chart
> - Almost $3 billion in total franchise sales . . . not counting merchandise [9]

Terminator Genisys, Ex Machina, Tomorrowland, Crimson Peak, Captain America, Guardians of the Galaxy, and many more are raking in top dollar.

No, most of these aren't set off world. Only *The Expanse, The Martian, Interstellar,* and *Gravity* have made the sort of waves that directly translate to more interest in off-world adventures. Games have done rather better, bringing *Halo, Destiny,* and *Mass Effect,* among many other titles, to a mainstream popularity that rivals leading movies in terms of money and influence. Games have the added bonus of giving the player agency and investment, allowing them long-term immersion in the world.

"[Producer/director Alfonso] Cuarón has stated that *Gravity* is not always scientifically accurate and that some liberties were needed to sustain the story. 'This is not a documentary,' Cuarón said. 'It is a piece of fiction.'" [1]. "According to NASA Astronaut Michael J. Massimino, who took part in the Hubble Space Telescope Servicing Missions STS-109 and STS-125, 'Nothing was out of place, nothing was missing. There was a one-of-a-kind wirecutter we used on one of my spacewalks and sure enough they had that wirecutter in the movie'" [2].

Celebrity scientists like Neil deGrasse Tyson give those hooked by popular media a step up from pop culture, a way to begin learning the ropes of science, while still keeping it fun and relatable.

These are the things that prime the next generation to not only continue the current efforts, but to launch new ones. They are accessible to almost everyone, on some level or another.

Where we need to focus our efforts is on the step between consumption and involvement. Not just for the kids, but for everyone. Getting off-planet will involve a lot more than just hard science.

Compare it to the Internet. Developed as a military tool, it burst out of its specialized box and became ubiquitous when it became accessible to everyone. But it went through its awkward phase—and is just beginning to leave that phase now—when movies like *The Matrix* hit it big and well-known authors like Tom Clancy had their best-selling novels translated into videogames.

But its journey was full of illegal, unpleasant, dirty bits. Much of what we have today is the result of black hats and commercial ventures, most of them failed. It was a frontier that claimed fewer lives than previous frontiers, but still did its damage. It was a Wild West. It still is, and it's still growing and evolving even as the US government attempts to rein it in.

We need the same sort of innovation and ownership over the future, too. The next battery breakthrough might be a kid on summer break. The software needed to develop a ship-managing AI might come from the games industry. A biomedical advance could come from a kid in the projects.

The future could come from anywhere. Pop culture is priming their imaginations and ambitions. We just need to make sure that the building blocks are there for everyone, regardless of their age, gender, origin, or education.

We need to open the future to everyone if we ever want to leave the earth's surface.

References

1. Wikipedia: *Gravity* (film). https://en.wikipedia.org/wiki/Gravity_(film)

2. Wikipedia: *Gravity* (film). https://en.wikipedia.org/wiki/Gravity_(film)

3. Wikipedia: *Interstellar* (film). https://en.wikipedia.org/wiki/Interstellar_(film)

4. Wikipedia: *Interstellar* (film). https://en.wikipedia.org/wiki/Interstellar_(film)

5. Wikipedia: *Gravity* (film). https://en.wikipedia.org/wiki/Gravity_(film)

6. Wikipedia: *Gravity* (film). https://en.wikipedia.org/wiki/Gravity_(film)

7. McNichols, Joshua. Why NASA Called Northwest Indian College Space Center. May 11, 2015. http://kuow.org/post/why-nasa-called-northwest-indian-college-space-center

8. McNichols, Joshua. *Why NASA Called Northwest Indian College Space Center.* May 11, 2015. http://kuow.org/post/why-nasa-called-northwest-indian-college-space-center

9. Murphy, David. *Halo's Final Statistics: 235,182 Years Played, 136 Billion Kills.* April 1, 2012. http://www.pc-mag.com/article2/0,2817,2402479,00.asp

What Would E.T. Believe? Anticipating the Contours of Extraterrestrial Religion

Peter M. J. Hess, Ph.D.

Let's imagine a group of space farers relaxing in the lounge of their starship heading toward Gliese 581-d, Kapteyn-b, or some other relatively nearby star with a potentially habitable planet in the "Goldilocks zone." As they hurtle through interstellar space, our astronauts speculate with a mixture of uncertainty, apprehension, and excitement about a host of practical and theoretical issues facing them. Will they find the planet already inhabited by intelligent life? What physical, chemical, or meteorological conditions might have shaped the contours of that species' experience? Will their intellectual life be based on biochemistry similar to ours, or so different from our own that it will scarcely be recognizable to humans?

Of particular interest to me is the question of extraterrestrial religious experience. Some might argue that this is so entirely hypothetical as to be of no practical value: why speculate about the putative spiritual experience of beings whom as yet we have no reason even to believe exist? It is important to consider the implications of the possible religious experience of an extraterrestrial species for two reasons: (1) in preparing for first contact we should naturally be interested in and sensitive to every aspect of an extraterrestrial's physical, emotional, intellectual, moral, and spiritual life. (2) Recognizing that understanding the universe does not revolve around adherence to the archaic terrestrial worldview of one tribe of Homo sapiens is essential to the continuing relevance of religious belief on Earth in the third millennium.

1. ET in Western Thought: Speculation and Clichés

Speculation about life on other worlds dates back at least to the Greek and Roman Atomists—Leucippus, Democritus, Epicurus, and Lucretius. They argued for a plurality of worlds from their premises of the "uniformity of nature" and of an infinity of atoms moving randomly in infinite space. However, this atomist idea was submerged for nearly two millennia by the dominant Aristotelian-Ptolemaic cosmology, based on a physics of elements having a "natural place" within the sphere of fixed stars. Aristotle (384-322 BCE) held that only one world is possible, a view reaffirmed by the medieval Catholic philosopher Thomas Aquinas (1225-1274).

But overconfident Aristotelians provoked a backlash when they declared on the authority of Aristotle that other worlds could not exist. In 1277 Bishop Stephen Tempier of Paris issued a series of condemnations limiting the self-proclaimed authority of arts masters of the University of Paris, declaring that "one may not hold that the first cause cannot make more than one world." The effect of this condemnation was to permit speculation in non-Aristotelian directions, including the theoretical consideration of a plurality of worlds, leading ultimately to the modern debate.

Cardinal Nicholas of Cusa (1401-1464) speculated that "in the area of the sun there exist solar beings, bright and enlightened intellectual denizens, and by nature more spiritual than such as may inhabit the moon—who are possibly lunatics—whilst those on earth are more gross and material." Nehemiah Grew (1641-1712) thought that every fixed star is a sun surrounded by planets, "which we assume to be inhabited in order for their respective suns to serve some purpose." Countless other figures joined the debate in the modern period.

It would of course be extraordinarily anthropomorphic to import the details of human religion into Imagining extraterrestrial life. Nevertheless, a reasonable working assumption is that if extraterrestrial religion exists it will be in some essential respects analogous to our own experience. Terrestrial religion is, after all, a product of human evolutionary history, and presumably ET religion would likewise be a product of the evolutionary history of its respective planets. Ritual, belief, prayer, doctrine, artistic creation, and other aspects of what we broadly conceive of as "religion" are artifacts of human experience and culture. Theologian Linda Gibler, O.P. offers a fine example of this in her exploration of the scientific and religious history of oil, fire, and water central to the Roman Catholic sacrament of baptism. Her work integrates the liturgical and theological significance of the sacramental elements into their 13.7 billion year cosmological and evolutionary history.

We need at this point to dispel one impediment to thinking theologically about ET, namely, the clichéd categories imposed on alien beings by science fiction. Too often ET is portrayed stereotypically either as evil or demonic on the one hand, or as benign or morally superior on the other. Examples include cardboard-cut-out evil aliens such as those in Independence Day, the arthropodic Sarris or reptilian Xindi in Galaxy Quest, the "chest bursters" in Alien, or the ant-like invaders in Battle, Los Angeles. On the opposite extreme we have the benevolent Klaatu from The Day the Earth Stood Still, the angelic healer E.T., the peaceful genetic shape shifter in Starman, and the wise teacher Yoda in Star Wars.

However, a more nuanced reading suggests that if extraterrestrial life has followed a similar evolutionary trajectory in the heavens as it has on earth, it will be characterized similarly by moral ambiguity. Examples include the Klingon Worf in Star Trek, The Next Generation, the Drac Jeriba Shigan in Enemy Mine, and the alien beings derogatorily referred to as "prawns" in District Nine. For theologians, the broadly accepted Neo-Darwinian evolutionary framework necessarily changes our theological understanding of "sin" to include the evolution of moral ambiguity.

Let us consider one of the foundations of morality to be the physical parameters of a planetary environment, and organisms' response to the constraints imposed by that environment. All planets have finite energy resources that are competed for by any evolving life forms. As life expands into every available ecological niche, predation between species and competition within species inevitably arises. Prior to the evolution of self-consciousness, competition and predation are ecological but not moral issues. However, with self-consciousness comes moral awareness: the recognition of the possibility that I could choose a different path; that, for example, rather than hoarding all our wooly mammoth meat in a time of famine, we could share it with a neighboring tribe.

Life developing in an ecological web of predator-prey relationships will thus reflect an evolutionary morality in which a range of moral ambiguities compete within a gradually widening circle of ethical inclusion. Robert Russell has summarized the probable character of ET nicely: if and when we meet alien intelligent life, it will be neither angelic nor demonic, but in theological terms *simul iustus et peccator* —at "once justified (redeemed) and a sinner." Human space-farers must be alert to the importance and the difficulty of recognizing the stages of moral development in alien species.

An example of such a morally ambiguous landscape is painted by Mary Doria Russell in her fine but darkly disturbing novel The Sparrow, recounting the story of a Jesuit mission to the fourth planet in the Rakhat system. The moral world of Father Emilio Sandoz, S.J., crumbles when he learns that on this planet two species—evolutionarily locked in a primordial predator-prey relationship—have co-evolved into rationality. The Runa species that historically has been the prey species continues to serve as food for the predator Jana'ata species. From within his Roman Catholic moral tradition Fr. Sandoz has no theological language through which to interpret this situation so opposite to the refined morality of most human religious traditions. The context of Rakhat 4 seems to contradict the assumed moral ends designed for rational creatures on the part of a loving creator

2. The Contours of ET Religion

It is in this context of the evolution of religion that I will reflect on the possible religious experience of ET. If we discover that life has evolved elsewhere in the universe, might we find that intelligent beings like us have also evolved a spiritual and ritual imagination? Might this imagination include any of the following elements found in different forms throughout terrestrial religions?

- Creation stories, narratives of salvation from floods or other primal catastrophes
- Rituals to celebrate planting and harvest, birth and death, coming of age and marriage
- God or gods, spirits and angels, demons or devils, saints
- Relationship between spirit, soul, mind and body
- Saviors, sacrifice, salvation
- Logos, Incarnation, divine involvement with the world
- Prayer, worship, consecration, commemoration, sacraments
- Morality, sin, original guilt or perfection
- Forgiveness, retribution, expiation of sin
- Problem of natural evil, redemptive suffering, theodicy
- Church, temple, priesthood, theocracies, evangelization
- Eschatology, heaven, hell, reincarnation, the ultimate fate of the universe

I propose that the evolution of morality, religion, spirituality, or theologies in an extraterrestrial context raises a host of questions that fall into a number of broad categories. Each of these sets of questions naturally constitutes a research area in its own right, but together they convey a sense of the magnitude of the task of evaluating theology in a context that takes the rest of the universe seriously. This essay will conclude with a briefly expanded consideration of two of the following categories as examples of the task of inter-planetary theology.

1. *Belief, faith:* Is belief in a transcendent dimension of reality an artifact of the evolution of human consciousness?

2. Theism: Will God(s) be thought of as acting in the world? Will monotheism be found to have replaced polytheisms in ET religions? Is monotheism an artifact of our own religious experience in a mono-solar cosmology? How might God be conceived of in a binary star system?

3. *God as Trinity:* God is triune in Christian experience and theology, but not in other human religions, although Hinduism addresses plural manifestations of the One. How might a multi-faceted experience of the divine be expressed elsewhere in the universe?

4. *Incarnation:* Could the Logos or "Christ principle" that is foundational to Christianity become incarnate/enfleshed/embodied elsewhere in the universe than Earth, perhaps multiple times?

5. *Salvation:* What does "salvation" mean in an extraterrestrial context, and is the idea relevant to every biological species? From what and to what should the word "salvation" be understood as referring?

6. *Salvific death:* Is the Christian idea of salvation through the death of a savior unique to human religious experience? If salvation through a salvific death is necessary, would multiple such deaths throughout the universe be necessary? Could salvation be accomplished on other planets without a sacrificial death?

7. *Competing or complementary experience:* if "the Christ" became incarnate a million years ago on Planet X, would the members of an expedition to Earth from that planet recognize Jesus of Nazareth as God incarnate? What might be the interface between an established ET theological doctrine and a human theological system imported from Earth?

8. *Gender roles:* Would the Roman Catholic argument that the priest—as an "icon of Christ"—must be male be undermined by the discovery of a female-dominant ET species?

9. *Suffering and theodicy:* How might the problem of unmerited suffering and the search for a cogent theodicy play out in different rational species? The tragedy of 9-11 is a theodicy problem for humans, whereas Thanksgiving would be a theodicy problem for turkeys if that species were rational.

10. *Religion and evolution:* Does religion carry either necessary or sufficient survival value to evolve on another planet?

11. *Comparative truth claims:* What is the relationship between religious belief as a tool with survival value on Earth, and the claimed truth values of comparative religious doctrines should we someday encounter ET?

12. *Future of religion:* Is religious belief something a species outgrows, or does it merely increase in sophistication? Would religious belief continue to play a role in the lives of a technologically advanced space-faring people?

3. The Task of Inter-Stello-Planetary Theology

The discovery of extraterrestrial intelligence would probably not in itself constitute a crisis for any but the most fundamentalist sects in the three Abrahamic traditions. However, the discovery of extraterrestrial religion certainly would challenge our terrestrial religions' claims of exclusivity and the concept of religion itself. But such challenges would be salutary for us. Let me illustrate this with a brief look at two theological challenges that would be raised by ET religion. Both of these issues point to the importance of reexamining our own terrestrial faiths in a broader and deeper cosmological light.

Example 1: Incarnation

Speaking for only one terrestrial religion, Christianity, the central claim is that eternal God, the ground of being and creator of the universe, worked to reconcile the world to Godself by becoming incarnate ("enfleshed") in a particular human person, Jesus. The particular locus (first century Palestine) but universal application of this incarnational claim is already a problem for inter-religious dialogue among humans. How much more problematic might it become in interplanetary religious dialogue? In light of the possibility of encountering ET religion, how might we envision Christianity thinking more universally about the idea of the "Incarnation"?

In *Christ in Evolution* Ilia Delio engages precisely this question about the universal significance of incarnational theology. Of the "Christ principle" she writes

On the terrestrial level, Jesus Christ assumes a bodily nature by which all of creation (that is, material reality) is assumed into relationship with God. Similarly, on an extraterrestrial level, incarnation must assume a form that includes the material reality of that creation, in whatever way that creation is constituted.

She notes that although the term "incarnation" might not be appropriate to another world order, since literally it means "taking on flesh," what it means more broadly speaking is embodiment of the divine Word in created reality. Delio sees the primary theological reason for incarnation not so much as the "forgiveness of sin" as the completion of creation in its relationship to God:

An understanding of the incarnation as an act of love rather than a condition of sin may be more fitting to an evolutionary universe where the understanding of human original sin is under revision.

I find compelling Delio's conclusion that no matter how many times the Cosmic Christ might become incarnate throughout the universe, there is only one spirit and one Christ.

Example 2: Eschatology

The second theological gauntlet would be thrown to any human faith that speaks about the fate of humans beyond this life. Eschatology is the branch of theology that deals with the last things: death, judgment, the end of the world, reincarnation, and heaven or hell. All eschatological ideas—including the recreation of a "new heavens and new earth"—were developed in a prescientific context that assumed a young, small, geocentric, and anthropocentric cosmos. We now know from physics and astronomy that although our sun will be around for another five billion years, that macro-life on earth (plants and animals) will end in about half a billion years, when the increasing luminescence of the expanding sun will increase the solar winds that gradually strip away our atmosphere and evaporate the oceans. Many galactic nebulae will continue to spawn stars with new planetary systems long after our death for billions of years into the future, until the gradually metalized galaxies pushes the universe toward an entropic heat death.

How can terrestrial religions speak meaningfully of "the end of the world" or "the new heavens and the new earth," when we recognize that stars and planets that have not yet even been formed may spawn extraterrestrial religions in our own or other galaxies? What does such religious language mean in a universe in which some ET religions may have gone extinct many millions of years ago, and others may not be born for a billion years to come?

Conclusion

If some day we encounter indigenous religion on Gliese 581-d or another planet, we must be prepared to engage them with a theological sophistication that transcends our earth-bound experience. And even if we never do make the momentous discovery of extraterrestrial religion, we will have been forced to take the critical examination of our earthly faiths to a meta-level to which they have not historically been subjected. We forget that our own religions—like the rest of our social structures—have evolved in a brief interglacial window of relatively hospitable conditions on earth. They also developed in a prescientific era characterized by relative ignorance of the vastness of the universe. Discovering ET religion would be a very important moment in humanity's "coming of age," and if religious faith is worth maintaining in the third millennium, it will profit from such theological self-examination.

References

1. http://io9.com/the-closest-known-potentially-habitable-planet-is-13-li-1585896900

2. Huston Smith intriguingly sketches how geography and climate might have influenced the three great families of terrestrial religious philosophies: the Western, the Indian, and the Chinese, "Accents of the World's Philosophies," in *Philosophy East and West*, Vol. 7, No. ½ (1957), 7-19.

3. Edward Grant, "Science and Theology in the Middle Ages," in David C. Lindberg and Ronald L. Numbers, eds. *God and Nature: Historical Essays on the Encounter between Christianity and Science* (Berkeley: University of California, 1986), 53-58. Steven J. Dick notes that "It is one of the great ironies of the history of ideas that while the passionate Atomist espousal of the idea of plurality of worlds fell on deaf ears with the rest of the Atomist system until the end of the sixteenth century, the rejection of the idea by the dominant Aristotelian worldview inspired critical discussion of plurality of worlds as early as the thirteenth century in the Latin West." Plurality of Worlds (Cambridge University Press, 1982), 23.

4. Nicolas Cusanus, *De docta ignorantia*, or *Of Learned Ignorance*, trans. Germain Heron (London, 1954), 114-115.

5. Nehemiah Grew, *Cosmologia Sacra* (1701).

6. Michael J. Crowe, *The Extraterrestrial life Debate, Antiquity to 1915: A Source Book* (University of Notre Dame Press, 2008), passim.

7. Frank Herbert seems to have done this in *Dune* (Penguin 1965). Religion is treated syncretistically and not in a particularly evolutionary way. "Zenshiism" is a hybrid of the religious principles of Zen Buddhism and Shia Islam; "Navachristianity" is a blend of Navajo and Christian spirituality.

8. Joseph Bulbia, et al., eds. (2008) *The Evolution of Religion: Studies, Theories, & Critiques*, Collins Foundation Press.

9. Current scholarship does not support the primacy of either myth or ritual, but sees them as sharing common paradigms. Eleazar Moiseevich Meletinsk, *The Poetics of Myth* (trans. Guy Lanoue and Alexandre Sadetsky, 2000 Routledge.

10. Linda Gibler, O.P., *From the Beginning to Baptism: Scientific and Sacred Stories of Water, Oil, and Fire* (Michael Glazier, 2010).

11. Daniel Dennett notes that the theory of evolution is transformative of every traditional concept, an idea certainly not lost on theologians. *Darwin's Dangerous Idea: Evolution and the Meanings of Life* (New York: Simon & Schuster, 1996), 63. The theory of evolution is a universal acid that "eats through just about every

traditional concept, and leaves in its wake a revolutionized world-view, with most of the old landmarks still recognizable, but transformed in fundamental ways." This is not news to theologians sophisticated enough to place theology in its scientific context, but it is worth bearing in mind.

12. Robert J. Russell, "Life in the universe: philosophical and theological issues," in *First Steps in the Origin of Life in the Universe*, proceedings, Sixth Trieste Conference on Chemical Evolution, Julian Chela-Flores, Tobias Owen and François Raulin, eds. (Dordrecht: Kluwer Academic Publishers, 2001).

13. Mary Doria Russell, *The Sparrow* (Villard, 1996).

14. Note that it is not even characteristic of all human religions, but only of those who interpret the tragic death of a religious leader in the language and sacrificial economy of First Century Judaism.

15. Desmond Stewart's sobering short story "The Limits of Trooghaft" deftly addresses the problem of unspeakable suffering imposed by rational beings on their sentient planet-mates. http://www.animal-rights-library.com/texts-m/stewart01.htm

16. Ilia Delio, O.S.F. *Christ in Evolution* (Maryknoll, NY: Orbis Books, 2008), 169. See also Delio, *The Emergent Christ* (Orbis Books, 2011).

17. Delio, Christ in Evolution, 168.

18. Peter D. Ward and Donald Brownlee, *The Life and Death of Planet Earth: How the New Science of Astrobiology Charts the Ultimate Fate of Our World*. (Holt Paperbacks, 2004).

19. For an engaging panoramic sketch of the human discovery of the vastness of our cosmos, see Timothy Ferris, *Coming of Age in the Milky Way*. (William Morrow & Co., 1988).

Memory Issues for 100YSS Missions:
Lifelogging, Social Media, and Virtual Time Capsules

Jo Ann Oravec, MA, MS. MBA, PhD

Professor, Information Technology and Supply Chain Management

University of Wisconsin at Whitewater

Abstract

In long-term space missions, issues relating to the storage and upkeep of human memories are becoming increasingly salient. The importance of memory for personal identity has been underscored in recent research on such conditions as Alzheimer's and Post-traumatic Stress Disorder (PTSD). In everyday Earth-based interactions, physical and virtual objects as well as spatial locations can remind individuals of past events and assist in organizing and refreshing personal memories over time; some individuals are already engaged in initiatives in which their entire lives are taped ("lifelogging") so that any slippage in memory could be compensated. In long-term space missions, such efforts to refresh memories and record new ones take on a different and decidedly complex character. This paper begins by exploring long-term memory issues, outlining research trends that relate to such fields as education, medicine, and gerontology. It continues by proposing specific personal and collaborative strategies for renewing and protecting long-term memories during 100yss space travel, strategies rooted in recent research involving information technologies, social media, and crowdsourcing. The paper also explores the development of virtual "time capsules" that would afford space mission participants the means to organize their schedules of memory refreshment at pre-determined intervals. The paper includes a discussion of the privacy and transparency issues involved as space mission participants store detailed aspects of their personal lives in various digital formats.

Keywords

memory, space missions, time capsules, memory support, privacy, social media

Your memory is a monster; you forget—it doesn't. It simply files things away. It keeps things for you, or hides things from you—and summons them to your recall with a will of its own. You think you have a memory; but it has you!
--John Irving [1]

1. Introduction

Our personal memories have often been a taken-for-granted part of everyday life, despite the kinds of frustrations described by US author John Irving in the epigraph. We may enjoy a few minutes looking back at a family vacation or perhaps wince at a lost opportunity or past mistake. However, human memory research has greatly expanded in the past several decades, expanding our insights about how short- and long-term memory operate and interact with each other [2]. Substantial advances in the biological dimensions of memory formation and retention (see Martinez & Kesner, 2014) have been coupled with new understandings of memory's social and cultural aspects [3]. As the average ages of the populations of many Western nations increase along with an expansion in the number of individuals with Post-traumatic Stress Disorder (PTSD), diseases and syndromes that serve to debilitate memory functioning have elicited concern and catalyzed extensive research efforts. This paper explores the support of "mission memory" at the individual and collective levels using some insights from recent memory research as well as perspectives from the developers of advanced information technologies. Mission memory support involves the requirements in terms of long-term memory to preserve a sense of personal identity for mission participants as well as maintain the integrity of 100YSS and other space missions.

The biological and social substrata for mission memory may certainly face challenges as individuals spend decades on lengthy space missions. Many of the difficulties involved in the support of long-term memory over the course of a lifetime (the length of some potential 100YSS initiatives) may not be directly foreseeable, despite efforts to extrapolate them from shorter space missions or from Antarctic exploration initiatives (the latter described in Love & Harvey, 2014) [4]. Many of the reinforcements of and refreshments for long-term personal memory will be missing or may be somehow compromised during long-term space missions. For example, the kinds of everyday reminders about the past that reinforce memory-related processes (such as Earth-based physical buildings and personal acquaintances) will not be directly available to an individual during long space flights. Social and physical environments (such as those available in random walks through one's neighborhood) have been shown to play considerable roles in how long-term memories are stored and refreshed [5]. As spacecrafts travel through many thousands of miles and as early memories have few anchors or sources of refreshment, even the most efficient and effective personal memory-related processes could become less fully operative. The use of information technologies in storing virtual aspects of these memory anchors in digital format is described later in this paper (following strategies outlined in John, 2014; Oravec, 1996, 2012; Stawarz, Cox, & Blandford, 2014, and others) [6], [7], [8], [9].

2. Why Be Concerned About Personal Memory During Space Missions?

The importance of personal memory in supporting individual functioning as well as collective interactions during space missions has been underscored through many recent research efforts. Declines in autobiographical memory have been associated with loss of personal identity [10], which could provide especially difficult issues for long-term space travel. Identity as an astronaut and mission participant could certainly deviate even during a relatively short space mission as roles expand and deflate, but some kind of a core identity may be essential. Langston and Pell (2015) outline some of the variations of perceived identity of space mission participants, including such labels and related roles as "astronaut, cosmonaut, taikonaut, yu hang yuan ("space navigating personnel"), vyomanaut, as well as citizen astronaut, civilian astronaut, space tourist and spaceflight participant" [11]. Individuals who cannot retrieve early memories of how they arrived in a spacecraft may not be able to recall basic dimensions of their on-board roles and their craft's mission; ultimately, they may possibly even rebel or otherwise sabotage the craft. The question of what "memory" itself constitutes can help us in our analysis of these mission-related issues. Analyses of the biology of memory are beyond the scope of this paper: some excellent compendiums of current research results are provided in the academic journals listed at the end of this section. Memory loss can be associated with various diseases and syndromes (such as Alzheimer's), with accidents (such as those resulting in concussions), and with the use of statins [12], as well as the result of certain addictions [13]. A former astronaut, Duane Graveline, wrote of the experiential dimensions of a severe loss of memory (possibly a result of the use of a particular statin) in the book Lipitor, Thief of Memory: Statin Drugs and the Misguided War on Cholesterol [14]. Graveline's account of what it feels like to lose substantial portions of his long-term memory (in "transient global amnesia"

242

syndrome) is chilling, with a detailed exposition of how unsettling it is not to be able to connect to one's past (even to forget one's family members and basic profession). Some researchers have speculated that increases in radiation received on long-term space missions could also have disabling effects on memory-related processes [15], [16]. Whether these radiation-related impacts will be as severe as the transient global amnesia described by Graveline remains to be seen; however, building preparation for mitigating some kinds of memory problems into 100YSS missions would be a prudent strategy, given the broad assortment of potential, emerging memory concerns. Memory research could also enhance our knowledge of addiction processes [17], which could themselves become problems during long space flights.

Memory can be construed as embodied in human brains as well as embedded in objects, sounds, and images (along with other sensory data):

Human memory is "embodied" in living personal memories and "embedded" in social frames and external cultural symbols (e.g., texts, images, and rituals) that can be acknowledged as a memory function insofar as they are related to the self-image or "identity" of a tribal, national, and/or religious community. Whereas the social or "collective" memory comprises knowledge commonly shared by a given society in a given epoch, cultural memory in literate societies includes not only a "canon" of normative knowledge but also an "archive" of apocryphal material that may be rediscovered and brought to the fore in later epochs.

--J. Assmann [18]

The human "embodiments" of memory may become problematic on long-term space missions as specific individuals sustain incapacitating illnesses or even die. The memories of space participants can indeed be captured in some format through oral history preservation techniques [19] and even "lifelogging," which would involve the continuous recording and subsequent storage in video format of individuals' everyday interactions and encounters. The embeddedness of memory can also be problematic in some dimensions, given the fact that many of the objects associated with participants' long-term memories may not be directly accessible on board the 100YSS spacecraft. For individuals who are born on board these issues and problems may take other forms, which will be explored in future 100YSS research.

Varieties of memory investigations that are relevant to space issues include longitudinal research that examines particular subjects over time. For instance, some individuals have been shown to have "superior autobiographical memories," being able to retrieve in organized fashion minute data about incidents years distant in their pasts [20]. One notable subject was observed in a clinical context to have the "ability to recall accurately vast amounts of autobiographical information, spanning most of her lifetime, without the use of practiced mnemonics" [21]. This individual described her personal experiences involving memory as dominating her life in many ways, some of which were negative; the memories often intruded into current aspects of her experience. Generally, individuals have less domineering relationships with their memories, with long-term memories residing in the background of their consciousness until they are triggered either through conscious attempts at recall or as a result of an encounter with a particular trigger. Triggers for memory recall can include sights and sounds; the other senses can also be involved, such as tastes, touches, or smells.

Emotional responses can play roles in the storage and retrieval processes of memories with a mechanism labeled as "tagging" [22], with some memories being enhanced and others diminished in relation to various emotional contexts. Not being able to encounter sights and sounds that are somehow linked with one's past may be disabling in terms of memory as well as a variety of other emotion-related ways: "Long-duration space travel can produce a sense of isolation and separation from family and friends." [23] Exploration of memory issues may assist in dealing with these other psychological and social aspects of space travel and in connecting individuals with their personal identity origins and past relationships on Earth.

This paper can only begin to map some directions for "mission memory" research: sustained interdisciplinary efforts will be required to understand the complexities of memory well enough to ensure the preservation of mission memory at the individual and collective levels. Academic journals such as *Memory Studies* (Sage Publications), *Memory* (Taylor and Francis Publishers), Memory & Cognition (Springer Publications) and the *Journal of Applied Research in Memory and Cognition* (Elsevier) are pioneering in the examination of memory from a systems perspective, incorporating social and environmental perspectives as well as neurobiological insights. Historical and cultural dimensions of memory on a collective scale are explored in the journal *History & Memory* (Indiana University Press).

Artistic and humanistic expressions can play pivotal roles in preserving and triggering long-term memory both on the personal and collective levels, potentially leading to expanded roles for arts and humanities specialists in relation to 100YSS missions. Another important aspect of recent memory research involves the role of forgetting or deletion; the role of forgetting in diminishing the effects of extraordinarily painful memories can itself be of critical importance in 100YSS missions in which some upsetting events will certainly occur, such as the deaths or injuries of loved ones [24]. Understanding forgetting as well as memory retention could be essential for space mission participants in efforts to manage and maintain their personal identities and sustain their wellbeings during lengthy missions.

3. Retention and Protection of Mission Memory of 100YSS Participants

The notion that personal memories of space participants would need to be "protected" over time (as well as retained) may seem quite strange. Interaction with individuals on Earth and discussion of space-related matters via social media and other means may not be entirely a positive experience for space travelers, however, and may disrupt their sensitive and nuanced understandings about the meanings of space missions. Space travel has a political and social component that can serve to alter memory both at the social and personal levels, even to the point of selecting which space-related artifacts to preserve [25]. For instance, Russian space-related memories have undergone various challenges as potential historical inaccuracies are examined [26]. British space science has been associated with the rise of scientific research and technologically themed identity in the UK [27]. Participants during 100YSS missions could find that the societal perspectives on their missions that are generated by observers and supporters on Earth could begin to vary dramatically from their own accounts and narratives, leading to difficulties if they rely on external, non-mission sources for refreshment of their memories. Such online activities as "crowdsourcing" (looking for answers through eliciting social media discussion) can be empowering in brainstorming the solution to various problems (as described in Howe, 2008) but indeed could be unsettling if it questions or distorts basic space mission perspectives [28].

On Earth, the US National Aeronautics and Space Agency's (NASA) history has been preserved through the years by the specific efforts of historians who specialized in space history and maintained particular museum artifacts [29]. However, how space missions are framed on Earth can vary as a result of a wide assortment of social, economic, and political shifts. Missions that seemed appropriate in some eras can be perceived as wasteful and even useless in others. Parker (2009) writes, "Though there are many ways in which the space race might be deemed politically suspect, it represents a triumph of a modernist concatenation of progress, technology and organization. In contrast to the introverted couches of the virtual, the sublime space between the stars might suggest a much more expansive relationship between technology and the human." [30] Although Parker projects a "political suspicion" around space missions, he is still relatively positive; however, other commentators may not be so supportive. The social media communications from Earth to space missions could thus serve to recast the missions in some way if precautions are not taken. Having digital and social supports for their own personal memories preserved on board in100YSS missions is one mechanism to retain the identity of the individuals involved and the integrity of the 100YSS mission itself; even online mental health counseling could be problematic if attitudes toward space travel on the part of Earth residents shift [31]. Completely autonomous memory support (support that does not require Earth contact and cannot be adulterated through Earth influence) would be a critical aspect of strategies to support and protect mission memory.

Challenges concerning the storage and retrieval of negatively charged or painful memories will loom large for space mission participants. Generally, the act of recalling items in distant memory or in linking current observations to the past can indeed have pleasurable dimensions, which have often been explored in psychology as well as in the humanities [32]. Indeed, some "memory games" have been marketed in many Western countries as pleasant ways to enhance one's brain functioning while also having an enjoyable experience [33]. However, negatively charged memories can often affect personal wellbeing as well as be problematic in terms of long-term storage [34]. As long-term space missions transpire, many of the memories associated with the mission could be negative in tone, including deaths, accidents, and other unfortunate occurrences. Post-traumatic stress disorder (PTSD) is often associated with such negative occurrences, with subsequent difficulties in memory storage and recall [35]. Painful and negatively charged memories may be hard to deal with, but valuable insights about missions as well as personal identities can be lost when only information about pleasant events and successes is retained and less desirable insights are forgotten or even erased. Future memory research may provide clues as how to handle such negatively charged memories in productive and functional ways.

4. Varieties of Memory Support Systems for 100YSS Missions

Memory support systems can vary in their styles and effectiveness; choice of a support system for personal use can reflect one's personal abilities as well as one's own values and eccentricities [36], [37]. Furthermore, needs for memory support can vary throughout individuals' lifetimes, increasing the support systems' requirements for flexibility and adaptability over time. Scalability (the capacity for handling growth in size) will also be important as memory troves grow through the years. Requiring every 100YSS mission participant to use a particular kind of memory support system could thus be problematic. Individualizing and fine-tuning a roster of different support systems could be the most effective way of ensuring that individuals retain sufficient "mission memory" to be able to complete successfully their 100YSS venture.

The styles of memory support that could be adapted for 100YSS use include those that are being manifest in today's mobile device market, applications that generally require relatively little in terms of user monitoring and upkeep [38]. Özkul and Humphreys (2015) make the following contentions about these emerging mobile media applications:

These new ways of preserving the past could be in the form of sharing locational information (e.g., geotagging, camera phone photos, check-ins), which would remind our future selves where we come from and how we used to be. We sometimes consciously create our everyday life narratives intending to hang onto a moment, or simply because the technology automatically saves our experiences, we unconsciously preserve our pasts. [39]

Many individuals are already using digital memory reminders to keep their busy schedules on track. Specific activity reminders are often built into everyday operations on the job or in households in efforts to ensure that memory lapses do not have a negative impact. For instance, Stawarz, Cox, and Blandford (2014) describe apps for mobile devices that remind individuals to perform such everyday functions as taking a pill or getting exercise [40].

The "checklist" (whether on paper or in digital form) has been a part of many successful professional environments as a memory support device, "making sure we apply the knowledge we have consistently and correctly" [41]. Many doctors, airline mechanics, engineers, and other individuals engaged in critical operations have utilized checklists to make sure that every step is followed in its correct sequence. Long-term memory refreshment may benefit from a kind of checklist structure, with the varieties of memories that are particularly important for personal identity reinforced on a regular schedule. Space mission participants could adopt a kind of "buddy system" in which they ensure that their partners review and acknowledge the various personal items enumerated on the checklist, thus helping to refresh their memories of certain important personal events and historical items (such as marriages, births of children, etc.). If the checklist fails to serve its purpose, these space participants could be made more immediately aware of potential memory losses and work to ameliorate these difficulties.

In Earth-based memory processing, locational elements can serve critical roles. In his dissertation Chance memories: Supporting involuntary reminiscence by design, Fennell (2015) maps some of the ways that individuals can be connected with long-term stored memories that are associated with particular locations [42]. The notion of "homesick tourism" was even coined by Marschall (2014) to refer to the kinds of travel that are associated with the refreshment of long-term memories of émigrés related to their homelands [43]. Such venues for reminiscence may be reproducible in some digital forms through virtual travels, although how successful such simulated experiences will be in refreshing memories is as yet unproven. Scheduling regular "virtual visits" to particular locations such as one's early school or college settings may potentially serve such location-based roles as virtual reality technologies improve. The stars and planets themselves could eventually serve as new virtual location reminders as 100YSS participants reminisce about occurrences during their lengthy flights.

As previously mentioned, lifelogging initiatives use video recordings of everyday life to store details of existence for subsequent retrieval and review [44]. Items associated with various events could be replicated through 3-D printing on demand in order to renew particular tactile experiences. Some researchers have been a bit skeptical about the prospects for lifelogging: for example, Whittaker (2014) asserts that many lifelogging efforts can be criticized for "overgenerating data and failing to focus on what is important" [45]. Space mission participants who engage in lifelogging may want to annotate the digital files of their experiences so as not to be lost in long streams of hard-to-access video.

Personal and group "virtual time capsules" that are designed to be viewed at certain junctures during 100YSS missions (or viewed by the children of mission participants) could also serve to renew and refresh memories. Oravec (2013) describes the roles of time capsules in the following way:

Time capsules are designed to remove selected objects (both physical and virtual) from the streams of everyday use and destruction, toward the goal of placing them in the reach of individuals in the future … items are sequestered from present applications and transported to a physical and conceptual space in which they will be received at a particular time in the future with minimal alteration and modification. [46]

Determining what is relevant to include in such virtual time capsules, as well as scheduling time capsule openings, would be a critical pre-launch effort requiring some "mental time travel" efforts as individuals project what will be relevant in the future [47]. The contents of memories can themselves be factors in memory storage and retrieval, whether virtual time capsules, lifelogging, or social media are involved as support. Some memories may be easier to store and retrieve than others as well as more critical to personal identity. The term "sticky" has been related to information in an assortment of ways, including the concept of "sticky ideas" (ideas that are especially memorable) as portrayed in Heath and Heath (2007) [48].

Substantial privacy and transparency issues will emerge in the development of memory support, especially since some of the memories involve may be unpleasant and perhaps even reflect societally-stigmatized or illegal behavior [49], [50], [51]. Davies, et al. (2015) outline the security and privacy implications of some of the pervasive memory enhancement and support systems just described, outlining strong potentials for abuse [52]. The aspects of our lives that lead us to perform heroic or exemplary activities may not be dimensions we want to share with others or even recall ourselves. If space mission participants fear that aspects of their personal memories will somehow lessen their societal status, they may be reticent to become involved in storage processes for these memories; subsequently, they may be less well equipped to maintain their personal identities over the course of lengthy 100YSS missions. Mission participants may thus need to be informed about the memory and identity-related insights described in this paper that increase the apparent need for memory support.

5. Some Conclusions and Reflections

This paper has focused on the instrumental and pragmatic aspects of long-term memory, specifically, aspects that relate to the "mission memory" of 100YSS participants. Memory issues have gained importance for mission-oriented strategies in part because of their couplings with personal identity maintenance. Personal identity issues have close ties with how individuals relate to the team and institutional collectives with which they are associated. Relying on Earth-based sources for memory support could be problematic; online collaboration platforms can change features and participation over time, so their use in refreshing long-term memory can be challenging. Such social media strategies as crowdsourcing [53] could be utilized to some extent to provide a kind of memory "backup" to space mission participants. However, autonomous memory support that cannot be adulterated through Earth contact may be required, whether in the form of on-board memory games, checklists, lifelogging, or virtual time capsules.

Designing space missions with memory-related perspectives in mind could engender new insights about the nature of human memory and personal identity, whatever the Earth or space contexts. Recordings and subsequent retrievals of on-board memories can also serve the role of teaching devices, equipping future space mission participants to understand more fully what is involved in long-term endeavors. Memory research presents exciting prospects for new ways to construe basic mental processes and to equip our human intellects to travel to the stars.

Notes

1. Irving, J. (2014). *A Prayer for Owen Meany*. New York: HarperCollins, p. 35.

2. LaRocque, J. J., Eichenbaum, A. S., Starrett, M. J., Rose, N. S., Emrich, S. M., & Postle, B. R. (2015). The short-and long-term fates of memory items retained outside the focus of attention. *Memory & Cognition*, 43(3), 453-468.

3. Martinez Jr, J. L., & Kesner, R. P. (Eds.). (2014). *Learning and memory: A biological view*. Elsevier.

4. Love, S. G., & Harvey, R. P. (2014). Crew autonomy for deep space exploration: Lessons from the Antarctic Search for Meteorites. *Acta Astronautica*, 94(1), 83-92.

5. Hirst, W., & Stone, C. B. (2015). Social aspects of memory. In *Emerging Trends in the Social and Behavioral Sciences: An Interdisciplinary, Searchable, and Linkable Resource*, 1–12.

6. Mols, I., Hoven, E. V. D., & Eggen, B. (20October). Making memories: A cultural probe study into the remembering of everyday life. In *Proceedings of the 8th Nordic Conference on Human-Computer Interaction: Fun, Fast, Foundational* (pp. 256-265). Association for Computing Machinery (ACM).

7. Oravec, J. A. (1996). Virtual individuals, virtual groups: Human dimensions of groupware and computer networking. New York: Cambridge University Press.

8. Oravec, J. A. (2012). Digital image manipulation and avatar configuration: implications for inclusive classrooms. *Journal of Research in Special Educational Needs,* 12(4), 245-251.

9. Stawarz, K., Cox, A. L., & Blandford, A. (20April). Don't forget your pill!: Designing effective medication reminder apps that support users' daily routines. In *Proceedings of the SIGCHI Conference on Human Factors in Computing Systems* (pp. 2269-2278). Association for Computing Machinery (ACM).

10. Jetten, J., Haslam, C., Pugliese, C., Tonks, J., & Haslam, S. A. (2010). Declining autobiographical memory and the loss of identity: Effects on well-being. *Journal of Clinical and Experimental Neuropsychology,* 32(4), 408-416.

11. Langston, S., & Pell, S. J. (2015). What is in A name? Perceived identity, classification, philosophy, and implied duty of the "astronaut." *Acta Astronautica,* 1185–1(Quote, p. 185).

12. Moyer, M. W. (2015). It's not Dementia, It's your heart medication. *Scientific American,* 38-39.

13. Hyman, S. E. (2014). Addiction: A disease of learning and memory. *American Journal of Psychiatry,* 162(8), 1414-1422.

14. Graveline, D. (2004). *Lipitor, thief of memory: Statin drugs and the misguided war on cholesterol.* Boston, MA: Infinity Pub.

15. Cucinotta, F. A., Wang, H., & Huff, J. L. (2012). Risk of acute or late central nervous system effects from radiation exposure. *NASA Human Research Program Roadmap.*

16. Hsia, R. E. T. M. (2015). Biological and psychosocial effects of space travel: A case study (Doctoral dissertation, Alliant International University).

17. Hyman, S. E. (2014). Addiction: A disease of learning and memory. *American Journal of Psychiatry,* 162(8), 1414-1422.

18. Assmann, J. (2011). *Communicative and cultural memory* (pp. 15-27). Springer Netherlands.

19. Willox, A. C., Harper, S. L., & Edge, V. L. (2012). Storytelling in a digital age: Digital storytelling as an emerging narrative method for preserving and promoting indigenous oral wisdom. *Qualitative Research,* 13(2), 127-147.

20. LePort, A. K., Mattfeld, A. T., Dickinson-Anson, H., Fallon, J. H., Stark, C. E., Kruggel, F., ... & McGaugh, J. L. (2012). Behavioral and neuroanatomical investigation of highly superior autobiographical memory (HSAM). *Neurobiology of Learning and Memory,* 98(1), 78-92.

21. Parker, E. S., Cahill, L., & McGaugh, J. L. (2006). A case of unusual autobiographical remembering. *Neurocase,* 12(1), 35-49.

22. Richter-Levin, G., Kehat, O., & Anunu, R. (2015). Emotional tagging and long-term memory formation. In *Synaptic Tagging and Capture* (pp. 215-229). Springer New York.

23. Kanas, N. (2015). Countermeasures for space travel. In *Humans in Space* (pp. 83-95). Springer International Publishing.

24. Spear, N. E. (2014). *The processing of memories: Forgetting and retention.* Psychology Press.

25. Gorman, A. (2005). The cultural landscape of interplanetary space. *Journal of Social Archaeology,* 5(1), 85-107.

26. Gerovitch, S. (2011). 'Why are we telling lies?' The creation of Soviet space history myths. *Russian Review,* 70(3), 460-4doi:10.1111/j.1467-9434.2011.00624.x

27. Sanford, P. (2014). Collaborations in space: Memories of British space science, 1960-19Britain & The World, 7(2), 261-2doi:10.3366/brw.2014.0151

28. Howe, J. (2008). *Crowdsourcing.* New York, NY: Crown Business.

29. Launius, R. D. (1999). NASA history and the challenge of keeping the contemporary past. *The Public Historian,* 63-81.

30. Parker, M. (2008). Memories of the space age: From Apollo to cyberspace. *Information, Communication & Society,* 11(6), 846-8(Quote p. 846).

31. Oravec, J. A. (2000). Online counselling and the Internet: Perspectives for mental health care supervision and education. *Journal of Mental Health,* 9(2), 121-135.

32. McLelland, V. C., Devitt, A. L., Schacter, D. L., & Addis, D. R. (2014). Making the future memorable: The phenomenology of remembered future events. *Memory,* 23(8), 1255-1263.

33. Martinovic, D., Whent, R., Adeyemi, A., Yang, Y., Ezeife, C. I., Lekule, C., ... & Frost, R. A. (2013). Gamification of life: Playing computer games to learn, train, and improve cognitively. *Journal of Educational and Social Research,* 3(8), 83-89.

34. Zaragoza Scherman, A., Salgado, S., Shao, Z., & Berntsen, D. (2015). Event centrality of positive and negative autobiographical memories to identity and life story across cultures. *Memory,* 23(8), 1152-11doi:10.1080/09658211.2014.962997

35. Rathbone, C. J., & Steel, C. (2015). Autobiographical memory distributions for negative self-images: Memories are organized around negative as well as positive aspects of identity. *Memory,* 23(4), 473-4doi:10.1080/09658211.2014.906621

36. Abdalla, A., & Frank, A. U. (2014). Designing spatio-temporal PIM tools for prospective memory support. In *Principle and Application Progress in Location-Based Services* (pp. 227-242). Springer International Publishing.

37. Mols, I., Hoven, E. V. D., & Eggen, B. (20October). Making memories: A cultural probe study into the remembering of everyday life. In *Proceedings of the 8th Nordic Conference on Human-Computer Interaction: Fun, Fast, Foundational* (pp. 256-265). Association for Computing Machinery (ACM).

38. Migliardi, M., & Gaudina, M. (2014). Memory support through pervasive and mobile systems. In *Inter-cooperative Collective Intelligence: Techniques and Applications* (pp. 239-271). Springer Berlin.

39. Özkul, D., & Humphreys, L. (2015). Record and remember: Memory and meaning-making practices through mobile media. *Mobile Media & Communication, 3*(3), 351-3doi:10.1177/20501579145658(Quote, p. 351).

40. Stawarz, K., Cox, A. L., & Blandford, A. (20April). Don't forget your pill!: Designing effective medication reminder apps that support users' daily routines. In *Proceedings of the SIGCHI Conference on Human Factors in Computing Systems* (pp. 2269-2278). Association for Computing Machinery (ACM).

41. Gawande, A. (2010). *The checklist manifesto: How to get things right.* New York: Metropolitan Books.

42. Fennell, J. E. (2015). Chance memories: Supporting involuntary reminiscence by design (Doctoral dissertation, Goldsmiths College (University of London)).

43. Marschall, S. (2014). "Homesick tourism": Memory, identity and (be) longing. *Current Issues in Tourism, 18*(9), 876-8

44. Ogata, H., Hou, B., Li, M., Uosaki, N., Mouri, K., & Liu, S. (2014). Ubiquitous learning project using life-logging technology in Japan. *Journal of Educational Technology & Society, 17*(2), 85-1

45. Whittaker, S. (20April). Technology and memory: From lifelogging to strategic reminiscence. In CHI'14 Extended Abstracts on Human Factors in Computing Systems (pp. 1-2). *Association for Computing Machinery (ACM).* (Quote, p. 1).

46. Oravec, J. A. (2013). Not now, perhaps later: Time capsules as communications with the future. *Leonardo Electronic Almanac, 19*(2), 72-(Quote, p. 72).

47. Macrae, C. N., Miles, L. K., & Best, S. B. (2012). Moving through time. Mental time travel and social behavior. *Social Thinking and Interpersonal Behavior,* 113.

48. Heath, C., & Heath, D. (2007). *Made to stick: Why some ideas survive and others die.* New York, NY, Random House.

49. Oravec, J. A. (2003). The transformation of privacy and anonymity: beyond the right to be let alone. *Sociological Imagination, 39*(1), 3-23.

50. Oravec, J. A. (2004). The transparent knowledge worker: Weblogs and reputation mechanisms in KM systems. *International Journal of Technology Management, 28*(7), 767-775.

51. Mols, I., Hoven, E. V. D., & Eggen, B. (20October). Making memories: A cultural probe study into the remembering of everyday life. In *Proceedings of the 8th Nordic Conference on Human-Computer Interaction: Fun, Fast, Foundational* (pp. 256-265). Association for Computing Machinery (ACM).

52. Davies, N., Friday, A., Clinch, S., Sas, C., Langheinrich, M., Ward, G., & Schmidt, A. (2015). Security and privacy implications of pervasive memory augmentation. *Pervasive Computing,* IEEE, 14(1), 44-53.

53. Howe, J. (2008). *Crowdsourcing.* New York, NY: Crown Business.

Additional References

54. Habermas, T., & Köber, C. (2014). Autobiographical reasoning is constitutive for narrative identity: The role of the life story for personal continuity. *The Oxford Handbook of Identity Development,* 149.

55. Jackson, M. C., Linden, D. J., & Raymond, J. E. (2014). Angry expressions strengthen the encoding and maintenance of face identity representations in visual working memory. *Cognition & Emotion*, 28(2), 278-297. doi: 10.1080/02699931.2013.816655

56. Kanas, N., Sandal, G., Boyd, J. E., Gushin, V. I., Manzey, D., North, R., ... & Wang, J. (2009). Psychology and culture during long-duration space missions. *Acta Astronautica*, 64(7), 659-677.

57. Kohsuke, Y. (2015). The importance of autobiographical memory for self/identity achievement. *Japanese Journal Of Developmental Psychology*, 26(1), 70-77.

Reconciling Religion and Science through Eternity

Rex Pay, M. Sc.

498 Drexel Drive, Santa Barbara, CA 93103

rex.pay@seti-setr.org

Abstract

An extrasolar civilization in which religion and science developed in harmony would very likely have eternity as one of its fundamental scientific concepts. It would probably see spacetime emerging from an aspatial, atemporal eternity containing formless energy that undergoes a phase change, so that minimum elements of spacetime with uniform properties in every direction condense out of it and form a spacetime lattice. Spacetime would not be infinitely divisible but would be made up of elemental quanta that give it a digital character. Formless energy would create the universe when it entered the lattice and gained form. For interstellar communication, extrasolar civilizations would derive their units of space and time from the dimensions of spacetime quanta, set by universal constants. They would assume the spacetime quantum to be a sphere, and may choose the Planck length to be the diameter of this sphere. Their Planck length could therefore be 13 per cent larger than ours. This could be a problem in communicating terrestrial dimensions to locations light-years away.

Keywords

interstellar, eternity, religion, cosmology, Planck length, non-locality

1. Introduction

This paper was prepared in response to a request for papers on the topic "Becoming an Interstellar Civilization" for presentation at the 2015 100 Year Starship Symposium held in Santa Clara, CA. A previous paper [1] showed that because becoming such a civilization depends on the presence of other civilizations within communication distance, it requires that planetary civilizations have lifetimes of one million years or more. Even to have just one such civilization present somewhere in the Galaxy over the period when civilizations could have emerged (the last 5.5 billion years) requires 5,500 civilizations to rise and fall within that period, leaving ourselves alone in the universe surrounded by the remains of 5,499 expired civilizations. For us to become an interstellar civilization would require, of course, at least one other civilization to be present, doubling the required rate of civilization production. This would be of little value if the other civilization were the other side of the galaxy, as one message could take up to 50,000 years to arrive. So the number of civilizations has to be increased to ensure that on average civilizations are spaced no more than, say, fifty light-years apart. This requires an enormous number of concurrent civilizations of great stability.

The previous paper [1] identified organized religion as a stabilizing force in the past for civilizations on Earth. In recent years, investigation into the social mechanism for this has been underway [2], [3], [4]. It appears that the size of a social group is an important determinant. In small groups, where everybody knows who can be trusted and who not, formal religions do not appear. But when group size gets large and an individual has to deal often with strangers, a formal religion appears in which spirits of the dead, karma, or a vengeful god will mete out mischance in this life or great discomfort in the next to those who cheat the social system. This has nothing to do with the words of sages like Confucius, the Buddha, or Jesus of Nazareth, who were concerned with the way moral individuals can live together in harmony and support each other in facing the hazards of existence. But it is possible that declining belief in the supernatural surveillance revealed by organized religion is leading to an increased reliance on electronic surveillance to maintain a functional social order.

Given the value of religion as a source of social cohesiveness and stability, it is important for an aspiring interstellar civilization to mitigate internal and external sources of religious instability. One external source is an antagonism that has developed between science and religion [5], [6]. And while such antagonism is understandable, given the history of persecution of scientists by religion, reconciliation between science and religion would be beneficial if it helped religious leaders accept scientific findings regarding the future of a planet with limited resources. One avenue for this is to revisit the overall picture of eternity and creation found in many religions to see if there can be a scientific account consistent with this and able to throw light on some of the current questions in theology.

It is not often recognized that religion on Earth has played a major role in shaping science by turning it away from certain areas of study. In the past, scientists who disagreed with religious doctrine on material matters have been silenced by threats, torture, and murder. As a result, the remainder have shunned forbidden topics, nurturing a resentment, conscious or unconscious, that the search for truth should be limited in this way. Fortunately, the situation is changing, with some religious leaders beginning to promote critical findings of science [7]. But the damage has been done: scientists continue to avoid investigating areas that are a sensitive matter for an established religion. One of these is eternity. Neglect of this dimension is regrettable because it means that science has an incomplete view of the nature of space and time.

Part of the reason for neglect is that the content of eternity is confused with the dimension of eternity. If one considers another dimension, such as volume, this enters the domain of religion if it contains holy water, but it is not sacred when it contains gasoline. Eternity is a similar dimension. But religion has sought to dominate the discussion of eternity by focusing on its content, which varies among religions. This makes it a subject for disputation and violence. So it is no wonder that scientists have steered clear of it.

However, it may be unwise to allow this neglect to continue, because phenomena such as non-locality in quantum effects are suggestive of the presence of eternity as a dimension. It is therefore worth looking at the concept more closely to understand its properties better.

There is another motivation as well. Over the past sixty years there has been increasing interest in the prospect of communicating with civilizations on planets orbiting other stars in the galaxy. It is possible that the sad history of persecution of scientists did not occur in some of these civilizations. In such cases, religion and science would have developed in harmony. Rather than having forbidden subjects to avoid, scientists would be free to explore concepts like eternity and incorporate them into mainstream science. As a result, extrasolar science may be different from our own in important respects. I will show that this can give rise to incompatible units for dimensions, preventing us from correctly interpreting communications from other civilizations.

First, of course, we need to see what concepts of eternity we have available. Ironically, but not surprisingly, the discipline that has investigated alternative concepts of eternity extensively is Christian theology. Theologians filled the gap left by the absence of scientific interest, drawing valuable distinctions concerning different types of eternity.

2. Eternity in Terrestrial Theology

In terrestrial civilizations we can trace the beginnings of concepts of eternity in Indian and Greek thought that speculated on an everlasting mode of existence very different from transitory life in this world. This led to eternity viewed as a dimension of time that extends forever, in both the past and future. We refer to this as sempiternity. With the standard three dimensions of space, it is the current view of eternity in science, modified by the assumption that there was nothing before emergence of the universe 13.8 billon years ago.

Sempiternity was already being questioned by Parmenides, who denied the existence of change (which requires sequence in time), and by Plato, who proposed timeless ideal forms. In the 6th century CE, the Roman philosopher Boethius saw eternity as God's mode of existence, with the attributes of life, timelessness, and perfection [8]. Timelessness is a new type of dimension. Boethius viewed it as an everlasting now without beginning, or end, or succession in time. This is referred to as atemporal eternity.

Archbishop Anselm of Canterbury in the 11th century added aspatial eternity, an absence of beginning, or end, or succession in space. In addressing God, he remarked, "For yesterday, today, and tomorrow have no existence except in time; but You, although nothing exists without You, do not exist in space or time." [9] This atemporal, aspatial eternity can be referred to as ultimate eternity. In some religious thought it appears as a dimension in which a divine mind contemplates the eternal truths and through creative energy manifests them in space and time.

Today, eternity remains an active subject of debate in theology, particularly in how it relates to sempiternity. The proposition that all times are simultaneously present in the mind of God (atemporal eternity) has been described as incoherent by theologians Anthony Kenny [10] and Richard Swinburne [11]. Their temporalist branch of theology emphasizes the activity of God in this world. It may allow that God created the world out of eternity, but after that He turned to existing in time, enjoying a past, present, and future.

Eleonore Stump and Norman Kretzmann have attempted to reconcile eternity and temporalism by arguing that timelessness and temporality may exist simultaneously in two different reference frames [12]. William Lane Craig [13] disputes the concept of eternal-temporal simultaneity, remarking that such a concept relies on incoherent notions like atemporal duration or conceptually indivisible extension. The perceived incoherence may arise however from the assumption that sempiternity and ultimate eternity must occupy the same region.

3. Eternity in Terrestrial Science

This debate is important to science. The theologians are faced with the problem whether a deity existing in eternity can act in space and time. My view is that the appearance of non-locality in quantum effects has raised a similar problem: whether eternity, as the dimension that does not recognize locality (the location in space and time recognized by temporalists), can co-exist with the spacetime that provides locality.

I suggest that the phenomenon of non-locality in physics signals the presence of eternity. Non-locality appears as the instantaneous interaction between particles separated in space and time. Particles are chosen that are created in pairs and move off in opposite directions. Typically, the experiment is done with photons that have related spins. A later change of spin induced in one will cause an instantaneous change in the spin in the other, no matter how far apart they are. This effect has been demonstrated by Alain Aspect and colleagues over a distance of 25 meters [14], but the same effect should be achievable over distances of light-years, if entangled particles can maintain their coherence over such intervals.

Under current physics, where the speed of light is the limiting speed for interactions over distance, instantaneous interactions are impossible. This is why when such interactions were observed, Einstein referred to them as "spooky." Their appearance means that in some interactions there is no concept of particles being separated at different locations in space and time. This non-locality is a description of ultimate eternity.

Indications of ultimate eternity also appear in the wave-particle duality attributed to fundamental particles like electrons and photons. In quantum mechanics, their behavior is accurately described mathematically by a wave equation. The solution to this spreads instantaneously throughout space and time and predicts the probable behavior of particles. The instantaneous spread has been confirmed in an experiment by Maria Fuwa and colleagues between Australia and Japan [15]. This instantaneity implies a spacelessness, timelessness, and lack of sequences that corresponds to atemporal, aspatial features of ultimate eternity.

These are other hints in physics of dimensions underlying familiar ones and causing a uniformity throughout the universe, so that every time acts like the same time and every space acts like every other space. In quantum mechanics there is an intimate relationship between time and the law of the conservation of energy [16]. This shows that the same experiment can be made at different times and produce the same result. Similarly, there is comparable relationship between space and the conservation of momentum. The same experiment can be made at different places and produce the same result. These and similar theoretical findings are consistent with scientific observation that show all parts of the universe are much the same on a large scale and appear the same in every direction (the cosmic principle). While this remarkable finding does not necessarily demand a concept of an ultimate eternity that precedes the locality imposed by spacetime, it suggests the idea is worth exploring.

4. Types of Eternity

So there is some common ground between religion and science in that both describe phenomena involving events that appear to occur in dimensions beyond the familiar four of space and time. They differ in the entities they study within these dimension.

In science, the dimensions of eternity can be reached by reduction of four-dimensional concepts of spacetime. If you retain the time dimension but take away one space dimension, you get two-dimensional forms of space, found in mapping, atomic monolayers, and concepts of membranes and event horizons. Take away two space dimensions and you reach the one-dimensional space at the base of information processing and communication theory.

The alternative of removing the time dimension yields atemporal eternities, regions where there is no progression in time, where every time is the same time. Such atemporal eternities can exist with various dimensions of space. At the event horizon of a black hole, space is two dimensional and time stops, yielding an atemporal, two-dimensional spacetime.

Extending atemporal eternities further, complete removal of space dimensions yields a zero-dimensional eternity where both space and time have no meaning, or every place is the same place and every time is the same time. This is a region without boundaries. It is comparable to Anselm's eternity.

5. Extrasolar Cosmology

I think extrasolar scientists developing a creation narrative with eternity as a starting point (consistent with many religions), would focus initially on atemporal, aspatial eternity. As the ultimate blank slate it offers the simplest starting point for a cosmology and has the greatest potential for creating a spacetime free of prior assumptions. However, the creation of a universe involves work and this requires energy. So there is a dilemma. In this eternity, there are no dimensions, so the energy that makes creation possible cannot be any particular form of energy. It has to be formless, with no dimensions in space or time.

How can there be an energy without form? Many types of energy exist in spacetime, including gravitational energy, kinetic energy, potential energy, heat energy, elastic energy, electrical energy, radiant energy, nuclear energy, pressure energy, and mass energy. All these can change from one form into another. Any scientific civilization can calculate the energy transferred in the conversion. They can confirm their calculations by measurements. They can verify that energy is conserved, that none disappears in a transfer. What they will not know initially, as Richard Feynman pointed out in our own case, is what is conserved or its nature [17].

Properties of energy reveal themselves in the context of a particular form of energy. Each form is tied to a set of dimensions. An extrasolar science will figure out that when energy is separated from its many forms, what remains is a dimensionless energy that can move from one form to another, picking up the necessary dimensions on the way. We can see the process taking place in the movement of a pendulum, where potential energy, described within the dimensions of mass and height, is transformed, in full view, into motion energy, described within the dimensions of mass and velocity.

For the energy of eternity to gain familiar forms requires the presence of spacetime, in which energy fills selected dimensions in a particular way. If a science assumes an initial condition of the universe to be a formless energy existing in eternity, then it will require spacetime to emerge from the energy of eternity in order for particular forms of energy, including matter, to appear and create a universe. To create a stable universe, however, the forms of energy and matter created must follow universal laws, such as conservation of energy. An extrasolar science will investigate how these laws might operate as spacetime is formed.

6. Emergence of Spacetime

We observe an irreducible level of random fluctuations in actions that occur in our present universe. I suggest that a fluctuation among eternities could have caused a phase change that precipitated spacetime and moved the universe towards immanence. Since in ultimate eternity all locations are the same location and all times are the same time, this fluctuation would occur instantaneously throughout all energy. It would initiate a phase change in which bubbles of spacetime condense out of eternity in the way that droplets of dew condense out of water vapor in a planet's atmosphere, or ice crystals grow in a planet's pools of liquid water.

Phase changes like this are familiar in our own accounts of evolution of the universe, so we should not be surprised that they occur in extrasolar accounts. In our cosmology, the universe starts with a sudden expansion (big bang) that initiates many phase changes. In one set, the four or five main forces emerge in succession: perhaps a unified force, then gravity, the two nuclear forces, and electromagnetism. In another set of phase changes, a succession of matter particles emerges—quarks, protons, neutrons, nuclei, atoms, and molecules.

To formulate a cosmology that starts with eternity, an extrasolar science would argue for a similar set of phase changes to ours, but add a prior one that yielded elements of spacetime. These form spacetime quanta. When packed together they are the building blocks that form four-dimensional spacetime. For simplicity, the building blocks would be considered as uniform and of minimum size. There would be no dimensions smaller than their dimensions. Once they combine into a four-dimensional spacetime, formless energy could enter and take up form. The evolution of the universe would then proceed through the phase changes we postulate for subsequent stages of our own explanation of material evolution.

7. Spacetime Quanta

With the first phase change, universal constants could be expected to gain expression. From a proposal that space and time emerged in the form of spacetime bubbles, the conclusion might have been drawn that a random foam appeared. However, this would not lead to an orderly, stable universe, because in every instant it would be impossible to foresee in what stretch of past or future time the next instant would be. A more reasonable assumption would be that in these bubbles the relationship between space and time is constant and expressed in the velocity of light, which later relates energy to mass. Other assumptions would be that the effect of mass on spacetime would be expressed in a gravitational constant, which relates energy to attractive force, and that the relation of energy to radiation would be expressed in a Planck constant of action (energy acting through time). These three constants would determine the size of the quanta.

As spacetime quanta have the minimum possible dimensions they are indivisible. That is, this extrasolar approach abandons the notion of infinitely divisible space. As the speed of light is constant, if no spatial interval can be smaller than the size of a spacetime quantum, then the quantum also expresses the smallest interval of time. The speed of light is the smallest interval of space traveled by light in the smallest interval of time. An extrasolar civilization would also conclude that spacetime quanta would have no preferred direction for their minimum dimensions, expressing the same interval in every direction. This would arise from the view that on a large scale the universe acts the same in all directions. For this reason, the quanta would be spherical.

When they come together, the quanta provide the extended dimensions required to provide potential for action. They make possible movement from one location to another and permit action between different locations. But they restrict movement through space and time to discrete, finite steps. This extrasolar spacetime would act like a digital medium.

8. The Spacetime Lattice

The initial phase change in an extrasolar science's account of emergence and evolution of spacetime would invoke universal constants and laws that determine the nature of the spacetime quanta. They would not expect their nascent universe to immediately evaporate. But they would have an inherent, and perhaps explosive, contradiction in a population of spacetime quanta with dimensions of space and time dispersed in the dimensionless energy of eternity. It is likely that they would assume the presence of a law that requires the minimization of kinetic and potential of energy differences (comparable to our principle of least action). This would lead to their newly born spacetime quanta being pulled into a closely packed lattice that minimized the remaining energy surrounding the quanta. In effect, an extrasolar science would probably see formless energy between quanta exerting a pressure on the quanta that packs them into arrays occupying least energy. The effect will be to pull the spacetime quanta together by a negative pressure, forming a three-dimensional lattice of closely packed spheres.

In the extrasolar cosmology, the presence of the spacetime lattice would allow the entry of formless energy into spacetime to take up form. Probably the first form anticipated would be radiation at an extremely high temperature. Given the lack of locality within the proposed ultimate eternity, this energy would be considered to enter throughout spacetime uniformly. The creation of specific forms of energy would be initiated homogenously in a uniform flash of energy.

9. Randomness in the Lattice

We can assume that if the proposed action of formless energy were fully effective in pulling the spacetime quanta together, it would yield a regular lattice in which the maximum space, 74 per cent, would be taken up by the spacetime spheres and 26 per cent would be occupied by the formless energy of eternity [14]. However, as there could be no regular pattern in eternity to seed formation of the most efficient packing, the spherical quanta might be pulled together in a random lattice. Then the quanta would occupy about 64 per cent of the space, increasing the space occupied by formless energy to 36 per cent.

At question here is the degree of randomness in the lattice. We can assume particles travel through the lattice by skipping from quantum to quantum, across the points of contact. No other movement through space between quanta would be possible, because that would require a passage through the energy of eternity, and the effective distance would not conform to the limit enforced by the assumption of a minimum size for intervals in space and time.

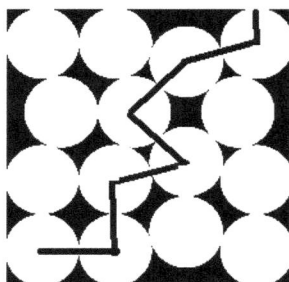

Figure 1: An artist's impression of a cross-section of the spacetime lattice with some irregularities. Particles can only transition across points of contact of spacetime quanta, so that a random pattern of movement around an average trend develops.

Instead, the particle would simply disappear from one spacetime quantum and appear at the next, in an interval no shorter than the smallest unit of time. The direction of movement would be set by the contact points between quanta. In a fully packed regular lattice, a quantum would have 12 contact points with other quanta.

A particle would therefore have to transfer between quanta in a restricted number of directions and those directions may not conform to the direction in which it is traveling. It would have to find a zig-zag route with an average direction corresponding to, for example, the straight line path expected for a photon of light. So the spherical quanta envisaged by extrasolar science, and their limited number of contacts, would bring a degree or randomness to particles, even when moving through a regular lattice. Dislocations in the lattice (Figure 1) would add further randomness to particle paths. The mathematics describing particle interactions would have to include an inherent randomness, rather like our own quantum mechanics.

An extrasolar science, noting that randomness does not appear at the macroscopic level, would probably decide that the random disturbances caused by the lattice would be overcome internally in some way so that light and other particles do travel in straight lines. Such randomness as remained would be recognized only over very small distances. It would be observed when a light beam is restricted to a very small size and very low intensity. They might agree with Richard Feynman, who remarked "...the idea that light goes in a straight line is a convenient approximation to describe what happens in the world that is familiar to us" [19]. The small random deviations that occur would have to be taken into account at scales found within an atom.

The contiguous lattice of interstices surrounding spacetime quanta would cause eternity and spacetime to be locked in an intimate embrace. In formless energy, all space would be the same space and all time would be the same time. When this energy is locked in the four-dimensional interstitial space, it might acquire some aspects of locality, even though it remains continuous, in contrast to the discrete nature of spacetime. It may therefore offer a path through which entangled particles might interact instantaneously, or enable a wave function to cover an immense region with no delay.

10.   Universal Units at Earth

On Earth, science has accepted, for the most part, that there is indeed a minimum length appropriate for our own description of the physical world. The accepted size for the minimal length is the one offered by the natural units proposed by Max Planck [20]. These units will also have been discovered by an extrasolar science because they are of fundamental interest as far as units of measurement are concerned. They are also the obvious dimensional units for any civilization seeking to communicate with other planetary civilizations in the galaxy. This is because the units are based on the universal constants that determine natural properties of space, not on physical objects kept in a standards bureau on a particular planet.

Each of these constants will be expressed in the local units of a planetary civilization. For example, one extrasolar civilization might express the speed of light in "oodles per twinkle," and we would not know how these units relate to our own. Informing us that they stood a few "megaoodles" high would mean nothing to us. But the universal constants are universal. And it turns out that when you choose the correct ratio between constants, you cancel out the local units and obtain universal units.

On Earth, the universal unit of length, the Planck length, is found by identifying the smallest region from which we can gain information. Quantum mechanics and general relativity each provide an approach to this minimal region. Where the two approaches meet, a joint value for the Planck length is obtained. The method has been described by Frank Shu [21].

In quantum mechanics the length of a probability wave gives the information needed to find where a particle is likely to be. The length of the wave is the Planck constant divided by the particle momentum, which is the particle mass times its velocity. When the speed of light is entered for the velocity, the particle shrinks to its Compton wavelength, the smallest wavelength associated with its mass. Further decrease in wavelength depends on increasing the particle mass. But as more mass is squeezed into a smaller space, the gravitational field rises the level for formation of a black hole, which also hides information. So, when a particle's mass creates a black hole about the size of its Compton wavelength, we can be sure we have found the smallest region from which we can gain information.

There is a problem in deciding how to equate the diameter of a gravitational hole to the length of a probability wave, which is a somewhat fuzzy variable. The solution is that the same smallest interval is obtained by both approaches when the black hole diameter is four times the Compton wavelength. On this basis, the interval can be found from two equations. The diameter of the black hole for a given mass is calculated from the Schwarzschild radius equation. The particle wavelength for that mass is calculated from the Compton wave equation. Combining these two equations shows that the Planck length can be obtained by taking the square root of the product of the Planck constant and the gravitational constant divided by the cube of the speed of light. This length divided by the velocity of light yields the Planck time.

$$\text{Planck length} \qquad = 1.62 \times 10^{-35} \text{ meter}$$

$$\text{Planck time} \qquad = 5.39 \times 10^{-44} \text{ second}$$

These intervals are extremely small. The Planck length is over ten million trillion times smaller than the diameter of the proton, the smallest length we have measured.

On Earth we calculate a Planck area as the square of the Planck length. That gets rid of the square root. So, multiplying the gravitational and Planck constants together and dividing by the cube of the speed of light gives

$$\text{Planck Area} \qquad = 2.61 \times 10^{-70} \text{ square meter}.$$

We calculate the Planck volume as the cube of the Planck length.

Planck Volume = 4.22 x 10^{-105} cubic meter

In essence we have selected a cube as our minimum volume, our equivalent of an extrasolar spacetime quantum. It is difficult, but not impossible, to imagine cubic spacetime quanta condensing out of eternity. And if they do, then each one comes with several minimum lengths, provided by their sides and diagonals. Again, it is possible to conceive of particles moving different minimum lengths according to the direction in which they move from one minimum volume to another. And the time intervals have to be self-adjusting to make sure that the intervals in the direction of the different minimum lengths continue to yield a constant velocity of light. But the complexity of this arrangement would not make it a first choice on the basis of aesthetics and simplicity. So, it is not surprising that we treat spacetime as infinitely divisible, even though we have found a minimum interval from which information may be acquired.

11.   Extrasolar Planck Universal Units

A civilization that assumes spacetime is made up of spherical quanta could approach the Planck length in a different way. After the spheres pack into a lattice with minimal spacing, movement of a fundamental particle from one to another will always represent a change of position corresponding to the same minimum length, independent of the direction of movement. Movement through the lattice will always be by the same incremental steps. The ratio of the minimum length to the minimum time will always yield the speed of light.

From the values of the three universal constants, the extrasolar Planck area will be the same as ours, namely, 2.61x 10-70 square meters. However, whereas we have assumed this is a square area, an extrasolar civilization is likely to assume it is a circular area -- the cross-section of a spacetime quantum. They would then calculate the Planck length as the diameter of that area.

Extrasolar Planck length	= $\frac{2 \times \text{square root of Planck area}}{\text{Square root of pi}}$	Eq. (2)
	= 1.13 square root of Planck area	
	= 1.13 of Earth Planck length	
Extrasolar Planck length	= 1.83 x 10^{-35} meter	
Extrasolar Planck time	= 6.09 x 10^{-44} second	

That is, the extrasolar Planck length may be 13 percent larger than the terrestrial Planck length.

12.   Problems in Communication

If we make use of the Earth Planck length in interstellar communications, we may confuse our extrasolar communicants. For example, suppose they are orbiting the bright star Canopus and decided to visit us. If we were to tell them (using the conversion factor correct for us) that we are located 310 light years from their star, they would convert our Planck units using their own conversion factor and set a lay line to arrive within our orbit. Unfortunately, they would overshoot by 40 light years. Not finding us, they might question our concept of accurate navigation. And, until we heard from them, we would be wondering for 40 years what happened to our visitors. We should be prepared to resolve this ambiguity at the earliest possible time.

13.   Conclusion

An extrasolar science that has enjoyed a congenial relationship with religion will have a very different physical cosmology to ours. Even though it may have the same mathematics of quantum mechanics and relativity, its description of the origins of the universe and phenomena such as spooky action at a distance may differ dramatically from ours. Extrasolar physics may consider spacetime to be a lattice of spherical spacetime quanta interspersed by the formless energy of eternity, providing a basis for the phenomenon of non-locality. In contrast, we have a picture of spacetime made up of cubic quanta, with no interstices for the energy of eternity and consequently no explanation for non-locality.

This extrasolar cosmology is similar to religious accounts of creation of the universe by a divine being existing in eternity. Because it aligns with various religious scenarios, I think this extrasolar cosmology will be more acceptable to an extrasolar religion than our concept of a big bang expanding from nowhere followed by a transient inflation to smooth over irregularities. There will be a view that the presence of eternity is pervasive everywhere, offering a solution to the theological dilemma of the coexistence of eternity and temporality,

Civilizations accepting the concept of a lattice of spherical quanta may have a Planck length that is 13 per cent longer than ours. This might cause confusion in our progress towards becoming an interstellar civilization.

References

1.   Pay, R. G. (2014). "Religion in SETI communications." *Journal of the British Interplanetary Society,* 67(193).

2.   Norenzayan, A. (2013). *Big Gods: How Religion Transformed Cooperation and Conflict.* Princeton, NJ: University Press.

3.   Slingerland, E. (2015). "Big gods, historical explanation and the value of integrating the history of religion into the broader academy." *Religion,* 45(585).

4.   Wade, L. (2015). "Birth of the moralizing gods." *Science,* 349(919).

5.   Dawkins, R. (2006). *The God Delusion,* Boston: Houghton Mifflin Company.

6.   Carroll, S. M. (2012). "Does the universe need god?" Chapter 17, *The Blackwell Companion to Science and Christianity,* Eds. J. B. Stump and A. G. Padgett, Chichester, UK: Wiley-Blackwell.

7.   Pope Francis. (2015). *"Laudato Si', mi' Signore* -- On Care of our Common Home." Encyclical Letter. The Vatican: Vatican Press.

8.   Boethius, A. *The Consolation of Philosophy,* Chapter 5, book 6, translated by H. R. James, London: Elliott Stock. Available on Project Gutenberg.

9.   Logan. I. (2008). *Reading Anselm's Proslogion,* p. 50, par 19.4, Farnham, England: Ashgate Publishing Ltd.

10. Kenny, A. (1979). *The God of the Philosophers,* p. 38. Oxford: Clarendon Press.

11. Swinburne, R. (1977). *On the Coherence of Theism.* Oxford: Clarendon Press.

12. Stump, E. and Kretzmann, N. (1981). "Eternity." *Journal of Philosophy,* 78(429).

13. Craig, W. L. (1999). *American Catholic Philosophical Quarterly,* 73(521).

14. Aspect, A. (1999). "Bell's inequality test: More ideal than ever." *Nature,* 398(189).

15. Fuwa, M. et al. (2015). "Experimental proof of nonlocal wavefunction collapse for a single particle using homodyne measurements." *Nature Communications 6*, Article number: 6665.

16. Feynman, R.P., et al. (1963). *The Feynman Lectures on Physics,* Vol. 1, 4-2, Vol III 17-3, Conservation Laws, Reading, MA: Addison-Wesley Publishing Company.

17. Ibid., Vol 1. 4, Unknown Energy.

18. Wikipedia. "Sphere packing": https://en.wikipedia.org/wiki/Sphere_packing.

19. Feynman, R.P. (1988). *QED: the Strange Theory of Light and Matter,* p. 53, Princeton, NJ: Princeton University Press.

20. Wilczek, F. (2008). *The Lightness of Being*, p. 156, New York: Basic Books. See also http://www.arxiv.org/pdf/0708.4361v1.pdf

21. Shu, Frank H. (1982). *The Physical Universe.* Mill Valley, CA: University Science Books, p. 397.

Acknowledgement

I thank my wife, Elsa Pay, for proofreading this document and for providing valuable suggestions for improvement.

A Roadmap to Interstellar Travel: The Societal Challenges

Abigail Sherriff

International Space University

Abigail.Sherriff@community.isunet.edu

Chris Welch

International Space University

Chris.Welch@isunet.edu

Abstract

An interdisciplinary roadmap is developed to show the preparations needed over a nominal 100-year timeframe to launch a slower-than-light, self-sustaining worldship to carry humans over many generations to other star systems. The project was carried out by Astra Planeta, one of two team projects for the International Space University's 2015 Master of Science in Space Studies. This paper discusses the societal aspects, particularly the challenges of society on and off the worldship and necessary mission planning leading to launch.

The project, as a whole, draws on information from past studies of interstellar missions, and a preferred concept for the worldship is identified. Strategies are outlined for the necessary development and follow a logical progression of enabling missions and projects. An interdisciplinary approach considers the technological and societal challenges leading up to the launch of a worldship, as well as operating it over hundreds to thousands of years.

Through this framework, many major topics were cultivated, including: onboard infrastructure, operational concepts, necessary ship subsystems, onboard societal issues, and mission planning. For civilization in particular to grow and flourish, planning for the mission and the onboard society must be taken into special consideration. The onboard society research examines topics such as education, religion, language, and ethics. Mission planning takes into account the conditions for enabling the construction of a worldship including political considerations, cultural changes, financing, and international partnership.

Since expanding civilization and humanity is arguably the central reason for interstellar travel, this paper also addresses these issues in the context of the proposed mission. The end result is a framework to achieve interstellar travel in the near future with an approach that is financially feasible, technically capable, and culturally desirable.

Keywords

Culture, education, financing, language, religion, society

1.   Introduction

This paper addresses the societal challenges as part of a project to create a conceptual design and nominal 100-year roadmap of a slower-than-light, multigenerational worldship to begin establishing humanity across the universe. The project as a whole is the result of the work of the Astra Planeta team, composed of twenty-two students from the International Space University (ISU) 2015 Masters of Science in Space Studies class. Each student comes from a different part of the global community, with the entire team representing four continents and seventeen different disciplines. The project made an effort to leverage these professional and intercultural differences inherent in the team to help aid the creation of an interdisciplinary roadmap. This paper provides a snapshot of the societal aspects drawn out by the work performed by the team. The full report is the culmination of a dedicated effort to critically evaluate existing literature to develop a roadmap for the worldship and should be consulted for a more comprehensive view of the project [1].

Before delving into the specifics of the societal aspects of the research, general background, overview, and outline of the project are provided. In the sections following, a more in depth discussion of the planning and onboard society is provided. References to the 'project' refer to the team project as a whole and following sections detail the societal challenges addressed.

1.1   Definition of a Worldship

An "interstellar worldship," or "worldship" for short, is a spacecraft designed to travel over vast distances at non-relativistic speeds (slower than light: < 0.10c). The term is reserved for spacecraft designed to carry a population of 100,000 people or more to a destination, which may take hundreds to thousands of years to arrive. A worldship is a generational ship in which many generations of people live their whole lives and die onboard. It is not a ship where the occupants are asleep or hibernating and it does not only carry human "seeds" for growth at the destination.

1.2   Aims and Objectives

An attempt has been made to offer an interdisciplinary approach to tackling the challenge of interstellar travel. The concept and roadmap provide a step towards overcoming some of the developmental challenges of interstellar travel. This project aims to contribute to this effort by furthering the state of knowledge of interstellar travel and establishing the foundation for future work and research to be accomplished. This requires that several challenges be discussed in addition to the technical ones. These include societal challenges onboard the worldship as well as throughout the developmental and construction phases. Political, economic, and legal aspects hold barriers for success that must be discussed, in addition to strategies for education and public outreach. The roadmap and suggestions are meant to showcase a preferred mission concept that is technologically and financially feasible, as well as culturally desirable. Given that the future cannot be predicted, the final goal of this project was not to create a detailed design of a worldship, but to identify key technologies and milestones that should be pursued for a preferred mission concept.

1.3   Constraints and Assumptions

In order to provide some context into which the preferred worldship concept could fit, some constraints and assumptions were discussed. Speculative technology such as warp drive, wormholes, and faster-than-light travel were not considered. Furthermore, it was assumed that the energies involved with fast travel (velocities greater than 10% the speed of light: 0.10c) would be prohibitive to achieve with near-future technology. Also, the population onboard the worldship is assumed to be alive and fully awake; thus, dedicated seed ships and cryogenic sleep are not considered. Finally, one of the project aims was to identify a preferred concept for a worldship, not to develop a detailed design.

1.4   Why 100 Years?

To most people a hundred years is a relatable period of time. One can certainly conceive of living that long. Thus, a hundred year time frame was chosen in part because it is approachable to people. It is not so far in the future as to be uninteresting, yet it is far enough that it seems a feasible timeframe. It also serves as a constraint, limiting the use of technology to those that currently exist or will in the near- future.

While the future cannot be predicted, logical and useful statements may be made about it. This allows humans to forecast alternative futures. One of these alternative futures could be the construction and launch of a worldship to explore our stellar neighborhood. Significant changes can happen in a hundred years, as is evidenced by the exponential growth in population and technology over the last one hundred years. To someone living at the beginning of the 20th century, the progress over the following hundred years would have seemed fantastical at the time. While this project makes no effort to predict the future, it considers the realm of what may be possible in the next hundred years. This gives the project scope and forces a realistic consideration of what a worldship may look like without relying on tenuous physics or incredible breakthroughs. The use of the word nominal when describing the timeframe was chosen carefully as it expresses uncertainty. It is a goal, meant to provide scope and boundary to the project as a reasonable starting point.

1.5   Mission Concept

In order to define a preferred concept for a worldship, an interstellar mission must first be conceptualized. That mission is then used to set the boundaries and focus efforts on creating a roadmap to the conceptual worldship. Thus, the proposed interstellar mission is as given:

> *The interstellar mission will consist of a single worldship directly travelling to a single stellar system within 25 ly from the Sun [2].*

Due to the large separation between stars, a voyage to a single stellar system is considered. As this will take hundreds to thousands of years, it is outside the scope of this project to consider what the worldship inhabitants will do when they arrive. A self-sustaining worldship would have the capability to leverage a stellar system's resources to manufacture the necessary components to settle a planet, construct a space station, or continue the voyage to another stellar system. Furthermore, due to the high energy required to accelerate a worldship, distances further than 25 light-years (ly) were considered to be unreachable in hundreds to thousands of years and are not considered. A fleet of several starships was considered as it offered several advantages, such as redundancy. However, the extra complexity and costs of constructing and managing a fleet was seen as prohibitive.

2.   Planning for the Worldship

The road to the launch of the worldship is full of hurdles and challenges both technological and societal. The societal challenges range from economical to cultural. There are many steps along the way that must be taken to become the interstellar society necessary to travel to a distant stellar system. Some of the major challenges and possible solutions are outlined.

2.1   Governance

Determining the political governance structure onboard the worldship is a major concern that must be developed during the planning and construction of the worldship to ensure that it is fully primed. Hill discusses three motivations for future space exploration, including a corresponding governance model for each [3]. First is a global emergency, such as an approaching large asteroid. The corresponding governance model would be similar to an economically controlled government where decision-making is highly centralized. The second is driven by free enterprise and follows the market economy model, which will be chaotic with private companies and multinational organizations working towards personal gain or prestige. In this scenario, each company must be independent of governmental support in order to be a true free enterprise, complicating the development of a worldship governance structure. Third, and most likely according to Hill, will be a hybrid approach between the first two scenarios. Governments will handle the first and most costly long-term efforts, while commercial organizations take up the short-term activities through incentivized financing. This is similar to the mixed economic and governance model that has been successful in the creation of the Internet or air transportation. In the beginning, the government and military took on the costly and risky endeavors, and once the technology was proven and costs were reduced, the commercial market took over. Each of these scenarios offers a possible catalyst that will lead us off of Earth and into interstellar space. According to Crawford, a space-faring society must have the preconditions of "survival, geopolitical stability, resource availability, and moral justification," so a world government must be in place before the human race can become truly space-faring.

While the idea of an effective world government may seem utopian to some, the idea of an Interstellar City where world governments, commercial enterprises, and the international community work together to develop the worldship may also prove to be a model worth exploring. An Interstellar City would ensure flexibility, as cities are well known for their ability to grow and change and can significantly contribute to all factors relating to the worldship.

The Interstellar City would be a centralized community that can assist in stimulating open and rational debate in a highly politicized area. It allows the centralization of an economic system and labor and would be central hub for major worldship experimentation, research, manufacturing, and governing. The city would be a place where all participating nations and nationals could live and contribute to the creation of an interstellar culture and provide a platform to facilitate collaboration. This city could be an autonomous, sovereign city-state that would be separate from the direct supervision of any single nation-state. Instead, it would be bound by its own constitution, rules, legal structures, and even economy. The city would be religiously and ethnically mixed, have its own currency, and be a center of technological and scientific expertise. Examples of autonomous cities in the past include Danzig, Tangier, or Trieste. These cities inspired creativity and fostered innovation [4]. The present day example of the Vostochny Spaceport that is being built in the Russian Far East is intended to be a state-of-the art "science city" [5] and could serve as a primary model for developing the Interstellar City. An Interstellar City on Earth will also serve as an analog for an Interstellar City on the moon, and finally an Interstellar City on the worldship. More importantly, it will serve as an important base to test governance models.

While the continuation of the human race, exploration, and expansion may be motivation enough for some countries, others may require bigger motivations. The idea of an exclusive, technologically advanced, and scientifically driven Interstellar City will serve as an additional motivation for the worldship project. All participating countries will be allowed to send a certain number of citizens to the City every year. All of the patents, technology, and spin offs that result from the innovation within the city will be the property of the city and the individuals that reside in it. There will be an internal economy that trades with other nations. Participating nations can expect to have a portion of their financial contributions returned to them in contracts, similar to the European Space Agency's Industrial Policy of geographical return [6]. In summary, national and personal prestige, access to technology, and generous return on investment schemes will all serve as tools to motivate nations and citizens to join the Interstellar City.

2.2 Financing the Worldship

Before humanity can realistically achieve interstellar travel, the society of Earth must adapt. One of these key societal aspects is economy. No single nation can realistically fund a worldship on its own, and no project of this scale has ever been attempted. For the construction of the worldship to go ahead, appropriate funding must be collected through international partnerships. This financing must take into account a global change towards the stabilization of world politics and a lessening of economic and social instabilities. Utilizing existing financing models paves the way to devise realistic approaches of achieving financing an interstellar worldship.

There are three complementary ways in which a project of this magnitude could be financed and developed. The following three-pronged approach aims to tackle both the financing and technology development aspects of the project:

1. International Interstellar Fund (IIF) similar to International Monetary Fund (IMF) Quota system
2. Incentive financing to further technological progress

3. Additional sources of funding:
   A.   Crowd Funding Platforms
   B.   Leveraged Finance

The International Interstellar Fund (IIF) being the main driving force through which the bulk of financing for the project will be carried out. The IIF will serve as the primary body for coordinating financing efforts for the duration of the project. The IIF will use a proportional equality system, or "criterial equality," which stipulates that though the voting rates and funding requirements may vary for each participating state, the calculation method for such rights will be the same for all, depending on distinct, predetermined criteria. An example of this is the allocation of votes in the Board of Governors of the IMF. This is the organization's highest decision-making body and allocates votes to participating countries' decisions based on their financial contributions to the Fund in terms of IMF quotas [7]. To keep the IIF mechanisms simple, the payment requirements will be based on this quota system.

Considering that there are 188 countries that are members of the IMF, applying the IMF model not only simplifies the process for member states, but will also ensure a smooth transition by taking advantage of existing protocols. Initially, the fund will consist of twenty founding members, which include the leading and emerging space nations. After initial founding, these countries will form a committee to establish a governance structure similar to the IMF. Using this model, the IIF body could require members to pay a contribution based on the weighted average criteria outlined below. Thereafter, an open invitation to all nations will subsequently be made allowing new members to be added upon payment of the required membership quota. Aiming for a $500 billion fund within 20 years of founding could be a realistic proposition if membership numbers reach two-thirds of the IMF. The benefits and incentives for participation will need to be emphasized to ensure successful financing, and this is where lobbying and outreach will play an active role. One of the incentives for participation in the IIF could include the production of energy in space. A portion of the output of this energy would correspond to the level of investment made. Space-based solar power systems could revolutionize energy consumption and help drive the economic case for the participating nations. The case for this is made unequivocally in the 1975 NASA Summer Study [8].

IMF's quota system is used to determine a country's financial and organizational relationship, based on these three aspects:

1. Subscriptions or quota share – the maximum financial resources a nation is required to pay upon joining
2. Voting power or voting share – member's voting power
3. Access to financing – the level of available financing for the nation usually capped at 200 percent of the nation's quota

Whereas IMF countries contribute funds to a pool through the quota system from which countries with payment imbalances can borrow, the IIF will serve as a body that utilizes this pool to directly invest in the infrastructure and technological development needs of the interstellar worldship project. The IMF aims to improve the economies of its members through the provision of loans. The IIF will aim to facilitate and foster economic activity for the member countries, advancing their technologies and delivering societal progress in pursuit of its long-term mission of interstellar travel.

Second to the IFF will be incentive financing, which is a way of enticing the development of technologies through award programs. This financing approach can play a strong supporting role that will help drive forward innovation. It is through incentive financing that much of the cost reductions and technological innovations have historically been achieved. An example of this is the Orteig Prize which was a $25,000 ($340,067 as of 2015) prize awarded to Charles Lindbergh for the first non-stop flight across the Atlantic [9]. There is also the modern equivalent of the Orteig Prize known as the X-Prize. These types of prizes help promote competition and drive technological innovation. Involving the commercial space ventures in this way adds to the compelling case for the mission and ensures there are added economic benefits of which can be taken advantage. Prize money for these awards could come from the IIF, with a focus on ensuring equal involvement from member states.

Additional funding sources, though miniscule in comparison, are also necessary. One type of additional funding that has seen recent proliferation is online Crowd Funding. A Crowd Funding Platform (CFP) is a form of funding that takes advantage of the general public or "the crowd" to raise funding for projects, otherwise known as campaigns. CFPs can be utilized to help support development costs and, in addition, foster public outreach.

A second type of alternative funding is leveraged finance, or borrowed capital. Utilizing leveraged finance to support technology development programs could support the interstellar project through difficult periods of the project. By leveraging debt financing, the project can offset any future costs associated with the project and increase the return on investment. The IMF model, for example, supplements its quota resources through borrowing if it believes that they might fall short of members' needs. A number of member countries and institutions stand by to lend additional resources to the IMF. This type of financing could support additional, unforeseen costs and help support the continuation of technology developments.

In order to fulfill the dream of interstellar travel, many economic issues must be resolved. A full-scale solar system-wide economy should be taking shape, in which the international community plays an integral part in its development and growth. A society capable of building a starship will likely already be a space-faring society, with a solar system-wide economy to draw upon, utilizing all the resources and energy sources available in that system [10]. It is therefore difficult to imagine a situation where interstellar worldships are being developed prior to the existence of a solar system-wide economy.

2.3   Public Outreach and Education

To successfully develop and construct an interstellar, multigenerational worldship, Earth's population must be supportive and educated. Gaining this support will be one of the major challenges on the road to becoming an interstellar society, so effective public outreach is of the utmost importance. Once the public is onboard with the project, they can be educated on the mission objectives and thus influence their governments and officials to support the interstellar project. Overall, public cooperation will be necessary, as in the days of Apollo, where outreach and educational programs reached people of all ages.

There are several challenges associated with an interstellar worldship that are otherwise not present in other space programs. The vast scale of the project gives the need for global participation from many nations. Reaching target audiences with contrasting cultural motivations, languages, and practices will make this difficult. Luckily, the worldwide society is becoming increasingly interconnected through the Internet and social media, but these platforms require new methods and strategies. Finally, retaining the public's interest over the lengthy project timeline will be difficult. The public must remain engaged to maintain financial and technical support. Luckily, there are opportunities to be found in the outreach challenges. For example, younger generations feel the need to be active participants, which can be leveraged by including interactive and individually driven outreach programs. NASA does this through programs designed to crowd source technical solutions from employees and the public alike. This increased focus on individuals and connectivity will also help to solve many international issues.

An effective outreach program for the worldship mission must be multifaceted to ensure that the entirety of the global community has access to participate and access information. Outreach must be pursued consistently and cleverly, and must adapt with technology and societal norms. There should be facets of the program that reach individuals in all fields. There may be design contests that reach members of the science, technology, engineering, and math (STEM) fields and there may be digital design competitions to attract individuals that are not so obviously connected to an interstellar mission.

The Interstellar City will be integral in enacting outreach and education programs. The governmental systems, ways of living, and technological advancements within the city will be fascinating to people on the outside. Visiting and learning about the city will pique the interest of individuals and encourage a different way of thinking. As projects such as the Biospheres become more sophisticated, members of the general public may have the opportunity to visit and live within the structures for short periods to experience life as a future explorer. The Interstellar City will also be the hub for reaching the global community.

In the world of today, social networking has become ubiquitous and can be utilized for a broad reach in a short amount of time. To do this, surveys and design contests may be utilized to ignite interest and even generate innovative new technologies. Additionally, social media can be utilized to monitor data trends that may prove vital in evaluating the effectiveness of outreach programs and interstellar campaigns themselves. Another popular outlet is interactive media, or computer games. Massive multiplayer online games (MMOs) such as World of Warcraft and Minecraft are hugely successful and have enticed people to create entire digital worlds. Games with similar mechanics could be utilized to explore mission concepts, psychological tendencies, and encourage new ideas. Computer game design and competitions will encourage passionate people to explore and gather knowledge on interstellar travel. One such game could be based on designing an interstellar ship, its environment, habitats, and preparing tools to survive. Media outlets such as these that work through the medium of the Internet are some of the most effective ways to reach much of the world's population with minimal expense.

By the launch date of the worldship, the world will be very different from today, and maintaining post-launch outreach activities will be essential. This will ensure strong interaction between Earth and the worldship while communication is possible. Also, this will inspire future generations of explorers to venture deeper into the universe and ensure that this is not a one-time project.

In addition to outreach, strong educational systems on Earth leading up to a launch of the worldship are essential to create a society capable of implementing the vision and to bring value to the purpose of the mission.

In order for a mission such as this to take place, an interstellar-minded society willing to take on the challenge must exist. This will require individuals who believe in the value of the worldship mission and are willing to devote their work to a pursuit they may never see carried out. This means that from a very young age students should be encouraged to think in the very long term. This mindset can be created through an interstellar education integrated into national curricula and extensive interstellar engagement programs.

The first priority will be to ensure that there are enough technically knowledgeable personnel on Earth to physically design and create the worldship. This means that education in STEM fields should continue to have an increasingly strong presence in school curriculum. Nebergall notes that a special focus should be placed on the general problem-solving methods rather than learning to solve specific problems. Creative thinking and innovation will be essential to ensure that the people on Earth are quickly adaptable and able to create what is need with the available resources [11].

An interstellar society requires a change in many ways of thinking in addition to traditional knowledge. In order to prepare students to think in this sustainable way, non-interstellar specific courses would also be vital. Examples of this would include project-based learning with regards to recycling and up-cycling. Interdependencies could be modeled through analog examples such as permaculture and biosphere design. Curriculum would include team-building activities, as well as create an environment in which students will learn to be resilient and self-confident.

A growing trend in the information age is the availability of free online education for those who want to learn for the sake of it. An online course initiative will be an effective way to provide information and create awareness between the enthusiast and interested people in our society. Free online short-term courses will gather and transform the approach of interstellar education. Students, scientists, professionals would have opportunities to participate in such type of courses. Educating individuals would provide a nice avenue to leading them directly to participate in many of the outreach programs in a productive way.

Adequate public outreach and education programs that reach the global community of individuals of all ages and backgrounds is essential. An interstellar, multigenerational worldship will require the majority of humanity's support, understanding, and effort to be a success. Fundamentally, this means that the outreach and educational programs should be a prime focus for this mission.

3. Worldship Society

The societal challenges for the citizens of Earth leading up to the launch of the worldship are daunting. Many considerations must be taken into account for what the society on the worldship will be, from the population to language to religion. Specific decisions about the makeup of the worldship society cannot be made at this early stage, but an idea of the challenges and options facing the society as the launch approaches is investigated.

3.1 Population

The size of the population is a very significant determining factor for the size of the worldship and scope of the roadmap, and there are a number of differing views on how to accommodate the size of the population. The current definition of a worldship is 100,000 individuals, but the specifics of achieving that value are flexible. The launch population size and age distribution are two key factors. Decisions about the population must be made as the launch date approaches.

The worldship must eventually be populated with 100,000 individuals, but the population size at launch may vary. At one end of the spectrum is to launch with the smallest population feasible for genetic diversity, usually defined as 10,000 individuals. This would allow the population to grow at a natural rate and apply constraints to that growth as necessary. However, if there were an event early in the journey that wiped out a large number of the population, the remaining population would struggle with genetic diversity and ship maintenance and upkeep. In addition to genetic diversity, another limiting factor on a ship starting out with a smaller population size is knowledge transfer. With fewer individuals at the start, less knowledge will be available to share with the following generations. At the other end of the spectrum is to launch the worldship with the full population size of 100,000 individuals. There would be little issue with genetic diversity or knowledge transfer, but the population would be faced with maintaining population control from the onset of the journey.

No matter the launch population size, there will eventually be a convergence of population versus space. There are a number of methods to achieve the population control necessary at that point. One method includes a reproduction lottery in which a couple can enter for the chance to have a child when a member of the population passes away. Another method is to limit the number of offspring each couple can produce, which would also help maintain genetic diversity. Each of these limited methods would introduce ethical issues of their own, which must be considered in depth by the population as a whole.

A second key factor is the composition of the launch population. The age distribution of Generation Zero must be considered. First, the population can be made up of individuals with the same age distribution as on Earth, with a specific number of youth, middle-aged, and elderly individuals. This would be the best option in terms of knowledge transfer and range of knowledge. Second, the population can be composed of individuals from one age range, who then self-distribute as subsequent generations are born. In particular, one option is a Generation Zero comprised of the youngest individuals feasible, which would allow for less emphasis placed on reproduction and knowledge transfer at the start.

In addition to age distribution is genetic and cultural composition. Creating a homogenous Generation Zero would lead to a less confrontational society, but would also limit genetic diversity. To mitigate this, a larger population size is advantageous. A heterogeneous Generation Zero would ease the genetic diversity, no matter the population size, but would introduce a number of possible inter-group confrontations. Ultimately, the population selection criteria must be stringent enough to weed out potentially troublesome individuals and create a Generation Zero that best meets the needs of sending humanity to the stars.

3.2 Culture

Humans are fundamentally social beings, so culture is engrained in every part of life. Culture can be seen as the manifestation of human intellect, reflected by social habits, cuisine, values, art, music, language, and religion. A single person is a part of many cultures, from a local community culture to a professional culture. With the rapid interconnectedness of the 21st century, the lines between different cultures are blurring. Where immigration, travel, and social media have brought cultures into contact, these cultures are adapting and changing. Furthermore, as the world's population grows, so does its cultural diversity and interconnectedness.

However, despite the interconnectedness of the world there is still a large amount of disconnect and contention between the world's nations and religions. This will undoubtedly continue to be inherent onboard the worldship, and strife and contention must be dealt with without causing more friction between groups that may lead to growing gaps over multiple generations. A mission of this scale will require a vast change in the global society's outlook and operations. With the multitude of nations participating in the planning and construction of the worldship, there will be nations working together who have never before done so. This will cause a clashing of cultures. Culture is a very fluid concept and therefore is capable of change at a much quicker pace, potentially causing geo-political complications if not properly mitigated [12].

On Earth, there are many examples of people of different geographical cultures coming together in one place to live peacefully. As the society becomes more homogeneous, aspects from the different cultures "melt together" in what is often called a melting pot. For example, the United States has experienced many occurrences of large immigration, and over time, the various cultures start to change and share characteristics. Generation Zero on the worldship will have a wide range of different cultures, and over time, those cultures will melt together. In addition to geographic cultural differences are generational cultural differences. Older generations are generally more rigid in their methods and beliefs whereas younger generations adapt more quickly. If Generation Zero is composed of a wide age distribution, this will come into play as well. Various cultural factors will affect the continuity of cultures over time through cultural reproduction, which will be significant for the worldship.

The establishment of the Interstellar City prior to launch will start the process of cultural melting to mitigate cultural strife onboard the worldship. Here, individuals from around the world will come together and start the cultural melting process. This will push forward the collective thinking and cohesiveness necessary for the worldship. An example of this is Star City in Russia, where most cosmonauts and astronauts from all over the world train. In addition to the training equipment, this facility has its own high school, shops, post office, and much more. In the Interstellar City, people will not just receive their training; they will also build connections, get used to each other, and start to create their own culture.

Ultimately, the culture that develops onboard the worldship will be affected by everything leading up to launch. Culture is an important discussion point in population selection to determine what traditions, religions, ethics, and language are carried onto the worldship. Choosing which cultural identities are allowed and disallowed in the selection process is a point that must be handled with delicacy, and it must be understood that a combination of cultures will be brought onto the worldship and no one cultural identity is inherently superior to another.

3.3   Ethics

The ethical implications to be considered for the worldship are truly global issues of concern and will affect the mission before launch and continue throughout the mission. Ethics and morals are difficult to separate and are different for each individual. For the purposes herein, ethics are defined as concrete global, social rules for how to act [13]. It should be noted that the ethics onboard the worldship will vary from the earthly ethics, and the divide will grow as the mission continues. At this point in the process, some of the implications should be parsed out and discussion should be started, but no finalized decisions can be made.

First to be considered are the ethical implications of sending off a sizeable group of individuals on a mission that only their fairly distant descendants would complete. Not only would Generation Zero be subjected to the remainder of their life lived entirely on a worldship, the next couple of generations would never even have the opportunity to live life on a planet. Although volunteers will populate the first generation, their descendants will have no choice. However, all the descendants are inherently affected by the actions and decisions of their ancestors, so this ethical consideration is manageable.

Another possible ethical contention is what to do with the deceased onboard the worldship. Different cultures handle their deceased in very different ways. Onboard the worldship, the options could include a funeral, cremation, or jettison into space. For a generational worldship, the loss of nutrients and carbons following generations of deaths would not be maintainable, so the bodies of the deceased must be recycled into the ecosystem of the ship through a controlled method. This will not sit well with many of Generation Zero, but they must board the worldship knowing this unavoidable aspect of worldship life. Once the ship is launched it will be about the worldship community's global survival and certain things will need to be done that may seem inconceivable to society now but will be necessary in the new "Wild West" of the galaxy [14]. Further delving into this particular topic are the issues of whether to jettison the bodies of those that have passed away due to some contagion. Considerations such as this must be taken into account additionally to ensure the continuation of the mission and the health of the population.

These considerations bring into focus one of the most important questions for the mission: will Earth continue to support a mission that must carry out deeds necessary for the survival of the mission that are not considered ethical and humane by Earth's standards? An example may be if the population must euthanize an infected portion of itself to continue a healthy existence and not waste resources on members of the population that may in turn endanger the wellbeing of the population as a whole. Along those lines, a long-term prison would be a major strain on the worldship resources, so a death sentence may be a necessary punishment for a severe enough crime, such as a serial killer.

The essence of humanity cannot, at this point, be conclusively determined, but is at the center of all ethical considerations. The ethics of a population are intrinsically tied to the humanity of that population. Although, determining what makes us "human," we have the ability to reflect on that capacity of being human, and therefore will retain the essence of what makes us human, even if it is guided by a different set of guidelines [15].

3.4   Religion

There are few other thoughts and traditions that have persevered as long as religion, making it a topic of great importance with respect to the success of the worldship. Religion can be either a great asset to the mission or one of the most preventive factors from success.

Onboard the worldship, religion may have a positive or a negative influence. The religious options for Generation Zero must be discussed at a global scale before the initiation of the selection process. The results of this discussion will in turn be implemented into the selection criteria. In general, there are three foreseeable options to be discussed, each with its own advantages and disadvantages:

1. Maintain the demographic of religions as on Earth
2. Allow only one religion onboard
3. Allow no religious beliefs and have an atheistic population at launch

First, the current demographic of religions on Earth as they exist may be maintained. The major advantage is a simpler selection process with one less criterion and will enable more breadth in the makeup of Generation Zero. Additionally, the health and psychological benefits found through religious practices will be retained. The major disadvantage is continued religious strife and contention, as found on Earth. This can lead to division onboard the worldship, particularly between the religions.

Second, one religion may be chosen, likely one of the major world religions: Christianity, Islam, Hinduism, etc. Selecting one religion will decrease interfaith strife, if not eliminate it, and maintain the health and psychological benefits of religion. However, it will add to the selection criteria and narrow the possible pool of applicants.

Third, Generation Zero can consist of all atheists who claim no religious beliefs. This follows the lines of the second option of limited, and likely eliminated, interfaith strife and increased selection difficulty. However, both the second and third options will produce strife during the discussions on Earth and during the selection process. It can be assumed that each religion will abdicate to be selected. In addition, if either one religion or no religion is selected, there is a high probability that individuals that have a strong wish to be included in the worldship population will simply lie about their religious beliefs, complicating the selection process. These individuals will harbor their beliefs until they can pass them down to their descendants, therefore cause the interfaith strife at some point following launch or in a later generation.

Weighing the pros and cons of each option and having the foresight to examine the possible outcomes of each decision is imperative. Regardless of which option is chosen, it will have a direct impact on the population selection. Therefore, such a decision must be made before the start of that process and perhaps before any population selection. The world must come together on some platform to discuss the implications and choose the most advantageous path.

3.5 Language

Language is another major societal consideration. Each individual will bring their own language, which will in turn give each individual their own outlook on life. These different languages could serve as a detriment, yet should also be cherished. Similar to other psychological and cultural issues of in-group and out-group biases, the population could become divided and have difficulty communicating across multiple different languages, especially those that do not have an etiological link, such as English and Mandarin [16]. Therefore, a common language is necessary to aid in collaboration and cohesion throughout the population.

The chosen common language will depend upon the most influential languages of the time. A globally influential language must not only be widely spoken, but also the centrality and connectedness to other languages. Figure 3 depicts the current global language network with book translations as the major data source. Each node depicts a specific language with the size representing the number of speakers; the connections between the nodes represent the number of translations. If the common language for the worldship were to be selected now, the most obvious choices due to their popularity are English, Chinese, or Hindi. Of those three, English has the strongest centrality to the other global languages. This would lead to the selection of English as the common worldship language, similarly to the selection of English as the common language of the aviation industry.

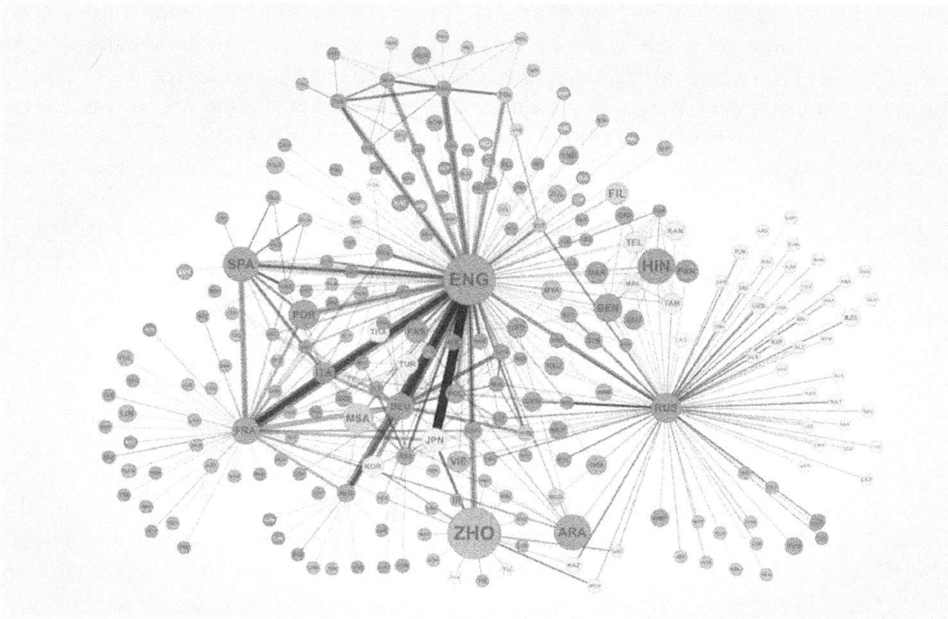

Figure 3: Global language network of book translations [17]

Language drift is another concern. Overall, languages are fairly structured, but the Linguistic Relativity Hypothesis dictates that one thing that will inevitably change will be vocabulary and therefore the population's outlook on the world [18]. For example, an individual from a tropical climate will not have a large vernacular for cold weather. Much of the earthly vernacular people use every day will be irrelevant onboard the worldship. With continued communication with Earth, the drift in language will not be large. However, as the communication delay grows, or if communication stops entirely, the worldship language will likely evolve into something incomprehensible to those speakers not a part of the worldship community. Ultimately, language is not only an important factor for the population selection, but also for the continuation of future communication of an interstellar humanity.

3.6   Education

Education is a key factor in producing competent and effective citizens, so the onboard education must be addressed. As time has progressed, humans have made a continuous record of knowledge in law, agriculture, philosophy, philology, natural history, politics, zoology, medicine, geography, theology, and departments of spiritual, intellectual, and social activity that lie in the range of human knowledge or speculation [19]. A record of human knowledge must be encapsulated in the worldship and accessible to the citizens as, in the words of Cicero, "Those who have no knowledge of what has gone before them must forever remain as children." In addition, an interstellar worldship will have special educational requirements to remain resilient and self-reliant. The educational system onboard must be developed to not only educate citizens about the past, but also prepare them to be effective citizens onboard. The limited population size means that there is a major risk of knowledge loss or skill shortage over time, which is a problem that educational methods must address.

As an interstellar mission would be carried out over multiple generations, many unexpected challenges are likely to arise. It is then argued that the focus of education should not just rely on teaching facts from history, but also on how to think logically to solve problems. Nebergall points out that human nature is accustomed to accepting and using technology without fully understanding the intricacies behind it [20]. As the population's survivability will rely on their capability to repair, recycle, and reuse the resources available, blind reliance on the technological systems of the worldship without a technical understanding could lead to mission failure.

More studies are essential in determining the best systems, configurations and groupings of early childhood educational institutes, primary and secondary schools, research centers, trade schools, laboratories, and universities. Summerford notes that diversity should be encouraged throughout the colony and used the example of a medium sized university on Earth, where, while each dormitory is different, certain halls are all known for something unique [21].

271

Requirements for technical, cultural, and practical aptitude dictate that a combination of classroom and hands-on settings for learning are needed. Students will need to be put through basic education that will teach them necessary life skills to survive and general knowledge about the world around them.

With the increased use of technology in the classrooms and integrated computer based learning, it is likely that education both onboard the worldship and on Earth will boost the usage of digitally based learning platforms. Additionally, virtual reality will likely be used to expand the learning environment beyond the worldship. This would help to decrease the risk of knowledge loss in vital areas that may otherwise be lost. One such example is an understanding of planetary geology. Unless samples are taken from Earth, the practical applications of this science could be lost and this may be vital to later stages of the worldship mission that require planetary manipulations and identifications. Virtual reality areas could help people onboard stay connected with Earth-based information and environments, as well as expose them to virtual scenarios and environments for the purpose of stimulating creative thinking and adaptive skills. The rate of information exchange and interconnectedness is unlikely to slow down; therefore information accessibility should not be a problem for worldship citizens. Instead, the challenge will be to make useful information available and presented in a way that is effective and interactive for learning purposes.

As they grow older, students will be assigned apprenticeships to learn from leaders who have mastered specific skills vital to worldship operation. Given the availability of and need for certain skill specializations, students would take part in apprenticeships geared towards their natural abilities and interests. The specializations will depend on the worldship's current needs in that particular timeframe. Furthermore, after some predetermined time period of work, individuals may have the option to complete another short apprenticeship and enter into a new profession.

Education is a vital part of the success of a worldship mission. The role and responsibility of education on a worldship will be a community-based effort to relay technical knowledge with craftsmanship, ethical concerns, and teamwork. It will act as a mechanism through which the original Earth culture and intent is preserved. Furthermore, education will play a role of catalyst in conveying knowledge transfer to deal with issues that people may not have dealt with before. Various constraints will still exist but people must be able to adapt to the changes in the new environment.

4. Conclusion

Humanity's destiny and vision has always lain amongst the stars. While early visionaries such as Verne, Goddard, and Tsiolkovsky dreamt and wrote about human travel into interstellar space, current organizations now work to make this a reality. This paper has made an attempt at contributing to this effort by outlining many of the societal challenges to be faced. The societal aspects from the planning and construction stages of the worldship to the actual flight itself have been discussed.

During the road to launch, some aspects include governance, financing, and outreach. Developing a governance system before launch is important. To do this, an Interstellar City can be established. This city will not only be useful to develop a governance system, but also to aid financing efforts, lessen cultural conflict, and begin education and training of future interstellar citizens. To finance the worldship, the primary method will be the establishment of an International Interstellar Fund to gather funds from the earthly nations with additional funding from incentive financing, crowd sourcing, and leveraged finance. Third, a majority of humanity must be invested in this grand project for it to be a success. To do this, an ambitious, multifaceted outreach and education system must be enacted.

There are many layers to society, and special considerations must be taken to each layer of the worldship society. Many of the societal aspects have different options that must be decided closer to launch by the whole global community. First, the starting population size and make-up of Generation Zero must be determined. The starting population may be the maximum of 100,000 or the minimum of 10,000. Additionally, the population may be homogeneous or heterogeneous with respect to age. The individuals that make up this population will come from a diversity of backgrounds, so cultural differences must be taken into account to ensure the smoothest transition possible into the new worldship society. The ethics onboard the worldship will be vastly different from the ethics found on Earth, from death sentences to burials. With respect to this, Earth may have a difficult time continuing to support such a possibly unethical mission. Differences in religion may also have an effect on the global community's support. Before the population can be selected, the religious backgrounds accepted must be considered, whether the population has the religious demographic found on Earth, one religion, or no religion. Each individual of the population, whatever their religion and culture, will not share the same first language. Therefore, a common language must be selected to ensure cohesiveness within the population, but the various languages brought onboard must also be cherished to keep the future generations onboard the worldship connected with their earthly ancestry. Finally, the educational system onboard the worldship should be outlined before launch, and it can be enacted and tested in the Interstellar City. All of these facets and more make up the considerations that must be taken into account for the worldship society. These considerations must be made before the launch of the worldship and are key to the success of the mission.

This paper covers just a small part of the effort to critically evaluate literature in order to create an interdisciplinary roadmap for the development of a slower-than-light, multigenerational worldship for interstellar travel. The worldship concept is one of a self-sustaining ship capable of transporting thousands of humans into interstellar space over many generations. Parameters included non-relativistic speeds and a population of 100,000 individuals living their lives. The full work was created by the Astra Planeta team, composed of twenty-two students from the International Space University's (ISU) 2015 Masters of Science in Space Studies class [22].

Acknowledgements

The author would like to acknowledge the entire Astra Planeta team whose efforts brought this project to fruition:
- Kyle Acierno
- James Bevington
- Shambo Bhattacharjee
- Chaitra
- Guang Chen
- Daphne De Jong
- Lei Geng
- Avishek Ghosh
- Karan Gujarati
- Micah Klettke
- Florin-Cristian Lazar
- Olivier Leblanc
- Yang Liu
- Hameed Monoharan
- Hamza Ragala
- Brian Ramos
- Jean-François Rococo
- Fergus Russell-Conway
- Mansoor Shar
- Abigail Sherriff
- Anderson Wilder
- Peng Zuo

Additionally, the author would like to thank Chris Welch, the advisor for the team and project, and the International Space University for the support throughout the development and completion of the project.

References

1. Astra Planeta, 201Astra Planeta Final Report. [online] Available from: <https://isulibrary.isunet.edu/opac/doc_num.php?explnum_id=731>

2. Astra Planeta, 201Astra Planeta Final Report. [online] Available from: https://isulibrary.isunet.edu/opac/doc_num.php?explnum_id=731

3. Hill, T., 200Space: What Now? The Past, Present, and Possible Futures of Activities in Space. Baltimore: PublishAmerica, LLLP, pp. 4-22

4. Hansen, M.H., 200A comparative study of thirty city-state cultures: An investigation. Kgl. Danske Videnskabernes Selskab.

5. Zak, A., 201The real rocket to nowhere. [online] Available from: <http://www.russianspaceweb.com/vostochny_soyuz.html>

6. European Space Agency, 201ESA, An intergovernmental customer. [http://www.esa.int/] European Space Agency. Available from: <http://www.esa.int/About_Us/Business_with_ESA/Business_Opportunities/ESA_an_intergovernmental_customer>

7. Tsagourias, N., 200Transnational Constitutionalism: International and European Perspectives. Cambridge University Press.

8. Johnson, R.D. and Holbrow, C., 197Space Settlements. A Design Study. NASA, Washington, SP-41[online] Available from: <http://www.unbc.ca/assets/history/courses/201101_nasa_sp_413.space.settlements.a.design.study.pdf>

9. "Raymond Orteig - $25,000 Prize." Charles Lindbergh - An American Aviator. Spirit of St. Louis 2 Project, 201[online] Available at: <http://www.charleslindbergh.com/plane/orteig.asp>

10. Martin, A., 19World Ships-Concept, Cause, Cost, Construction and Colonisation. Journal of the British Interplanetary Society, 37, p.243.

11. Nebergall, K., 20Becoming an Interstellar Culture. 100 Year Starship 2012 Symposium Conference. Houston, TX: CreateSpace Independent Publishing Platform, pp.291–299.

12. Wooldridge, B., 20Preparing an Interstellar Civilization for the 100 Year Starship: Don't Hold Your Breath! 100 Year Starship 2012 Symposium Conference. CreateSpace Independent Publishing Platform.

13. Arnould, J., February 20Religious & Ethical Questions. Internal Communication.

14. Arnould, J., February 20Religious & Ethical Questions. Internal Communication.

15. Arnould, J., February 20Religious & Ethical Questions. Internal Communication.

16. Kondo, Y., Bruhweiler, F.C., Moore, J. and Sheffield, C. eds., 20Interstellar Travel & Multi-Generational Space Ships: Apogee Books Space Series Burlington, Ontario: Apogee Books.

17. "Global Language Network." Global Language Network. MIT Media Lab Macro Connections Group, 20[online] Available at: http://language.media.mit.edu/visualizations/books

18. Swoyer, C., 20The Linguistic Relativity Hypothesis. [online] Stanford Encyclopedia of Philosophy. Available from: http://plato.stanford.edu/entries/relativism/supplement2.html

19. Kern, A., 20Classical Education In Ancient Greece.

20. Nebergall, K., 20Becoming an Interstellar Culture. 100 Year Starship 2012 Symposium Conference. Houston, TX: CreateSpace Independent Publishing Platform, pp.291–299.

21. Summerford, S., 20Colonized interstellar Vessel: Conceptual Master Planning. 100 Year Starship 2013 Public Symposium Conference. The Dorothy Jemison Foundation for Excellence.

22. Astra Planeta, 20Astra Planeta Final Report. [online] Available from: <https://isulibrary.isunet.edu/opac/doc_num.php?explnum_id=731>

100 YEAR STARSHIP™

Data, Communications, and Information Technology

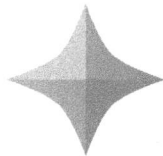

Data, Communications, and Information Technology

Chaired by Ron Cole

Colorado, USA

Track Description

Whether enroute or having arrived at a destination, robust capabilities to gather, analyze, compile, store, retrieve, transmit and receive information is essential to any deep space journey.

Information, communication and data transmission capacity are constrained by vast distances, signal degradation, energy availability, bandwidth, data management, time delays, direction and pointing accuracy, instrumentation as well as existing earth based and deep space networks.

Presenters in Data Communications and Information Technology are asked to present research, concepts and systems that facilitate the process of finding earth analogues and that are actionable within the next 5, 10, and 15 years. Areas for discussion include ground and earth orbiting equipment and systems; advances in artificial intelligence (AI), as well as solar and extra-solar system networks and capabilities. In addition discussion of software, hardware, syntax and design techniques that aid in or result from the discovery of Earth 2.0 are welcome.

Track Summary

Sending and receiving information by interstellar travelers or robotic vehicles requires development of new methods to traverse the vast emptiness between stars. Additionally, in the absence of routine and timely communication with Earth, a probe or traveler must be self-sufficient in gathering, generating, compiling, storing, analyzing and retrieving data while ensuring these systems are operational over the lifetime of the mission and beyond. For this year's symposium, we had four excellent presentations that address the concerns of this technical panel. Following are gist of those presentations and email addresses of the authors for your follow-up for more details.

The First 72 Hours: Communications in Crisis

by Jaym Gates (jaym.gates@gmail.com)

Last year, my presentation focused on the overall needs and problems in communications during a crisis or disaster. I used an event, the massive King Fire to illustrate the uses of social media and community empowerment to mitigate the effects of a disaster. I then used that to demonstrate some of the communications challenges facing us in space, and some of the basic tools and education space programs will need to address.

My topic this year is focused on a much smaller scale. Communications are the nervous system that holds any massive project together; however, it is also often taken for granted, or downright disregarded. Lack of communications in peril is all too common; the more dangerous the situation, the higher our risk of not making it out, the more we need our last words to be heard.

In September, a convoy of trucks, stock trailers, and firefighters was attempting to bring a load of livestock and pets out of one of the towns in imminent danger from the Butte Fire when they had a terrifying brush with a horrible end. "We're stuck on Highline Road," one of the drivers posted on Facebook (FB). "The fire jumped the road ahead of us, and we don't know if we'll make it through before the fire overtakes us. "They made it out, thanks to the firefighters who were with them, and immediately updated the FB status. These weren't social media-obsessed millennials, just ranch wives and retirees in crisis.

Last year, I presented the need for the development and empowerment of community leaders and crisis-management training. But those leaders and systems will be useless if we don't pay attention to the techniques they use to build and manage their communities.

We have more methods of communicating with each other than ever, and yet less awareness of the language we use. Critical reading and discussion is seldom taught in schools. The ability to speak and write clearly is rare, and poorly understood, to say nothing of critical listening.

The science of communication should be studied, but so should its very practical application, especially if we ever hope to send settlers and civilian parties of any sort off world. Even the relatively short trip to Mars will require social management and tact. Community managers, discussion moderators, writers, and even editors need to be brought into the loop, to help develop the tools and resources necessary to maintain a calm and productive environment in general, and to manage and direct communication in event of trouble.

Yet another thing that is all too often disregarded: women frequently form the communications, response, and recovery load of a community crisis. Women are taught early on to communicate, and we rely on it to ensure our most basic safeties. The women of Katrina, Sandy, and Syria, and other traumatized areas may be one of our most valuable resources.

In a massive crisis or disaster on Earth, the first 72 hours are critical. That is generally the length of time it takes to mobilize any large-scale response, and the time in which everything is in the most upheaval.

In space, resources may not be available from outside. Rescue might not be able to get there fast enough. Communications resources may vanish during a crisis. Those on board whatever vessels we use will be utterly responsible for their own survival. The emotional stress may be exponentially greater than what we are used to in high-risk Earth environments, even in times of non-crisis.

Additionally, the people on board a large-scale off-world mission will likely be of incredibly diverse backgrounds, mixing, among others, scientific, technological, military, and social communication styles, many religions and cultures, and language differences. Finding a way to foster and support clarity on a basic level between the groups in a steady, reliable environment will be key to effective communications in crisis. Because in the end, we're all alone, and nobody is coming to save us.

Recent Progress in Interstellar LINKS Exploiting the FOCAL Space Mission

By Dr. Claudio Maccone: clmaccon@libero.it
Major reference: Maccone, C., "Mathematical SETI", a 724-pages, published by Praxis-Springer in 2012. ISBN-10: 3642274366 ISBN-13: 978-3642274367 Edition: 2012. See, in particular, Chapters 12 through 16.

In this paper we describe some recent lessons learned regarding the Sun Gravity Lens:
1. First of all its use for keeping the telecommunication link between the Earth and any future truly Interstellar Mission by exploiting the Sun as a huge radio antenna, the ONLY antenna capable of assuring robust radio telecommunications across interstellar distances up to 10 light years away or more.
2. Another use of the Sun Gravity Lens in the optical frequencies is to visualize hugely magnified picture of extrasolar planets lying on the other side of the Sun with respect to the spacecraft position. This is a rather new and recent field of mathematical modelling

Sun Gravitational Lens: The geometry of the Sun gravitational lens is easily described: incoming electromagnetic waves (arriving, for instance, from the center of the Galaxy) pass outside the Sun and pass within a certain distance r of its center. Then a basic result following from General Relativity shows that the corresponding deflection angle (r) at the distance r from the Sun center is given by (Einstein, 1915): $\alpha(r) = 4GM_{Sun} / c^2 r$

For any spherical celestial object (namely both the Sun and all the Planets) the FOCAL GRAVITATIONAL SPHERE has a radius from the Sun center given by:

$$d_{focal\_object} \approx c^2/4G \times r^2_{Object} / M_{Object}$$

Replacing into this formula the MASS and RADIUS of each planet, and sorting the results out in increasing order, one can create a table of the radiuses of the focal spheres of all the planets. (See the reference for details of this table).

The "gravity lens" concept means that the Sun and any other massive celestial body is an antenna since they can increase the intensity of the signal, by virtue of its deflection. The Gain associated to any celestial body is the ratio between the intensity of the signal in presence of the celestial body compared to the intensity of the signal without the celestial body. The gain is constant along the focal axis but is wavelength-dependent.

Radio Bridge Between Any Two Stars: Electromagnetic waves path are deflected by the gravity field of each star, and made to focus on the two FOCAL space crafts placed on opposite sides with respect to the two stars. These two FOCAL space crafts must be strictly positioned along the axis passing through the two stars' centers. Then, the transmission powers between the #1 FOCAL spacecraft and the #2 FOCAL spacecraft are greatly reduced. In fact, the huge (antenna) gains of the two stars, plus the modest (antenna) gains of the two FOCAL space crafts combine to yield a huge total (antenna) gain of the whole system, meaning that the transmissions powers between the two stellar systems become quite affordable. This is the key to build a GALACTIC INTERNET, that might already have been created in the Galaxy by Aliens capable of sending FOCAL space crafts around. But we, Humans, still are cut off from all that since we have not yet sent any FOCAL space craft 550 AU away from the Sun, namely we have not yet learned about how to use the Sun as a gravitational magnifying lens.

Conclusion: This author believes that a radio link still remains the best choice, that could work, however, ONLY if the gravitational lens of the Sun is used as an amplification tool. We thus suggest that:

A first FOCAL 1 spacecraft is launched away from the Sun up to 1000 AU in the direction of the sky opposite to Alpha Cen B. This spacecraft will then act as a relay satellite to insure robust telecommunications across the 4.37 light year distance between the Sun and Alpha Cen B.

A second FOCAL 2 spacecraft is launched towards Alpha Cen B as the truly interstellar spacecraft going to Alpha Centauri B. It may take quite a few years to get there, of course, (50 years at least?), depending on the propulsion system, but its robust telecommunications with the Earth are insured by the gravitational lens of the Sun and by the FOCAL 1 relay spacecraft. After FOCAL 2 reached Alpha Cen B, we envisage positioning it in the direction opposite to the Sun beyond the minimal focal distance of 452 AU from Alpha Cen B. Thus, the Sun and Alpha Cen B, plus the two FOCAL spacecrafts 1 and 2 on their opposite sides, will make up a radio bridge insuring cheap (i.e. modest in power) and robust telecommunications for all future space missions to Alpha Cen to come.

In conclusion, this paper goes beyond traditional papers only, all concerned with propulsion for the first interstellar mission, in that it pointed out that a robust telecommunication system has first of all to be created by exploiting both stars' gravitational lenses plus two FOCAL space crafts on opposite sides: SUN FOCUS COMES FIRST. INTERSTELLAR COMES SECOND.

Managing Technical Debt in Deep Space
by Jesse Warden jesse.warden@gmail.com
There are 3 challenges writing software for interstellar space travel:
- We don't know what needs to be changed before the need arises;
- those who can change it might not be on the flight; and,
- no deep space network exists to upload new software changes to an in-flight craft.

How do we solve these problems? By both managing technical debt (software written to get something working "well enough" vs. good, scalable quality in space) cover managing technical debt through various strategies for self-writing software and developing deep space networks.

To do that, we must leverage the two opportunities before us that will help enable the technology needed: Consumer space technology and continued advances in quantum computing.

The Internet of Things, cheap hardware, and shrinking size of that hardware has allowed consumers to build their own satellites. They do this by leveraging existing, open source software as well as creating new software libraries & technologies platforms. The open source space software ecosystem is huge, accessible, and growing. Middle schoolers are sending software they wrote into space on commodity hardware and contributing back to the internet lessons learned.

Various forms of Artificial Intelligence have been used for years to solve large data problems. The biggest problem of all, navigating the unknowns of space, will need more powerful AI. They'll not only update our navigational code, in seconds, but they'll make it better. How that process happens is the utmost importance for the mission, and for the continued existence of the human race.

Communications' Requirements on a Generational Starship

By Alexander James Sweetman asweetma@msudenver.edu.

This research focuses on the communication needs and social health aspects of a community inhabiting a mass-locked generational starship in conditions that may not allow for media up-link from Earth.

Comparing the effective transfer of data of numerous future technologies and the projected population diversity requirements of a generational starship, recommendations can be formed for the transmission of select media, or the possibility of a self-sustaining media on board. Using calculations of possible data up-link based on current, near-future, and hypothetical means of communication to a long distance starship, a hard boundary can be reached on up-link bandwidth. Recommendations on media transmission, or the requirement of self-propagating media channels, will be based on these calculations. Current and past analog scenarios, such as polar research expeditions, or cases of "cabin fever," will shed light on human need for media and communication, and will further inform communication requirements, media modalities and frequency recommendations.

Track Chair Biography

Ron Cole

Ron finished a 50-year career with the US Intelligence Community in May 2013. Ron received the Civilian Defense Meritorious Award upon leaving the NSA to begin working with Scitor Corporation as a systems engineer technical advisor to the NSA and the National Reconnaissance Office (NRO) on system development and data processing. Ron left Scitor to go to work for Riverside Research as a senior advisor for five years to the National Geospatial-Intelligence Agency on technology developments for mission execution and then moved to support the NRO on policy and management issues.

The First 72 Hours: Crisis Communications on the Edge

Jaym Gates

Editor, War Stories; Communications Specialist

1. Introduction

In 2014, I explored the potential risks and weak points of the communications needed to deal with space-based disasters, drawing on well-known Earth disasters such as Katrina, Sandy, the King Fire, and even the Challenger. I discussed the need for off-world efforts to understand the role of social media, popular opinion, and fraud in these hypothetical disasters. From there, I explored the possibility of using those future needs to develop a much stronger crisis communications science here on Earth, as well as empowering and training communities to lessen their reliance on federal and nonprofit disaster-relief efforts in wide-scale emergencies.

We have seen demonstrated, repeatedly, breakdowns between the many organizations and interests in disaster response, within one government, much less when governments begin cooperating. Command centers unable to communicate with each other because of outdated, incompatible technology, administrators and politicians struggling for power, commercial and humanitarian interests colliding head-on with each other and military/government efforts.

Few of these operations take into account the established communications systems within the communities they are striving to help, or the role those communities might play in alleviating disaster.

The communities bearing the brunt of disaster seldom know how to cooperate fully with outside efforts.

In September, 2015, two major fires, the Butte and the Valley, broke out within days of each other in Northern California. The state was already stretched thin on resources due to other fires, as well as supporting the efforts to suppress the massive Chelan Complex Fire in Washington, and general exhaustion at the end of an unusually long and brutal fire season. Two major fires in populated areas stretched those thing resources to the breaking point.

In any disaster, the first 72 hours are essential. It takes time to mobilize any large-scale response, much less to implement it, and the remote nature of the Butte Fire meant that simple travel allotments took nearly half of that time. Too, the fire was deceptive, and seemed to be under control. Resources were released, and attention waned. People on the verge of evacuating stood down and settled back in.

When the fire exploded into a fast-moving disaster, bearing directly toward a cluster of isolated towns, it took the region off-guard. Towns were evacuated to shelters, only to have those shelters threatened by the fire. Most of the local resources were used up in trying to help people evacuate. Stock and pets had to be left behind, a devastating blow for a county that relies primarily on ranching and agriculture.

However, freshly off of the King and Sand Fires of the previous year—both of which had occurred in the neighboring county—the communities were able to adapt quickly and use social media to create a crisis response plan. Many residents from the neighboring county had also taken the previous year's lessons to heart and spent the time to certify themselves with the local authorities for disaster response. Resources descended on the small towns of Amador and Calaveras Counties and began coordinating with the local authorities.

Almost none of these resources were official. Most were farmers, ranchers, retirees, and business people. The local contractors hauled in their bulldozers and began cutting firebreaks. Horse owners brought their trucks and trailers and set up schedules with the firefighters, beginning evacuation efforts. Those not able to go into the active zones organized extensive food and supply donation campaigns. The local casino opened its doors and kitchens to refugees, and sent its employees out to help evacuate those who couldn't make it out on their own.

By the time the first resources began showing up from Southern California and Nevada, a massive, effective system was already in place, reducing a potentially lethal situation for a community already on the edge to oen that is survivable.

Now, months later, they are still functioning, helping restore homes, coordinating donations from hay brokers and feed stores for ranchers who lost the means to make it through the winter, and sheltering homeless families until they can figure out how to move forward.

Most of this work was done through social media, primarily Twitter and FB. There were the inevitable miscommunications and overlaps, but overall, it brought thousands of people together in a desperate environment and mitigated an untold amount of loss and destruction.

That all began to disintegrate when the Red Cross and other relief organizations arrived. Turf wars and tensions drove away successful volunteers by the score, leading to a sharp breakdown between what was already happening, and what needed to happen going forward. Resentment built, and vital resources were diverted to deal with the tension. Scheduling and communication issues routed duplicate resources to issues, and the smooth system began floundering.

These are the sorts of situations that we should be studying in great detail if we want to understand how to prepare for the first off-world disaster. Not catastrophic failure, where nothing can be done, but smaller disasters that leave people on their own, struggling to survive. We need to understand how communications systems have evolved, and how to prepare ahead of time for the inevitable.

Whether that is on a generation ship, a Mars colony, or a space elevator doesn't matter, they'll be waiting a lot longer than 72 hours, and they'll have a fraction of the resources any on-Earth community could rely on.

Early preparation is key, but before we can prepare, we need to understand the bigger picture…and the elements that comprise that picture. Both the forest and the trees, as it were.

2. Institutional Communication

Social media has been one of the most commonly-used tools for 'popular' public relations campaigns for big brands and celebrities. Recently, as social media has broadened its user base from 'super users' to everyone from grandparents to high-powered politicians, those efforts have been met more and more frequently with impressive backlash.

Twitter Q&A sessions, in particular, are bad news for anyone without an iron-clad reputation and public image. JP Morgan was one of the few PR departments that learned very quickly from its mistake, pulling their scheduled Twitter chat the day before it was scheduled, but many high-profile politicians and CEOs don't learn until too late, as REI's CEO learned through a disastrous Reddit AMA (Ask Me Anything) that highlighted questionable performance metrics and tanked the goodwill earned through their Black Friday anti-consumerism initiative.

Others are learning the necessity of having 24-hour social media management. British Airways watched its public support drop after a disgruntled passenger bought promoted positions to air his grievances. It was hours before British Airways caught the problem and began trying to smooth ruffled feathers, but the damage had already been done.

But the biggest, most egregious and indefensible social media disasters have been in the wake of tragedy. The ability to automate tweets and posts is a social media manager's god-send, until a disaster and a promoted tweet meet head-on. The occasional bleak irony does result, such as when the NRA accidentally asked shooters what their weekend plans were, mere hours after the Aurora shooting, while others are merely blatant, unresearched cash-ins, like the Celeb Boutique's co-opt of the Aurora hashtag to promote a Kim Kardashian-inspired dress. Others are even more blatant: Epicurious used the Boston Bombing tragedy to market its products.

These aren't small companies, but they're safe in one respect: they've done something idiotic, but it didn't involve the direct loss of human life.

Now imagine if a commercial space company makes an error that leads to the loss of human life, or costs tax-payers billions? It's no big deal, it happens regularly, right? Here's the other thing protecting the companies making these mistakes though: they're in the background. We're used to making small moral sacrifices to get our jeans and oil, and the people effected are very seldom of much importance to the average person.

Another example of this is highlighted by the recent Paris terrorist attacks, and the coverage those received, versus massive Boko Haram and ISIL massacres of African or Middle Eastern victims. There are acceptable losses, as long as they aren't visible. As long as they aren't possibly someone we know.

The first fifty years of space exploration, at least, will be one of the most visible projects in human history. The massive attention to the Mars Rover campaign, the celebrity of certain ISS astronauts, these are foreshadowings of what we will need to be aware of when we begin to send humans out of atmosphere on any regular basis.

Imagine what social media would have looked like if it had been around during the Moon Mission. America was already, as a nation, obsessed with every detail of the mission, when such coverage was minimal, and there was no opportunity to engage outside of the immediate community. Negative feedback certainly existed, but it had little power to touch the people responsible for the mission. The media hounded every possible detail, but the transparency then was a joke compared to what is required now.

Now, the more popular something is, the more necessary it is, the more of a negative backlash develops against it. In and of itself, a negative backlash can be mitigated in the public opinion if there is already a strong crisis-response plan in place and a sharp PR team to implement it. The bigger danger is that for an effort as controversial as off-world exploration and development, any public outcry will be used by opponents to try to cut funding, cut support, and demonstrate to easily-bought politicians that there's not really any support for off-world ventures.

The scenarios above are predicated on a series of missions going perfectly smoothly, which is unlikely. SpaceX has been an excellent example of how damaging even a single disaster can be, even if it doesn't involve a loss of life. The media was full of headlines like "Explosion! SpaceX CRS-7 Mission Ends In Disaster!", "Falcon 9 Disaster Messes Up SpaceX's Launch Schedule", and "SpaceX Launch Fails: How Safe Is Rocket Travel?".

When NASA's Antares exploded, the images saturated social media. They were saved, oddly enough, by the fact that the photos were spectacular, leading to a somewhat more friendly social media response. (The lesson here, PR managers, is that if you have to have a disaster, make sure the photos are beautiful.)

It was only a shuttle, taking supplies to the ISS, but it spread quickly and made people question the safety of space travel. Add at least one human casualty and the higher-visibility effort of sending someone out of atmosphere again, and the blow to public support and opinion could be catastrophic.

NASA and SpaceX, fortunately, have handled these set-backs with significant skill and aplomb. The major players in the space race, so far, know their public, and know how to keep the shareholders, politicians, and general fans at least somewhat mollified.

But to really get the new space race going, it isn't going to just be the major players. The field is widening all the time, and with NASA's recent announcement that they're moving their focus to goals farther out of the atmosphere, the commercial interests will begin to bring a new element of risk into the game of public and political opinion.

We place high value on the lives of those most visible to us, and we can see from our past how much celebrity will likely be attached to the explorers and risk-takers of the future. That notoriety and attention could be the best thing for the future of the off-world programs, and it could be the worst.

It only takes one death for all the optimisim, all the hope, and all the glee to go away, issues that will intensify by orders of magnitude if we find a way to send large numbers of humans off-world. Proper handling can mitigate the long-term effect of such a tragedy, but only with proper planning, proper support, and an understanding of the power of social media and social communication.

3. Social Communication

With the advent of the internet, humans found a new and delightful freedom of communication. Chat rooms and private groups formed almost immediately, places for people to talk, gather, and be heard. A new village waterhole began to develop, one divided sharply along the lines of interest, rather than location or profession. It was an excellent development, but one that would highlight just how poorly humans communicate.

The internet's history has been plagued with outbursts, illegal endeavors, and security issues. In many ways it is still the wild west, a frontier we're still discovering the rules and boundaries of, while realizing how very dependent we are on its services.

We are conditioned now to expect immediate information and connection. A brief service or power outage can wreck havoc on the day-to-day function on an individual. A simple city-wide power outage can snarl essential services and cost lives. Customer service must be immediate, response time measured in minutes, even for the most non-essential problems.

As we venture off the Earth's surface, however, we may not have these immediacies. We will need to understand that we are indeed alone, without a network of support. The farther from Earth we go, the more we will need to have ways for those adventurers to communicate with the ones left behind. It will be one thing for the scientists and soldiers, but if the goal is to develop off-world colonies, ordinary people will need to be able to cope with long stretches of isolation.

They will also need to be able to communicate efficiently in an environment that will amplify every word and miscommunication. Off-world projects will be intensely claustrophobic, creating a pressure-cooker of potential miscommunication, resentments, and delays.

Social media has demonstrated, repeatedly, that it is a powerful tool for communication…and for misunderstanding.

As mentioned above, it is highly likely that the first people out of orbit will be celebrities in their own right. They will be under intense scrutiny, and will need to have an understanding of how social media and communications work.

Unfortunately, we are doing little to educate people in proper communications from an early age. Knowing how to be clear and concise in any length is as important as knowing how to address a letter, but these are not skills being taught to kids, or even teenagers and young adults. Many will pick it up from their peers and immersion in the environment, but as seen from recent events, that may not be enough even for daily life, much less a closed environment like a generation ship.

The use of the internet to coordinate hate speech, bullying, and large-scale harassment is being widely documented at this time. Terrorists are using social media to recruit and organize. Many people are stepping up to find ways to combat such targeting, but little incentive exists to stop it before it starts.

Riot Games recently ran a study on the rampant harassment within their popular League of Legends MMO (mass multiplayer online). Instead of simply banning players who harassed, they clearly spelled out the issue that had caused the complaint and action. Most players immediately apologized, many stating that they hadn't thought their words through, or didn't understand the impact their words had. Since this study, Riot says it has lifted over 280,000 'Offensive' gamer statuses and has put into place community groups to monitor and educate gamers who offend their fellow players.

But let's go back to the most important statement above: "they hadn't thought their words through, or didn't understand the impact of their words".

More than ever, the world relies on words to function. Many of us maintain our relationships online, communicate with clients and managers online, fall in love online, play and shop and plan online. As we move farther from the Earth's surface, the necessity of understanding how words work will expand exponentially. It's one thing to order the wrong part when replacing it will only take a couple of days.

To ensure that we will be able to support life in the long run in space, we will need to know how to talk to each other, to our lifelines on Earth, and to the people hoping we fail.

4. Conclusion

It is easy to discount words. We use them every day. They are free, and so ubiquitous that we tend to disregard or forget their power. As critical debate and thinking training fades from schools, we need to begin addressing how the next few generations will learn to communicate efficiently, clearly, and in a way that does not cause more problems than silence.

This will need to include social media training, higher-level social development, and conflict resolution. Understanding why we need these skills in everyday life will develop a robust communications system for our military, our social situations, and our explorers. We will be able to prepare for communication breakdowns, misunderstandings, and the stress and strain of off-world isolation and pressure.

We use words every day. Now we need to learn to use them good.

100 YEAR STARSHIP™

Designing for Interstellar

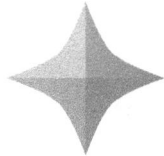

Chaired by Karl Aspelund, PhD

Assistant Professor, University of Rhode Island

Visiting Assistant Professor, University of Iceland

Track Description

Presenters in Designing for Interstellar are asked to consider parameters for designing probes, and vehicles or habitats for robotic or human crews –Earth, Earth orbit and deep space based—that will actively accelerate finding an analogue Earth and which may be implemented within the next 5, 10, 15 and 25 years. What aspects of design will be impacted by the various methodologies instituted to discover a planet outside of our solar system capable of supporting terran life? What are the design parameters that should be met to optimize the chances and rapidity in which such a planet may be identified?

Designs for probes and crewed vehicles must address the unique characteristics and extreme conditions of isolated research bases, deep space and interstellar space. The equipment, structures, tools, materials, cleaning and maintenance processes—the accoutrements of life and work— surround and create an operating environment or habitat. Such an environment protects, nourishes and facilitates daily activities. For living things, the environment must support the myriad physical needs. For higher order creatures, physical, mental and emotional requirements must be met as well.

Understanding, optimizing and manufacturing design for sustainability are critical for success—with a living crew or robotic probes.

Track Summary

The design track this year was a multifaceted, interdisciplinary affair, with a welcome turn toward a more specific and pragmatic approach to design problems –even when veering toward the theoretical and philosophical. The five presentations were each followed by excellent question-and-answer sessions, driven not in the least by the energetic and enthusiastic participation of students in the audience. (The increased numbers of students in attendance is a very welcome development, which should be encouraged in future symposia.)

Five presenters (including the chair) delivered presentations on design problems ranging on the psychological implications of design decisions, the systematic methodology for interstellar design, the actual conceptualizing of interstellar missions, the implications of current terrestrial best practices in ecologically sound architecture, and the future of the design process itself as human culture moves off Earth. The breadth and diversity of these topics is in itself testimony to the need to explore our approach to space-based communities from not only the practical points of designing, but also the cultural and experiential.

Hank Hine (The Salvador Dali Museum, USA) and Charles Hine (Hine Inc. Consulting) began the session with "Solid Ground in Interstellar Space: Mission and Metaphor." The presentation wove together observations from a variety of fields to show a beneficial use of commonly understood historical and natural metaphors and symbolic forms in the designing of physical environments and habitats in space. They showed how such an approach could psychologically ground interstellar crews in an aesthetic understanding that would allow their hu-

291

man experience, and their place in relation to the Universe, to be bridged from Earth to the ultimate destination. Here was a much needed reminder that even as we are in awe of our inventions and technological capabilities, that the language of human experience is rooted deeply and often expressed without words.

Christopher Andrew Corner (Steersman Technology Ltd, UK) followed with "Advanced Design Methodology for Deep Space Systems." The design of complex vehicles for interstellar travel may rely on unproven or undefined technologies in addition to more technically mature areas. This presentation demonstrated a methodology and structure that allows advanced system designs to be captured and advanced in a coherent manner. Risk areas, and areas where further research and development are required, can also be easily identified. Designs are captured and considered at three discrete levels. This allows the dependencies within and between layers to be captured, and then further –as the design matures— allows the cascade effects of changes to be analyzed across the design: The *Conceptual Layer* captures high level concepts including likely processes and requirements in areas such as Propulsion Systems and Spacecraft Electronics. The *Logical Layer* is described in a qualitative way, and captures initial logical designs of physical and software systems and the processes that involve their use through a formalization of the Conceptual Layer. Then, the *Implementation Layer* contains quantitative data, and captures the detailed design that realizes the Logical Layer.

Pete Swan (International Space Elevator Consortium) then presented his and Bruce Mackenzie's (Mars Foundation, USA) conceptual model of "Leapfrogging World Ships." Conceptualizing on a grand scale, they showed the benefits of designing missions on the basis of convoy ships that would, rather than aim for a specific destination, drop off "seed ships" en route to an ultimate destination of a distant star. This would establish a culture of continuous expansion and create a path for other, faster, travelers to follow. The simple, but elegant concept, prompted much discussion, questioning, and speculation on the nature and purpose of interstellar travel and the large-scale technical hurdles involved.

T   om Hootman (MKK Consulting Engineers, USA) then brought us back to Earth and presented: "We Are All Astronauts: Net Zero Design on Earth and Beyond." He showed examples of the state of the art for high performance and net zero architecture on earth today, where energy, water, air and other flows through building are all monitored and analyzed. He explored how these current best practices could be applied to preparation for space travel and planetary colonization. The implications of how net zero building practices could be expanded as a test bed for interstellar design were many and exciting, and the case was made clear for an immediate need for dialogue between Earth-bound architects and designers for space, to benefit the future of building practices on Earth as well as transfer established knowledge to space-based systems. It was generally agreed that a whole new field of discussion had been opened by this talk, and that future symposia must include follow-up.

Finally, the undersigned track chair, Karl Aspelund (University of Rhode Island, USA, University of Iceland) considered "Designing for Earth 2.0: New Design Philosophies and Processes." Here, the case was made for future design thinking (both *for* space and *in* space) to shift away from modern industrialized framings. The different needs and constraints, as well as the size and social structure of space-based communities will demand a less product-oriented design field and a more component-based and ecologically sound approach. A model was presented of a fusion of engineering-design and art toward a process-oriented design culture that is less of the modern age, and more reminiscent of the ancient Greek Techne: The "knowing-how" where design, art, and craft meet.

Track Chair Biography

Karl Aspelund, PhD

Assistant Professor, University of Rhode Island
Visiting Assistant Professor, University of Iceland

Karl Aspelund, PhD, is Assistant Professor at the Department of Textiles, Fashion Merchandising and Design at the University of Rhode Island. His research interests lie in examining the role textiles and design play in the creation of identity, the impact of the textile life-cycle on the Earth's environment, and how the design community can contribute to the goal of environmental sustainability. He is now turning toward investigating the design and cultural needs and constraints of apparel in long-term space exploration. After graduating from the Wimbledon School of Art (1986,) with a degree in 3d design, Karl worked as an artist and designer for theater and film for 20 years and has taught design since the early 1990's. Before coming to URI in 1996, he was head of the Department of Industrial Design at the Reykjavik Technical College in Iceland. Karl completed a Ph.D. in 2011 in Anthropology and Material Culture from Boston University's University Professors Program, where his dissertation was awarded the University Professors Edmonds Prize as the best dissertation of the academic year 2010-2011. Karl is the author of two design textbooks, "The Design Process," (2006) and "Fashioning Society,"(2009.) The third, an introduction to designing, is due in early 2014.

Advanced Design Methodology for Deep Space Systems

Christopher A. Corner, BEng (H).

Steersman Technology Ltd, Unit 29, Basepoint Business Centre, Caxton Close,

East Portway Business Park, Andover, Hampshire, UK, SP10 3FG.

Abstract

The design of complex deep space probes suitable for finding an analogue Earth may rely on as yet unproven or even undefined technologies in many areas, in addition to other areas which are more technically mature. A methodology and structure which allows such advanced system designs to be captured at the Conceptual, Logical and Implementation layer of maturity is required. The Conceptual Layer captures high level concepts including likely processes and requirements in areas such as Propulsion Systems, Electronics, etc. The Logical Layer captures initial logical designs of physical and software systems and the processes which involve their use through a formalization of the Conceptual Layer and is described in a qualitative way. The Implementation Layer captures the detailed design which realizes the Logical Layer and contains quantitative data. Importantly the model structure allows the layer dependencies to be captured. This allows the cascade effects of changes as design matures to be analyzed across the design. Risk areas and areas where further research and development are required can also be easily identified. An example of the approach is a re-design using the methodology of the NASA Innovative Advanced Concepts Realistic Interstellar Explorer spacecraft design. The approach allows advanced system designs to be captured and advanced in a coherent manner and allows useful progress to be made in a wider context. Alternative technical approaches can be assessed for their impacts on the system design as a whole and research and technical development can be focused on key areas.

Keywords

systems, interstellar, design, qualitative, risk

1. Introduction

1.1   Background

The design of complex vehicles such as deep space probes suitable for finding an analogue Earth may rely on as yet unproven or even undefined technologies and theories in areas such as advanced propulsion [1] and instantaneous communications [2], in addition to other areas which might be considered to be more technically mature such as thermal control [3] and guidance systems such as star trackers [4]. Thus technologies on which the deep space probe design is dependent can be described as being at different Technology Readiness Levels (TRLs) [5].
A design methodology and structure which both allows such advanced system designs to be captured and matured in a coherent manner and supports a managed and understood risk approach to system development is required. This paper presents such a methodology and structure using a layered architectural model using Conceptual, Logical and Implementation layers which are linked by a set of dependencies defined in the Logical layer by a qualitative model. Examples using the framework and methodology are shown in the context of the NASA Innovative Advanced Concepts (NIAC) Realistic Interstellar Explorer spacecraft design.

1.2   Case Study Selection

In 1999 NIAC proposed a spacecraft called the Realistic Interstellar Explorer [6]. This was a proposal for a small probe that would be capable of achieving a significant penetration into the Very Local Interstellar Medium, perhaps out to ~1000 Astronomical Units (AU) within the working lifetime of the probe developers (i.e. less than 50 years). The science goals of the Realistic Interstellar Explorer were to:
*   Explore the nature of the interstellar medium and its implications for the origin and evolution of matter in the Galaxy.
*   Explore the structure of the heliosphere and its interaction with the interstellar medium.
*   Explore fundamental astrophysical processes occurring in the heliosphere and the interstellar medium.
*   Determine fundamental properties of the universe, e.g., big-bang nucleosynthesis, location of gamma-ray bursts, gravitational waves, and a non-zero cosmological constant.
*   The Realistic Interstellar Explorer concept provides an ideal example to use the proposed framework and methodology to model the spacecraft architecture and demonstrate its utility through worked examples.
*   The Realistic Interstellar Explorer architecture comprises three main systems:
*   Launch vehicle Delta III-class.
*   Perihelion manoeuver system providing high-thrust propulsion (e.g. solar thermal propulsion).
*   Interstellar Probe.
*   The focus of the system model is the Interstellar Probe system.
*   Although the system laydown was amended during Phase II of the Realistic Interstellar Explorer work [7], the example design of the Interstellar Probe is that from Phase 1 and is described as follows.
*   The probe mechanical design consists of three main mechanical elements, a Radioisotope Power Source (RPS) assembly, a central support mast, and an optical dish.
*   The RPS is placed at one end of the mast in order to minimize the radiation dose to other spacecraft components.
*   At the other end of the mast is the large optical dish, which faces away from the direction of travel and back toward the solar system.
*   Instruments and spacecraft electronics boxes are placed on the back of the optical dish and along the mast.
*   Four booms, used for field measurements, are mounted orthogonally to one another at the perimeter of the optical dish.
*   The spacecraft secondary battery and power control system are placed inside of the mast roughly at its mid-section.
*   The communication system laser is also placed inside the mast and points out the end of the mast, through a small hole in the 1-m optical dish, toward the hyperboloid reflector.
*   This physical architecture is illustrated in Figure 1.

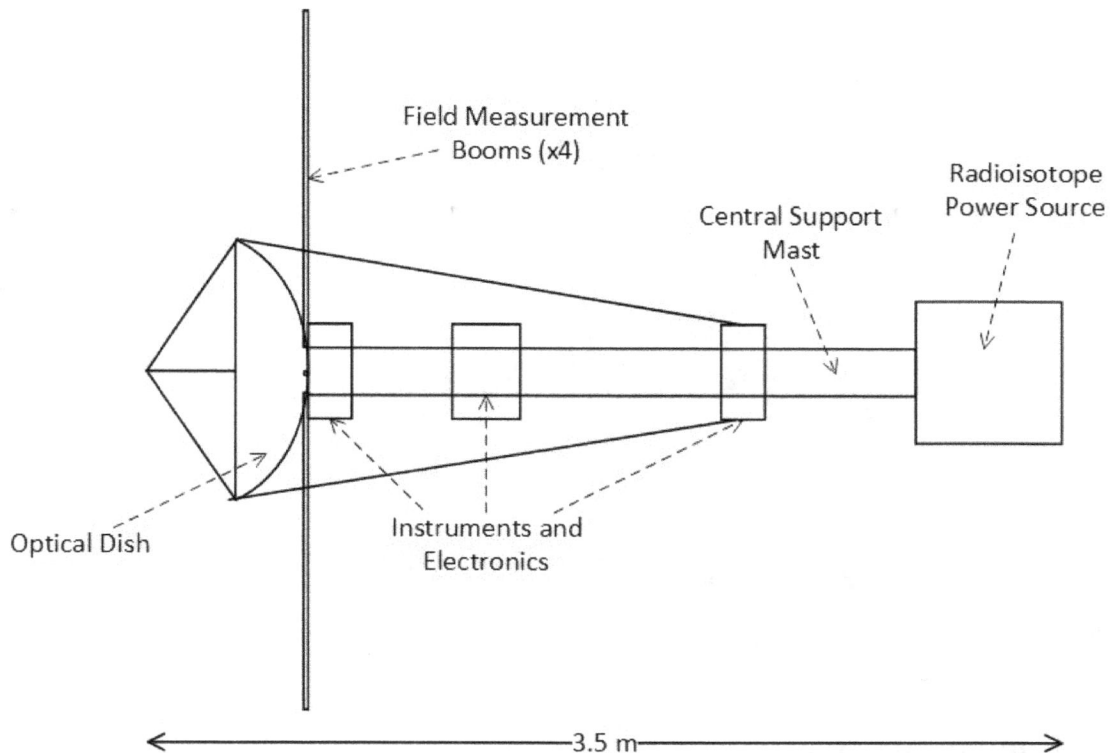

Figure 1. Phase I Interstellar Probe

1.3   Problem and Hypothesis

The proposed systems engineering framework and methodology allows us to explore a number of critical issues which must be resolved in order to successfully progress designs of systems comprising of subsystems at different TRLs. These critical issues include.

- How do we manage subsystems from concepts through to detailed designs?
- How do we manage the dependencies between subsystems?
- What's the impact if a concept doesn't prove successful?
- Do the details of a design affect concepts in other areas?
- How do we focus research and development?

2. Layered Architectural Framework and Methodology

2.1   Layered Architectural Framework

The layered architectural framework comprises three layers Conceptual, Logical and Implementation. These are linked by a set of dependencies defined in the Logical layer by a qualitative model. Each layer typically contains more elements and information than the layer above it. Figure 2 illustrates the layered architectural framework.

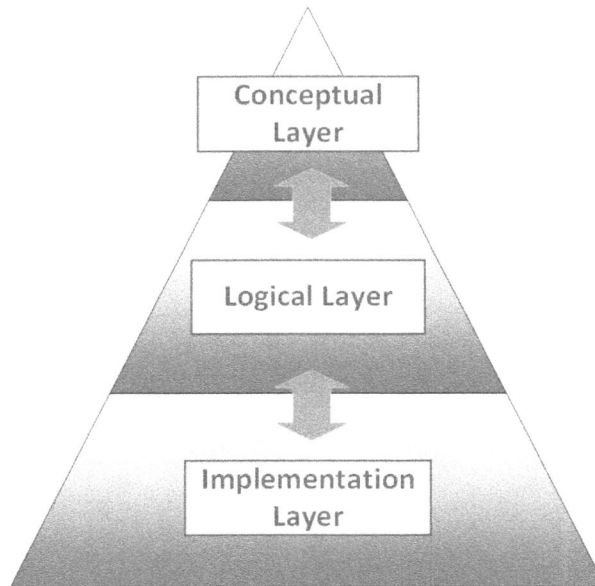

Figure 2. Layered Architectural Framework

2.2   Conceptual Layer

The Conceptual Layer is a meant to be understood by all stakeholders. It describes the entities and their relationships that form a domain of interest – for example the design of a deep space system. It typically contains fewer objects than Logical Layer and is more free form than the Logical Layer, being described "boxes and lines". It allows the high level concepts to be thought about and designed in formats similar to Visio/PowerPoint.

2.3   Logical Layer

The Logical Layer is formed by a set of qualitative objects which have formally defined attributes and relationships and thereby form a qualitative model. It is designed to be readable and understandable by stakeholders but contains formalised object types that define key logical and semantic information.
A key aspect of the Logical Layer is that it links the Conceptual Layer to the Implementation Layer.

2.4   Implementation Layer

The Implementation layer is defined through quantitative objects that form the design of specific implementations including systems, subsystems, software applications, and so on. These designs are captured in application specific tools, such as CATIA for hardware computer aided designs, Matlab for numerical data analysis and modelling, and Enterprise Architect for software designs.

3. Qualitative Models

Quantitative simulation based on quantitative models aims at producing precise numerical results as answers to user questions about the problem domain. Such precise numerical answers are often overly elaborate, contain much more information than it is actually needed, and may actually prove difficult or impossible to quantify [8]. In everyday life, humans use common sense to reason about problems qualitatively, without numbers. A qualitative domain model allows a domain to be represented qualitatively. This model can then be used as a basis for analysis and simulation. Qualitative models may be better suited for tasks such as designing new systems, supporting diagnostics and explaining complex system behaviour than either static or qualitative models [9].

The domain model allows qualitative formal simulation of system elements such as:
* Functions.
* Dependencies.
* Structure.

Qualitative not quantitative:
* For example, a motor provides rotation function not specific rpm values.
* Supports quick and simple simulations even of complex system of systems.
* This can include functional chains through a system.

It has been found that correct understanding and simulation of system behaviours can be accomplished with parameters which take on only a few values (e.g. high speed, low pressure) [10]. Such qualitative models are generally less time consuming to develop than quantitative models and can be developed when hard and fast quantitative values may not be available or, if available, may not be accurate or trustworthy.

4. Averiti Notation

4.1   Averiti Overview

The Averiti domain modelling language is an open source (MIT License) notation to describe qualitative models which is implemented in XML. It allows Component Information, Structural Information, Functional Information and Operational Information to be formally captured which in turn allows complex systems, or entire systems of systems, to be defined and modelled.

The Averiti notation has been designed to allow the qualitative modelling of equipment focused systems. Such systems are formally structured, their interactions are well defined, they are generally deterministic, and any available blueprints, UML/SysML models, or other design documents can be used as the source information. The Averiti language focusses on the logical structure and function of systems. Spatial and material properties are more difficult to capture at the qualitative level of detail and the modelling of these properties is less comprehensive. Systems are described as collections of subsystems.

The domain modelling language allows the following types of information to be described:
* Structural information about the subsystems (part/whole relations and connections between items at the same level)
* Properties of the subsystems (e.g. the shape of gears, the normal operating pressure of fluids)
* Functional information about the subsystems (the inputs and outputs of operating components, and the function they perform)
* Operational status information about the subsystems (OK/Damaged/Failed).
* Information about the Averiti Component and Structural information is presented below, for full details the Averiti specification [11] should be consulted.

4.2   Averiti Component Information

Component information describes the basic elements and concepts which form an Averiti domain model:
* System: a technical system which accomplishes tasks on demand. All systems are processors of something. Human actors in a system of systems can also be described as systems.
* Function: the function performed by a system.
* Media: Systems transmit their effects through various media which themselves have properties (e.g. pressure, temperature). These media include:
  o   Gases
  o   Liquids

- o Mechanical movement
- o Electric current
- o Control signals (digital or analog)
- instance_of: The system is an instance of a particular generic system from the domain model library (e.g. instance_of diesel_engine).
- TRL: The Technology Readiness Level (TRL) of the system.

4.3   Averiti Structural Information

Structural information allows the structure of a system of systems or a system and it subsystems to be described. This information is critical for analysing queries against the model. For example it should be able to answer questions such as which subsystem components are part of a particular subsystem? Or are two specific subsystems part of the Solar Array Subsystem? This level of detail supports hypotheses of new facts which would make the reported facts into a coherent whole. Query rules can be applied against structural information, for example "Solar Array Subsystem must contain Solar Cells". Models can be tested against such rules and variances flagged up for analysis and rectification.

Structural information includes the following elements:
- part_of: a pointer up in the system hierarchy; the subsystem belongs to the system/subsystem to which the part_of notation refers.
- made_of: a pointer down in the system hierarchy; the subsystem to which the made_of notation refers is a sub-component of the system/subsystem.
- attached_from: a pointer across the system hierarchy; the subsystem to which the attached_from refers is attached from the subsystem.
- attached_to: a pointer across the system hierarchy; the subsystem is attached to the subsystem to which attached_to refers.
- adjacent_from: a pointer across the system hierarchy; the subsystem receives an input from the subsystem to which adjacent_from refers.
- adjacent_to: a pointer across the system hierarchy; the subsystem provides an input to the subsystem to which the subsidiary element adjacent_to_name refers.
- adjacent_to_name: a subsidiary element of adjacent_to; specifies the name of the subsystem that the subsystem is adjacent_to.

The structural information defined above also lets a model be queried for completeness. If a subsystem is detailed as being adjacent_to another subsystem, but that subsystem either does not exist in the model or does not hold the equivalent attached_from relationship then the model is incomplete.

5. Case Study – Realistic Interstellar Explorer

5.1   Overview

The Realistic Interstellar Explorer is now used as a case study to illustrate some of the uses of the layered architectural framework and methodology.

5.2   Realistic Interstellar Explorer Power System Example

The Realistic Interstellar Explorer has been modelled at the system and subsystem level in the Logical Layer of the layered architectural framework. An example mapping from a Conceptual Layer "boxes and lines" view to a subsystem described using Averiti notation in the Logical Layer is shown in Figure 3.

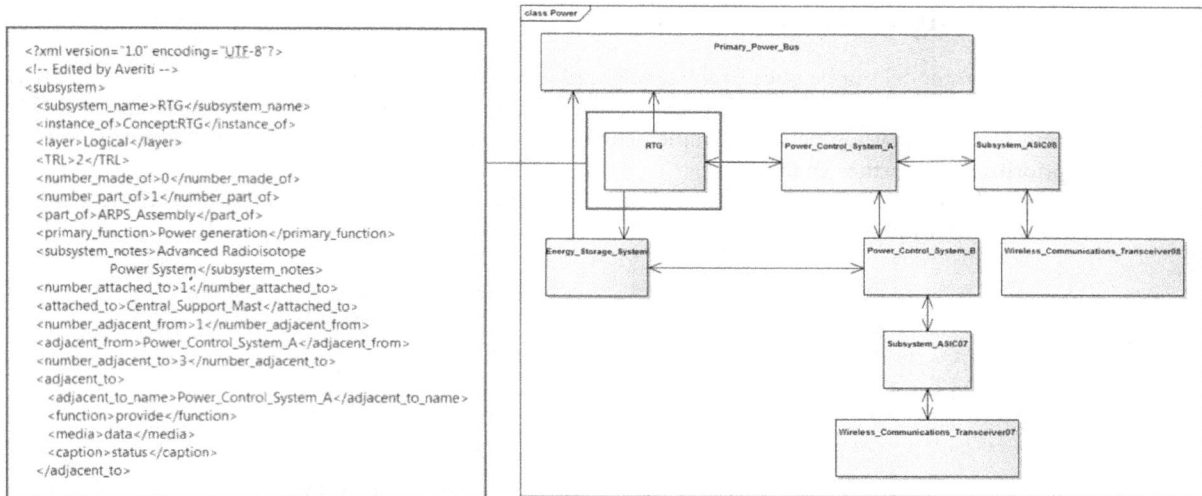

Figure 3 – Power System Conceptual - Logical Layer

5.3 Subsystem Dependency Analysis

The structure of the qualitative domain model in the Logical Layer allows dependency analyses to be undertaken. For example a change to one subsystem may drive changes to other systems through their dependency relationships such as adjacent_to and adjacent_from.

For example, an increase in the mass of the Electrical Power system may drive changes to the Main Propulsion subsystem with more thrust being required; an increase in the ionizing radiation from the Electrical Power system Radioisotope Thermoelectric Generator (RTG) may require changes to the spacecraft Structure subsystem to increase shielding; and an increase in thermal radiation from the RTG may require improvements to the spacecraft Thermal Control subsystem.

Likewise, an increase to the electrical power requirements for the Instruments and spacecraft Electronics will drive changes to the Electrical Power system, which in turn could have the additional dependency impacts on other spacecraft subsystems described above.

The concepts of subsystem dependency analysis are shown in Figure 4.

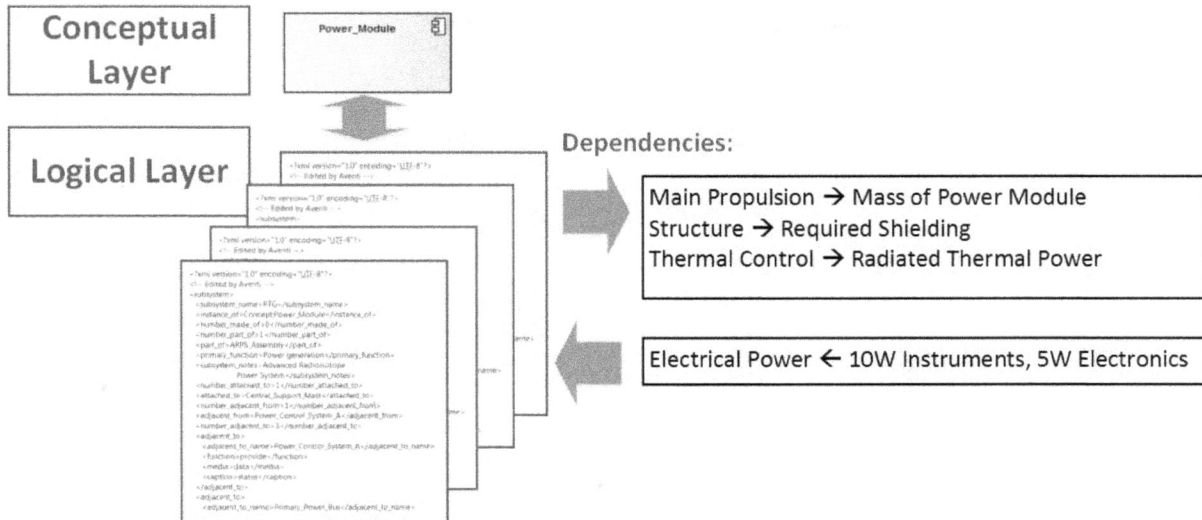

Figure 4 – Subsystem Dependency Analysis

5.4   System Design TRL Assessment

The qualitative domain model can be queried to survey attribute values across multiple subsystems that comprise the model. For example, the TRL of each subsystem can be queried to give the analyst a comprehensive overview of the subsystem research, development and implementation status. Subsystems at a low or unquantified TRL may then be prioritised for further analysis.

An example TRL assessment across a subset of the subsystems in the domain model for the Realistic Interstellar Explorer is shown in Figure 5.

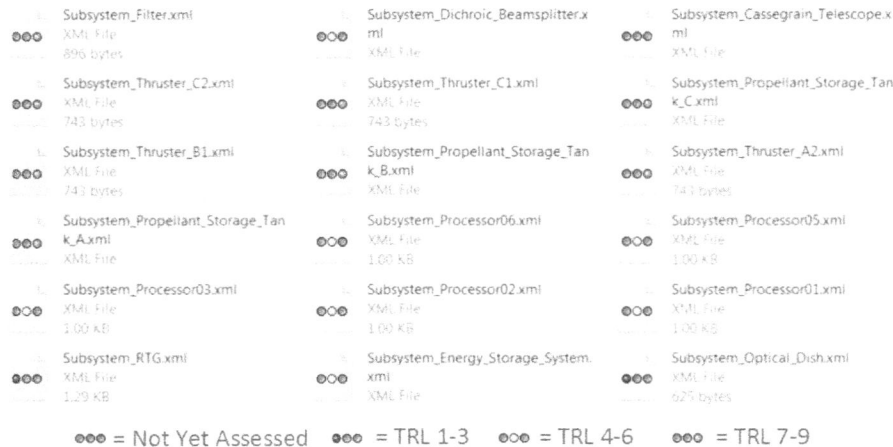

Figure 5 – Subsystem TRL Assessment

5.5 Research and Development Prioritisation

To further analyse a subsystem which has been identified as having a relatively low TRL with respect to the other subsystems, its dependency relationships, as defined in the qualitative model, can be analysed.

A subsystem that has many dependency relationships – e.g. many subsystems being dependent on Electrical Power to operate – and has a low TRL would clearly be a focus for further research and development. That is because ongoing changes to that subsystem will impact the detailed design and implementation of all those subsystems that are dependent on it, potentially leading to wasted effort and major redesign in areas that had been considered to be stable.

An example of a low TRL dependency analysis is shown in Figure 6.

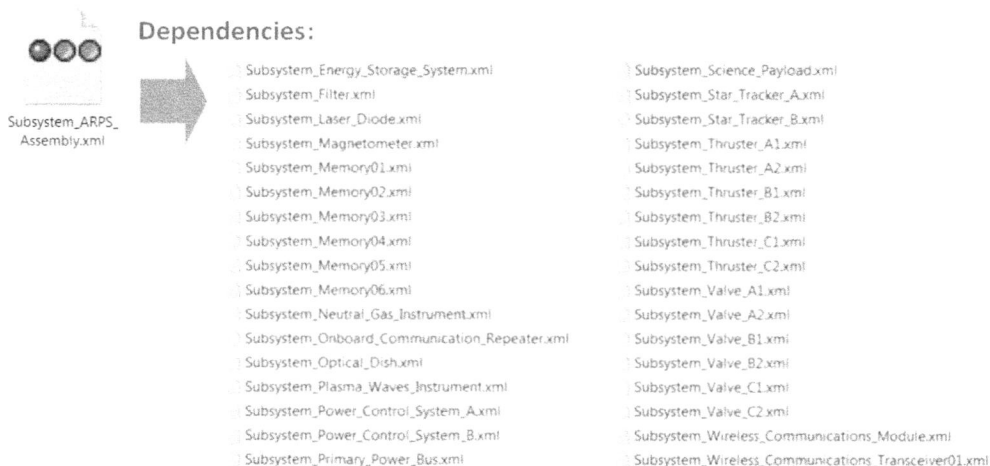

Figure 6 – Low TRL High Dependency Analysis

302

6. Conclusions

6.1 Summary

This paper has presented design methodology and structure which both allows advanced system designs to be captured and matured in a coherent manner and supports a managed and understood risk approach to system development. The design methodology and structure provides a layered architectural model using Conceptual, Logical and Implementation layers which are linked by a set of dependencies defined in the Logical layer by a qualitative model. The design methodology and structure allows:

- Understanding of the impact on the system if a concept does not prove successful.
- Management of functional needs from concepts through to detailed designs.
- Understanding of how the details of a design affect concepts in other areas.
- Management of dependencies between functional needs.
- Focussed and prioritised research and development.

6.2 Applications

The layered architectural model has been developed and tested using the Realistic Interstellar Explorer as a case study, by effectively reverse engineering the Logical Layer domain model to the Realistic Interstellar Explorer design. The next stage of development and refinement of the model would be to use it to support and manage the development of a complex deep space system design.

References

1. Alcubierre, M., "The warp drive: hyper-fast travel within general relativity", Classical and Quantum Gravity 11 (5): L73–L77 (1994).

2. Benford, G., Book, D.L., Newcomb, W.A., "The Tachyonic Antitelephone". Physical Review D 2: 263–265 (1970).

3. Gilmore, D.G., "Spacecraft Thermal Control Handbook, Volume I: Fundamental Technologies". The Aerospace Corporation, 2002.

4. "Star Camera". Space Technology NASA Jet Propulsion Laboratory.

5. Mankins, J.C., "Technology Readiness Levels: A White Paper". NASA, Office of Space Access and Technology, Advanced Concepts Office (1995).

6. McNutt, Jr., R.L., "A Realistic Interstellar Explorer Phase I Final Report NASA Institute for Advanced Concepts". The Johns Hopkins University, 1999.

7. McNutt, Jr., R.L., "A Realistic Interstellar Explorer Phase II Final Report NASA Institute for Advanced Concepts". The Johns Hopkins University, 2003.

8. Bratko, I., "Qualitative Modelling". University of Ljubljana, 2005.

9. Forbus, K.D., "An Introduction to Qualitative Modeling". Northwestern University, 2003.

10. Ksiezyk, T., Grishman, R., "An Equipment Model and its Role in the Interpretation of Nominal Compounds". HLT '86 Proceedings of the workshop on Strategic computing natural language: 81-95 (1986).

11. "Averiti Qualitative Domain Modelling Language and Environment", http://www.averiti.com/, retrieved 11/17/15.

Solid Ground in Interstellar Space: Mission and Metaphor

Dr. Hank Hine

The Dali Museum, 1 Dali Blvd, St. Petersburg, Fl 33701

hhine@TheDali.org

Charles Hine

Hine Inc. Consulting, 140 M Street #0642, Washington, DC 20002

charles.h.hine@gmail.com

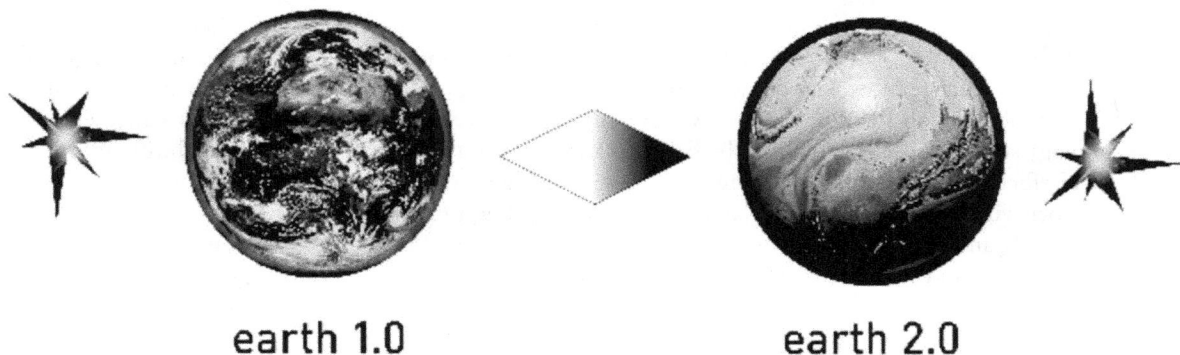

Figure 1: Earth 1.0 and Earth 2.0 graphic, Credit: The Dali Museum

Abstract

Early extraterrestrial vehicles have been intricately tied in their design to elements practical for escaping earth gravitation. Missions to the solar and interstellar mediums require habitations that are larger and designed to provide traveler wellbeing.. These environments may be designed to amplify the voyagers understanding and commitment to mission. The human mind rehearses action by its imaginings. Certain designs of terran and even universal reference should be included among those spatial configurations to remind us of our mission. Images in the environment that elicit thoughtfulness around the mission – origin, necessary tactics, and target should be indelibly present in the habitations created for extended space missions. The spaces we inhabit become the diagrams by which the universe is decoded, the source of our sense of relation and difference, our maps of intimacy and otherness.

Keywords

metaphor, prolonged habitats, design, mission, space, interstellar

1. Introduction

A human-populated starship to Earth 2.0 stirs our ambitions, activates our problem solving capacity, and promulgates technological advances. It gathers hundreds to conference, as here in San Jose. Recent inquiry into the incremental steps to accomplishing this great transplantation of human culture has articulated two principal concepts: first that the initial stellar probes are likely to be robotic using nano- and pico-technology, and secondly, that way stations in the solar and Interstellar Medium (IM) will be necessary to mature our capabilities for interstellar travel.

The incremental stages proposed by Cohen and Becker [1] posit semi-permanent staging centers positioned at a Lagrange point between solar bodies successively at greater distance from the earth. The rich observational opportunities present in the IM from the Kuiper Belt to the Oort Cloud are cited by Friedman and Garber. [2] Manned missions to establish these stagers at distances from a few AU to 5,000 AU have this in common: they will require great periods of time in travel.

The craft designed and constructed for these voyages will be home for a long time, a significant fraction of the voyager's life, and in the greater mission, the entirety of a life and even that of successive generations. But as "home," the abode offers rich environmental possibilities.

In fact, the house and the rooms it comprises become the diagrams by which the universe is decoded, the source of our sense of relation and difference, our maps of intimacy and otherness. They provide us our first sequestering comfort and then our windows and our threshold to world beyond. As the philosopher Gaston Bachelard writes in the Poetics of Space, "…thanks to the house, a great many of our memories are housed, and if the house is a bit elaborate, if it has a cellar and garret, nooks and corridors, our memories have refuges that are all the more clearly delineated. All our lives we come back to them in our daydreams." [3] Bachelard's words assume that the subconscious mind may be stimulated through both supraliminal and subliminal stimuli in design. [4]

This paper is concerned exclusively with manned missions and with conceptual habitat designs that may enhance the wellbeing and stability of the crews, amplifying their sense of mission. It assumes a synthesis of art and technology such as those proposed by the Bauhaus and by innovators in the beginnings of the digital era. The viability and efficiency of future missions are the potential dividends of these enhancements.

This paper will examine: the history and assumptions guiding prolonged extraordinary habitats such Polar stations, Sky Lab, and the ISS (International Space Station): the role of metaphor in human cognition and psychology; and images and structures that can convey and amplify the mission and the progressive steps toward it.

2. The Design of Prolonged Extraordinary Habitats

In preparation for the design of Skylab in an analogous environment, Werner von Braun appointed a NASA representative to crew with the team of Jacques Piccard's Ben Franklin submarine to study the effect of proximity and isolation for the six man crew drifting in the Gulf Stream for four weeks.

Figure 2: The Ben Franklin, Credit: Giovanola/Grumman, builder

Skylab's basic structure, a cylinder 24 feet long and 22 feet in diameter, was that of a Saturn V fuel tank. It was inhabited for 171 days by three successive three-man crews. As an extended experience in space it provided astronauts and land-based observers insights into the challenges of prolonged weightless habitation. Part of its mission was to assess the needs and develop strategies for extended space travel. Feedback from the crews identified aggravating design errors which included: insufficient human docking strategies for sleep and work in weightlessness, lack of flexibility in repurposing area functions, noise and light pollution, and insufficient areas of privacy for maximum wellbeing. [5] NASA hired Raymond Loewy, with extensive industrial design experience to recommend features.

Figure 3: Raymond Loewy's 1953 Studebaker Commander Starliner, Credit: Studebaker

Many of Loewy's proposals foresaw the challenges of prolonged and crew-intensive habitation and were sensitive to the ergonomic needs of the astronauts.

Figure 4: Skylab, Credit: NASA

The construction of Halley VI Antarctic research station, which became operational in 2012, provides us with several practical lessons that are applicable to the design of extended mission habitats. [6] Weather often forces the station personnel to remain indoors for months a time, [7] so extended periods of confinement were taken into consideration when designing this new station. Visual stimuli, a sense of space and even smells were incorporated into interior features. [8]

307

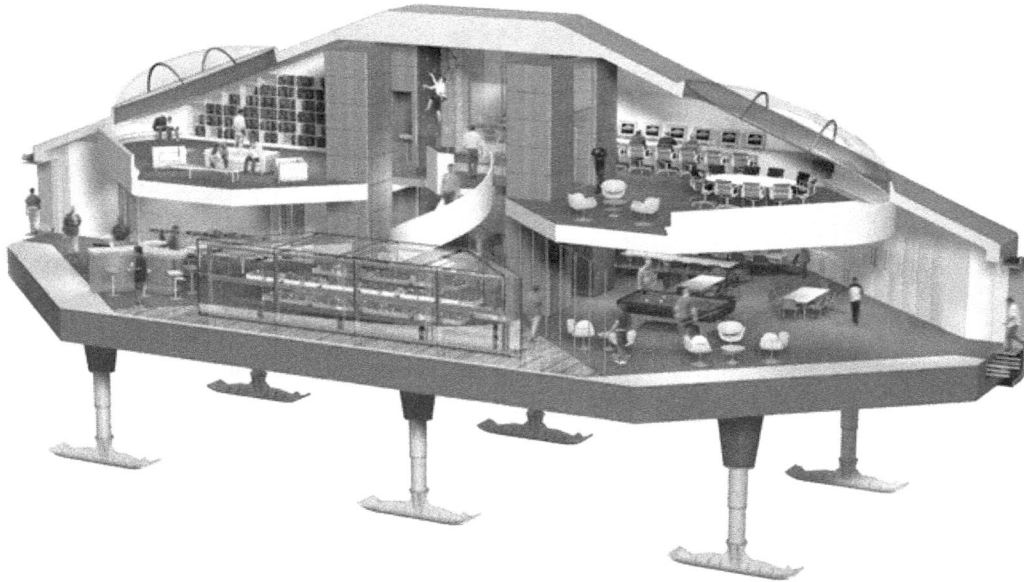

Figure 5: Halley VI Antarctic Research Station Common Area, Credit: British Antarctic Survey

Wood veneering was chosen to bolster an aesthetic appeal and stimulate the sense of smell. Cedar, provides the occupants of Halley VI with both, rather than the sterile futuristic man made facades that Hollywood usually adorns the rooms and halls of its futuristic sets. Halley VI considers social interaction in his design. The Bar in the Halley V was the central feature and communal area. The Bar area in Halley VI has access from multiple points and provides varying lines of sight, encouraging its occupation by multiple social groups. [8]

Figure 6: The International Space Station, Credit: NASA

Improving the functionality and comfort of SkyLab, the International Space Station (ISS) sought enhanced habitability through sight lines, lighting, circulation, privacy, proximities, and food supply, though still within the governing outline of successive cylindrical structures. NASA'S Habitability Design Center had as its goal "to make the astronauts feel at home, and in shaping the module it has to reconcile the social functions of architecture and design with a mind-boggling array of engineering and safety constraints." [9]

The design of the ISS benefited from the realization of space architects that the conventions of terran architecture were based on humanity's experience with 1-G environment, where massing and orthogonal configurations prevail. The gravities of Earth's orbit, the moon, or Mars allow for distinct guiding concepts.

However, constraints remain. The limitations due to environmental hazards are multiple. For long-term durations, radiation shielding will be essential as will be structural integrity, and energy conservation, so viewing ports/windows will need to be severely limited. [10]

Artificial viewing ports will be necessary to provide a sense of space. Screens project images of space, our solar system, the current star positions, or even artificial representations of space.

These early space environments were designed, of course, to provide temporary habitation – support in an environment hostile to terrestrial life. Consider a space station or ship as a sensory deprivation chamber. They provide for support for basic human functioning. For the psychological wellbeing of the occupant the designer must add sensory stimuli. In missions of prolonged habitation, missions which may constitute the final environment of the voyagers, much more than life support is needed: we seek human wellbeing. Wellbeing may be a useful term because of its comprehensiveness. The concept includes protection, access to social engagement, privacy, and the ability to initiate these conditions. A sense of self-possession, energy and mission – senses that gives life meaning - are depletable resources in a non-terran environment and sources of renewal are essential.

Researchers continue to explore the dimensions of human performance beyond the technical. The psychological health of crew and motivational factors have been acknowledged in the emerging literature such as in H. Jones' "Starship Life Support" in which the emotional amelioration from cultivating food, for instance, is balanced with the greater efficiencies of mass with a dehydrated stored-food supply from earth. [11]

These further elements have been called in the literature, "human factors." What are these human factors? What is necessary to promote wellbeing? Higher order creatures require a sense of place and purpose to function optimally. Living crews may be sustained in their purpose by a physical environment reflective of the mission and of their relation as a part to the whole. Long-term habitation in space requires a driving force, a sense of purpose and origin. NASA mission patches have been the emblematic sense of purpose for all of their missions. [12] Long-term missions will need to incorporate a sense of origin as well and the design of the work and living environments will need to incorporate these emblems overtly into their designs. These conscious and subconscious design features will instantiate the sense of origin and purpose necessary for prolonged duration missions.

Let's assume the needs of stagers or starship voyagers to be co-extensive with our needs on earth. As identified by the National Institute of Health, wellbeing includes interpersonal, psychological, esthetic, philosophical, and spiritual, domains. [13]

To what extent can the designed environment influence these domains and thus, wellbeing? In fact there is an extensive literature on the impact of the built space on the human psyche beginning with the Roman architect Vitruvius, who wrote of the breadth of knowledge required of a good practitioner. The architect, he said should be acquainted with "astronomy and a theory of the heavens," [14] in order to create buildings of durability, convenience and beauty. From Alberti to Frank Lloyd Wright, architects have asserted the power of built structures to guide good character and have sought how structure might influence psychological, emotional, healing,

Following the philosophic positions of Kant and Hagel, the realms were expanded to an architecture of ideas, built environments as expressions of abstract ideas and metaphysics. A totally determinist position is unreasonable, but evidence indicates that architecture can channel thought and behavior in a predictable manner. Charles Montgomery cites a 2008 study that shows that "a ten-minute walk down a South London main street increased psychotic symptoms significantly, while a sense of awe reduces the prevalence and severity of mood disorders." [15]

Evolutionary biologists argue that the ways in which living creatures choose, adapt and construct their habitats become a potent evolutionary agent. [16] Thus, the constructed environment by influencing the inhabitant may influence the mission.

3.   Metaphor

Recent work in linguistics, neurology, and cognitive neuropsychology has demonstrated that human reasoning is mediated by metaphor. This work postulates that metaphorical meaning functions as a map from a concrete space (the source) to an initially more abstract space (the target) that is made more concrete by the metaphoric connection. [17] The metaphor transfers associations from the source domain to the target domain. The topology of the source domain is retained and maps its structure of image patterns into the abstract reasoning patterns in the target zone. Thus, higher level meanings at the target are constructed by images at the source. If this is so, then the nature and relation of experienced images can determine meaning of the constructs (metaphors) targeted by the source images.

Research also suggests that other sensory experiences may play a role in construing meaning as images do. Brain structures in the sensory-motor regions are exploited to characterize abstract concepts. Gallese and Lakoff argue that conceptual knowledge is mapped within our sensory-motor system. "Imagining and doing use a shared neural substrate." [18] Thus, seeing and imagining also share that pathway. Beyond the scope of this paper is the consideration of the full range of our sensory capability to prompt high order conceptual structures, yet it can be inferred that in that other sensory prompts exist to heighten the near sense of mission.

4. Metaphors of Place—Images and Structures

Living, working, and communal spaces need to stimulate a sense of purpose and combat isolation and the loss of mission focus. By incorporating two fundamental elements— an aesthetic proportion, and a supraliminal feature coalescing the mission and a sense of origin— the habitat may contribute to a sense of mission.

One may reasonably assume that a starship capable of achieving 0.1 C might also accommodate more ample space, asymmetries, and successive reconfigurations beyond those of our early forays into space. Apartments, laboratories, workshops, storage areas, libraries, command rooms, animal bays, medical facilities, and recreation spaces will be comprised. We assume the creation of a biologically and ergonomically appropriate gravity environment. Presumably, experiments like Biosphere 2 will provide paradigms for sustainable food supplies with only partially open loop systems.

Such capability would allow the creation and recreation of a series of environments. These spaces could correspond to an instructional curriculum to inculcate a deeper sense of a terran origin and a novo-terran future. How can a variety of spatial configurations be achieved in a space that is confined? Recent work at Houston has proposed stereo-lithographic fabrication of spaces.

Figure 7: Concept of 3D Printing of a house, Credit: Contour Crafting Robotic Construction System, University of Southern California

Our process of envisioning a future is an analogical process, just as is our creativity. We draw relations from the source (our past and present) and extend those relations to the new context, our future, our target.

Our bank of source images provides many that suggest order and growth. We suggest that some of the following may provide useful visual metaphors for extended missions in interstellar space.

...for a sense of awe, Sagrada familia
Figure 8: Sagrada Familia, Barcelona, Exterior, Credit: Xavi Gracia, Flickr Creative Commons

Figure 9: Sagrada Familia, Barcelona, Interior, Credit: Wikimedia Commons user Miki-pons, Creative Commons Attribution-Share Alike 3.0 Spain

It is possible to generate ideas through visual metaphors as in this structure of poured concrete, glass and steel in which the classical and the fantastical are merged and hard materials in a geodesic format are made to seem ductile and elastic,

Figure 10: The Dali Museum, St. Petersburg, FL, Exterior, Credit: The Dali Museum

or make material and ever-present the concept of the multiple perspectives that constitute our vision of the world as in The Dali Museum's glass vista.

311

Figure 11: The Dali Museum, St. Petersburg, FL, Interior, Credit: The Dali Museum

Figure 12: The Dali Museum, St. Petersburg, FL, Interior, Credit: The Dali Museum

This geodesic structure of 1064 triangles distinct in shape and size at The Dali Museum in St. Petersburg, Florida affirms the mission of the art and the museum to provide different perspectives.

This helical staircase references the structure of DNA, a fascination held by Salvador Dali. Its extension beyond the floor of the museum instills in the mind the concept of upward aspiration.

Figure 13: The Dali Museum, St. Petersburg, FL, Interior, Credit: The Dali Museum

We derive aesthetic concepts by looking at both nature and history. In many of the creations that humanity considers beautiful, a mathematical commonality pervades. One such is the golden ratio, an infinite scalable relation of the small to the large.

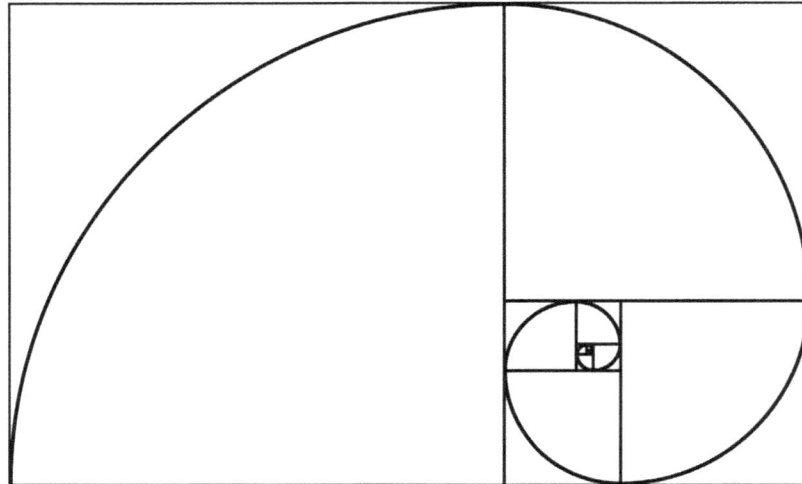

Figure 14: The Fibonacci Spiral, Credit: Raiana Tomazini, Wikimedia Commons

The Fibonacci Spiral visually demonstrates this ratio and can be seen in many naturally occurring patterns: snail shells, flowers and plants, and weather phenomena. This ratio appears in the design of great architectural icons.

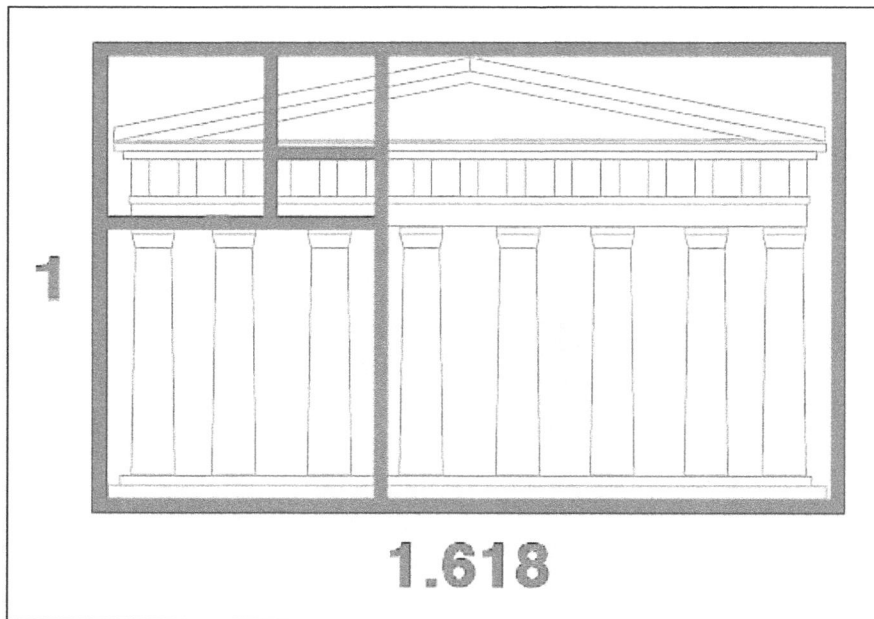

Figure 15: Golden Ratio Overlaid on the Parthenon, Credit: Gray's School of Art

The Parthenon, a Greek icon, is an exemplary achievement. The height and width of the building are in proportion to the golden ratio. The disbursement of the columns and the adornments to the frieze above the capitals similarly follow this proportion. Visually, the use of this fundamental relationship instills a sense of harmony, perhaps one based in the similar proportions of the human face.

Aside from the grand architectural achievement, let us remember Bachelard's comment about the generic house as a model of the mind and the source of memory: that house with attic and basement and its rooms filled with the potential to register the entire range of human emotion.

Figure 16: House, Credit: Flickr Creative Commons

Figure 17: House, Credit: Flickr Creative Commons

Some elements that have such fundamental importance on earth rise to the level of the sacred. The configuration of atoms in the water molecule might be used as a defining image or as a template for habitation layout to secure in the voyagers' minds their destination.

Water Molecule

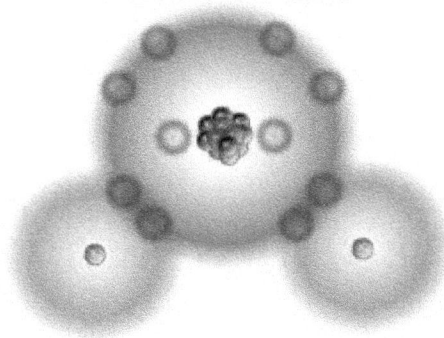

Figure 18: Water Molecule, Credit: Professor John Blamire, Brooklyn Collection

Similarly, molecular oxygen might provide comfort and a reminder of the destination.

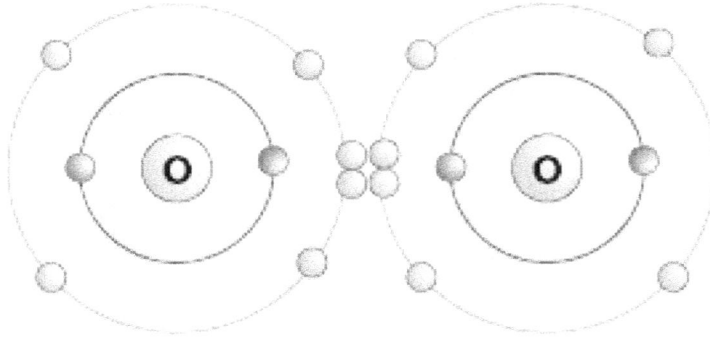

Figure 19: Molecular Oxygen, Credit: Danielle Fong

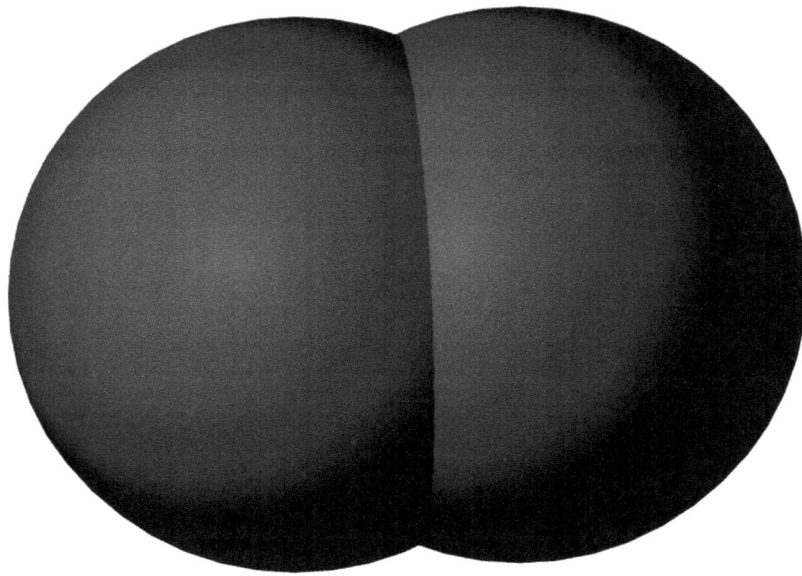

Figure 20: Molecular Oxygen, Credit: Ulflund (user), Wikimedia Commons

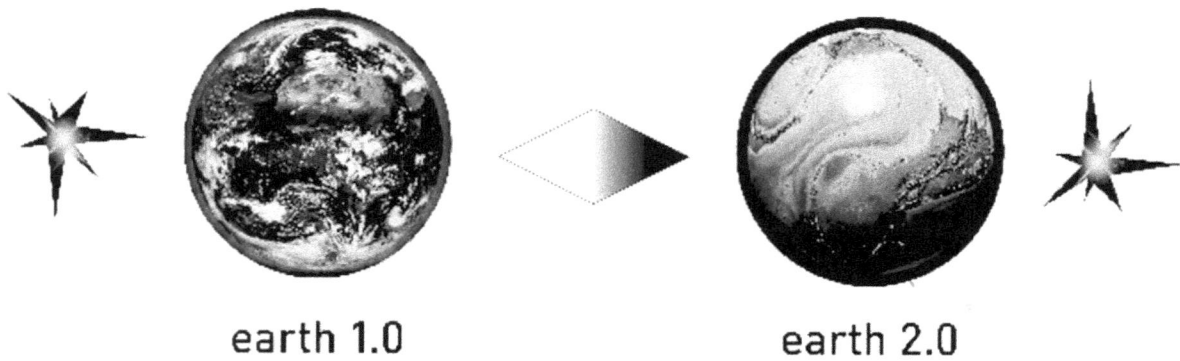

earth 1.0   earth 2.0

Figure 21: Earth 1.0 and Earth 2.0 graphic, Credit: The Dali Museum

This form embodies the perfection of engineering and a metaphor of the infinite unfolding from the smallest point.

Figure 22: Fibonacci Spiral Shell, Credit: Flickr Creative Commons

Figure 23: Fibonacci's Spiral Appearing in Nature, Credit: Flickr Creative Commons

The terran world provides a myriad of examples of this universal form.

317

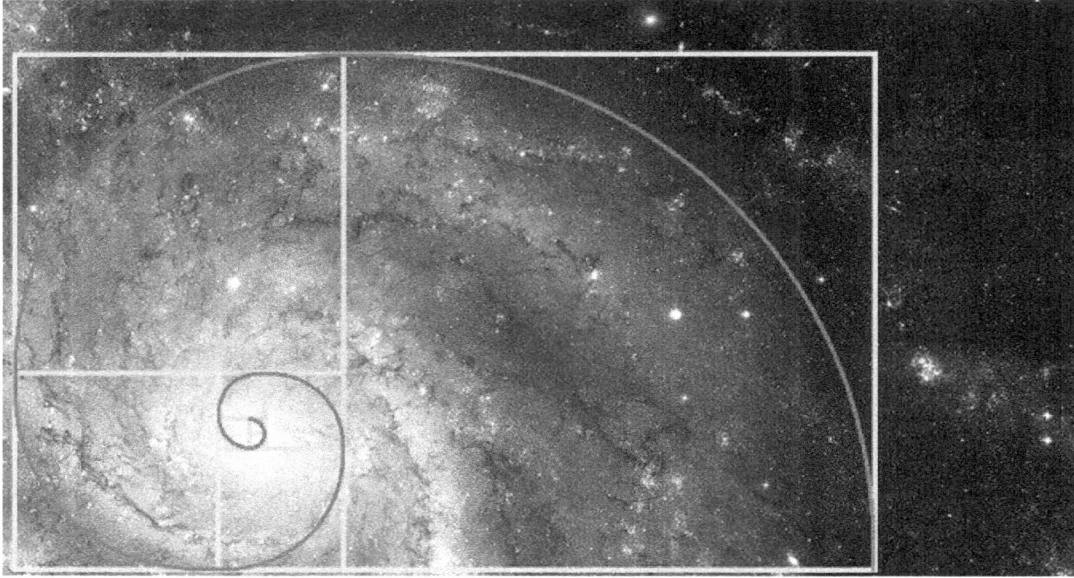

Figure 24: Fibonacci's Spiral describing the configuration of stars in a galaxy, Credit: AMCC-MCEO ARCHIVE

As we take those forms that have had the compelling meaning on earth with us into space and equip ourselves to reconfigure our spaces towards the meanings that may grow through our mission, perhaps we will see the harmonies of our own small world present in the grand expansion of our understandings that awaits us.

5. Conclusion

Environmental signs that elicit mission—origin, tactics, and target—should be indelibly present in the habitations created for extended space missions. The design of early extraterrestrial vehicles has been intricately tied to elements practical for escaping earth gravitation. In missions based in the solar and interstellar mediums, habitations that are more ample and more designed to provide the conditions of wellbeing are possible and efficacious. These environments may be designed to amplify the voyagers' understanding and commitment to mission. Certain designs of terran and, perhaps, universal references are among those spatial configurations to remind us of mission. Ultimately though, we will need iterative, flexible habitation designs that can be articulated by the evolving needs and imaginations of the voyagers. As a resource for them we will attempt to gather the experiences with which to inform the minds of a second or third generation without terran experience. These will include selections from the encyclopedia of human experience formative of higher mental and physical functioning. What these experiences will be and how they will be delivered are subject, at least initially, to the designs of the team that sets this project in motion.

References

1. Cohen, Marc M. & Becker, Robert E. "Infrastructural Development Approach to the 100 Year Starship," In M. Jemison, A. Almon (Eds.), "100 Year Starship 2012 Symposium Conference Proceedings," CreateSpace Independent Publishing Platform, 2013.

2. Friedman, Louis & Garber, Darren. "Science and Technology Steps Into the Interstellar Medium." Keck Institute for Space Studies Publications. Keck Institute for Space Studies, 2014. Web. October 2015.

3. Bachelard, G., "Poetics of Space," Beacon Press, Boston, 1969.

4. University College London. "UCL study: Subliminal Messaging 'More Effective When Negative.'" UCL News. UCL, 28 September 2009. Web. October 2015.

5. Fairburn, S. "Retrofitting the International Space Station." Book Chapters (Art). Open Air at Robert Gordon University, 2009. Web. October 2015.

6. British Arctic Survey. "Halley VI Research Station - British Antarctic Survey." British Arctic Survey, n.d. Web. October 2015.

7. Hugh Broughton Architects. "Hally VI Antarctic Research Station." Projects. Hugh Broughton Architects, n.d. Web. October 2015.

8. Broughton, Hugh. "Behind the architecture of the UK's Antarctic station." Voices. British Council, 5 September 2013. Web. October 2015.

9. Vanderbilt, Tom. "International Space Station: Johnson Space Center, Houston, Texas." Interiors 161.2: (2001). 36-41. Art Source, EBSCOhost. Web. October 2015.

10. NASA. "Fire Away, Sun and Stars! Shields to Protect Future Space Crews." Topics: Space Stations. NASA, 14 January 2014. Web. October 2015.

11. Jones, H. "Starship Life Support." SAE Technical Papers. SAE International, 12 July 2007. Web. October 2015.

12. Space Center Houston. Mission Patch. Space Center, n.d. Web. 24 October 2015.

13. National Institute of Health. 1990. "Quality of life assessment: Practice, problems, and promise." Proceedings of a Workshop, Bethesda, Md. 1990. In Boschi, N. & Pagliughi, LM. "Quality of Life: Meditations on People and Architecture." Aufsätze aus Sammelbänden. Fraunhofer IRB, 2002. Web. October 2015.

14. Vitruvius, P., De architectura. Trans. Morgan, M.H., Harvard Press, Boston, 1914. 10. E-books Directory. Web. October 2015.

15. Golembiewski, J. 2014. In Montgomery, C., "Happy City: Transforming Our Lives Through Urban Design," Farrar, Straus and Giroux, New York, 2013.

16. Odling-Smee, F.J, "Niche Construction: The Neglected Process in Evolution," Princeton University Press, Princeton, 2003.

17. Lackoff, G. & Johnson, M. 1980. In Turney, P.D., Neuman, Y., Assaf, D. & Cohen, Y. "Literal and Metaphorical Sense Identification through Concrete and Abstract Context." ACL Anthology. Association for Computational Linguistics, 2011. Web. October 2015.

18. Gallese, V & Lakoff, G. "The Brain's Concepts: the Role of the Sensory-Motor System in Conceptual Knowledge." Cognitive Neuropsychology 22.3-4. 3 January 2007: 455-479. Taylor and Francis Online, 2007. Web. October 2015.

100 YEAR STARSHIP™

Life in Space: Health, Astrobiology, Earth Biology, and Bioengineering

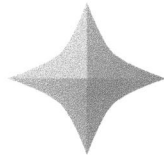

Life in Space: Health, Astrobiology, Earth Biology, and Bioengineering

Chaired by Terry Mulligan, MD

Clinical Associate Professor in Emergency Medicine,

University of Maryland School of Medicine

Track Description

How will the myriad fields making up the life sciences impact and be impacted by finding an indisputable Earth 2.0? Presenters are asked to consider the following areas for discussion from the perspective of experiments, projects and work that may reasonably be started/achieved within the next 15 years.

Biology and Astrobiology. Most life sciences contributions to finding exoplanets have targeted the ability of a planet to support life with origins on Earth. Yet, as "Earth-evolved" or terran humans, plants and other life forms travel deeper into space, farther away from Earth and eventually our solar system, greater understanding of the fundamentals of life mechanisms is demanded. Concurrently, as the search for life beyond the Earth continues, a re-evaluation of what is defined as "life" may be needed.

Once an exoplanet is identified as within the "goldilocks zone" what markers should be used to evaluate life that may have evolved outside of Earth? What questions about Earth ecosystems should be asked and answered? Also, how might the interstellar environment itself be used to advance life science research?

Health and Bioengineering. Assuming the search for Earth 2.0 will require humans in deep space for extended periods of time, how can the issues surrounding support of human crews be addressed? The health care needs to sustain human life over long periods of travel and colonization in unfriendly environments must be met. Preparation for radical shifts in nutrition, potential therapeutics, growth and development, physiology and ethics must be made.

Biomedical engineering advances knowledge in engineering, biology and medicine, and improves human health through cross-disciplinary activities that integrate the engineering sciences with the biomedical sciences and clinical practice. Furthering knowledge and understanding of living systems through innovative and substantive application of engineering sciences based experimental and analytical techniques exist side by side with the development of new devices, algorithms, processes and systems that advance biology, and medicine.

Biomaterials, tissue engineering, personal protection, human and agricultural waste material and recycling are other examples of biological related systems engineering that will have to be designed and built in space.

Track Chair Biography

Terry Mulligan, MD

Clinical Associate Professor in Emergency Medicine, University of Maryland School of Medicine

Dr. Mulligan is an assistant professor of emergency medicine at the University of Maryland School of Medicine and the director of the department's International Emergency Medicine (IEM) Program. From 2006 to 2010, he lived in the Netherlands and directed two emergency departments and EM residencies. He is an Extraordinary Senior Lecturer/Visiting Assistant Professor at Stellenbosch University in Cape Town, South Africa. He is a board member of the International Federation for Emergency Medicine (IFEM) and the immediate past-chair of the ACEP Section for IEM. He is a co-editor of the peer-reviewed journal The African Journal of Emergency Medicine. Over the past 12 years, Dr. Mulligan has initiated and participated in emergency medicine and acute care system development programs in over 20 countries.

Neurocognitive Behavioral Health Monitoring During Space Missions

Curtis Cripe, Ph.D.

NTL Group, Inc. 23623 N Scottsdale Rd. Suite D3-414, Scottsdale, Az. 85255

ctcripe@att.net

Abstract

Previous mission reports suggest that long duration space crews are at risk for psychosocial/behavioral issues that are affecting crewmember cognitive performance, moods, work performance and overall mission success. Behavioral health evidence indicates crewmembers are experiencing a myriad of stressors related to reduced neurocognitive performance that include lapses of attention, mood changes, cognitive performance decline, sleeping problems, and waning motivation. When viewing these reports from a behavioral medicine lens, there is a need to monitor, predict, and remediate reduced brain based neurocognitive changes in real time, especially during long duration space missions. With this in mind, a need exists for a space qualified Neurocognitive Predictive Performance Surveillance - Screening and Remediation tool (NPPS-SR) to monitor onboard behavioral health changes. The NPPS-SR tool can be used to predict neurobehavioral performance changes aiding mission control teams in structuring onboard mission activities based upon crewmember's cognitive resiliency. Further, the NPPS-SR tool can be used to strengthen neurocognitive resiliency changes based upon real-time neurocognitive performance markers. To date, brain-based neurometric literature contains over 6,100 publications related to cognitive performance (listed on Pubmed), all derived from earth based analog clinical populations. More critically, it is vital to note such publications have common NASA roots stemming from original early astronaut selection research. More importantly, clinically focused similar NPPS-SR tools exist and are presently used to successfully monitor, predict and aid in remediation of neurocognitive performance in earth based analog populations that include concussion, TBI, depressions, substance abuse, and learning issues.

Keywords

Behavioral Health, Radiation effect, Microgravity effect, Cognitive Rehabilitation

1.Introduction

An astronaut's resiliency while performing during the complex and dynamic conditions in the extreme environment of space is critical to the success of any manned space mission. Equally crucial, though less intuitive to harnessing the full benefits "stemming" from space exploration spin-offs, is the continued engagement of the credible leadership qualities of each crew member upon their return to earth and their subsequent reintegration into society.

This is achievable only by way of continued evidence of their exemplary performance in post flight leadership roles back on earth. Preserving opportunities to successfully demonstrate such leadership qualities by proactively protecting and preserving neurocognitive performance, will further reflect the value of manned space travel in the eyes of the world. Such due diligence would include accounting for potential neurocognitive declines due to exposure to long term space environmental effects.

From a systems engineering perspective, mission success requires many mission subsystems to function together flawlessly as an integrated unit. Many of these subsystems are mechanical in nature (i.e. spacecraft infrastructure) that support the mission. However, the most mission critical and vulnerable subsystems are the astronauts themselves; their requirements and needs must be addressed with care.

When viewing an astronaut from a systems' perspective, two major categories musts be addressed by the systems engineering and mission planning/operations teams: 1) physical health; and 2) behavioral health. With this framework in mind, this article will focus on an astronaut's onboard behavioral health from a behavioral medicine perspective. Behavioral medicine is an interdisciplinary field that integrates knowledge about human behavioral expressions by incorporating the knowledge from the fields of neuroscience, psychology, social sciences, occupational sciences, and medicine that are relevant to healthy behavioral expressions, which in our case are the astronaut's ability to perform both mission tasks and their interrelationship abilities with each other in space.

For long duration missions the impact of negative behavioral health responses is a concern, especially those due to space related brain changes affecting neurocognitive performance. These concerns include impacts on mission performance during critical tasks, fragmentation of crew cohesiveness, and ground team/crew communication disconnects. Additionally, these concerns include possible reductions in an astronaut's cognitive thinking abilities (reasoning, decision and judgmental thinking, perception, memory, attention and general information processing), and/or possible brain based emotional regulation disturbances affecting an astronaut's sense of self-agency (sense of self control) and utility, leading to potential disruptions in attention and performance. For example, disruptions resulting from radiation effects could impact mission safety and success by inducing distractions to learning and one's affinity to access memory, (including speed of memory recall and/or speed of cognitive processing), especially during time-critical reactions and tasks such as a docking sequence, EVA, deorbit burn, or landing sequences.

Deep Space Readiness Consortium, (DSR 30), is a focus group of subject matter experts dedicated to assisting in characterizing the scope and strategic positioning of potential countermeasures to assess, track, and reinforce a healthy and safe operational environment conducive to sustaining productive and thriving behavioral performance. The anticipated 2030 Mission to Mars (DSR30), with its focus of discussion centering, in part, around brain adaptation during long duration missions have contemplated questions around the effects of exposing crews to altered gravity environments. During the discussions several questions have arisen as they relate to on board behavioral health during long duration space missions:

- Is there a real possibility that neurophysiological changes might occur?
- Would negative neurophysiology changes place the mission and crew in life threatening conditions?
- If an Astronaut's visual perceptual abilities were distorted during docking or landing what might happen?
- If an Astronaut's memory retrieval abilities were distorted, slow or delayed what might happen?
- If an Astronaut's social perception abilities were distorted or dysfunctional in a multicultural, confined space during high pressure situations what might happen?
- How can the risk of a Crew-to-Ground communication disconnect be minimized and managed?
- What would happen if the ground Mission Planning Teams and/or the Mission Crew were unaware of neuro-cognitive changes and are there potentially behavioral markers that might link or map back to neurocognitive effects that could serve as an" early warning system"?
- Can we anticipate these possibilities and protect against them?

2. Psychosocial Difficulties from Mission Reports

From articles summarizing previous mission reports, it is clear that many of the above questions are possible and need to be addressed for long term space missions. During the early years of manned space exploration, missions were comparatively short and crews were mostly comprised of males, generally from similar cultures and backgrounds [1]. Accordingly, the impact on astronaut behavior and performance due to psychological and interpersonal factors was considered minimal and generally not addressed by either mission planning or operations teams. However, recently with the introduction of longer term missions on orbiting space stations, missions have become longer with crews drawn from multinational sources, often consisting of mixed genders and cultural backgrounds.

American and Russian anecdotal mission reports are indicating the presence of psychosocial issues occurring in space crews. When viewing these issues from a psychosocial model, it appears that these issues are impacting space missions by inducing detrimental effects on cognitive performance, crewmember moods and work performance [1–14]. However, behavioral health evidence from near earth short and long term mission reports also suggests many crewmembers are experiencing a myriad of stressors that are related to reduced neuro functional performance, especially those involving longer-durations of space environment of four to six months that might be intensifying observed psychosocial difficulties in its place.

These difficulties include lapses of attention, mood changes, cognitive performance decline, work performance decline, emotional lability, psychosomatic symptoms, sleeping problems, irritability toward crewmates and/or mission control staff, and a decline in robustness and motivation [4, 10-14]. When viewing many of these difficulties from a neurobehavioral lens, the evidence suggests that they might be derived from exposure to microgravity and space related radiation or other environmental factors, which in turn is effecting a crew member's neuro-cognitive performance abilities [1-14], therefore modifying crew member's emotional behavioral responses.

The astronaut health literature contains several articles addressing various aspects of crew related psychosocial difficulties. A recent meta-analysis by Kanas and Manzey [3] summarizes four potential sources of possible reduction in neurocognitive performance that might induce related neurobehavioral problems. These sources include: (1) physical factors, derived from acceleration, microgravity, radiation, and changes in light/dark cycles; (2) habitability factors, that are derived from vibrations, noise, temperature, light, and air quality; (3) psychological factors, derived from isolation, danger, monotony and workload; and finally (4) social or interpersonal factors, that may include gender issues, cultural effects, crew size, leadership issues and personality conflicts. Nevertheless, little attention on how these sources of possible reduction in neurocognitive performance exists and how it might affect mission critical events or how to monitor these effects in flight.

3. Why Is Neurocognitive Performance Important--

Self Agency, Ego Strength, and Cognitive Resiliency

Neurocognitive performance is intimately related and correlated to self agency, ego strength and cognitive resilience. Self agency refers to the subjective awareness that one is controlling their actions, volitionally. It is the "I" inside of ourselves that is executing our movements and thinking our thoughts. It is the source of how we creatively access our internal reserves to uniquely make things happen in the world. Ego strength, as a concept, depicts how an individual preforms in life by managing internal and external stresses and maintaining emotional stability, without loosing self agency. Yet, ego strength, as is self agency, also is dependent upon cognitive resilience [15]. Clinically, ego strength is frequently used to described a patient's ability, in the face of distress and conflict, to perform and maintain their internal sense of self integrity (i.e. be themselves). High or strong ego-strength individuals approach challenges with an attitude that they are able to overcome problems and grow stronger as a result. In our case, an astronaut's strong internal sense of ego-strength will allow him or her access to natural internal resources required to cope with problems, including finding new ways of dealing with struggles "in the moment" - all without losing a sense of self agency (i.e. won't panic under pressure). This is an extremely important quality needed in space. In contrast, weak ego-strength individuals view challenges as something to avoid and in many cases, reality can seem too overwhelming to deal with "in the moment". Weak ego-strength failures during moments of high reconciliation demands, such as during mission critical task or those affecting crew group cohesiveness could compromise crew performance.

Baumeister [15], in his Ego-Resource Depletion model, described how one's ego strength fluctuates depending upon the amount of cognitive effort exerted during the day. The model offers an explanation on the interrelationship between ego strength and ego-resources, and neurocognitive constructs that include self-regulation, executive function, and control interact. Baumeister proposes that self-regulation, like many other cognitive domains, will fatigue with extended effort. In his model, ego-resource capacities are not fixed, but fluctuate throughout the day and rely on internal resources that by their nature are limited and finite. Cognitive neuroscience studies support this view by demonstrating that many neurocognitive constructs, such as self-control and executive function, are mediated by fatigue [16-21]. Subjectively, cognitive fatigue is often observed not only as a decrease in ego strength, but with a corresponding increase in negative neurobehavioral expressions such as impulsivity, grumpiness, and/or narrowed myopic executive decision-making processes (reduced executive function capacity). Furthermore, evidence suggests that ego depletion detrimentally affects not only self-control and executive function, but also our general cognitive resilience as well, thus affecting performance on many tasks. From resiliency and coping studies in earth based analog populations that include learning challenged children, adults with Traumatic Brain Injury (TBI), and those in substance abuse recovery, the author has found this concept to be not only true, but also a key ingredient in the treatment remediation plan for individuals with brain-based challenges [22,23,24].

Cognitive resilience is a psychological construct that depicts the capacity to overcome the negative effects of setbacks involving associated stresses on cognitive functions that affect performance. There is a direct effect on an individual's ego strength and self agency depending on their cognitive resilience [18]. Intrinsically, cognitive resilience can be understood as a continuum of cognitive functionality. Cognitive processes that become overwhelmed and ineffective are on one end of the continuum. In contrast, the other end of the continuum represents an area where there are few or no negative effects due to stress on cognitive performance. Matching where a person lies along the cognitive resilience continuum with the task at hand is significant.

For long duration missions, gaining a measure of an astronaut's cognitive resilience is vital for mission success: 1) to predict performance ability (or risk) during higher levels of mission demands; 2) to predict when cognitive resources might need to be replenished (rest, or performing a battery of cognitive enhancement exercises); 3) to provide a quantifiable measure of the astronauts' general Ego Strength. Presently, in clinical practice the author has successfully applied measures of cognitive resilience, based upon specific brain measurements as real time neuro responses, to both child and adult earth based analog populations. These methods have consistently and quantifiably aided these earth analog populations in identifying and in recovery from learning and brain development issues, TBI, and substance abuse [22,23,24]. If similar objective brain response measures were obtained during a mission, this added data will give mission control teams a window into how to assist structuring onboard crew mission activities as well as assigning similar brain based exercises as used in earth based analog populations to help ensure mission success.

4. What are Possible Neurocognitive Performance Markers

From a classic neuropsychological model perspective, neurocognitive functions are considered fundamental building blocks that make up neurobehavioral expressions or behaviors. Behaviors range from task execution to interpersonal or mood expressions, to general or specific thinking abilities. Classically, neurocognitive functions are classified into categories such as executive function, perceptual function, memory function and self regulatory function, each with corresponding interconnected neurological structures that define a specific neuro network. It is important to note that the brain operates as an integral unit with each network either receiving or transmitting information to another network. In other words, most neurocognitive functions interact and influence each other to receive information, perceive, store, learn, recall, feel and create a thought and then to execute a task. This action occurs in a precise sequence with precise timing.

When using the right instrument, this action can be observed with quantifiable variables through measurements of intensities (strengths) that include speed, power, connectivity, timing and phase relationships within neural network elements via the EEG. Thus from a behavioral medicine or neuroengineering perspective, all neurobehavioral expressions or behaviors can be considered performance expressions derived from the interaction between individual brain subsystems. Yet, operating as an integral unit to produce the desired behavior. Moreover, each subsystem and subsequent expression is measurable. For this reason, to predict astronaut performance and onboard behavioral expressions, it is important to monitor neurocognitive performance of many brain structures and predict their performance ability effect on task performance. In other words, if the neurocognitive functions that support an astronaut's memory system should become slow to respond or fail to respond, it is important to know how it will impact other aspects of their mission performance.

To date, changes in neurocognitive and neurobehavioral responses, due to microgravity effects, sleep circadian cycle disruption, and radiation exposure, are not well understood, but have been observed and reported in the space literature. For example, recent ISS experimental reports indicate that an astronaut's visual perceptual processing abilities are affected during medium duration missions [25]. These effects include: orientation illusions, errors in sensory localization, postural imbalance, changes in vestibular, reflexes, space motion sickness, face recognition errors, negative impacts on mental rotation abilities, and errors in judgments of direction and special orientation. Report summaries indicate that effects on visual perceptual processing abilities are due to the brain's reduced ability to adjust to a microgravity field.

Specifically, as can be seen in figure 1, event-related recordings of EEG signals in the frontal and occipital brain areas showed that phase-locking (a measure of neuro network connectivity) of specific oscillations were suppressed in weightlessness for 3D tunnel images but not 2D images. Meaning that in a microgravity field, the brain processes 3-D images differently, but not 2D images. More importantly, the report demonstrated the brain on Earth processed 3D information using a directional top-down control from the frontal to the occipital areas. However, in weightlessness, the fronto-occipital top-down informational control was altered into a diverging flux originating from the central brain areas towards the frontal and occipital areas almost equally. Further when observing astronaut 3D drawing from mission experiments, they resemble drawing from children with learning issues know to originate due to cerebellar issues, either structural or with lack of primitive reflex suppression [25]. In our case, a more likely conclusion would be as a result of reflex disturbances due to microgravity effects.

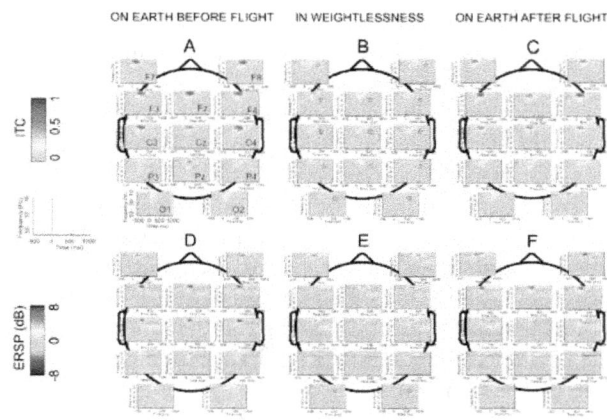

Figure 3. Effect of microgravity on the topographical representation of ITC and ERSP. ITC are represented in the upper part (A–C) and the ERSP in lower part (D–F). Grand average (n = 5) triggered by the 3D tunnel presentation on Earth before flight (A, D) in weightlessness (B, E) and on Earth after flight (C, F). Each map corresponds to a single recording channel (from F7–F8 to O1–O2) disposed on the scalp. Statistical significance (Friedman ANOVA) p<0.05 is indicated by an asterisk. doi:10.1371/journal.pone.0082371.g003

Figure 1.

Clinically, similar observations have been reported in adults with blood flow reduction due to heart issues, and children with learning issues. Of concern within these populations is that similar neurocognitive performance changes have been shown to impact both memory recall (precursor to dementia) and the brain's natural self-regulatory abilities as they relate to information processing abilities. These compromised neuro abilities effect objective thoughts and subjective emotional interpretations, moods, and cognitive resilience [25-35]. For long term missions, where there is a chance of the brain to naturally adapt to the microgravity environment, these adaptations need to be considered for crew onboard performance and reintegration upon Earth's return.

Equally important, if not more important are possible changes to the brain structure and its performance due to radiation effects. Radiation is known to effect hippocampal performance by reducing its neurogenesis ability and thus affecting cognitive resiliency [41-46]. Hippocampal neurogenesis aids in the ability to adapt to environment changes by providing the necessary building blocks for neuroplasticity. Neuroplasticity is used to expedite and take shortcuts in accelerating the use of amount of information acquired in every second by learning from what has transpired from past actions. Neuroplasticity through neurogenesis is required to make new neuro connections and adapt to changes that are deeply rooted in existing neuronal connections. With this in mind, the hippocampus is one of the key brain structures responsible for learning and memory and is involved in reaction times (or delays) required to react to an event [36-46]. Both positive and negative experiences exhibit notable affects on the hippocampus, altering memory, tendency towards anxiety, and an individual's perception of their environment and how they handle stress [36-40].

It is important to note that hippocampal performance is considered a key structure in mediating stress and supporting cognitive resilience [36-40]. Compromised hippocampal performance has been implicated in affecting the ability to cope with changes in the environment, as well as creating anxiety, depression, and other performance-inhibiting disorders [41,42]. Of concerns for the 2030 Mission to Mars are routine behaviors that occur during the 3 years in space, which to some may be considered negative and stressful. More importantly, not often accounted for, are indirect hippocampal performance effects that either directly or indirectly affect task performance that ultimately has an effect on an astronaut's cognitive reserve, cognitive resilience, emotional resiliency and general ego strength.

5. Is Neurocognitive Performance Measureable in Space?

The short answer is yes and the instrument to do so is presently onboard the ISS. What is required is the development or a modification to existing clinical tools to an on board software package that monitors targeted neurocognitive systems and subsystems. The preferred methodology that the author is proposing is based upon a neurometric methodology which had its original roots in the selection of the astronauts in the original NASA space program.

During the late 1950s, a UCLA team led by Dr. Ross Adey, pioneered a numerical method of using raw EEG measurements as part of a NASA study directed at astronaut selection [46-48]. Similar to our needs today, the thrust of the study was to determine the strength of the neuro-performance of astronaut candidates. Study results were used in initial astronaut selection, as well as in future NASA pilot training studies, which is where the author was first introduced to the methodology in graduate school.

It is important to note, Dr. Adey's team's initial study developed many of the essential quantitative foundations required to digitize and use the EEG as an analytical tool, still used today in clinical settings. The initial UCLA database included several hundred subjects based upon UCLA faculty and students and those who were candidates for the early NASA space exploration. In the 1970s, two Swedish neurologists, Drs. Milos Matousek and Ingemar Petersen, published the first clinically oriented normative database which allowed clinicians to compare a child's brain wave activity to a normative standard [49,50]. This study was replicated and verified by E. Roy John's lab at New York University in 1975 [51-56]. Beginning in the 1990s, co-registration of qEEG measurements with other instruments that include: Positron Emission Tomography (PET), Magnetic Resonance Imaging (MRI), CT Tomography, Single Photon Tomography (SPECT) occurred providing equivalent numerical markers that allows cross registration between instruments.

From these efforts, EEG neurometric measures have grown to include several non-invasive neuro-electric neuroimaging methodologies such as EEG Tomography (tEEG), that provides recordings with a time resolution of less than 1 millisecond with a 3-dimensional spatial resolution of about 1 cubic centimeter. To date, this methodology has been applied to many clinical conditions such as ADD, ADHD, schizophrenia, compulsive disorders, depression, epilepsy, TBI and a wide number of clinical groupings of patients, with well over 6,100 studies currently listed on Pubmed. The tEEG methodology, which includes insights in to the neuro network Functional Effective Connectivity is proposed as the primary analysis methodology.

Within the last ten years, tEEG measurements have been applied to many practical applications outside of the clinical realm that include many Department of Defense (DOD) projects. These projects are focused on teamwork, workload monitoring, emotional content processing, and others, all applicable to onboard neurobehavioral monitoring.

For example, Stevens and colleagues [21] showed that team work neurophysiological synchronies measuring individual member workload and engagement measures are interlinked with each team member. More importantly, these measures can be used as a general measure of team cognition and have been demonstrated as EEG based neurophysiologic indicators of team communication metrics that are complementary to conventional subjective paper measures of team cognition. Equally important, researchers from around the world are demonstrating that there are several effective EEG based neurophysiologic indicators of emotional and task performance engagement that are consistent across populations and cultures.

6. How Can We Monitor Neurocognitive Performance?

As noted earlier, cognition functions from various cognitive domains that act as an integrated system. These domains include, but are not limited to, arousal, perception, attention, memory, learning, thinking, mental organization, affect (feeling) and expression, plus executive functions. Recent cognitive neuroscience has begun to establish how many neurocognitive functions operate to produce neurobehavioral expressions in terms of cognitive endophenotypes describable in neural network terms [29-33]. Core cognitive endophenotypes of brain-behavior models are broadening our understanding of healthy brain function and many mental disorders by explaining the importance of unconscious schemas, motivational processes or learning and reinforcement principles, and how they relate to both proper brain function and psychopathology [36-42].

At present research is primarily focused on identifying dysfunctional cognitive endophenotypes found to underlie common symptoms in many mental disorders. Impulsivity is an example of a meaningful cognitive endophenotype that has been recognized in several categorical disorders such as ADHD, anti social behaviors, substance use and more [57]. In this manner, cognitive endophenotypes offer clinicians the ability to target specific cognitive dysfunctions that contribute to the mental disorder rather than treating symptoms, by using targeted brain exercises to strengthen the areas of the brain (neuro network) that define the endophenotype [30–33,57-63]. Another example of a neurocognitive endophenotype is working memory, which is correlated with hippocampal performance. Fig 2 is a slide presented at the 2015 100 YSS Symposium depicting working memory operations and the network level (Brodmann areas are specific locations in the brain that define its neuro network) and its possible radiation affects.

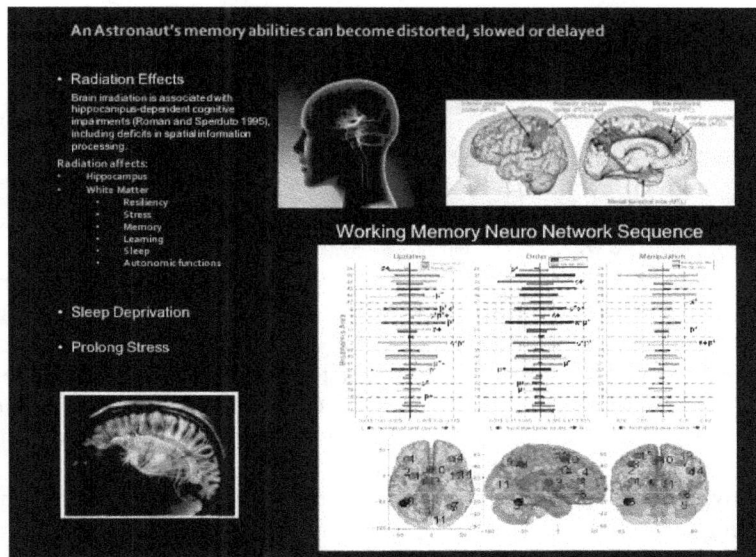

Figure 2. Depicts working memory neuro network and areas of the brain that may be affected by radiation.

Based upon functional connectivity neurometrics Figure 3 depicts an image that highlights the processing strength of the functional connectivity of the hippocampus with 44 Brodmann areas. This image was derived from a patient who was presenting with working memory issues and mood dysregulation that was affecting not only his cognitive resiliency, but his willingness to remain employed as a managing executive at his work place. Note the the light blue colors in the figure indicating weak or low Functional Connectivity between the Brodmann area and the hippocampus from each hemisphere and the brighter orange colors indicating over active or hyper connectivity. Green areas represent age appropriate levels of functional connectivity.

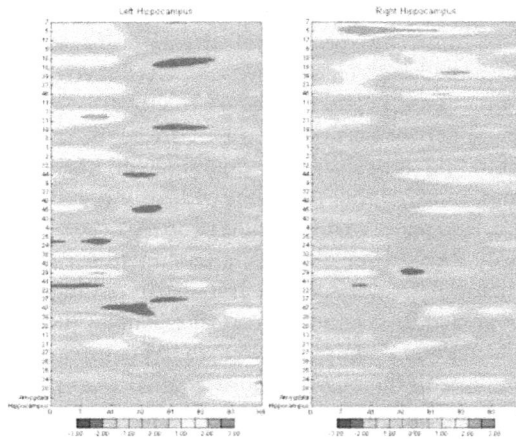

Figure 3. Functional Connectivity Neurometrics depicting hippocampus functional connectivity between 44 Brodmann brain areas. Green areas represent normal response connectivity; Blue areas represent hypo connectivity; Orange areas represent hyper connectivity.

7. In Conclusion--

Neurocognitive Predictive Performance Surveillance - Screening and Remediation tool (NPPS-SR)
Onboard Crew Monitoring and Cognitive Remediation Tool

In conclusion, the success of any manned space mission depends not only upon an astronaut's ability to perform and cope in space, but also equally to continue to demonstrate their natural leadership qualities upon return to earth and reintegration into society. Research suggests that long duration space crews are at risk for psychosocial/behavioral issues within the crew environment that may affect cognitive performance, crewmember moods and work performance. Behavioral health evidence conveys that crewmembers are experiencing a myriad of stressors that are related to reduced neurocognitive performance. These difficulties include lapses of attention, mood changes, cognitive performance decline, work performance decline, emotional lability, psychosomatic symptoms, sleeping problems, irritability toward crewmates and/or mission control staff, and a decline in robustness and motivation.

When viewing many of these issues from a neurobehavioral lens, there is a need to monitor, predict, and remediate reduced neurocognitive changes in real time, especially during long duration space missions. To support the viability of such a tool, to date, EEG based neurometric related literature contains over 6,100 publications listed on Pubmed, all derived from earth based analog populations. More critically, it is vital to note all such publications have common NASA roots stemming from original early astronaut selection research. Additionally, clinically focused neurocognitive performance tools have now evolved to successfully aid in the remediation of earth based analog populations including concussion, TBI, depressions, substance abuse and learning issues.

With this in mind, the author is suggesting there is a need for the creation of a space qualified Neurocognitive Predictive Performance Surveillance - Screening and Remediation tool (NPPS-SR) to monitor onboard behavioral health. The NPPS-SR tool can be used to predict neurobehavioral performance in order to aid mission control teams in how to assist structuring onboard crew mission activities, which will aid crew members in strengthening possible neurocognitive resiliency based upon real time neurocognitive performance markers.

References

1. R. North, Human requirements for long-duration missions: Antarctic and Arctic stations, planetary surface operations, and space transportation vehicles, Revista Portuguesa de Medicina Militar 40 (1991).

2. J.R. Ball, Safe Passage: Astronaut Care for Exploration Missions, Institute of Medicine, National Academy of Sciences, Washington, DC, 2001.

3. N. Kanas, D. Manzey, Space Psychology and Psychiatry, second ed., Microcosm Press, El Segundo, CA, and Springer, Dordrecht, The Netherlands, 2008.

4. D.A. Shayler, Disasters and Accidents in Manned Spaceflight, Springer/Praxis, Chichester, UK, 2000.

5. P. Suedfeld, Canadian space psychology: the future may be almost here, Canadian Psychology 44 (2) (2003) 85–92.   J. Rivolier, R. Goldsmith, D.J. Lugg, A.J.W. Taylor, Man in Antarctica. The Scientific Work of the International Biomedical Expedition to the Antarctica, Taylor & Francis, London, 1988.

6. G.M. Sandal, T. Bergan, M. Warnche, R. Værnes, H. Ursin, Psychological reactions during polar expeditions and isolation in hyperbaric chambers, Aviation, Space and Environmental Medicine 67 (1996) 227–234.

7. P. Suedfeld, Invulnerability, coping, salutogenesis, integration: four phases of space psychology, Aviation, Space and Environmental Medicine 76 (2005) B61–B66.

8. Sweet, T, Panda,N., Hein, A, Das,S, Hurley,S, Olschowka,J. , Williams,J., & O'Banion, (2014) M, Central Nervous System Effects of Whole-Body Proton Irradiation, Radiation Research 182, 18–34

9. Denisova, N., Shukitt-Hale, Rabin & Joseph (2002) Brain Signaling and Behavioral Responses Induced by Exposure to 56Fe-Particle Radiation Radiation Research158 725-734

10. Heuer, H.; Manzey, D.; Lorenz, B.; Sangals, J. Impairments of manual tracking performance during space-flight are associated with specific effects of microgravity on visuomotor transformations. Ergonomics 2003, 46, 920–934.

11. Johannes, B.; Salnitski, V.P.; Polyakov, V.V.; Kirsch, K.A. Changes in the autonomic reactivity pattern to psychological load under long-term microgravity—twelve men during 6-month spaceflights. Aviakosmich-eskaia i Ekologicheskaia Meditsina 2003, 37, 6–16.

12. Manzey,D.;Lorenz,B. Poljakov,V. Mental performance in extreme environments: Results from a perfor-mance monitoring study during a 438-day spaceflight. Ergonomics 1998, 41, 537–559.

13. Manzey, D.; Lorenz, B.; Heuers, H.; Sangals, J. Impairments of manual tracking performance during space-flight: More converging evidence from a 20-day space mission. Ergonomics 2000, 43, 589–609.

14. Whitmore, M.; McQuilkin, M.L.; Woolford, B.J. Habitability and performance issues for long duration space flights. Hum. Perferm. Extrem. Environ. 1998, 3, 64–74.

15. Baumeister R.F., Heatherton T.F. (1996). Self-Regulation Failure: An Overview. Psychol Inq. 1996; 7:1–15.

16. Baumeister, RF., et al. (1994). Losing control: how and why people fail at self- regulation. Academic Press.

17. Muraven M, et al. Self-control and alcohol restraint: an initial application of the self- control strength model. Psychol Addict Behav. 2002; 16:113–120.

18. Lavie, N., (2010). Attention, Distraction, and Cognitive Control Under Load. Current Directions in Psycho-logical Science 2010 19: 143.

19. Thush, C., et al. (2008). Interactions between implicit and explicit cognition and working memory capacity in the prediction of alcohol use in at-risk adolescents. Drug and Alcohol Dependence, 94, 116-124.

20. Oberauer, K., Su, H., Schulze, R., (2000). Working memory capacity—Facets of a cognitive ability construct. Personality and Individual Differences, 29, 1017-1045.

21. Stevens, R., Berka, C. Behneman, A, Wolgeuth, T, Lanb, J, Buckles, R. (2011) Linking models of Team Neurophysicologic Synchronies for Engagement and Workload with Measures of Team Communication. 20th Conference on Behavior Representation in Modeling and Simulation (BRIMS), Sundance, UT

22. Cripe, C. (2014) "Psychology's New Design Science: Theory and Research, (Eds.) S. Imholz and J. Sachter, Univ of Illinois/Common Ground Publisher - Curtis Cripe, Chapter 6 "Neuroengineering – Brain Recovery Methods – Applied to Substance Abuse Recovery"

23. Cripe, C. (2006) 'Effective Use of LENS Unit as an Adjunct to Cognitive Neuro-Developmental Training', Journal of Neurotherapy, 10: 2, 79 — 87

24. Gunkelman, J., Cripe C. (2008) Clincial Outcomes in Addiction: A Neurofeedback Case Series. Biofeedback 36.4, 152-156

25. Cheron G, Leroy A, Palmero-Soler E, De Saedeleer C, Bengoetxea A, et al. (2014) Gravity Influences Top-Down Signals in Visual Processing. PLoS ONE 9(1): e82371. doi:10.1371/journal.pone.0082371

26. Donovan, M. H.; Yazdani, U; Norris, R. D.; Games, D; German, D. C.; Eisch, A. J. (2006). "Decreased adult hippocampal neurogenesis in the PDAPP mouse model of Alzheimer's disease". The Journal of Comparative Neurology 495 (1): 70–83.

27. Mu, Y; Gage, F. H. (2011). "Adult hippocampal neurogenesis and its role in Alzheimer's disease". Molecular Neurodegeneration 6: 85

28. Jacobs, B. L., H. van Praag, F. H. Gage (2000). "Depression and the Birth and Death of Brain Cells". American Scientist 88.

29. Emmelkamp, et. al. (2014). Advancing psychotherapy and evidence-based psychological interventions. Int. Jrl. Methods Psychiatr. Res. 23(Suppl. 1): 58–91.

30. Lambert M.J. (2007). What we have learned from a decade of research aimed at improving psychotherapy outcome in routine care. Psychotherapy Research 17(1), 1–14.

31. Insel T., et al. (2010). Research Domain Criteria (RDoC): toward a new classification framework for research on mental disorders. American Journal of Psychiatry.

32. Hofmann S.G. (2008). Cognitive processes during fear acquisition and extinction in animals and humans: implications for exposure therapy of anxiety disorders. Clinical Psychology Review 28, 199–210.

33. Hofmann S.G., Sawyer A.T., Fang A., Asnaani A. (2012). Emotion dysregulation model of mood and anxiety disorders. Depression and Anxiety 29, 409–416.

34. Prigatano, G.P. & Schacter, D.L. (Eds.) (1991). Awareness of deficit after brain injury: Clinical and theoretical issues. New York: Oxford University Press.

35. Heatherton, T. and Wagner, D. (2011). Cognitive Neuroscience of Self-Regulation Failure. Trends Cogn Sci. 15(3): 132–139.

36. Neves, G; Cooke, S and Bliss, T (2008). "Synaptic plasticity, memory and the hippocampus: A neural network approach to causality". Nature Reviews Neuroscience 9 (1): 65–75 doi:10.1038/ntn2303.

37. Becker S (2005). "A computational principle for hippocampal learning and neurogenesis". Hippocampus 15(6): 722–38. doi:10.1002/hipo.20095

38. Wiskott L, Rasch MJ, Kempermann G (2006). "A functional hypothesis for adult hippocampal neurogenesis: avoidance of catastrophic interference in the dentate gyrus". Hippocampus 16 (3): 329–43.

39. Aimone JB, Wiles J, Gage FH (June 2006). "Potential role for adult neurogenesis in the encoding of time in new memories". Nat Neurosci. 9 (6): 723–7.

40. Shors TJ, Townsend DA, Zhao M, Kozorovitskiy Y, Gould E (2002). Neurogenesis may relate to some but not all types of hippocampal dependent learning. Hippocampus 12 (5): 578–84.

41. Opendak, M; Gould, E (2015) Adult neurogenesis: a substrate for experience-dependent change Trends in cognitive sciences   DOI: http://dx.doi.org/10.1016/j.tics.2015.01.001

42. Opendak, M; Gould, E (2011) New neurons maintain efficient stress recovery Cell stem cell -

43. Sweet TB, Panda N, Hein AM, Das SL, Hurley SD, Olschowka JA, Williams JP, O'Banion MK (2014). Central nervous system effects of whole-body proton irradiation. Radiat Res. 182, 18-34.

44. Cherry JD, Williams JP, O'Banion MK, Olschowka JA (2013). Thermal injury lowers the threshold for radiation-induced neuroinflammation and cognitive dysfunction. Radiat Res. 180, 398-40

45. Cherry JD, Liu B, Frost JL, Lemere CA, Williams JP, et al. (2012) Galactic Cosmic Radiation Leads to Cognitive Impairment and Increased Ab Plaque Accumulation in a Mouse Model of Alzheimer's Disease. PLoS ONE 7(12): e53275. doi:10.1371/journal.pone.0053275

46. Adey, W.R., Walter, D.O. and Hendrix, C.E. (1961). Computer techniques in correlation and spectral analyses of cerebral slow waves during discriminative behavior.Exp Neurol., 3:501-524

47. Adey, W.R. (1964a). Data acquisition and analysis techniques in a Brain Research Institute. Ann N Y Acad Sci., 31;115:844-866.

48. Adey, W.R. (1964b). Biological instrumentation, electrophysiological recording and analytic techniques. Physiologist., 72:65-68.

49. Matousek, M. & Petersen, I. (1973a). Automatic evaluation of background activity by means of age-dependent EEG quotients. EEG & Clin. Neurophysiol., 35: 603 -612.

50. Matousek, M. & Petersen, I. (1973b). Frequency analysis of the EEG background activity by means of age dependent EEG quotients. In P. Kellaway & I. Petersen (Eds.), Automation of clinical electroencephalography (pp. 75-102). New York: Raven Press.

51. John, E.R. Functional Neuroscience, Vol. II: Neurometrics: Quantitative Electrophysiological Analyses. E.R. John and R.W. Thatcher, Editors. L. Erlbaum Assoc., N.J., 1977.

52. John, E.R. Karmel, B., Corning, W. Easton, P., Brown, D., Ahn, H., John, M., Harmony, T., Prichep, L., Toro, A., Gerson, I., Bartlett, F., Thatcher, R., Kaye, H., Valdes, P., Schwartz, E. (1977). Neurometrics: Numerical taxonomy identifies different profiles of brain functions within groups of behaviorally similar people. Science, 196, :1393-1410.

53. John, E. R., Prichep, L. S. & Easton, P. (1987). Normative data banks and neurometrics: Basic concepts, methods and results of norm construction. In A. Remond (Ed.), Handbook of electroencephalography and

clinical neurophysiology: Vol. III. Computer analysis of the EEG and other neurophysiological signals (pp. 449-495). Amsterdam: Elsevier.

54. John, E.R., Ahn, H., Prichep, L.S., Trepetin, M., Brown, D. and Kaye, H. (1980) Developmental equations for the electroencephalogram. Science, 210: 1255– 1258.

55. John, E. R., Prichep, L. S., Fridman, J. & Easton, P. (1988). Neurometrics: Computer assisted differential diagnosis of brain dysfunctions. Science, 293, 162-169.

56. John, E.R. (1990). Machinery of the Mind: Data, theory, and speculations about higher brain function. Birkhauser, Boston.

57. Robbins T.W., Gillan C.M., Smith D.G., de Wit S., Ersche K.D. (2011). Neurocognitive endophenotypes of impulsivity and compulsivity: towards dimensional psychiatry. Trends in Cognitive Science 16, 81–91.

58. Miller E, Cohen J. (2001). An integrative theory of prefrontal cortex function. Ann Rev Neurosci; 24: 167–202.

59. FitzGerald, M., Carton, S., O'Keeffe, F., Coen, R. & Dockree, P. (2012). Impaired self- awareness following acquired brain injury: current theory, models and anatomical understanding, The Irish Journal of Psychology, 33:2-3, 78-85.

60. Carroll, E. & Coetzer, R. (2011). Identity, grief and self- awareness after traumatic brain injury, Neuropsychological Rehabilitation: An International Journal, 21:3, 289-305.

61. Barco, P.P., Crosson, B., Bolesta, M.M., Werts, D., & Stout, R. (1991). Training awareness and compensation in postacute head injury rehabilitation. In J. Kreutzer & P. Wehman (Eds.), Cognitive rehabilitation for persons with traumatic brain injury. Baltimore: Paul H. Brookes.

62. Bandura, A., (1991). Social Cognitive Theory of Self-Regulation. Organizational Behavior and Human Decision Processes 50, 248-287.

63. Heatherton, T. and Wagner, D. (2011). Cognitive Neuroscience of Self-Regulation Failure. Trends Cogn Sci. 15(3): 132–139.

The Contributions of Occupational Science to the Readiness of Long Duration Deep Space Exploration

Janis Davis, Ph.D.

Stanbridge College, 2041 Business Center Drive, Irvine Ca. 92612. Janis.davis@stanbridge.edu

Maria Absi
Macy Burr
Howard Koh
Rochelle Telles

OTS Stanbridge College, Master of Science in Occupational Therapy Program

Abstract

This study introduces the contributions of occupational science (OS) to the preparation and support of astronauts during long duration space exploration. Given the hostile environment of space, it is not surprising that there is grave deterioration of both physical and mental health when off Earth. However, OS, through occupational therapy (OT), can identify strategies that maintain health and minimize disruptions in task performance for mission success. To determine the gaps in NASA's preparation of astronauts for long duration space exploration and the viable contributions of OT. Because occupational therapists are trained to address deficits and modify environments to support meaningful engagement in occupations, the OT practitioner is well suited to address the disabling conditions astronauts experience in space. A literature review revealing the challenges of deep space travel on humans was completed. A survey was also sent to (N=170) occupational therapists worldwide to identify opinions about the profession's involvement in deep space exploration. Ninety-seven percent (N=163) of the participants believed that OS can inform long duration space travel. Approximately ninety-eight percent (N=166) of respondents believed that OT interventions can be used on space travelers during long duration space flights. Occupational therapy interventions can be implemented in any phase of space flight to increase the likelihood of mission success and astronaut safety and well-being.

Keywords

Occupational therapy, occupational deprivation, meaning, astronaut long duration space exploration

1. Introduction

Long duration space exploration places humans in a unique environment never before experienced by earthbound souls. A significant body of research currently exists describing the various health effects of space travel on astronauts. Scientific knowledge and technology such as magnetic resonance imaging (MRI), computed tomography (CT) scans and positron emission tomography (PET) scans of focal areas in the body have allowed researchers to distribute the body into subcomponents to study the effects of space exploration on human health. [1-8] Unfortunately, space travel is not commonly viewed from a holistic and occupation-based perspective. This study is unique in that it sought to identify how Occupational Therapy (OT) with its foundation in Occupational Science (OS) could contribute to the well being and optimized task performance of astronauts.

Occupational science is multidimensional with its philosophy grounded in a holistic view of the individual. The science studies the components of function as well as occupations, encompassing the elements of cognition, psychomotor skills, emotions, social awareness, self-knowledge, and many other concepts associated with a variety of social science and health disciplines. Occupational science is primarily concerned with promoting health, well-being, and a higher quality of life through a balance of occupations. It is also used to inform all areas of OT practice. [9] By examining the transactional relationship between the individual (client factors) and skills needed for optimal task performance (performance skills) occupational therapists facilitate health, well-being and enhanced participation in life's roles. [10]

Given the extreme conditions in space (e.g. microgravity, radiation, isolation), there is degradation in many health domains that will influence well-being and occupational performance. [1-8] Occupational therapy practitioners are uniquely qualified to examine the intersection of person, environment, occupation, and performance during long duration space exploration through current practice models and frames of reference. A thorough review of the literature was undertaken to reveal the risk factors associated with space travel related to human health and functioning. In addition, ideas regarding interventions to promote wellness for life in space to ameliorate these risks were explored.

Another purpose of this study was to determine if occupational therapy clinicians and educators would agree that OT has a place in long duration space exploration. The researchers were also interested in what types of models and frames of reference and interventions OT clinicians and educators would use to contribute to the health and well being of astronauts.

2. Literature Review

Occupational therapists are trained in assessing the interrelated components of client factors such as body functions and body structures when considering a client's ability to engage in meaningful occupations. Body structures include anatomical parts of the body which support body functions. [10] Body functions which are dependent upon body structures include sensory, affective, cognitive, perceptual, cardiovascular, respiratory and endocrine functions. [10] Because astronauts engaged in long duration space exploration need to adapt to a hostile anti-gravity environment, are exposed to radiation, perform under extreme stress and are essentially isolated from the rest of humanity, there are many detrimental effects imposed on them. Table 1 displays the client factors affected by space travel. Furthermore, Table 2 shows the performance skills that are impaired by the extreme environment experienced in space.

2.1 Cardiovascular Effects

Space travel has extreme adverse effects on the cardiovascular system. [3] In the beginning of space flight the central blood volume increases while perfusion and hydrostatic pressure in the lower half of the body decreases. The slightly higher preload and stroke volume can lead to bradycardia, increased renal blood flow and polyuria. Over time the plasma volume and the efficacy of orthostatic reflexes regulating blood pressure decrease. These effects can lower blood pressure, cause fainting and make it difficult to stand upright. Other cardiovascular changes such as heartbeat and heart rhythm irregularities and reduced aerobic capacity may occur. [3]

2.2   Musculoskeletal Effects

Researchers have revealed that the anti-gravity environment also has negative effects on the musculoskeletal system. [3] Specifically, skeletal muscle atrophy is a major concern. [4] Research indicates that long-term missions could reduce overall muscle function by nearly half. [2] Study subjects who completed missions with a duration of up to two weeks showed alterations in muscle activation variability. Specifically, increased variability in ankle and knee joint motion, alterations in head-trunk coordination and impaired balance were noted. Osteoporosis is also a serious health consequence caused by the loss of mechanical stress experienced in space. [2,5] Canan [2] indicated there to be a continuous and progressive bone loss during microgravity exposure that can be identified as early as after one week in space. Furthermore, Fulford [5] found that the leg and hip bones lose their density most quickly in microgravity increasing the risk of: kidney stones, fractures, hip and spine issues and impaired healing ability.

2.3   Sensorimotor Impairments

In addition to the debilitating effects of low gravity on the cardiovascular and musculoskeletal system, radiation exposure may cause central nervous system damage. Researchers have found that changes in the sensorimotor system occur during space travel. [8] These impairments include: diminished control of movement, reduced ability to see and interpret information from the eyes, impaired spatial orientation, and difficulty walking. [4] Alterations in the sensorimotor system can contribute to impaired postural control, locomotion, gait and manual control. [4, 8] Astronauts can also suffer from an overall perceptual distortion, disorientation and reduced performance in tasks relying on a high level of sensory-motor and sensory-cognitive skills such as piloting a spacecraft.

2.4 Cognitive Deficits

Research supports the relationship between microgravity exposure and reorganization in the somatosensory cortex and the cerebellum. For example, a decreased number of synapses and degeneration of axon terminals have also been identified. [4] De La Torre [3] suggest that the following neurocognitive deficits may occur: somatosensory problems, self position accuracy problems, color perception and loss of acuity, sound localization and binaural hearing, impaired executive functioning, concentration problems, difficulty acquiring targets in voluntary movements, visual attention problems, irritability, and lack of motivation.

2.5 Vision Degeneration

Visual deficits have been identified as a major concern for astronauts undergoing space travel. [3, 14] This can be attributed to structural changes in the retina and optical abnormalities caused by cerebral-spinal fluid compressing on the optic nerve. [11] Cowing [11] reported that about 29% of astronauts who returned from two-week space shuttle missions and 60% who spent six months aboard the International Space Station (ISS) reported visual deficits.

2.6. Vestibular Dysfunction

The microgravity environment also disrupts the vestibular system. Motion sickness occurs because sensory input identified from the eyes, muscles and joints or vestibular systems conflict with each other to cause these phenomena. [3] De La Torre [3] indicated that approximately 70% of astronauts experience space motion sickness during the first week of the mission. Astronauts also can suffer from a loss of balance and spatial orientation due to the vestibular system degrading. [2]

2.7 Sensory Deprivation

Because astronauts have restrictions of the visual, auditory, olfactory, kinesthetic, tactile and gustatory systems, they might be at risk for experiencing sensory deprivation. [1] Additionally, many aspects of space travel such as eating the same foods, hearing the same sounds, continuously smelling the same smells and being around the same people for such long periods of time may become monotonous. Researchers have suggested that a lack of sensory input can impair how information is processed in the brain. [1] Furthermore, Rasmussen [12] linked sensory deprivation to neurological changes in the brain that can lead to psychological effects such as hallucinations and anxiety.

2.8   Sleep Deprivation

Circadian rhythms help maintain sufficient periods of sleep and wakefulness enabling people to perform occupations efficiently. However, this system is driven by variable amounts of daylight and darkness in different seasons and places. In space, the lack of environmental cues such as natural light and other factors such as anxiety, workload, stress and isolation result in sleep loss, poor quality of sleep and fatigue in astronauts. Astronauts usually have between 5 to 6 hours of sleep per night and even less in emergency situations. [3] Sleep deprivation is a major risk factor for astronauts because fatigue in the workplace has been associated with increased risk of accident and injury in several studies. [13, 14]

2.9   Physical and Psychological Effects of Stress

Stress is also a risk factor for astronauts undergoing long duration space travel that has physiological and psychological implications. [15] Not only do daily stressors lead to irritability, short temperedness, and a lack of concentration, but psychological stress can have long-term effects on an individual's sympathetic nervous system. [7, 9] With excessive activation of the fight or flight response, elements such as cortisol, catecholamines and neuropeptides can produce negative effects on the immune system and can possibly contribute to disease onset and progression. [16] Extreme stress can even induce cardiac arrhythmias. [17] Stressors in space that might trigger this response include: constant acceleration, heat, noise, vibration, problems with food and oxygen supply or waste disposal, radiation, and toxic fuels. [7] NASA Extreme Environment Mission Operations (NEEMO) have indicated the psychological effects include depressed mood, anger, irritability, anxiety, interpersonal tension and conflict with group members. [6]

2.10   Interpersonal Conflict and Ineffective Communication

Currently only analogue missions such as NEEMO, the inflatable lunar habitat in Antarctica and other research programs have been set in place to explore how humans tolerate and interact in deprived environments. In the polar analogue mission, group interaction, outside communication, workload, and recreation/leisure were found to be the most salient behavioral issues [18] Interpersonal conflict and ineffective communication is a major risk factor during space exploration A crew that cannot work together in a cohesive manner will have negative effects on occupational performance and mission success.

3. Occupational Therapy's Analogous Contexts and Interventions

The Occupational Therapy Practice Framework (OTPF) which outlines the domain and process of the profession of occupational therapy, lists the client factors and performance skills needed to perform most occupations. [10] It also outlines the types of interventions used in therapy to remediate or sustain optimal health and wellness. Occupational therapists routinely work with persons with neurological, central nervous system, visual, musculoskeletal, cardiac, vestibular, sensory and mental health disorders. Therefore, occupational therapists may have a role in addressing these deficits and enabling astronauts to work effectively and efficiently on long duration space exploration missions.

Occupational therapists have been identified as important members of the cardiac rehabilitation team. [19] Not only do they assist their clients in returning to meaningful activity and participation in life roles but they also have the knowledge and background to address psychosocial adjustment issues. For example, OT practitioners teach patients with cardiovascular disorders relaxation and control of breathing techniques that ultimately help improve their symptoms. [20] They also play a key role in noticing defense mechanism like denial and educating the client and family on coping strategies to mitigate the fear and anxiety that often accompanies a myocardial infarction.

Additionally, neurodevelopmental techniques may be beneficial in helping astronauts retain normal movement. Occupational therapists utilizing these techniques help their clients to restore normal movement and functioning by providing sensory information regarding natural muscle activation patterns. Researchers have even found that pressure on the soles of an astronaut's foot can promote a vertical sense and increase neuromuscular activation, suggesting the effectiveness of interventions based on neurodevelopmental theory. [8] By helping astronauts increase range of motion, strength and endurance, OT practitioners can facilitate optimal performance.

When planning for long duration space exploration, occupational therapists would suggest sensorimotor interventions. Helen Cohen, an occupational therapist, was part of a team that developed a training program aimed to improve the ability of astronauts to adapt to new sensory environments called the Sensorimotor Adaptability (SA) Training. [21] By assisting crew members in learning how to solve sensorimotor, balance, and/or locomotor challenges, adaptation to this extreme environment can be increased. Sensorimotor Ability Training involves walking on a treadmill that is mounted on a six degree of freedom motion base in front of a screen that provides visual stimuli. It has been proven to improve locomotor adaptability, stability which reduces energy expenditure. [22] Ironically, one of the very few occupational therapists involved with NASA, is involved in this training program.

Occupational therapy experts in vestibular rehabilitation offer interventions to reduce the negative effects that anti-gravity has on the vestibular system. Motion sickness has been controlled by a user-worn see-through display, utilizing a visual fixation target coupled with a stable artificial horizon and aligned with user movement for people experiencing motion intolerance and spatial disorientation. [23] Vestibular rehabilitation training has also been shown to improve other symptoms that are associated with space travel. [24] For example, Porciuncula, Johnson, & Glickman [25] identified vestibular rehabilitation training as an effective means of improving postural stability and gaze. Additionally, Cohen and Kimball [26] found that a vestibular rehabilitation program was effective in decreasing vertigo and improving independence in daily living. Cohen-Mansfield [28] asserts that occupational therapists play a vital role in vestibular rehabilitation because many patients don't follow their exercise regimens and it might be more effective to incorporate vestibular rehabilitation exercises into functional activities. [28]

Vestibular functioning is only one of the few sensory impairments OT practitioners are skilled in addressing. Therapists practicing in vision rehabilitation help their clients remain as independent as possible while dealing with the functional limitation that comes along with vision impairment. For example, they teach their clients how to compensate for visual acuity loss by using adaptive equipment or techniques to enhance functioning. [29] They determine how vision impairment has limited the person's ability to complete tasks and then modify the task and/or the environment accordingly. Furthermore, OT practitioners also apply their expertise with adaptive devices and assistive technology to enable people to use optical and non-optical devices to complete activities. [30]

Because occupational therapists are trained to address deficits and modify environments to support meaningful engagement in occupational endeavors, they have insight into ways the spacecraft can be modified to reduce sensory deprivation. Bachmand et al., [1] have revealed the influence of sensory input on cognition and the way in which information taken in from the environment influences information processed in the brain. The Biophilia hypothesis proposes that individuals have evolutionarily adapted to gather information through exploring their environment using the senses and this curiosity has been beneficial to the development of cognitive processes. [1] This implies that a lack of sensory input experienced on a space mission can have a negative impact on cognition. Mansfield-Cohen et al., [28] identified the positive effects that multi-sensory environments have on mood levels and cognition suggesting the importance of enhancing the environment of the spacecraft. Implementing a variety of stimuli can be beneficial to an astronaut's cognition. For example, video screens and virtual reality headsets may be useful modalities. Specifically, videos of hikes, trails or landscapes could make exercising on the treadmill more enjoyable. [1] Three-dimensional video games or videos could also offer crew members ways to explore the environment and increase sensory input. [1]

Cognitive rehabilitation may be necessary for astronauts experiencing the cognitive deficits induced by space travel. Researchers have found that repetitive cognitive training can result in macro and micro structural activity changes in the brain as identified by an MRI in people with brain injuries. [31] Additionally, Cho et al. [30] indicated that cognitive enhancement training improves the performance of daily living activities, cognition and decreases depression levels in older adults with dementia. [31] Similarly, Cochet et al [31] identified cognitive remediation to be effective in improving problem solving, memory and attention in individuals with schizophrenia.

OT interventions focused on stress management provide opportunities for astronauts to deal with the dynamic interactions of mental, physical and psychological stressors associated with the astronaut experience.

Smith-Forbes et al [32] describe occupational therapy's role in addressing intellectual, physical, emotional and or behavioral reactions of service members in combat or other military operations. Stressors during combat may mirror stressors experienced in space suggesting the need for OT interventions in optimizing occupational performance of astronauts. Training astronauts on the use of cognitive behavioral techniques, such as reframing and cognitive restructuring, may assist in minimizing the irrational thinking that could occur during long duration space exploration.

Sleep deprivation may be even more debilitating than operating under extreme stress. Fatigue in the workplace has been associated with increased risk of accident and injury in numerous studies. [16-17, 34] However, by modifying performance patterns such as habits, roles and routines, OT interventions can address the sleep problems experienced by astronauts. [10] Occupational therapists are trained to identify contextual, performance pattern activity demands and client factors that may interfere with restful sleep. Occupational therapists use stimulus control to address sleep problems such as adapting nightly routines to decrease stimuli. [35] They also incorporate sleep restriction/compression (e.g., setting a predetermined sleep schedule) techniques such as sleep education, light therapy and relaxation techniques. (35) Additionally, OT interventions to promote better sleep can include physical activity, earplugs, eye-masks, and sleep inducing music. [35]

Ball [18] states that the highest priority in optimizing performance and general living conditions will be the development of an effective systematic approach to the management of a compatible, productive team who must live together in an isolated, dangerous environment. Selection of personnel will be strenuous, requiring those intended to travel into deep space for longer periodic missions in order to go through a rigorous screening process. Occupational therapy practitioners are trained to identify barriers to optimal occupational performance at the personal and social level. For example, group work is a common treatment modality used by occupational therapists practicing in psychiatric settings. [27] In utilizing this treatment approach, occupational therapists must assess culture and societal factors that influence roles and group behavior. Because most OT practitioners are mandated to be culturally competent, they are knowledgeable about group leadership and dynamics. This implies that they may have insight into developing crews that can work together efficiently.

4. Method

4.1 Participants and procedures

Given the exploratory nature of this study, a mixed method approach was used. After the literature review was complied, a survey was sent to occupational therapists worldwide. The literature review was limited to English-language literature published primarily between 1990 and 2014. Databases were searched including: Proquest, Google Scholar, PubMed, Academic Search Complete, MEDLINE complete, ERIC, and Ebook Collection. Search terms used included: "National Aeronautics and Space Administration (NASA)", "space exploration", "neuroscience", "psychological effects", "occupational science", "cognition", "physiological effects", "context", "environmental conditions", "occupational deprivation", "sleep deprivation" and "sensory deprivation." Basic inclusion criteria were publications dating between 1990 and 2015 and articles pertaining to the domains of health affected by space travel. In addition, a drop-box with research articles and various resources provided by the Long Duration Space Exploration Behavioral Readiness Consortium, 2030 (DSR30) and advisement from American astronaut, Dr. Yvonne Cagle, was used to draw conclusions.

After approval from the Institutional Review Board for Protection of Human Subjects, researchers developed an eighteen-item questionnaire and random and snowball sampling was used to contact occupational therapists as participants. Demographic questions such as area of practice, education level, age, and area of residence were included. Questions regarding what types of OT frameworks, models, and interventions were included. Open-ended questions regarding thoughts about how OS and OT could contribute to long duration space travel were included. Google Docs was used to send the anonymous survey and completion of the survey indicated consent.

4.2 Analysis

Results of the qualitative information were coded to identify common themes and sub-themes. Each researcher read participant's responses thoroughly. Categories were identified and headings selected. Under each heading, phrases or words were listed. Further examination of the data yielded themes and subthemes by searching for patterns and connections among responses. To establish interrater reliability, each researcher coded the information until common themes and subthemes were identified. The findings revealed there is a high level, excellent, agreement among raters, k = 0.90.

5. Results

Ninety-six percent of the participants believed that Occupational Science can inform long duration space travel and 97.6 % of respondents believed that OT interventions can be used on space travelers during long duration space flights. When given a list of OT frames of references and asked which one would specifically contribute to a long duration space mission, 69 % responded with biomechanical, 67% responded with sensory integration, 64% responded with cognitive behavioral, 58% chose psychodynamic, 32% selected rehabilitation, 20% chose neurodevelopmental, 21% selected that another frame of reference could be used that was not mentioned, and 3% were unsure of a theoretical frame of reference that could be used (See Figure 1). The following question on the survey stated, "What performance skills are most important to focus on while providing an OT intervention to astronauts? (See Figure 2)." The responses included: 80% emotional regulation Skills, 64% motor and praxis skills, 64% sensory perceptual Skills, 46% cognitive skills, 35% communication, and 10% said other. The last question with detailed data asked participants to select an OT model that they thought would contribute to long duration space travel. The answers were as followed: 66% person-environment-occupational-performance model, 57% model of human occupations, 41% occupational adaptation, 33% ecology of human performance model, 24% occupational behavior model, 13% unsure, and 7% said other (See Figure 3). Many OT practitioners recommended interventions that aligned with the gaps in knowledge identified by NASA on the Human Research Roadmap. For example, psychosocial interventions were mentioned 52 out of 170 times (See Table 3).

6. Discussion

Inherent in the OS discipline is the link between engagement in meaningful activity and well-being. Therefore, it could have been expected that the one concept that came up most frequently in the participant's qualitative responses was meaning. With the inability to participate in meaningful occupations, one's well being is threatened. This implies that continued participation in meaningful occupations in addition to just work is crucial to optimize the health of astronauts.

While considering astronauts undergoing long duration space exploration, leisure may be subordinate to the completion of mission objectives. However, leisure might be a necessary means of restoration. From an OS perspective, leisure is crucial for optimal health and well-being. Numerous researchers have revealed the positive effects of leisure activity on health and well-being. [37, 38] Specifically, Kleiber et al [37] identified leisure as an important resource for dealing with stressful life events. Similarly, Garcia et al [38] identified the link between leisure-time physical activity and psychological well-being, which was operationally defined as self esteem and subjective vitality. Because humans have an intrinsic need for contact with nature, horticultural leisure activities might be especially beneficial and might even address sensory deprivation. [38, 39] Some occupational therapists who took our survey recommended virtual reality games with nature-scenes and the utilization of plants on the spacecraft. Future research should focus on enabling astronauts to participate in an optimal amount of leisure time for restoration. Additionally, ways to maintain occupational balance should be investigated.

Occupational therapists employing psychosocial techniques may present adequate suggestions for the gap regarding identifying and validating effective treatments for adverse behavioral conditions and psychiatric disorders during exploration class mission. [40] Specifically, interventions focusing on: coping with stress, depression and other behavioral consequences, emotional regulation, mindfulness and meditation were frequently mentioned.

The majority of therapists taking our survey believed sensory perceptual skills and the sensory integration frame of reference to be of most importance while providing treatment to people traveling to space. They believed that sensory deprivation, sensory integration, proprioceptive and vestibular exercises would be important. Because sensory and sensorimotor deficits have been identified as gaps in knowledge to be addressed by NASA, OT assessments and interventions might provide an opportunity for them to be addressed.

Due to the extensive knowledge regarding motor and praxis skills in regards to human performance, OT practitioners likely have the background to assist NASA with creating interventions to treat musculoskeletal inju-

343

ries that may occur during the mission. [40] They also might provide some insight into creating exercise regimens that assist in optimizing musculoskeletal health for maintenance of occupational performance. Specifically, physical activity, isometric exercises, range of motion (ROM) endurance and ergonomics appeared as major subthemes.

Team function has been identified as an important area to research by NASA. Additionally, understanding how personal relations/interactions affect behavioral health and performance is an area of concern. [40] Because occupational therapists have knowledge regarding social and communication skills and the influence of the social context on occupational performance, they might be able to assist with addressing these concerns. Specifically, social skills, group interaction, social connections with friends and family, journaling and communication logs appear to be potential countermeasures.

Performance patterns such as routines, roles, rituals, and habits appeared as a major theme in our qualitative analysis. Specifically, sleep and rest routines were mentioned frequently. The qualitative analysis indicated that therapists believe adaptation to be extremely important for astronauts undergoing space travel. Adapting to a new environment focused on adapting to a difference sensory environment. This includes adapting to different roles or activities or preferred activities for participants in artificial environment as well as changes in gravity to occupations and tools. Being experts in adaptation and environmental modification, OT practitioners might be able to provide significant insight into helping astronauts adapt to the extreme environment experienced during space travel.

Interestingly, when asked what thoughts participants had about the role of OS in deep space exploration a very small pool of participants included the term "occupational science" in their responses. These results indicate that the term or concept of OS does not easily come to mind for this population. It is unclear if this is a lack of knowledge or familiarity with the understanding of OS. However, participant responses to this questions that appeared multiple times were "occupational deprivation" and "meaning," two concepts closely associated with OS.

5.1 Limitations

It is important to note that the list of occupational therapy interventions suggested from the respondents is by no means complete or exhaustive. The responses suggest a role for occupational therapy as a means to address the many health risks of long duration space missions. The sample interventions collected by this survey are, at the very least, a foothold to occupational therapy's claim as an integral part of maintaining the health of people who participate in space missions. Another limitation of our study was the relatively small sample size and an underrepresentation of occupational therapists from Midwestern and Eastern areas of the United States. Furthermore, it is unknown what prior knowledge the individuals who participated in the survey have regarding space exploration, which may also compromise the validity of results. Despite these limitations, this study provides an excellent foundation for future studies. Based on these findings, relevant frames of reference, models of practice and interventions have been identified that can assist future researchers in helping astronauts optimize health and well-being, resulting in enhanced task performance and mission success.

6. Conclusion

The OT profession firmly believes OS can inform the long duration space mission. Moreover, the OT profession believes OT interventions can be used by space travelers on long duration space flights. Occupations can either facilitate or limit the capacity of an individual to succeed in adapting to environmental demands. Deep space exploration presents environmental demands not yet experienced by humankind. By exploring the many dimensions and inherent meaning of occupations, insight into meaningful occupational therapy interventions in space can be illuminated. These interventions can address the deficits associated with long duration space exploration.

Acknowledgements

Our team would like to thank Dr. Yvonne Cagle, American astronaut, Jan Kinsgaard, DSR30 Knowledge Specialist and all those involved with the DSR30 consortium for their guidance in completing this research.

References

1. Bachmand K, Otto C, Leveton L. Countermeasures to mitigate the negative impact of sensory deprivation and social isolation in long-duration space flight. 2012. NASA Johnson Space Center, Houston, TX: National Science Biomedical Research Institute and USRA Division of Space Life Sciences.

2. Canan JW. Health effects of human spaceflight. Aerosp Am. 2013; 24-30.

3. De la Torre G. Cognitive Neuroscience in Space. Life. 2014; 4(3): 281-294.

4. Koppelmans V, Erdeniz B, De Dios Y, Wood S, Reuter-Lorenz P, Kofman I et al. Study protocol to examine the effects of spaceflight and a spaceflight analog on neurocognitive performance: extent, longevity, and neural bases. BMC Neurol. 2013; 13(205): 1471-2377.

5. Fulford MH. To infinity and beyond! Human spaceflight and life science. The FASEB J. 2011; 25(9): 2858-2864.

6. Palinkas L, Keeton K, Shea C, Leveton L. Psychosocial characteristics of optimum performance in isolated and confined environments. NASA Center for AeroSpace Information. 2011.

7. Ruff E. Psychological effects of space-flight. Aerosp Med Hum Perform.1961; 32: 639-642.

8. Souvestre P, Landrock CK, Blaber AP. Reducing incapacitating symptoms during space flight: Is postural deficiency syndrome an applicable model? Hippokratia. 2008; 12(1): 41-48.

9. Henderson A, Cermak, S, Coster, W, Murray, E, Trombly, C, & Tickle-Degnen L. The issue is: Occupational science is multidimensional. American Journal of Occupational Therapy. 1991; 45: 370–372.

10. The American Occupational Therapy Association. The American Journal of Occupational Therapy, 2014: 68(1): 1-52.

11. Cowing K. Researching Changes to Astronaut Vision in Space-SpaceRef [Internet]. Spaceref.com. 2013 [cited 15 February 2015]. Available from: http://spaceref.com/space-medicine/researching-changes-to-astronaut-vision-in-space.html

12. Rasmussen J. (Ed.) (1997). Man in isolation and confinement. Chicago, IL: Aldine Pub.

13. Dawson D, Reid K. Fatigue, alcohol and performance impairment. Nature. 1997 July; 388(6639): 235.

14. Ferguson S, Lamond N, Kandelaars K, Jay S, Dawson D. The Impact of Short, Irregular Sleep Opportunities at Sea on the Alertness of Marine Pilots Working Extended Hours. Chronobiologia. 2008; 25(2-3): 399-411.

15. Leszczyńska I, Jeżewska M, Grubman-Nowak M. Dynamics of stress as a predictor of health consequences in Polish drilling platform workers. Longitudinal study: part I. Int Marit Health. 2014; 65(1): 33-40.

16. Hall J, Cruser D, Podawiltz A, Mummert D, Jones H, Mummert M. Psychological Stress and the Cutaneous Immune Response: Roles of the HPA Axis and the Sympathetic Nervous System in Atopic Dermatitis and Psoriasis.Dermatol Res Pract. 2012; 2012: 1-11.

17. Ziegelstein RC. Acute emotional stress and cardiac arrhythmias. JAMA. 2007; 298(3): 342-329.

18. Ball R, & Evans H. Safe Passage: Astronaut Care for Exploration Missions. Washington, D.C.: National Academy Press; 2001.

19. Tooth L, McKenna K. Contemporary Issues in Cardiac Rehabilitation: Implications for Occupational Therapists. Br J Occup Ther. 1996; 59(3): 133-140.

20. Tomes H. Cardiac Rehabilitation: An Occupational Therapist's Perspective. Br J Occup Ther. 1990; 53(7): 285-287.

21. Bloomberg J, Peters B, Cohen H, Mulavara A. Enhancing astronaut performance using sensorimotor adaptability training. Front Syst Neurosci. 2015; 16(90): 129

22. Krueger WWO. Controlling motion sickness and spatial disorientation and enhancing vestibular rehabilitation with a user-worn see-through display. Laryngoscope. 2010: 121(2): 17-35.

23. Trapp W, Landgrebe M, Hoesl K, Lautenbacher S, Gallhofer B, Gunther W, Hajak G. The effect of vestibular rehabilitation on adults with bilateral vestibular hypofunction: a systematic review. J Vestib Res. 2012; 22: 5-6.

24. Porciuncula F, Johnson CC, Glickman LB. The effect of vestibular rehabilitation on adults with bilateral vestibular hypofunction. J Vestib Res. 2012; 22(5-6): 283-298.

25. Cohen H, Kimball T. Changes in repetitive head movement task after vestibular rehabilitation. Clinical Rehabilitation. J Vestib Res. 2004; (18): 125-131.

26. Balance Centers. OT Takes Leading Role in Vestibular Treatment. [Internet]. United States: Florida Balance Centers; 2015 Nov [Cited 2015 Nov 12]. Available from: http://www.balancecenters.com/?page_id=52

27. Cole MB, Tufano R. Applied theories in occupational therapy: A practical approach. Thorofare, NJ: Slack, Inc. 2008.

28. Cohen-Mansfield J, Werner P. Management of verbally disruptive behaviors in nursing home residents. J Gerontol A Biol Sci Med. 1997; 52(6): 368-377.

29. Nordvik E, Walle K, Nyberg K, Fjelle M, Walhovd B, Westlye T, Tornas S. Bridging the gap between clinical neuroscience and cognitive rehabilitation: The role of cognitive training, models of neuroplasticity and advanced neuroimaging in future brain injury rehabilitation. Neurorehabilitation. 2014; 34(1): 81-85.

30. Cho M, Kim D, Chung J, Park J, You H, Yang Y. Effects of a cognitive-enhancement group training program on daily living activities, cognition, and depression in the demented elderly. J Phys Ther Sci. 2015; 27(3): 681-684.

31. Cochet A, Saoud M, Gabriele S, Borallier V, El Asmar C, Dalery J, D'Amato T. Impact of new cognitive remediation strategy on interpersonal problem solving skills and social autonomy in schizophrenia. Encephale. 2005; 32(2): 189-195.

32. Smith-Forbes ME, Najera C, Hawkins D. Combat operational stress control in Iraq and Afghanistan: Army occupational therapy. Mil Med. 2014; 179(3): 279-284.

33. Yazdi Z, Sadeghniiat-Haghighi K, Loukzadeh Z, Elmizadeh K, Abbasi M. Prevalence of sleep disorders and their impacts on occupational performance: A comparison between shift workers and non-shift workers. Sleep Disord. 2014: 1-5.

34. Leland NE, Marcione N, Schepens-Niemiec SL, Kelkar K, Folgberg D. What is occupational therapy's role in addressing sleep problems among older adults? OTJR. 2014; 34(3): 141-149.

35. Lloyd C, Maas F. Occupational Therapy Group Work in Psychiatric Settings. Br J Occup Ther. 1997; 60(5): 226-230.

36. Chen H, Tu H, Ho C. Understanding Biophilia Leisure as Facilitating Well-Being and the Environment: An Examination of Participants' Attitudes Toward Horticultural Activity. Leis Sci. 2013; 35(4): 301-319.

37. Kleiber D, Hutchinson S, Williams R. Leisure as a Resource in Transcending Negative Life Events: Self-Protection, Self-Restoration, and Personal Transformation. Leis Sci. 2002; 24(2): 219-235.

38. Garcia M J, Castillo I, Queralt A. Leisure-time physical activity and psychological well-being in university students. Psychol Reports. 2011; 109(2): 453-460.

39. Abraham A, Sommerhalde K, Abel T. Landscape and well-being: A scoping study on the impact of outdoor environments. Int J Public Health. 2004; 55(1): 59-69.

40. Foster J, Patel S. Human Research Roadmap [Internet]. Nasa.gov. 2015. Available from: http://humanresearchroadmap.nasa.gov/

Table 1: Client Factors Affected By Space Travel

Client Factor	Disruptions
Body Structures	Bone mineral loss, skeletal muscle atrophy, cardiovascular deconditioning, radiation effects on neuroplasticity, cerebral-spinal fluid compressing on the optic nerve
Body Functions	Alterations in muscle activation variability, orthostatic reflexes, decreased strength, balance problems, vision degeneration, sensory deprivation, the lack of sensory integration needed for optimal functioning including proprioception, sight, auditory and tactile restrictions, impaired manual control, impaired executive functioning, concentration problems, anxiety, depression

Table 2: Performance Skills Affected By Space Travel

Performance Skills	Disruptions
Motor and Praxis Skills	Delayed motor controls and spatial orientation leading to significant challenges completing tasks and meaningful occupations, motor coordination and movement timing problems, impaired postural control and self position accuracy, difficulty walking
Sensory Perceptual Skills	Vision degeneration, sensory deprivation, the lack of sensory integration needed for optimal functioning including proprioception, sight, auditory and tactile restrictions, visual attention problems, somatosensory problems, sound localization difficulties, vestibular dysfunction and binaural hearing
Cognitive Skills	Color perception, memory loss, impaired executive functioning, diminished attention, perceptual disorientation
Emotional Regulation SKills	Psychological effects of extreme working conditions, excessive activation of the stress response, depression, irritability, and lack of motivation.
Communication Skills	Ineffective communication, lack of team cohesion, interpersonal conflict, diminished social interaction

Table 3: Thematic Coding of Survey Fill in Responses

Themes with subthemes	Frequency (*f*)
Meaning	**62**
Psychosocial Interventions	**52**
Coping with Stress/Depression	23
Emotional Regulation	8
Cognitive Training/Retraining	7
Mindfulness/Awareness Training	6
Journaling/ Storytelling	4
Leisure	**51**
Sensory-Perceptual Interventions	**48**
Integration	16
Proprioception	5
Vestibular	4
Modulation	2
Biomechanical Interventions	**40**
Strengthening	14
Physical Activity	12
ROM/Stretching	8
Ergonomics	8
Endurance	2
Isometrics	2
Adaptation	**36**
Communication and Social Skills	**31**
Maintaining Social Connections	12
Group Interaction	9
Social Skills	5
Performance Patterns	**30**
Routines	21
Roles	7
Rituals	3
Habits	3
Occupational Balance	**10**
Occupational Science	**19**

The fill-in survey responses for "I believe the following OT interventions may be useful for space travelers who endure long periods off Earth," "what other thoughts do you have about the role of Occupational Science in deep space exploration?" and "what other thoughts do you have about occupational therapy's role in deep space exploration?" were combined and thematically coded and reflect high interrater reliability. Note: Frequency is the total number of responses but unique to only one respondent

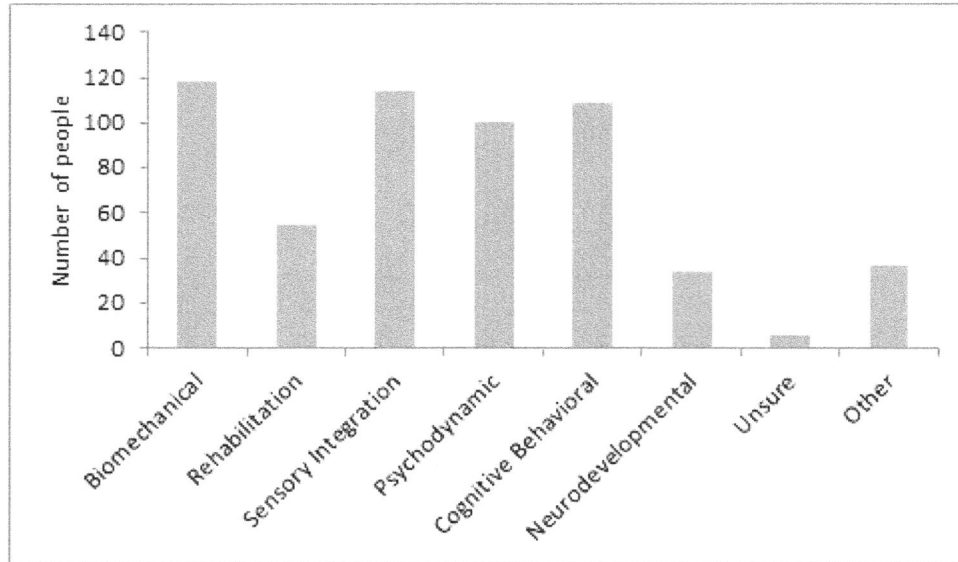

Figure 1. The Occupational Therapy Frames of Reference that could contribute to a Long.
Figure 1. Survey responses to "What specific occupational therapy frames of reference could contribute to a long duration space mission?" Bio mec = biomechanical, rehab = rehabilitation, SI = sensory integration, psy dyn = psychodynamic, cog behav = cognitive behavioral and neurodev = neurodevelopmental.

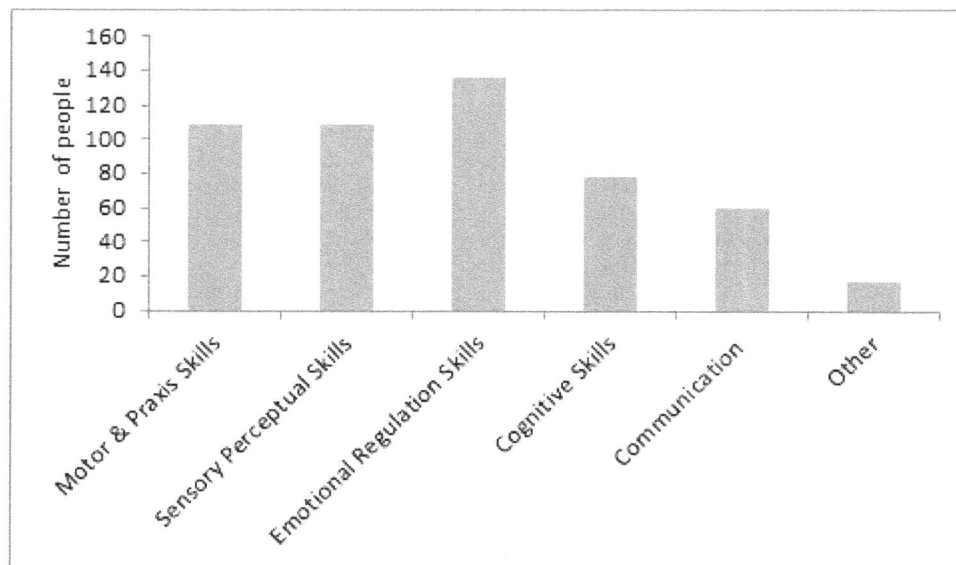

Figure 2. Targeted Performance Skills for Occupational Therapy Intervention.
Figure 2. Survey responses to "What performance skills do you think are most important to focus on while providing an OT intervention to astronauts?" Phy = physical, cog = cognitive, psychosoc = psychosocial, comm = communication.

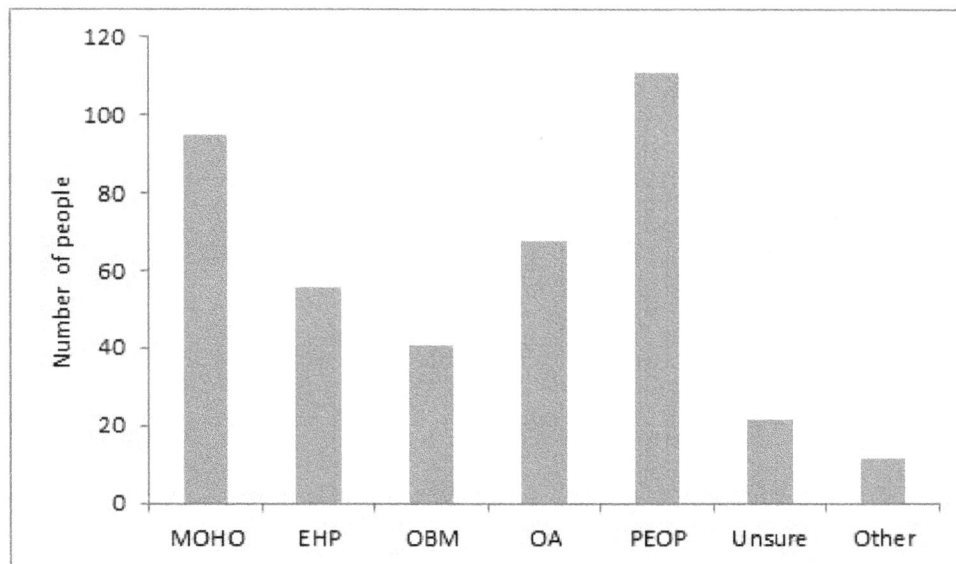

Figure 3. Occupational Therapy Models for Long Duration Space Missions.
Figure 3. Survey responses to "which of the following OT models do you think can contribute to long duration space travel?" MOHO= Model of Human Occupation, EHP = Ecology of Human Performance, OBM = Occupational Behavior Model, OAM = Occupational Adaptation Model, PEOP= Person- Environment- Occupational- Performance Model.

Astrosociology: Deviance Aboard a Long-Duration Spaceflight

Jim Pass, PhD

CEO, *Astrosociology Research Institute.*

P.O. Box 1129, Huntington Beach, CA 92647

jpass@astrosociology.org

Abstract

Life aboard a spacecraft on a long-duration mission, and perhaps one seeking an Earth 2.0 for settlement purposes, will prove to be difficult, if only because it will involve confinement and isolation. When adding additional social problems based on various other causes to the mix, the hardship and complexity of social life will become quite a bit more challenging. Here, coverage of the social and cultural forces that involve one category of social problems sure to develop serves as the main focus. The social conditions aboard a spacecraft will provide pressures to defy social values (important cultural ideas) and thereby produce defiance against social norms (acceptable rules of behavior). The result will often lead to forms of deviant behavior, from insignificant transgressions to serious criminal acts, potentially including homicide and sabotage, for example. Moreover, assuming children and teenagers are aboard, various forms of deviance will not be confined only to adults. Juvenile delinquency is also inevitable. As such, the social construction of a criminal and juvenile justice system must occur. This includes law enforcement, one or more courts, and a corrections system. Additionally, assigning personnel to fill the statuses of police officer, judge, and corrections officer require planning before launch. Social control and punishment mechanisms tailored to each type of deviant act will be required just as it is necessary within terrestrial societies. While such a long-duration voyage will require the construction of other institutions as well, the focus here is on the characteristics of deviance aboard a spacecraft on a long-duration mission and the responses by the social system to deviant acts. The inevitability of deviance in terrestrial societies strongly suggests its inevitability in space societies, including those within a spacecraft's isolated social environment and a settlement. Both types of social environments require the establishment of a social society when a large population is involved.

1. Introduction

Astrosociology is an academic field focused on the study of *astrosocial phenomena*, which include space-related (1) social patterns – relating to the life, welfare, and relations of people; (2) cultural patterns – involving ideas, including values and beliefs; and (3) behavioral patterns – or the direct actions of people. This author founded astrosociology as an academic field in 2004, initially as a subfield of sociology with a broad purview with space as the anchor.

> Recognizing that contemporary human efforts in space are best viewed as the tip of an iceberg and as possible precursors of grander future efforts, astrosociology proposes to move sociology into the space age… Astrosociology deals with the broad, societal contexts of activity pertaining to space, as well as actual space exploration including human space exploration and the search for extraterrestrial life (Harrison, 2005:14).

However, it very quickly became apparent that other fields and disciplines were relevant and non-sociologists wanted to participate in astrosociological education and research.

Thus, this academic field was adapted in order to bring all the social sciences and humanities into the space age, and indeed, now into the NewSpace age – characterized by commercial ventures in cooperation with and without contributions by government agencies – as well. As a result, astrosociology is a multidisciplinary academic field that includes the social and behavioral sciences, the humanities, and the arts – in collaboration with the physical and natural space sciences, including the STEM subjects. This field, at its very core, encourages interdisciplinary education and research, including within the social sciences and between the two branches of science. Nevertheless, this discussion provides mostly sociological insights and analyses based on this author's background and training. With this in mind, it is important for scientists and scholars from other disciplines and fields to contribute to this discussion in order broaden the understandings on a more holistic manner.

A central assumption of this paper is that a large population aboard a spacecraft on a long duration spaceflight must construct a *space society* aboard ship, which one may view as a collection of individuals with statuses (or positions) and roles (expectations) that organizes social life into a more or less structured and predictable reality. Whatever the overall social structure, the point is that structure is necessary for a collection of people to survive – and, importantly, thrive – on a long-term basis, whether a quasi-military or a democracy, or any other political system. The aim is not to create a utopian society, just to create one that is stable and not too harmful to the citizenry.

Social life among citizens in a large population during a long spaceflight is no different fundamentally from social life in a permanent settlement. Both involve isolation from the bulk of humanity and confinement in an artificial structure surrounded by deadly space or an inhospitable atmosphere. One must ask a fundamental question. What is necessary and sufficient for survival both in terms of the physical requirements and the sociocultural requirements? Each type of consideration requires construction, whether it relates to a physical structure or a social structure. Furthermore, one must also consider how the two interplay together, or even better, how to construct both types so that they work together most successfully for the benefit of the spacefaring population.

Another thing is certain. Any long-duration mission or permanent settlement will not have the luxury of assistance from Earth. It seems obvious, but these types of migration efforts must be self-sufficient. How is this possible without Star Trek style replicators? It seems that any mission to another star system needs to balance speed with practical resource collection efforts, perhaps traveling fast from point to point where resources are identified. Once again, this involves a holistic approach to maximize the likelihood of success, which is only possible if scientists from both branches participate in planning.

Major assumption: any long-duration migration to another space environment requires significant input from social scientists, humanists, and artists. Moreover, constructing a space society – a human ecosystem – on a suspected Earth 2.0 requires major planning that starts with the training on Earth, carries on during the spaceflight, and then after transferred to the space society (Pass, 2015). Astrosociology exists to organize "the other branch of science," which consists of social and behavioral scientists, humanists, and artists – so that interested scientists and scholars can participate in space education and research; in this case, to help make the voyage to another star system possible.

2. Why the Social Sciences, Humanities, and the Arts?

The logic behind the need for astrosociologists to participate in space missions, especially those that involve transporting a large population to another star system, involves the idea of a holistic approach to space exploration and settlement. The involvement of human beings alone should indicate the need for social sciences to understand, predict, and react to behaviors in closed isolated physical environments. The two branches of science consisting of the "hard" sciences and the "soft" sciences have long remained separated in the study of space issues, though this separation cannot continue. Scientific and technological change themselves indicate that their impacts on society will continue to produce broader social change in a number of different areas. A large part of this impetus is astrosocial change brought about by astrosocial phenomena.

Sociologists and other social scientists were required on Earth to understand tremendous upheavals that resulted in the transition into capitalism as part of the industrial revolution. Joining a collection of strangers without any common ground for expected behavior or cooperation would simply result in chaos. Therefore, a structure is necessary so that individuals can see where they fit into the overall system.

STEM (only the "Hard" Sciences, in addition to Technology, Engineering, and Mathematics) comprises a vital collection of fields and disciplines that are necessary for space exploration and settlement, but they have proven insufficient to accomplish it successfully. Focusing too much on technology fails to incorporate the human element, the very component that the STEM disciplines exist to serve and protect. The human dimension complicates matters for engineers, for example, but they can construct vessels and habitats that allow humans to live in harsh space environments. What STEM disciplines cannot accomplish, what they do not even attempt to address, is how the humans living inside their constructions that keep them safe from the elements interact with one another. They do not focus on what types of forces encourage conformity or how to deal with deviance and other social problems that are sure to arise.

The STEAM movement moves in the right direction for space education and research, but it fails to address the other important contributors. The "A" in the acronym refers only to the arts. While astrosociology focuses on space matters alone while STEM and STEAM represent broader foci, it is important to broaden the scope for all STEM-related investigations that focus on typically hard science subjects. Collaboration between the two branches is important in all scientific inquiries, though space research represents one of the most urgently needed areas. STEAM does, in fact, emphasize the traditionally underrepresented arts. However, the approach taken by the STEAM movement is not inclusive enough to cover all fields and disciplines related to space (Pass and Harrison, 2016), and collaboration between the hard sciences and so-called "soft" sciences is crucial to human space exploration and settlement. The social and behavioral sciences, along with the humanities, deserve as much attention as the arts, and arguably more.

3. Defining and Planning for the Space Society

A space society implies the replication of what one normally considers as the typical structure of a terrestrial society with a dominant culture and social institutions. Such an endeavor is a monumental task because it should exist before an expedition launches off the surface of the Earth. The implications of this relate to the massive undertaking necessary to organize a large collection of individuals into at least a semi-cohesive society with common objectives and expectations. Sponsors of such a massive project will need to determine how many resources they are willing to devote to it.

Spacecraft and space habitats represent extreme closed systems (Pass, 2011). Individuals will be isolated from all semblances of Earthly comforts and characteristics. Their new home will likely be more stark and cold. The best way to cope with such circumstances relates to the planning process, which involves forging these individuals with diverse backgrounds and belief systems into a common (and new) worldview that maximizes the likelihood of cooperation during a long – and perhaps lifelong – spaceflight. Training must involve assimilation to some extent without striving for a complete melting pot scenario, which is likely to result in resistance.

It may seem like the best way to achieve such an outcome is to find as many people as possible with the same or similar backgrounds and characteristics. That way, planners would create a head start in terms of involving people that share common characteristics from the beginning. However, sociological and anthropological research has shown that a population consisting of people with diverse backgrounds and skills is most advantageous. Biodiversity is important, which is based on a variety of biological factors; and similarly, social and cultural diversity are also advantageous. Heterogeneity is preferred over homogeneity in many areas such as sex, race, ethnicity, religious affiliation, occupation, and others. Space research by social scientists has confirmed these findings (see, for example, Williams, 2003).

A diverse and large population, necessary for survival and biological diversity in a space ecology, requires extensive cooperation along with a minimal level of conflict and disruptive behavior. Planners must consider what type of social system to construct. Assuming that a quasi-military structure is ruled out for a large population, two basic models exist when basically constructing a new society from scratch that must operate successfully from the beginning. These two models are conceptual rather than attempts to create utopias, and oversimplified. For example, modern societies consist of social organizational patterns that include traits of both extremes discussed here. However, they do serve adequately for the present illustrative purposes. The first is a *Gemeinschaft* society, which is a community structure based on a homogenous culture. Social ties tend to be informal, intimate, cooperative, and familiar. A strong commitment to the group exists and social cohesiveness is strong. In contrast, a Gesellschaft system is equivalent to modern society and characterized by heterogeneity. Social relationships are formal, goal-oriented, and often between strangers. Durkheim (1893) adopted these two concepts that were proposed by Tönnies (2001) in his work on explaining the transformations caused by the industrial revolution. While a Gesellschaft society possesses a very simple division of labor, a Gemeinschaft society possesses a complex division of labor. Durkheim focused on similar conceptualizations that he termed "mechanical solidarity" and "organic solidarity," respectively (Durkheim, 1893). (See Pass, 2015, for additional details in the context of planning for space societies).

For planners of the space society, the task becomes what type of structure is most preferable for a large population? One must keep in mind the "engineering" a social system is just as important as engineering a physical system (i.e., spacecraft and habitat). Three stages exist for implementation of the project – that is, the construction of the space society – as follows (Pass, 2015):

1. conduct terrestrial planning and training,
2. implementation aboard the spacecraft, and,
3. transfer of society to final settlement location (almost certainly a closed habitat system).

It represents an expensive and extensive project from the beginning, if the entire process is implemented. Those in charge of implementation must also keep in mind that social change will occur so that the original social system will adapt from stage to stage and due to social interactions and events that occur as time passes.

Another important question demands attention, though is outside the scope of this discussion. What type of ecology will exist? In other words, what types of life should be included besides humans? Which animals, flowers, plants, insects, microbes represent required additions to ensure human survival? A total absence of microbial life, for example, may prove unhealthy for humans.

4. Social Structure and Organization

A *social structure* is the organized system of relationships within a group such as a family, economy, or church community. From the smallest group, a dyad, to the largest organization or institution, social structure exists to provide organization. It exists outside of individuals. No population can survive long without the stability provided by an underlying organization of social structures. Two key elements of social structure are status and role. A status is a position. Status set includes all the different positions a person inhabits. A role is an expectation that relates to a particular status. A role set consists of all the expectations that define how a person should act and what he or she is expected to do when in that status.

A *social institution* is a system of relationships that provide structure and order to individuals through the creation and maintenance of a normative order (that is, normal and acceptable behavior within society). Social institutions regulate behavior within a society in order to achieve certain goals or to provide resources in tangible or intangible forms. Human beings cannot live together without social structure and specifically social institutions, and a society could not function stably and predictively without them.

Even with this simplistic discussion about social structure, it should be clear that the construction of a space society that involves structure even among all the groups and institutions represents a monumental task. Much of it will involve trial and error before getting things in acceptable order, but terrestrial societies exist as models that provide structure that are favored and others that are avoided. Much thought must be put into the construction of a space society in order to prevent excessive deviance and other social problems from occurring.

5. Social and Cultural Forces

A social reality exists beyond the individual so that society is more than simply the sum of its parts (Durkheim, 1893). Planners must realize that society is emergent. That is, it takes on a character that one cannot explain based simply on the individuals involved or its component structures.

Durkheim (1982/1895), in arguing for sociology as a distinct science, emphasized the existence of what he termed "social facts." Social facts are things because they are separate from individuals, consisting of values, cultural norms, and social structures. They are just as real and transcendent as any physical fact such as the elements that make up a rock for geologists. We can sense the existence of social facts through social and cultural change. Our social structures adapt to change as do individuals. Moreover, our ideas alter over time. It could be due to adaptations, breakthrough ideas, or advancements in science and technology. Values and the rules protecting them change.

Social and cultural forces produce change that, while not always predictable, is nevertheless inevitable. Thus, in planning for a space society, one can only construct a starting point and take pains to maximize the possibility that change occurs on a positive track over time. Still, attempts to manage social change can prove to be difficult and even result in unforeseen or dangerous outcomes. Each of the three plan implementation stages will possess its own challenges. Controlling and reacting to deviance is one of the monumental of these as the space society evolves.

6. The Impacts of Confinement and Isolation

Spacecraft and habitats are confined spaces in the strictest sense because they do in fact consist of enclosed spaces, which is a fundamental element of their definition. They are closed systems because they present residents with a permanently enclosed structure. They must live within it, and rely on it, for their very existence. They represent the entire living space for all inhabitants.

"Isolation can cause behavioral anomalies that provide important information to researchers" (Pass, 2011:363). Anxiety and stress, related to and also unrelated to negative effects caused by the space environment, can result in problematic behavior such as deviance. The difference between the romanticized notion of living in space and the reality of doing so, if extreme enough, can contribute to abnormal behavior. Planners should take this into account when training potential spacefarers and designing the physical structures in which people must live. Thus, careful construction of both physical structures and social structures represent vital considerations.

7. Deviance and Other Social Problems

The foregoing discussion makes it clear that the process of sending a large population to another star system is a daunting prospect. The issues involved are many and complex. Much needs to be thought out before worrying about deviance and other social problems becomes a central concern. However, it is important to plan for coping with social problems as the following discussion strongly suggests. The lack of organization encourages the violation of norms based on values that favor harmful behavior.

On Earth, social problems have proven inevitable. Likewise, the subset of behaviors that comprise the category called "deviance" involves a cultural universal. That is, deviant acts exist in every known society. Deviance is defined by the larger culture of society, and different societies therefore can treat the same behavior as deviant of conformist in nature. Additionally, subcultures and countercultures within the same society define particular criminal behaviors as conformist. These differences could be based on religious values or simply those of a criminal or delinquent gang or syndicate.

Sociologist Diane Vaughan (1996) discussed the concept of the "normalization of deviance" in her analysis of the Shuttle Challenger accident. When patterns of behavior that were once considered too risky become entrenched into the culture of an organization due to a string of successes, decision making becomes based on a type of arrogance, a mindset that nothing can go wrong even when policies set in writing are violated. The launch decision of Challenger in weather that was too cold was known, and at least one manager pointed it out, but the decision to launch went forward anyway because the commission of deviance became a normal course of behavior; that is, until the tragedy occurred.

Deviance and social problems in general represent relativistic concepts. Their evaluation is in the eye of the beholder, so that a person committing the act views it as normal acceptable behavior while the witness or victim interprets it as deviant. This is important because it complicates matters. They are not simply right or wrong, deviant or conforming behaviors, and trying to understand their natures to prevent them in the future becomes more difficult.

8. Space Law and Definitions of Deviance

Law and astrosociology is defined by Christopher Hearsey as "the study of the nexus between law and astro-social phenomena" and furthermore, law consists of a system of rules, or legal norms, that require enforcement

(Pass, Hearsey, & Caroti, 2010). The legal system in the space society must provide the framework that defines acceptable behavior. Importantly, the type of government will determine what types of laws become implemented. A despot will design a legal system that favors personal advantages at the cost of less powerful citizens. It is important to construct the type of political system that would support the types of laws that favor the protection of average citizens

The constraints imposed by space law result in the placement of boundaries around what is acceptable behavior, but they also map out actions that could conceivably receive deviant labels. Space ecologies in spacecraft (and later habitats) produce unfamiliar social conditions despite terrestrial analogs. Thus, unanticipated actions may result in definitions of new forms of deviance. This is one of the causes of social and cultural change. The legal system adapts to new conditions just as other parts of society do. Creation of laws and other types of social rules in essence produces forms of behavior that become impermissible, or deviant, and which may have been permissible the day before they took effect. Planners will need to decide which laws to incorporate into the legal system as well as their relative seriousness, which in turn determines the type and level of punishment for their violation.

9. Crime and Delinquency

These types of behaviors are subsets of deviance. They are formal in nature because legislators enacted them. Therefore, crime and delinquency are violations of formalized laws or social rules. The difference between the two boils down to the age of the offender and the type of law involved. Additionally, while everyone may steal a candy bar from a store, a juvenile commits delinquency while an adult commits a crime. Other differences exist, however, such as with statutory rape and drinking in which being under a certain prescribed age, either as victim or perpetrator, makes the act criminal or delinquent depending on the circumstances.

Explaining crime and delinquency is a difficult task because so many variables exist, and thus cause and effect become mired in a great many potentially important variables. In fact, sociologists, those that specialize in criminology for example, have a long history of proposing and testing theoretical models to explain crime and delinquency. Theories include those that focus or blame the individual, those that focus on societal conditions or blame society, critical theories that blame capitalism and the elites of society, and functionalist theory that views deviance as a disruption in the status quo, or the equilibrium (for example of a discussion involving a space context, see Hermida, 2006). No one theory or approach can account for all types of deviant behavior. Each one has been thoroughly tested and can explain some forms crime and delinquency. Such knowledge is valuable to planners. Nevertheless, crime and delinquency will occur in space societies just as they do in all terrestrial societies.

10.   Health and Safety Violations

Health and safety violations are deviant acts. They are violations of legal, administrative, and organizational rules. These rules can come from public and private entities. When planning a space society, the question become what rules will be necessary to protect both the physical structure and the social structure, the latter of which involves humans interacting with one another.

One might think that because the commission of health and safety violations could result in devastating injuries, illnesses, and even the destruction of the habitat, serious harmful actions would never occur or represent rare occurrences. However, once again, Vaughan's (1996) concept of the "normalization of deviance" demonstrates how routinized procedures can result in complacency and even the inability to recognize violations or their dangers, which can play a devastating part in the commission of health and safety violations. Thinking about these possibilities represents a vital type of exercise.

Thus, rather than a waste of time, the study of health and safety violations in the space habitat will provide planners of the physical and social elements with an invaluable arsenal of theoretical and practical knowledge that will result in a safer and healthier ecology (Pass, 2011:356).

The very action involving the acknowledgment of the inevitability of health and safety violations will likely result in methodologies, policies, and practices that could reduce their likelihood of occurrence. Doing nothing will likely have the opposite effect.

11. Other Forms of Deviance and Types of Social Problems

Various forms of deviance not characterized as crime or delinquency exist, as do various degrees of seriousness. They range from piercing one's nose or violation of taboos. Less serious forms of deviance not considered the violations of criminal or delinquent laws may not attract much attention or formal punishment, but they may result in less formal negative sanctions.

Other forms of deviance include the violation of administrative rules or regulatory law. Examples of administrative rules, which have the backing of legal standing and are created by state agencies, include environmental and housing regulations. Policies of organizations consist of a collection of rules, principles, procedures, and guidelines for proper behavior. Compliance is expected and the organization considers violations as deviant and subject to punishment such as reprimands, demotions, pay cuts, and even firings.

In addition to deviance, many other types of social problems exist that can somehow harm an entire citizenry, such the failure of the life support system aboard the spacecraft or within the habitat. Many social problems, such as the effects of environmental pollution harm some segments of the population more than other segments. For example, the poor are most likely to feel the effects of pollution because they can only afford to live in areas in which harmful conditions exist. In the spacecraft or space habitat, will a class structure exist in which the second and third class citizens live in less safe and healthful locations? How will the leadership treat maintenance workers in terms of their job risks and living conditions? Forcing workers to take unreasonable risks can be seen as a type of deviance, especially by the victims of the harm it causes and their families.

12. Criminal and Juvenile Justice Systems

Because deviance exists, it becomes necessary to cope with it in ways that can ultimately protect society. Like any social problem, crime and delinquency each require formal attention by the government and other agents of social control. Police represent the most common source of law enforcement in terrestrial societies. Construction of a space society must take this into account.

A criminal justice system consists of practices and institutions that enforce the law by apprehending, prosecuting, defending, sentencing, and punishing those suspected or convicted of criminal offenses. The juvenile justice system does the same for juveniles. Thus, criminal and juvenile justice systems represent formal government social control. Although deterrence, or the prevention of crime and delinquency through the threat of punishment, represents a potential goal, as does rehabilitation, most of the effort falls to punishment after the fact. Police departments enforce the law normally by capturing suspected criminals and delinquents. The courts determine the guilt or innocence of the suspects. Correctional institutions punish the convicted criminals or delinquents, usually by incarcerating them.

A permanent space society must construct some type of criminal justice system – and juvenile justice system if younger individuals are part of the population – in order to cope with the inevitable crime and delinquency that will occur. Like with other social institutions, the formalization of this system of relationships is critical. Otherwise, citizens will not know what to expect or how to behave properly. It is beneficial in terms of minimizing the rate of deviance if they can internalize the rules.

13. Social Control

Social control consists of all of the methods and mechanism a society employs to maintain a normative social system; that is, the ways in which a society enforces conformity in line with its social norms. These norms, or social rules, range from informal unwritten expectations to formal written laws and regulations. There are many ways in which norms are enforced from a number of different types of individuals. The bottom line is that a society needs to ensure a certain level of conformity in order for it to function properly, however this normative state is defined. "In the most fundamental terms, 'social control' referred to the capacity of a society to regulate itself according to desired principles and values" (Janowitz, 1975:82). Laws emerge that protect a certain type of value, but non-legal norms exist to protect important values that are not codified into law. The overall purpose is to induce conformity to the most important social rules of a given society.

The question that arises in this context is the following: Who creates the laws that are upheld utilizing methods of social control? Sociologists have conceptualized social control in a number of different ways, from critical and Marxist theorists who argue that social control exists to protect the elites over the lower social classes to the functionalists who focus on social control as a means of protecting social order. In reality, both extremes have merit and the many sociologists who fall in the middle of the continuum have much to offer as well, but these concerns are beyond this discussion's focus even though they require attention as we move forward.

Mainstream sociology, and criminology, which is a subfield, has a long tradition of studying social control in addition to attempting to explain it nature and predict its effects. Various sociologists and others have proposed different types of social control. Each one potentially contributes to increasing the likelihood of conformity for most individuals in society. The brief treatment herein includes a subset of some of the most commonly theorized types found in the literature. These dichotomies from various theorists overlap somewhat in terms of their criteria.

One dichotomy involves direct and indirect social control (Mannheim, 1940). Direct social control directly regulates and seeks to control individual behavior so as to encourage conformity. Families and schools are influential agencies of social control. Common agents of direct social control include parents, teachers, classmates, and clergy. Indirect social control is exercised most commonly by secondary groups through such means as customs, traditions, and public opinion.

Two types of sanctions, positive and negative, are applied in the attempt to control behavior (Young, 1934). Positive sanctions, or rewards, exist to reinforce "positive" behavior through methods such as praise, appreciation, and monetary gain. Negative sanctions, or punishments, exist to prevent further deviant behavior by the perpetrator through incarceration, fines, criticism, and other means, and to hopefully deter others who learn of the offense and punishment.

Another important dichotomy consists of formal and informal means of social control. Government/written rules versus private/unwritten rules. Informal social control comes from informal authority such as a primary institution and it prevents the most deviance since most interactions with authority figures do not involve official law enforcement actors. Primary groups exercise informal social control. Examples include family members, clergy, teachers, and peers. In contrast, secondary groups exercise formal social control, and social control agents of this type include police officers, corrections officers, and FBI agents. Informal guidance by family members (e.g., parents, siblings, aunts and uncles) and friends is a major source of social control in any society. Nevertheless, modern societies utilize formal agencies and formal agents of social control such as police, the National Guard, army, as well as traditional media and social media. Formal control increases steadily as a society modernizes because secondary ties become increasingly influential, and the threat of punishment becomes a last resort attempt to induce conformity when informal social control fails.

Thus, various ways to attempt to control crime and delinquency exist. A few are discussed here to demonstrate the complexity of implementing a criminal justice system and social control mechanisms; and despite all of these controls, deviance remains a common feature of society. Potentially, social control strategies that involve indirect, informal and positive efforts produce the best outcomes, meaning that they are more likely to prevent or at least discourage crime and delinquency, and deviance generally. The problem is that these mechanisms do not exist in a social vacuum so that other influences weaken their impact on some individuals.

Also of great importance is that the same strategies to prevent deviance can also encourage its commission (Marx, 1981). Subcultures that promote crime and delinquency strongly influence their members to embrace deviance as an acceptable lifestyle while lessening the impact of societal sources – including laws and "positive" role models – that continually stress the wrongfulness of committing "acts against citizens and society."

14.   Training of Astrosociologists and Interested Social Science Students

Today, there are far too few students in the social sciences and humanities involved in the study of space-related issues. Astrosociology was founded on the principle that the education and training of social scientists, behavioral scientists, and humanists is absolutely necessary for the advancement of space exploration for both humans living on Earth, and those who choose to travel into space, for a limited time, or those who migrate into space for a long period of time, or even permanently. For those who wish to think outside the box – and literally outside the influence of their home planet – astrosociology presents the social science student of any age the opportunity to participate in something unconventional and exciting.

Whether or not a particular social science student decides to become a pioneering astrosociologist, development of the *astrosociological imagination* – which allows him or her to look beyond the surface of social and cultural events in one's personal life to recognize the underlying influences of astrosocial phenomena – presents a different lens for looking at the world. It is patterned after Mills' (1959) concept of the sociological imagination, which of course focuses on terrestrial society. In the present case, it provides a deeper understanding of the relationship between humankind and outer space. A worldview such as this prepares even those pursuing STEM careers for the increasing influence of space in the daily lives of individuals, social groups, institutions, and societies as they potentially move toward the spacefaring ideal type (Pass & Harrison, 2007). This ability to recognize the influence of space will result in a more comfortable existence for those who remain on Earth as well as those who decide to leave it for long-duration spaceflights or permanent residence in another space environment, whether that takes place aboard a spacecraft or on the surface of another cosmic body.

Astrosocial phenomena already affect human lives and the progress of societies for a relatively small cost. Understanding them is important to science and society. The future of humankind may well depend on its ability to live in space environments, as well as on Earth. The training of astrosociologists and related science students is quite late indeed – and thus, we need to catch up. While similar attempts to create an academic field such as astrosociology have occurred, such as the creation of exo-sociology as a subdiscipline of sociology (Mejer, 1983), none have succeeded until now. Past failures due to disinterest or distain of the study of space issues, possibly due to their inclusion into the category of pseudoscience and the broad paintbrush applied that included UFOs and alien abductions, have stifled social scientific input since the dawn of the space age. Nevertheless, an intrepid collection of social scientists has carried the torch for the social sciences even while their disciplines have largely ignored space issues (see Harrison et al., 1988, as a good example).

As astrosocial phenomena become more pervasive and impactful, failure to study them will result in decreasing social science input and more (or continued) speculation. It makes coping with deviance less effective and thus harmful aboard a starship, especially over the long course of a voyage to Earth 2.0 in another star system. The time has passed to welcome social scientists, humanists, and artists working on space issues to the table, as their input will provide new layers of understanding to traditional analyses.

15. Conclusion

The proposed project presented here – including the idea to construct all the pertinent parts of a space society from the training/resocialization process on Earth, to transferring the society to the spacecraft, and finally transfer to the habitat if Earth 2.0 is located – represents one admittedly grand means to increase the likelihood that a large population can survive and thrive in space on long duration expeditions. The focus on deviance in the context of larger issues associated with the construction of an entire space society has made it impossible to provide details about all the important institutions that must exist. However, it can be stated that significant decisions need to be made about the functions of each institution, including its orientation and objective(s), and how they all interact as parts of society. As examples, what type of political system and economic system are installed, and how are religious groups handled? The family structure encouraged and how other types are tolerated undoubtedly reflects important decision that requires tremendous thought. This planning process is daunting, so it is important to start thinking seriously about them now

Thus, the foregoing discussion should make clear that sending a large population on an extrasolar expedition involves much planning, training, and resocialization. It also involves luck in terms of how the society actually functions over time and how citizens react to their new environments and how they behave. While the focus on preventing and reacting to deviance is an important element of constructing a space society aboard ship or within the space habitat is quite important, many other elements that comprise it require attention. Replication of a society in space is a daunting proposition. On the other hand, failing to put in the necessary work will result in chaos and disorganization.

Deviance is inevitable and any settlement in a space environment must construct social institutions to respond to it, especially those acts that harm the social order as a whole, social groups, or individuals. Utopias are ideal types and thus unrealistic. Thus, the goal should be to cope with deviance rather than to attempt to create a system characterized without deviance. Responding to crime and delinquency is a complex matter. Planners need to implement social structures to cope with it. The process begins on Earth with the inclusion of social and behavioral scientists, humanists, and artists in collaboration with physical and natural scientists.

While the STEM disciplines and fields represent vital contributors to the success of any long-duration space expedition or settlement, it is necessary but not sufficient for that success. The other branch of science, which includes the social and behavioral sciences, the humanities, and the arts, complements the physical and natural sciences. Both branches must work together in formal collaboration to maximize the likelihood of successful survival in confined and isolated space environments. The "other" branch of science has much to contribute to space education and research.

Social life aboard a spacecraft heading for Earth 2.0 would be rather chaotic and probably unpleasant without an overall well-structured societal structure. This includes an identifiable culture and social institutions. A criminal justice system – and perhaps a juvenile justice system – is required for a large population. Potential citizens of the space society are much better off if they understand what is expected of them and the difference between acceptable and unacceptable behavior.

An important fact to consider is that criminals and deviants of all sorts will remain in the spacecraft or space habitat unless planners decide to construct alternative structures such as prison ships or separate prisons on the surface of a planet, asteroid, or moon. Therefore, they will need to be confined or otherwise dealt with.

The migration to Earth 2.0, a star system beyond our own solar system, represents a monumental challenge for humankind. As such, it is wise that the effort has started already. Moreover, volunteers can learn much within our solar system even as astronomers and planetary scientists search for another habitable planet that is friendly to human survival. Settlements on the Moon and Mars can provide invaluable knowledge and knowhow in terms of how to survive off the Earth's surface. Interstellar migration will involve a long voyage aboard a starship, so realistic terrestrial analogs for the interstellar mission could occur. It can provide those who lead farther missions in the future to get a good handle on what types of issues will arise in actual space missions. Long voyages within our own solar system would place voyagers in position to study the impact of deviance and other types of social problems.

Finally, the *astrosociological imagination* allows the researcher to look beyond personal and subjectively important matters so that the true aspects of social reality become uncovered and available for scrutiny. In the area of deviance in space ecosystems, the ability to strip away the subjective meanings and see things objectively allows for a more accurate understanding of how astrosocial phenomena impact on human behavior and lead some individuals to break social rules of various types. Furthermore, the acquisition of the astrosociological imagination provides insights on how to respond to deviant acts in objective and thus more effective ways. It is a learned skill that every astrosociologist must master and is beneficial to anyone seeking deeper answers to questions dealing with human beings in space.

References

1. Durkheim, E. (1893/1964). *The division of labor in society,* New York: The Free Press.

2. Durkheim, E. (1982/1895). *The rules of the sociological method.* Translated by W. D. Halls. New York: The Free Press,

3. Harrison, A. A. (2005). "Overcoming the image of little green men: astrosociology and SETI." Paper presented at the 2005 California Sociological Association conference in Sacramento, CA. Retrieved from the Astrosociology Research Institute (ARI) Virtual Library [online archive], http://www.astrosociology.org/Library/PDF/submissions/Overcoming%20LGM_Harrison.pdf.

4. Harrison, A. A., Billingham, J., Dick, S. J., Finney, B., Michaud, M. A. G., Tarter, D. E., Tough, A., & Vakoch, D. A. (1988). In Allen Tough (ed.) "The role of the social sciences in SETI." Section V, Paper 1 in *When SETI succeeds: The impact of high-information contact,* 71-86.

5. Helmreich, R. L. (1983). "Applying psychology in outer space: Unfilled promises revisited." *American Psychologist:* 38(4), 445-450.

6. Hermida, J. (2006). "A legal and criminological approach to criminal acts in outer space." *Annals of Air and Space Law,* Vol. XXXI.

7. Janowitz, M. (1975). "Sociological theory and social control." *American Journal of Sociology,* 81(1), 82-108.

8. Mannheim, K. (1940). *Man and society in an age of reconstruction.* London: Kegan Paul.

9. Marx, G. T. (1981) "Ironies of social control: Authorities, as contributors to deviance through escalation, nonenforcement and covert facilitation," *Social Problems*, 28(3), 221-246.

10. Mejer, J. H. (1983). "Towards an exo-sociology: Constructs of the alien." *Free Inquiry in Creative Sociology*, 11(2), 171-174.

11. Mills, C. W., *The sociological imagination*, Oxford, England: Oxford University Press.

12. Pass, J. (2011). "Deviance in space habitats: A preliminary look at health and safety violations." *Physics Procedia*, 20(2011): 353-368. Retrieved from the Astrosociology Research Institute (ARI) Library [online archive]: http://www.astrosociology.org/Library/PDF/SPESIF2011--Deviance-in-Space-Habitats.pdf

13. Pass, J. (2015). "Astrosociology and the planning of space ecosystems," AIAA Space 2015 Conference & Exposition, AIAA 2015-4650. Retrieved from the Astrosociology Research Institute (ARI) Virtual Library [online archive]: http://www.astrosociology.org/Library/PDF/Space2015-JPass-PlanningSpace Ecosystems.pdf.

14. Pass, J., and Harrison, A. A., "Shifting from airports to spaceports: An astrosociological model of social change toward spacefaring societies," AIAA Space 2007 Conference & Exposition, AIAA 2007-6067, 2007, Retrieved from the Astrosociology Research Institute (ARI) Virtual Library [online archive]: http://www.astrosociology.org/Library/PDF/Contributions/Space%202007%20Articles/Airports%20to%20Spaceports.pdf.

15. Pass, Jim, and Harrison, Albert A. (to be published in 2016). "Astrosociology: Outer pace, the convergence of the social sciences, and beyond," chapter in *Handbook of Science and Technology Convergence*, edited by Bainbridge, W. S., & Roco, M. C., New York: Springer Science + Business Media.

16. Tönnies, F. (2001). *Community and civil society*. Cambridge: Cambridge University Press.

17. Vaughan, D. (1996). *The Challenger launch decision: Risky technology, culture, and deviance at NASA*, Chicago: University of Chicago Press.

18. Young, K. (1934). *Introductory sociology*. New York: American Book.

Starship Alpha: The Case for an Earth-based Proto-starship

Manuel Richey

29005 Orchard Road, Paola, Kansas, 66071

manuel.richey@honeywell.com

Abstract

This paper proposes the creation of a self sufficient earth-based colony to simulate the effects of long-term space travel as would be encountered on a generational starship or other space based colony. The purposes for creating such a colony are identified. Issues are identified that can be addressed though the creation of such a colony (as well as those that can't). It is shown that creation of an earth-based colony can address most of the unknowns of long term space flight, leaving only a small list of purely technical challenges.

This paper describes one approach to conducting such an endeavor, what can be learned from it, and how to finance it. It is not assumed that such an endeavor can be planned flawlessly up front. Rather the described approach assumes an initial plan with refinement over time during its execution. Its final outcome (after several years) will ultimately be a self-sustaining colony. Financing for the endeavor is discussed to demonstrate project viability. It will come via NASA sponsorship, corporate donations, volunteer colonists supporting themselves through their own labor, crowd-sourced fundraising, TV contracts and enthusiasm such as that created by the Mars One project. From this 1st generation colony, we'll learn what can be produced and what must be transported; we'll learn solutions to physical and social issues pertaining to colony wellbeing and self sustainment; we'll then build the 2nd generation colony off-planet.

Keywords

proto-starship, colony, prototype, generational, simulation

1. Introduction

When Earth 2.0 is finally identified, a journey for exploration or colonization to this planet will traverse considerable distance and require a significant amount of time. With future technology, this trip may still require several generations to complete. The issues that must be surmounted for the creation of a viable space colony or generational starship are numerous. Examples are: propulsion, shielding, construction, political organization, crew psychological wellbeing, agriculture, local manufacturing, information technology, communication, environmental control, recycling, financing, equipment maintenance, etc. Most of these issues should be modeled in a closed system on earth prior to the construction of an actual starship or large space colony. Attempts have been made to do this in the past, such as Biosphere-2 [1] or HI-SEAS [2]. However, no attempt has yet been made to simulate a large (100+ person) self-sustaining colony over an extended (multi-year) period of time.

The time is right to make an attempt at creating a self-sustaining space colony here on earth; to learn what can be learned before an actual colony of such size is attempted in space. When a self-sustaining colony can be maintained under favorable conditions on earth, whose only imported raw material is electricity, then many of the non-propulsion issues involved in the creation of a generational starship or space station will have been addressed.

Can such a colony be constructed today and maintained using existing technology with limited finances? Would such an endeavor be worth attempting? This paper says YES to these questions, describing who, what, when, where, why and how. The biggest of these questions are why and how, but let's start with what.

2. What is an Earth-based Proto-starship?

Before a generational starship or large colony is developed in space, it should be prototyped on earth where adjustments can be made and the consequences of failure and catastrophe are reduced. This prototype is an earth-based proto-starship (Starship Alpha). It is stationary facility that emulates the living area of a generational starship or space colony. Indeed, in this paper we use the term proto-starship and earth-based colony interchangeably. A truly successful self-sustaining proto-starship or colony is a pressure vessel that contains and supports a sufficiently large group of humans with only one imported raw material, and that is electricity. Everything else is recycled. An information link to the outside world would also be provided, but we won't treat information as a raw material in this paper.

3 Why Create an Earth-based Colony

The concepts of a self-sustaining space colony for off planet expansion, or of a generational starship to reach a distant home, are well known and popularly written about in fiction. But is either of these concepts really even feasible? We don't yet know. We won't know for sure until we have successfully built one. We cannot predict how many failures will be required before a successful approach is concluded. Even if we think we have solutions to all of the problems we can anticipate, our efforts will undoubtedly be hampered by problems we cannot anticipate.

The purpose for creating a colony on earth to simulate either a generational starship or a space colony would be simply to test out theories and hypothetical solutions to the kinds of problems such a colony would encounter in the hostile environment of space. It makes sense to first encounter these problems in a friendly environment, where they can be overcome with less risk. The advantages of construction cost, the risk to life and limb, and the reduced consequences of failure all decisively indicate that the very first space colony should be built and populated on earth.

There are two approaches used for mission planning in space. The typical approach is top down, where all possible problems are identified upfront, and possible solutions are brainstormed. The best of these solutions are then selected for implementation. A less favored approach is bottom up or empirical. Under this approach, a problem is presented to a heavily vested group, which solves the problem via an initial approach followed by iterations to correct deficiencies. Due to the nature and risk associated with space flight, the top down approach to mission planning is generally used. Though both approaches have strengths and weaknesses for addressing problems, the bottom up approach of solving problems empirically is seldom available for endeavors in space. An earth-based space colony presents the opportunity for a bottom up approach to solving many of the problems associated with long duration space flight.

The technology exists to create this colony today. We need not wait. For many issues, this won't be an attempt to test already arrived at solutions, but rather a test bed for creating solutions to problems both foreseen and unforeseen. Not all problems that the colony could encounter need be solved at the outset. Many of these problems will be solved as they are encountered by the colony itself. Both the problems and the solutions can then be documented and reused. As Werner Von Braun said, "One good test is worth a thousand expert opinions."

As an example, the technical aspects of atmospheric management are not addressed by this paper, but would be addressed during this colony experience. The colony provides the laboratory and the rats (colonists) for those who are working on the technical aspects of atmospheric management. This is done outside of the starship by the support team. This starts with normal temperature and humidity control as is found in any building. However, it gradually transitions to a completely closed atmosphere through means which are not described here. When the starship atmosphere finally becomes self contained and remains so for a long duration, then a solution for atmospheric management has been achieved. Much of this activity can be performed by external support personnel without actually disrupting the colonists or risking their health, and can be completed long after the colony is initiated.

So what aspects of long duration space flight are addressed by creating a prototype colony on earth? That question is better answered by asking what aspects cannot be addressed with a prototype colony. Here is that surprisingly small list:

- Starship propulsion. Creating a stationary earth-based colony will not help further our knowledge on propulsion in space.
- Power Generation. This problem may benefit from simulation with an earth-based colony, just to evaluate the maintenance capability of the colony and to provide measured data on electricity usage. However, the actual problem of power generation can efficiently be solved without the creation of such a colony. Commercial power would probably be used just to reduce the cost of supporting the earth-based colony.
- Construction in Space. Though much can be learned about the physical facilities required to support a long term colony in space by first building an earth-based colony, many aspects of space construction cannot be addressed by such an endeavor. After all, one advantage of creating a prototype colony on earth is the low construction cost as opposed to the extremely high construction cost of building a structure capable of withstanding the harsh environment of space.
- Shielding. An earth-based colony is already adequately shielded against radiation, so it is not a natural candidate as a platform for radiation testing. Shielding from micrometeoroids can likewise not be tested with an earth-based colony.
- Gravitational effects. An earth-based colony would naturally use simulated artificial gravity (i.e. real gravity) with the assumption that there are no significant differences between artificial gravity and the real thing. The validity of that assumption must be verified by other means.
- Atmospheric leakage. Though an earth-based colony could be constructed within a vacuum in an attempt to address this issue, doing so would raise the cost of the facility and would not be an economical approach to solving this technical issue.

Most other issues pertaining to long term human space flight can be studied and addressed with the creation of an enclosed long term earth-based colony.

3.1   What Will Be Learned From An Earth-based Colony?

A list of issues involved in creating a space colony on earth are listed in Table 1 with an indication of how an earth-based colony can help address the issue. Obviously, entire books (or even libraries) could be written on each of these issues, and we won't have room to describe any of them in detail. The table does however list the issues, identify them as technological, logistical or human related, and the last column is color coded to identify which issues can be addressed with an earth-based colony. In the third column of Table 1, red shading indicates that an issue cannot be helped by creation of an earth based-colony. Green indicates that the issue is almost identical for both a space colony and an earth-based colony. Yellow indicates that the issue is similar, but not almost identical between a space colony and an earth-based colony. We will start this table with the five previously identified issues that cannot readily be addressed by an earth-based colony.

Table 1: Issues with Space Colonization

Issue with Long-term Space Flight	Type(1)	Can an Earth-based Colony Help?
Starship propulsion	T	Cannot help (earth colony is stationary)
Power Generation	T	Some help, (maintenance & data on power usage)
Construction in Space	T/L	Shouldn't help (not cost effective)

Radiation Shielding	T	Shouldn't help (wouldn't test with real humans)
Long-term Effects of Artificial Gravity	T	Cannot help (cannot defeat real gravity on earth)
Atmospheric Leakage	T	Shouldn't help (not cost effective)
Colony Organization and Government	H	Almost identical to issue faced by space colony
Law Enforcement and Justice System	H	Almost identical to issue faced by space colony
Medical issues	L/H	Almost identical to issue faced by space colony
Crew Psychological Wellbeing	H	Almost identical to issue faced by space colony
Colony Economy	H	Almost identical to issue faced by space colony
Organization of Labor	H	Almost identical to issue faced by space colony
Agriculture	T/H	Almost identical to issue faced by space colony
Population Management	H	Almost identical to issue faced by space colony
Potential Life Threatening Catastrophes	T/H	Much lower risk than issue faced by space colony
Recycling	T/H/L	Almost identical to issue faced by space colony
Social Issues	H	Almost identical to issue faced by space colony
Colony Communication	T/H	Almost identical to issue faced by space colony
Physical Facility Maintenance	L/H	Similar to issue with space colony
Manufacturing	T/L/H	Almost identical to issue faced by space colony
End of Life Issues	H	Almost identical to issue faced by space colony
Religious Issues	H	Almost identical to issue faced by space colony
Documentation of Lessons Learned & Procedures	H	Would directly benefit a space colony
IT Infrastructure	T/L	Almost identical to issue faced by space colony
Atmospheric Management	T	Almost identical to issue faced by space colony
Family Life	H	Almost identical to issue faced by space colony
Ranching	H	Almost identical to issue faced by space colony
Colony Termination	H	Not similar to issue faced by space colony
Moral Issues	H	Similar to issues faced by space colony
Water Treatment	T	Almost identical to issue faced by space colony
Clothing and apparel	T	Almost identical to issue faced by space colony
Colonist selection approach	H	Almost identical to issue faced by space colony
How do you handle people who want to quit the colony?	H	Not applicable to a space colony
*Spillover to non-space industries	T	May help other types of colonies (ocean, arctic, etc). Lessons Learned from extreme recycling may benefit all of mankind.
Risk to life and health	H	Greatly reduced from that of a space colony
Criteria for measuring colony success	H	Different criteria for success from a space colony

Note 1: Each issue is categorized as to type as follows: T=Technical, L=Logistical, H=Human.

One may ask, "Why are so few of these issues color coded red?" In other words, "Why are so many of the issues of long-term spaceflight also applicable to an earth-based colony?" The answer is because the vast majority of the issues have to do with keeping the crew alive and prosperous while residing inside a pressure vessel over a long period of time. If you are a crew member living inside a pressure vessel, where that vessel actually resides has little impact on your daily life since you live inside the vessel. Thus, if you are on a starship in route to a distant solar system, or if you are in a simulator here on earth, for 99% of the voyage, your daily life would be the same.

Obviously, since the deltas for pressure, temperature and radiation are much less for a simulator on earth, it will cost a lot less to construct an earth-based simulator than it would a starship.

Table 1 presents a large number of closed system colony issues at a very high level, but does not solve them. As was mentioned previously, the proposed colony would serve as a laboratory for testing solutions to the problem of long duration space travel. The approach presented here is that the colonists and mission planners start with a solution to each of these problems and that the colony then be constructed and inhabited, and that colony life begin based on the initial solutions. As deficiencies are observed in colony life, adjustments will be made and implemented by the colonists with the help of an external support crew as needed. An earth-based colony probably won't start as a fully self sustaining colony, but after years of adjustment, it will hopefully achieve that status.

For an earth-based colony, failure is accounted for and recovery is built into the system. Furthermore, the consequences of failure are non-catastrophic. If one aspect of colony life does indeed fail, the simulation can continue and allow other aspects to succeed. To illustrate this (again using the atmospheric example), should a viable approach to atmospheric recycling prove too difficult to achieve during the lifespan of the colony, the colony can carry on using regular atmosphere and continue testing and refining other important aspects of space colony life.

To summarize, an earth-based self sustaining colony need not start with solutions to the problems of long term space travel. Rather, it develops and tests solutions, and ultimately ends with them.

4. When Should We Create An Earth-Based Colony?

Now that we understand why such a colony should be developed, let's move on to the question of when it should be developed. It should be developed at least two generations before a generational starship is actually launched. The first generation of a colony would be hand-picked from a large group of volunteers. The second generation of colonists will not be hand-picked, but will inherit their status. It is reasonable to assume that the second generation will not be of the same caliber as the first. Therefore, it is important to observe the second generation closely to determine colony viability when both motivation and capability have decreased.

5. How Should an Earth-based Colony be Developed?

Now let's focus on how an earth-based colony could be created, funded and managed. Bear in mind that there are myriad ways to create an earth-based colony. This paper describes one approach to demonstrate that creating such a colony is viable. The path we'll take to accomplish this is as follows:
• Estimate an appropriate colony size
• Select a starship approach
• Outline a top level design of a proto-starship and facility
• Price the facility and effort

We'll then present a table describing how numerous issues with long term space colonization can be addressed by creating an earth-based colony, and conclude with a discussion of financing.

5.1 Colony Size

The first question that must be addressed for an earth-based colony is its size. How many people are we talking about for creation of a space colony? That of course depends on the duration of the colony, the cost of construction, and the probable list of willing colonists. It also drives most other decisions made about creating such a colony.

Obviously, a colony on the order of 5,000 people or more would be best to assure genetic diversity and trade specialization [3]. However, the cost of moving that many people (and the materiel necessary to support them) into space would be astronomical. To date, economic expenditure has been the limiting factor on space exploration, and it is reasonable to assume that this will continue to be the case. So for the sake of economics, let's assume that a minimum sized colony is desired for a generational starship. Just to drive a nail in the coffin of the big colony approach though, let's point out that given the choice of sending one large colony on an interstellar voyage or numerous smaller colonies, the numerous smaller colonies approach would win to mitigate the catastrophic effect of failure en route.

From a purely biological point of view, anthropologist John H. Moore estimated that a population of 150–180 would permit a stable society to exist for 60 to 80 generations — equivalent to 2000 years [4]. We can safely assume that this colony size could be reduced through the use of artificial insemination and the transport of a genetically diverse supply of sperm from earth to a distant destination. But let's keep the 150 number in mind as we look at other approaches to arriving at a minimal sized generational starship.

A minimal colony must maintain sufficient skill set to both survive and thrive while isolated from the rest of the human population. Though much information will be available to the colony via the onboard intranet, thorough expertise is still probably required in several fields. The critical fields for maintaining high levels of skill are medical, engineering, manufacturing and environmental/agricultural. A minimum quantity of skilled professionals for these essential fields is three, to allow mentoring of replacement personnel and to accommodate the unexpected loss of these personnel. As an example, three people capable of performing surgery should be planned to be available to the colony at all times. Should one (or even two) of these people die, replacements can be trained by study and through mentoring from the remaining surgeon. These high-skilled personnel will not be deep experts in sub-fields of their discipline, but rather broadly proficient across their discipline. Specialized knowledge is available when needed via the colony intranet and via communication with the outside world.

Only a certain percentage of the colony can be considered productive workers at any given time since many will be attending school and some will be retired and beyond their working years. Let's take a stab in the table below at a minimum sized colony based on critical professions and see what number that leads us to.

Table 2: Critical Professional Roles

Occupation	Quantity
Surgeon/Doctor	3
Health Professional: Doctor/Dentist/Optometrist	3
Electrical/Computer Engineer	3
Software Engineer/IT Specialist	3
Manufacturing/Mechanical Engineer	3
Agricultural/Environmental Expert	3

Data from the 2010 U.S. census [5] indicates that 35% of the population is of educational age (younger than 24) and 11% is of retirement age (older than 70). The remainder of the population is of working age. If we make the assumption that only 25% of that population will have the capacity (motivational, mental, emotional, etc.) to achieve competency in these fields and then pad by 10%, a minimum sustainable colony size would be:

CP = minimum colony population = 1.1 * 18 / ((100 – 35 – 11) *.01 * .25) = 147 individuals.

Having reached approximately the same number twice, let's assume the use of a transported bank of genetic material to ensure genetic diversity, then round this last number up to 150 individuals and set that as our minimum target colony population.

5.2 Facility

We now need to discuss a colony facility. In fact, we will do a top level design and calculate the cost of such a structure, because that cost is important to determining the plan's viability.

Many approaches to space habitat design are possible, but we need to select one approach to demonstrate feasibility. So here is an example approach that we select for the facility for this colony. We call it the long hallway approach to starship design. We'll arrange four long hallways with rooms on either side of the hallway, three stories high. The four hallways are linked together with cables or structural ties in a cylindrical manner as shown in Figure 1. Artificial gravity is provided by a rotating cylindrical arrangement in space, and is simulated by real gravity in a parallel arrangement on earth. During acceleration or deceleration, such as during the departure or arrival phases of a generational starship, the hallways can be repositioned to provide artificial gravity by aligning the hallways with the axis of acceleration. The same figure shows how the hallways would be set up for an earth-based colony. Admittedly, this approach is taken because it maps well to an earth-based colony.

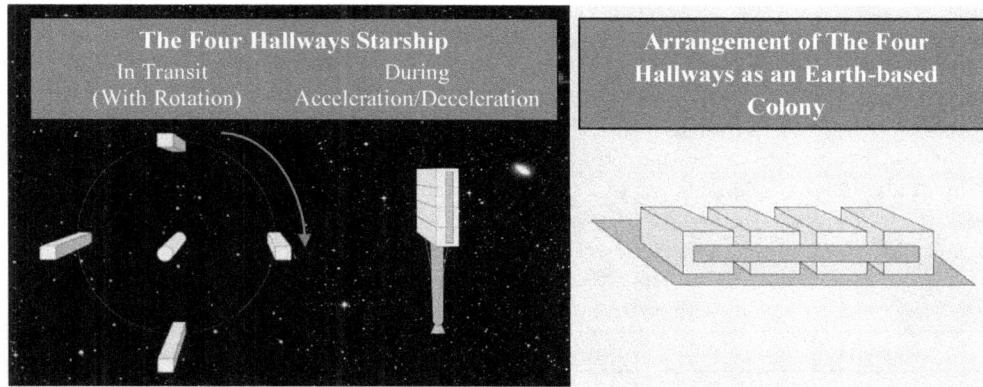

Figure 1: Space Station Approach, Credit: Background Image from European Southern Observatory
http://www.eso.org/public/images/eso0116a/

The hallways are redundant to accommodate disaster recovery. As an example, if a fire started in one hallway, the occupants could be evacuated to other hallways until the atmosphere is reconditioned and the damaged hallway reclaimed. Should a meteorite puncture the hull of a hallway and kill all of the occupants, then people from the other hallways may be able to repair the hull breach, reclaim the damaged hallway and increase the birthrate to eventually repopulate the empty hallway.

Another advantage of this approach is that in space the hallways could possibly be constructed from re-purposed launch vehicles. We typically build rockets in stages and then drop the stages as their fuel becomes exhausted. The fuel constitutes most of the weight of a rocket stage, so why not reuse the hull? It could then be repurposed in orbit to become a single hallway of a multi-hallway space station or starship.

Let's describe this approach just a little further. Each hallway would contain a generic facility and a specialty facility. Each hallway's generic facility would have the following services:
- Domestic living area
- Hygiene Areas
- Agricultural Area
- Gym/Exercise Area
- Recreational Area with inside 2 story high park facility.
- Cafeteria/Meeting Area
- Life Support Systems
- Two standard docking ports (one on each end) each with an attached escape vehicle capable of holding 1/6th of the colony (i.e. a little more than ½ the population of each hallway).
- A trolley docking port for transportation to other hallways.
- A small but well thought out trolley system will serve to transport personnel from one hallway to another. Each Hallway would also have one specialty facility for a total of four specialty facilities. Although actual equipment and supplies would be dispersed throughout the hallways to allow for disaster recovery, each hallway would focus its remaining area on creating a single specialty facility. These would be as follows:
- School/Social: Classrooms with a small theatre facing the meeting area.
- Hospital/Geriatric: A room for surgery, a warehouse for medical storage, offices for doctor/dentist/optometrist visits & a pharmacy.
- Factory/Recycling: Additive and subtractive manufacturing, recycling center, and a warehouse area
- Administration/Control : Administrative offices, IT and engineering labs, and a control room

To simulate the conditions that would be found in an actual space colony, the starship habitat should be as compact as possible, while still providing sufficient space for a contented and self sustaining crew. So what is that volume? How much room should we allocate for the inhabited area of our colony?

371

5.3 Size

Again, for economic reasons, this area should be as small as possible. However, we will not assume a level of compression as is typically found on a submarine, where techniques such as sharing beds (hot-bunking) are utilized to reduce domestic space, and cramped facilities are designed with knowledge that the crew's mission is short in duration. We will utilize slightly more space for psychological wellbeing, with the knowledge that the colonist's mission is essentially lifelong. The facilities will still be Spartan by terrestrial standards though. Since all light is artificial, work shifts can be utilized with hot-swapping of office and production stations. With that thought, we allocate space per capita as shown in the following table.

Table 3: Physical Volume per Colonist

Activity	Volume in m3	Comment
Agriculture	50	This number was taken from an analysis by So, et al. on evaluating the Mars One Project [6]
Living Quarters	8	Assumes a private dormitory for each colonist of 2x2x2 meters.
Public Space	8	Assumes a commons area for meals and meetings with 2x2x1meters/person and a tall park-like open natural area of 1x1x4 meters/person.
Manufacturing/Lab	2	Assumes manufacturing and lab facilities of 1x1x2 per person. These stations may be shared on work shifts.
Mobility	4	This is for a central hallway area of 0.5x2x4 per person
Kitchen and Gym	1	Assumes 1x1x2 per 8 people for a kitchen area and an equal amount for an exercise area.
Escape Vehicle Area	1	Should accommodate one person seated for a total area of 0.5x1x2 per person
Storage	2	Assumes 1x1x2 per person for storage of manufactured products and supplies being transported and emergency supplies
Office	3	Assumes a 2x2x3 meter office area for 1/4th of the colonists. This also includes a hospital area
Hygiene Space	1	Lavatories and Laundry Facilities. 1x2x2 toilet/8 people and 1x2x2 shower/8 people.
Maintenance	8	This includes electrical cabinets, waste water treatment, plumbing, electrical wiring, duct work, recycling, atmospheric management and soil storage and management facilities. Assumed 2x2x2 in area/person.
Total	*88*	*In cubic meters*

If you multiply this space by the total planned population, and add a 15% margin of error then the total internal volume for our target population is:

CV = internal volume for colony = 88 x 150 x 1.15 = 15180 cubic meters

With the assumption of three stories, each with an internal hallway that is 1 meter wide and two high, and rooms on each side of 2.5 meters deep and 2 meters high, you have a module width of 6 meters with a height of 6 meters. The total length for four such hallway structures can be calculated:

15,180m3 / (6m (width) x 6m (height) x 4 (hallways)) = 105.4 meters/hallway

5.4   Layout

The most appropriate external facility to house an earth-based colony would be a modern warehouse or man-ufacturing facility. These are typically divided into a front end office area and a back end warehouse or factory, consisting of a single large open area and loading ports for trucks in the back. The front lobby and office area would provide a public face to the colony project and house the small support staff. The large warehouse area would house the colony itself. This colony's living facility would be constructed within the warehouse area, which is typically open and usually about 20 to 30 feet high. The colony living facility could be constructed to resemble the layout of the inhabited area of a similar structure in space. The warehouse would provide a controlled environment that will ultimately bring down the cost of the colony structure housed within it. It would also serve to shield the colony from external influences.

The four hallways give a warehouse area of about 110m x 40m. If you add 20 meters on the front for admin-istrative space and 10 meters on the back for truck berths and storage, your warehouse would occupy 7700 square meters and could be laid out as shown in the following figure.

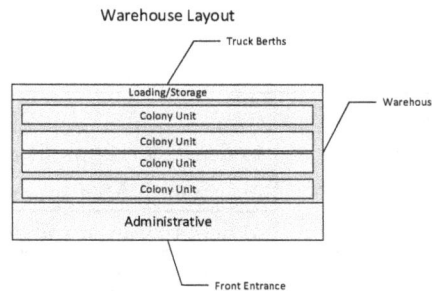

Figure 2: Layout of External Structure Housing Colony

The front lobby and administrative area of the facility houses the external management of the colony (i.e. Ground Control if you will). It would contain a male and female dormitory, a cafeteria, a recreation room, a front lobby and numerous workstation cubicles. The dormitories and cafeteria are used to house the college interns who will constitute the majority of the ground crew. This is elaborated on in the finances discussion later in the paper.

5.5   Facility Cost

We now need to provide a rough estimate for the cost of this facility for determining feasibility. Let us assume that warehouse space is leased and that an adequate existing location and facility can be identified. Warehouse leasing costs vary dramatically by location, but numerous facilities are available ranging from $3 to $10 per square foot per year. Numerous locations offering $5/sqft/yr are currently available in the Midwest [7]. We'll use that as our target.

Though the warehouse can be leased, the colony will need to be constructed. Units could be prefabricated and transported to the site where they can be assembled. If we use construction cost data from the Midwest [8], we would select a cost of from 70-$150/ft2. We'll stay on the cheap side of this, knowing that the inside facility need not stand up to the weather, that its foundation is already in place and that each hallway can be prefabricated. For this estimate, we'll assume that the indoor colony facility cost's $80/ ft2. This puts the total lease and upfront cost of constructing the colony facility at:

Annual Lease Cost = 7700m2 x 10.76 ft2/m2 x $5/ft2 yr = $413,490/yr

Upfront Internal Construction = 9m x 105m x 2 floors x 4 units x 10.76 ft2/m2 x $80/ft2 = $6.533 million dollars.

This is of course a rough order of magnitude calculation, but it is very useful for determining the feasibility of an earth-based colony. We'll use this dollar amount later when discussing the finances of the operation. This does give us an idea though that the plan is not at all unreasonable. When compared with the $6 billion [9] that is anticipated for the Mars One initial crew, this is quite economical and should have a higher probability of success.

The facility estimate doesn't include specialized equipment for recycling air, water and other materials. It also excludes needed medical and production equipment. Corporate donors will be sought for obtaining this equipment.

Now if we had made different choices for our space colony, then the corresponding earth-based colony would appear different. A spacecraft designed for constant acceleration (i.e. accelerating to the journey's halfway point and decelerating past that point) might have a large number of small stories as in a small high-rise building. A colony with a large population (>1000 people) might become a town constructed in a small valley that curves upwards at both ends and is domed overhead.

6 Approach to Starting an Earth-based Colony

So how should an earth-based colony be created? The answer is "in phases", with each phase requiring significantly more effort and resources than the previous. A next phase would not start until the previous phase had been successfully completed. This approach, outlined in Table 4, reduces risk and allows for early termination of the project should it become unviable.

Table 4: Colony Startup and Financing

Phase	Purpose	Activities
1	Organize and Sponsorship	1) Create a non-profit organization
		2) Obtain NASA sponsorship
		3) Select location and obtain state & local govt sponsorship
2	Raise Funds	4) Crowd-sourced fundraising
		5) Corporate donations
		6) TV contract (A Mars One idea)
		7) Colonist application fee (small) and sponsorship (larger)
3	Construct and Recruit	8) Select initial volunteers
		9) Plan and construct one hallway
		10) Operate initial hallway for a year with pioneer crew
4	Complete	11) Continue fundraising
		12) Construct three additional hallways
		13) Recruit additional colonists
5	Ongoing Operation	14) Continuous fundraising
		15) Facility maintenance
		16) Colony refinement and Analysis

6.1 Financing an Earth-based Colony

Of course, funding for basic research is currently difficult and government spending is particularly constrained. In fact, the advantages of this whole concept may seem pretty obvious, but no one is yet doing it (except in Hollywood). The reason it hasn't yet been attempted is primarily financial. So how can such an endeavor as that described by this paper be funded in today's economic environment? Here is a seven step process to achieve funding today:

1. NASA Sponsorship
2. State Sponsorship
3. Crowd-sourced Fundraising Campaign
4. Corporate Sponsorship
5. TV contract
6. Recruit Volunteers with Sponsorship
7. Donations

Let's now look at each of these steps in more detail.

6.1.1   NASA Sponsorship (This is step 1)

This is step 1 because this step provides legitimacy to the project. It costs the government little, but can be leveraged by the colony to obtain significant additional funding. Of course, what NASA gains by sponsorship is a significant voice in how the colony is constructed and operated without having to finance the endeavor itself. The fuel for this whole financial approach is enthusiasm (or hype), and that starts with NASA sponsorship.

6.1.2   State Sponsorship (Step 2)

Once NASA sponsorship is achieved, a suitable location for the colony facilities must be obtained.
- Place all 50 states in competition for this NASA project
- Possible donation of Land (maybe part of a space port)
- Selection process of U.S. state to host colony (e.g. no property taxes, inexpensive real estate, willing to exempt local laws such as standards for room occupancy, etc)

6.1.3   Crowdfunding Campaign (Step 3)

Armed with a noble cause, and exploiting the angle that the government would accomplish this if NASA were adequately funded, a well planned Kickstarter campaign should be able to raise a significant amount of money. Assuming that a successful Kickstarter campaign can be assembled and became a top performer, millions of dollars of funding could be brought in by this method. An average of the top 10 Kickstarter projects earned $8,678,228 [10]. None of those campaigns supported a cause as noble as the one described by this paper. None of them were endorsed by NASA. None of them would further knowledge and serve mankind the way this campaign would. Let's assume that Kickstarter and other crowdfunding campaigns can be mounted that will achieve the average of the top 10 to date.

6.1.4   Corporate Sponsorship (step 4)

While the crowdfunding program is proceeding, corporate sponsors should be sought. For corporations, supporting this project could be a big publicity windfall. This may involve bragging rights on TV commercials, as well as in other advertising endeavors. Donations could be monetary in nature, or could be made in the form of essential equipment. Corporate donors would be recognized as sponsors and have the right to advertise their sponsorship.
- Facility Donations (e.g. donations of mock B330 modules or mock Orion Cockpit for station use.)
- Hardware, IT and component donations from various companies
- Donations of medical and manufacturing equipment
- Corporate Partners could provide equipment for experimentation (e.g. air and water treatment companies). They would receive recognition as sponsors.
- Will create a monthly publication (A magazine for all sponsors) to keep them informed and provide free sponsor advertising. Will also maintain a website and Facebook page to keep everyone (inside and outside) informed of what is happening.

It should also be noted that corporations may wish to sponsor the project just to obtain some influence on its direction and test out their equipment. After all, this is a lab. A target amount for corporate sponsorship should be set and solicited. For the purposes of this effort, that target amount for corporate sponsorship is set for $1 million dollars. The Mars One project has raised close to this amount from corporate sponsors [11], and Mars One has a much higher probability of failure.

6.1.5   TV contract (step 5)

Speaking of Mars One, here is an idea borrowed from them. According to the US Bureau of Labor Statistics, we spend much more on the entertainment industry than we do on space exploration or research, about $2,500 per family per year [12]. A TV contract would provide a way to funnel some of that money back into productive activity. This activity would create celebrity, then use celebrity to generate funding.

The colony would be something new, highly unusual, extremely challenging, but with a noble purpose. These characteristics provide the ingredients for an excellent TV show. If well produced, a program based on this colony should easily appeal to a television audience who may wish to participate, but which can only do so by proxy. As a TV show, the project's celebrity impact will generate revenue from TV proceeds and from other sources related to its celebrity status. Some would even argue that the worse off the colony becomes, the better TV it makes, but we won't go there.

TV contracts for reality series vary from a small stipend to $100K per episode with the higher number for highly successful series [13]. For the purposes of this analysis, let us assume a TV contract would generate $50K/episode and run for 30 weeks/year for eight years. This would result in $1.5 million annual income from TV revenue. This is not an unreasonable assumption with good production and NASA sponsorship. After all, it was Alfred Hitchcock that said "drama is life with the dull bits cut out". There would surely be drama within the confines of a colony and ground crew, even if no one is being killed. This approach may entail providing cameras in the conference rooms and recreational facilities that would not be present in an actual space habitat. But this natural drama could be molded by a skilled editor into a successful TV show and the cameras would be useful for psychological analysis.

The possibility for licensing and merchandising may also exist. Some examples would be T-shirts, cups, toys and sponsor clubs. Once the main TV contract has expired, and due to the celebrity status of the project, publicity income will trickle in, albeit at a much slower pace. Let us assume a small amount of continuing income from articles, interviews, licensing and speaking fees of $100K/yr over the duration of the project. Maybe even a book deal or two would pop up. At the least, a TV contract would serve as a draw for volunteer colonists and ground crew.

6.1.6 Recruit Volunteers (Step 6)

Due to the actual size of the colony and support staff, this project would become extremely expensive unless volunteer labor is used. The proposed staffing approach in this paper relies on using purely volunteer labor for continuing operations. The obvious reason for using volunteers is to drastically control personnel costs. Due to the number of people involved, this project only becomes economically viable if personnel costs are kept low. Here is how this would work:

- No employees allowed. All volunteers are motivated solely by the cause of human advancement. The volunteer colonists are being asked to sacrifice numerous comforts for the sake of learning. Like the colonists, the support staff should also be making sacrifices.
- Volunteer colonists support themselves through their own labor within the station.
- Each candidate colonist would pay a small fee (>$40) for their application to be considered and processed, but each finalist would obtain a sponsor to pay a sponsorship fee ($5,000) to sponsor that colonist. The colonist is already making a significant sacrifice, so the financial sponsor would typically be a corporation or club or relative. That sponsor would keep sponsorship rights for that colonist for the duration of the colony unless they sell those rights. If the colony is successful, then individual colonist sponsorships could even be sold at a profit to other entities.
- The support crew is made up almost entirely of college interns who are offered free room and board at the facility, but no salary. They cook and clean up afterwards. They take out the trash and vacuum. They perform facility maintenance. They do technical analysis and staff the front desk. They even run the project alumni program to solicit funds from previous interns.
- Retired volunteers to provide professional services such as Project Administration, intern supervision, HR, facilities, accounting and legal.
- Personnel on-loan from corporate sponsors and from NASA may be paid, but would not be financed by the project.

The use of college interns for staffing the ground crew may appear to be a liability, but it will actually prove to be the opposite. The interns will rotate every semester and provide each intern with space program experience. In the long run, this will benefit the entire space effort. They will bring a constantly regenerating enthusiasm to the project that could not be obtained in any other way. Their long term support will spill over into financial support as they become project alumni.

Though it may seem important for continuity to have a core paid staff, having one would decrease the dedication of the volunteers. They will give more if everyone is giving, and no one is getting. For this reason a highly paid CEO is out of the question. It is better to settle for a qualified retired CEO who will donate his time for a worthwhile project, than to have the very best that money can buy.

Are there organizations that thrive based on this principle of volunteerism? Surely there are, but they are mostly religious [14]. What the religious organizations have that produce prodigious amounts of volunteer labor is a noble cause; a cause considered worthy of sacrifice by their volunteers. This project would likewise be a noble cause and considered worthy of sacrifice by many people. Volunteers will come pouring out of the woodwork to participate. The challenge will come in selecting appropriate volunteers.

Of course, an established project is required to recruit volunteers, but such projects must somehow become established. Some initial salaried personnel may be required to set up the project. These would be phased out though when the project is well established.

6.1.7 Continuing Donations (Step 7)

The project would be set up as a non-profit organization, and as such can easily solicit tax exempt donations. A website and Facebook page requesting sponsors similar to those supporting the 100 Year Starship program would be created and some money would trickle in. Benefits to sponsorship would be VIP visits to the facility, occasional phone calls with colonists, and other typical benefits that sponsors to non-profit organizations receive (i.e. T-shirts, coffee mugs, monthly magazine, etc.). An alumni association for interns would also be created with similar benefits.

Let us target a modest amount of $250,000/year in donations. This is not entirely unreasonable. After all, donations for the largest 100 non-profits ranged from 118 million per year up to over six billion dollars per organization [15].

6.2 Balance Sheet Overview

We now need to assemble all of this information and demonstrate financial feasibility. Here's is a back of the napkin balance sheet for the project:

Table 5: Colony Balance Sheet

Item	Expense	Income	Balance
STARTUP			
Expenses			
• Facility	$6,533,472		
• Initial Staff Salary	$1,250,000		
• Furnishings for Front Office	$500,000		
• Colony Equipment and Furnishings (See Note 1)	$1,500,000		
Income			
• Crowdfunding Campaign		$8,678,228	
• Corporate Donations		$1,000,000	
• Colonist Sponsors		$750,000	
Total Startup Balance	*$9,783,473*	*$10,428,228*	*$644,756*
CONTINUING OPERATION (per year)			
Expenses			
• Salaries	$0		
• Facility Lease	$413,490		
• Utilities (Electricity, Natural Gas, Water, Phone/Data, Waste removal).	$50,000		
• Cafeteria Expense	$70,200		
• Taxes	$0		
• Profit	$0		
• Miscellaneous Expenses	$50,000		
Income			
• TV Contract (See Note 2)		$0	

• Interest on Investments (return-4%/yr)		$265,790	
• Continuing Donations		$250,000	
• Continuing TV and Publicity Income		$100,000	
Total Continuing Operation Balance	*$583,490*	*$615,790*	*$32,100*

Note 1: Much of the furnishings and equipment should not be purchased, but rather manufactured by the colonists; thus ensuring that all essential parts can be replaced and recycled. This expense is only for raw material and specialty items.

Note 2: Half of this income together with the remaining startup income is invested, and an annual return is used to cover annual expenses. This return is shown in the row titled "Interest on Investments". The other half of this income is placed in an endurance fund that is provided to the surviving colonists on termination of the colony. Hopefully this endurance fund could be increased.

This rough balance sheet shows the project starting in the black and continuing in the black year after year. Admittedly, this balance sheet is only a ROM estimate, and the funding approach is unorthodox. But it is also creative and viable. We have demonstrated that finances need not hinder the founding of an earth-based colony. So what's stopping us?

7. Conclusion

In this paper we have described the reasons for creating an earth-based self-sustaining colony. In a nutshell, it would facilitate confronting and solving most of the issues of long term space flight. There are of course numerous approaches to creating a successful and useful earth-based colony. This paper has described one approach to creating and financing such a colony, thus demonstrating that it would be of value, and that it can be accomplished.

The enthusiasm generated by the Mars One project indicates that the time for such an endeavor is upon us. From this 1st generation colony, we'll learn what can be produced and what must be transported. We'll develop new approaches to environmental maintenance and recycling. We'll learn new approaches to production and farming in an isolated and confined space. We'll learn solutions to physical and social issues pertaining to colony wellbeing and self sustainment; we'll then build the 2nd generation colony (Starship Beta) off-planet.

About the Author

Manuel Richey (manuel.richey@honeywell.com) is a principal engineer at Honeywell International in Kansas where he has worked for over 30 years. He is also a computer science instructor at Fort Scott Community College and holds eleven U.S. patents.

Bibliography

1. University of Arizona, "Fast Facts," 2014. [Online]. Available: http://b2science.org/who/fact. [Accessed 15 August 2015].

2. E. A. Moore, "Six humans step into one-year isolation dome," USA Today, 31 August 2015.

3. C. M. Smith, "Estimation of a genetically viable population for multigenerational interstellar voyaging: review and data for project hyperion," Acta Astronautica, p. 16–29, April-May 2014.

4. CNN, "Report: Make Deep Space Travel a Family Affair," 9 March 2002. [Online]. Available: http://www.cnn.com/2002/TECH/space/03/09/family.spacetravel/. [Accessed 14 August 2015].

5.   United States Census Bureau, "Patterns of Metropolitan and Micropolitan Population Change: 2000 to 2010," 6 May 2013. [Online]. Available: http://www.census.gov/population/metro/data/pop_pyramid.html. [Accessed 15 August 2015].

6.   S. Do, K. Ho, S. S. Schreiner, A. C. Owens and O. L. de Weck, "An Independent Assessment of the Technical Feasibility of the Mars One Mission Plan," September 2014. [Online]. Available: http://dspace.mit.edu/handle/1721.1/90819. [Accessed 14 August 2015].

7.   LoopNet, "Search Properties for Lease," 15 August 2015. [Online]. Available: http://www.loopnet.com/xNet/MainSite/Listing/Search/SearchResults.aspx#/KS/Industrial/For-Lease/c!ARYC$BAQ. [Accessed 15 August 2015].

8.   Brown Wegher Construction, "How Much Does a Construction Project Cost?," 2015. [Online]. Available: http://www.brownwegher.com/cost-of-construction-chart/. [Accessed 15 August 2015].

9.   Mars One, "How much does the mission cost?," 2015. [Online]. Available: http://www.mars-one.com/faq/finance-and-feasibility/how-much-does-the-mission-cost. [Accessed 15 August 2015].

10. N. Zipkin, "The 10 Most Funded Kickstarter Campaigns Ever," 30 March 2015. [Online]. Available: http://www.entrepreneur.com/article/235313. [Accessed 14 August 2015].

11. Mars One, "Support Mars One's Human Mission to Mars," 11 March 2015. [Online]. Available: http://www.mars-one.com/donate. [Accessed 15 August 2015].

12. United States Departement of Labor, "Economic News Release," 9 September 2014. [Online]. Available: http://www.bls.gov/news.release/cesan.nr0.htm. [Accessed 15 August 2015].

13. R. Haidet, "Pay Scale For Reality Hosts & Stars," 24 August 2011. [Online]. Available: http://realitytvmagazine.sheknows.com/2011/08/24/pay-scale-for-reality-hosts-stars/. [Accessed 29 8 2015].

14. J. DiSanto, "Penn Research Shows That Mormons Are Generous and Active in Helping Others," 12 April 2012. [Online]. Available: http://www.upenn.edu/pennnews/news/penn-research-shows-mormons-are-generous-and-active-helping-others. [Accessed 15 August 2015].

15. M. Hrywna, "The NPT 2014 Top 100: An in-depth study of America's largest nonprofits," The NonProfit Times, pp. 1-8, 1 November 2014.

100 YEAR STARSHIP™

Propulsion and Energy

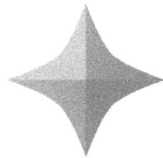

Chaired by Hakeem Oluseyi, PhD

Physicist, MLK Fellow, MIT, TED Fellow,
Chief Scientist Discovery Channel, Associate Prof. Florida Institute of Technology

Track Description

A major aspect of discovering details of exoplanets is to get closer to them, to take samples and test actual physical properties beyond our solar system. Profound breakthroughs in the generation, storage and control of energy for propulsion, as well as communications and data gathering instruments are required to get to the interstellar medium in 10-20 years, much less to reach another star. Such breakthroughs are accompanied by robust leaps in theory and technology paradigms, and also incremental advances in engineering technology deployable in the next 5, 10 and 15 years.

Presenters in Propulsion and Energy are asked to present research and supportable ideas on how to address the design and deployment of instruments, probes and vehicles that will accelerate travelling beyond our solar system and closer to exoplanets within the next 25 years.

Track Chairs Biography

Hakeem Oluseyi, PhD

Physicist, MLK Fellow, MIT, TED Fellow,

Chief Scientist Discovery Channel, Associate Prof. Florida Institute of Technology

---

Hakeem M. Oluseyi, PhD is an internationally recognized astrophysicist, science TV personality, and global science education activist. His research interests span the fields of astrophysics, cosmology, and technology development. He currently has 7 U.S. patents, 4 EU patents and over 60 scholarly publications in the areas of astrophysics, optics and detector technologies development; nanotechnology manufacturing; observational cosmology; and the history of astronomy. Dr. Oluseyi leads a group studying processes by which electromagnetic fields and plasmas interact in order to understand solar atmospheric heating and acceleration, which has resulted in a new in-space propulsion technology.

Photonic Laser Thruster Laboratory Demonstration towards Interstellar Photonic Railways

Young K. Bae, Ph.D.

Y.K. Bae Corporation, 218 W, Main St., Suite 102, Tustin, CA 92780, USA

ykbae@ykbcorp.com

Abstract

The spacecraft speeds required for interstellar travel would consume unprecedented amounts of energy that cannot be practically provided by onboard energy sources, even with nuclear fusion and antimatter. The Laser Sail beams thrust from the mother ship to the mission ship via a laser beam, eliminates the need of the onboard energy sources, and has been considered to be the best proven propulsion for interstellar travel. However, the Laser Sail has a very low energy to thrust conversion efficiency, thus requires too high laser power and too large spacecraft structures to be practical. Photonic Laser Thruster (PLT) has been developed and demonstrated to enhance the Laser Sail efficiency by factors greater than 1,000 by repetitively bouncing recycled photons between the mother ship and the mission ship. Under the auspice of NASA Innovative Advanced Concepts, PLT has been demonstrated for photon thrusts up to 3.5 mN and enhancement factors up to 1,500 using a 1-kW thin disk laser system, and to dynamically propel, slow, and stop a 1U cubesat in laboratory environment. We propose the Photonic Railway: a permanent energy-efficient reusable interstellar transportation structure, which consists of an array of PLTs. A 100-year four-phased developmental roadmap of the Photonic Railway towards interstellar commutes, which can provide return-of-investment for furthering development, is proposed. If successfully implemented, the Photonic Railway is projected to bring about a quantum leap in the human economic and social interests in space from explorations to terraforming, mining, and permanent habitation in planets, asteroids, moons and exoplanets.

Keywords

Interstellar Flight, Photonic Laser Thruster, Photonic Railway, Spacetrain, The Laser Sail, Photon Rocket

1. Introduction

One of the major requirements for interstellar flight is propulsion technology that is capable of propelling spacecraft to speeds at least 10 % of the light speed, v~0.1c = 30,000 km/sec. For more than five decades, photon propulsion, which generates thrust by physically bouncing light on the mirror or sail of spacecraft, had been researched for interstellar flight, because it is founded on established physics and technologies. [1,2] Recently, photon propulsion became more realistic because of breakthroughs that a successful recycling photon thruster capable of amplifying thrust by orders of magnitude, which increases the photon propulsion efficiencies by orders of magnitude, was demonstrated in a laboratory setting [3,4], and that a successful space deployment of solar sail, which is a form of photon propulsion, was achieved. [5] Such breakthroughs along with recent impressive development of high power lasers and associated hoptics now establish necessary technological foundations for photon propulsion towards interstellar flight.

Traditionally, the photon propulsion concept, the photon rockets, that utilizes photons generated by onboard photon generators, such as blackbodies or lasers, powered by solar or nuclear power. Here, the author developed a unified theory on such onboard photon propulsion that can be used for providing its maximum attainable velocities. In this unified theory, it is assumed that the propulsion system has a single stage. Suppose the total mass of the photon rocket/spacecraft is M that includes fuels with a mass of αM with $\alpha<1$. Let us assume that the fuel mass to propulsion-system energy conversion efficiency is γ, and the propulsion-system energy to photon energy conversion efficiency is δ that is much smaller than 1. Then the maximum total photon energy available for propulsion, E_{ph}, is given by

$$E_{ph} = \alpha\gamma\delta Mc^2.$$ (1)

If the total photon flux can be directed at 100 % efficiency to generate thrust, the total photon thrust, T_{ph}, is given by

$$T_{ph} = \frac{E_{ph}}{c} = \alpha\gamma\delta Mc.$$ (2)

The maximum attainable velocity, V_{max}, of the photon propulsion system for $V_{max}<<c$, is given by

$$V_{max} \approx \frac{T_{ph}}{M} = \alpha\gamma\delta c.$$ (3)

Examples of the maximum attainable velocities by onboard nuclear-powered photon propulsion systems are given in Table 1.

Table 1. The maximum velocity obtainable by photon propulsion with onboard nuclear photon generators with exemplary parameters.

Energy Source	α	γ	δ	V_{max}/c
Fission	0.1	10-3	0.5	5x10-5
Fusion	0.1	4x10-3	0.5	2x10-4

The maximum velocity limits by two nuclear power concepts are orders of magnitude less than the required 0.1 c for interstellar travel, thus they are impractical for interstellar missions.

The above theoretical limits posed by photon propulsion with onboard photon generators can be overcome, if the photon generators and the mission spacecraft are physically separated. In this concept, the photons are beamed from the photon source to the spacecraft using lasers as in the Laser Sail. In particular, Forward [6] worked on a wide range of interstellar propulsion concepts including photon propulsion and antimatter rocket propulsion, and proposed [2] The Laser Sail aiming at the goal of achieving roundtrip manned interstellar flight. However, according to the recent study by Millis [8] the power and engineering requirements to implement the original the Laser Sail are projected to be only achievable beyond the year 2500. Considering the unprecedentedly large world-scale investment required for such interstellar flight, unless there is enormous potential financial return from such endeavours, the chance of sustaining continuous return investment in such programs is highly unlikely.

An important theoretical understanding and development of the Laser Sail was obtained by Marx, [9] Redding, [10] and Simmons and McInnes [11] who calculated that the energy conversion efficiency of photon propulsion is approximately proportional to v/c at low speeds (v<0.1c), thus is very small at very low speeds (v<<0.1c). However, once the spacecraft reaches higher speeds (v>0.1c), owing to the favourable Doppler-shift energy transfer, photon propulsion becomes much more energy efficient, thus there is a need to bridge this energy efficiency gap. Meyer et al. [12] followed by Simmons and McInnes [11] proposed that recycling photons between the spacecraft and the photon beaming source would be a solution to this issue. Possible applications of photon recycling using passive resonant optical cavities (lasers are located outside of the optical cavity), the Laser Elevator, in launching and propelling spacecraft at higher velocities with higher efficiencies than those available by exiting rocket engines, was first proposed and extensively studied by Meyer et al. [13] They concluded that for missions requiring very fast transit times in the solar system or for interstellar fights, the recycling photon propulsion vehicles are much more energy efficient than that carry their own propellant, such as nuclear rockets.

However, the usage of such passive resonant optical cavities for recycling photon propulsion was questioned by the author, because they are extremely unstable against the motion of the cavity mirrors, thus unsuitable for propulsion. [14,15] Therefore, the author proposed the use of active resonant optical cavities, in which the optical gain medium is located within the cavity, and named the thruster with such optical cavities as the Photonic Laser Thruster (PLT). [15] The proof-of-concept PLT was demonstrated in laboratory environment by the author under the auspicious of NIAC/NASA, and its several spin-off space applications, including the usage in primary propulsion and satellite/spacecraft maneuvering were proposed and investigated. [3,4,16]

The author proposed a permanent energy efficient transport structure based on multiple PLT systems, the Photonic Railway, which aims enabling routine interstellar commutes via Space Trains, which can be used for profitable endeavours. [17] The Photonic Railway, if successful, would radically depart from the conventional spaceship concepts, in which a single spacecraft carries both an engine and a large quantity of fuel. Rather, the Photonic Railway would have permanent reusable space structures that propel Space Trains, which would consist of mainly crew habitats, and navigation and crew safety equipment.

This paper reports also the recent scaling up and the demonstration of PLT, which establish the technological foundation of the Photonic Railway. It is predicted here that the development of the practical Photonic Railway will further require development of x-ray lasers, advanced material science and technologies for radiation protection, low-weight super-strength spacecraft structures, and large long-lifetime super mirrors.

2. The Photon-Particle Unified Theory of Propulsion

Photon propulsion uses direct momentum transfer of photons to propel spacecraft and has been researched since the beginning of the 20th century. [1] The capability to propel spacecraft is largely governed by the rocket plume exhaust velocity, and its ratio, the specific impulse, I_{sp}, to the gravity acceleration constant, g. According to Special Relativity, the highest velocity of the rocket plume exhaust can have is the light velocity, c = 3 x10^8 m/sec. Therefore, photons are the ultimate rocket propellant that will produce extremely high specific impulse, I_{sp}, which is given by:

$$I_{sp} = \frac{c}{g} = 3.06 \times 10^7 \text{ sec} \cdot \tag{4}$$

In relativistic rocketry, the momentum, p, of a propellant is given by

$$p = \sqrt{\frac{E^2}{c^2} - m_0^2 c^2} \, . \tag{5}$$

where E is the kinetic energy of the propellant and m0 is the rest mass of the propellant. E in turn is given by

$$E = \frac{m_0 c^2}{\sqrt{1-\beta^2}}. \tag{6}$$

where β=v/c and v is the exit velocity of the propellant exhaust.

The specific thrust. F_{su}, is a parameter to indicate how efficient a thruster is in converting the input energy to thrust. The theoretical upper bound of specific thrust, F_{su}, the ratio of the relativistic momentum to the relativistic kinetic energy, is given by

$$F_{su} = \frac{p}{E - m_0 c^2}. \tag{7}$$

By using Eqns. 4-6, Eq. 7 can be further simplified into

$$F_{su} = \left(\frac{1}{v}\right)\frac{\beta^2}{1-\sqrt{1-\beta^2}}.$$ (8)

Eq. 8 is the unified relativistic equation that can be applied whether the propellant is mass particles or photons.

Fig. 1 shows the unified specific thrusts, F_{su}, of chemical rockets, electric thrusts that include Hall thrusters, and Pulsed Plasma Thrusters, in comparison with that of photon thrusters.

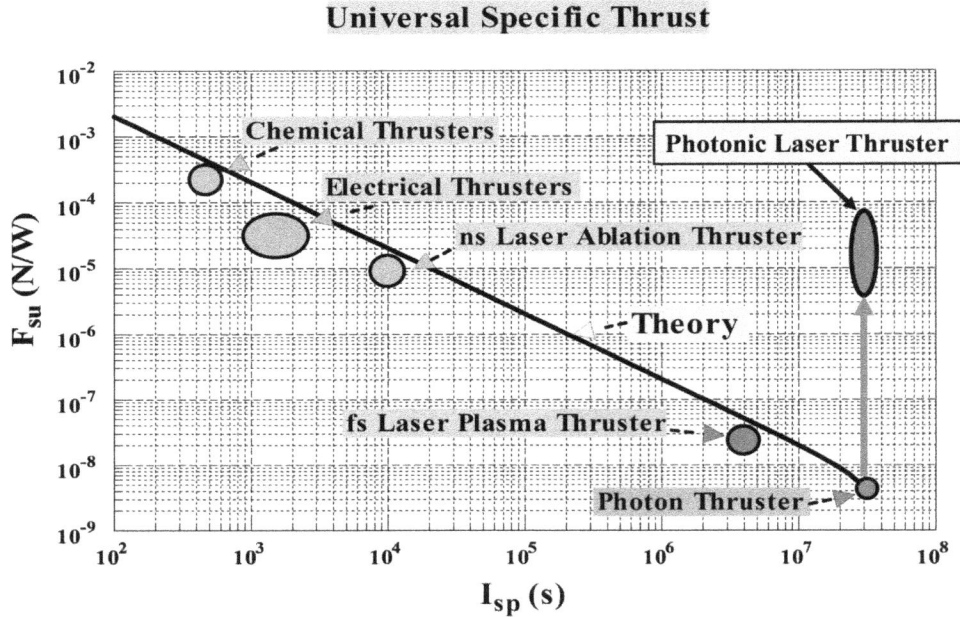

Figure 1. The overall specific thrust (thrust to power ratio) of representative propulsion systems. The solid back curve represents the unified theory expressed in Eqn. 8. With photon thrust amplification, Photonic Laser Thruster (PLT) has the ultimate I_{sp} as well as a high specific thrust, F_{su}.

The specific thrust of the photon thruster is several orders of magnitude smaller than that of conventional thruster, such as electrical thrusters, because it has the highest I_{sp}. The solid black line in Fig. 1 represents the unified specific thrust, F_{su}, given by Eq. 8. The inefficiency in producing thrust at extremely high I_{sp}, is for all thrusters, and it is not unique to the photon thruster. In other words, if conventional thrusters can be made to have $I_{sp} \sim 10^7$ sec, theirs specific thrust would be similar to that of photon thrusters. Therefore, the thrust efficiency does not depend on whether the propellant is made of photons or other particles, such as protons, in achieving relativistic velocities. With photon thruster amplification greater than 1,000, the specific thrust, F_{su}, of PLT can be comparable with that of LOX thrusters and Lightcraft, but the I_{sp} of PLT would be orders of magnitude larger than that of the latter. [15] Therefore, PLT and Photonic Railway are proposed to be the next breakthrough propulsion needed for expanding the scope of human space activities from the near earth activities to routine interstellar commutes.

3. Laboratory Demonstration of Photonic Laser Thruster (PLT)

In his theoretical work, Marx [9] derived for the first time the energy transfer efficiencies from the photon energy to the kinetic energy of the spacecraft of photon propulsion. For the laser sail, the instantaneous efficiency η_i and the total efficiency η_t, are given by: [9,11]

$$\eta_i = \frac{2\beta}{1+\beta},$$ (9)

$$\eta_t = 1 - \sqrt{\frac{1-\beta}{1+\beta}},$$ (10)

where $\beta = v/c$. At low speeds, $\beta \ll 1$, $\eta_i \sim \beta$ and $\eta_t \sim \beta/2$, therefore, at non-relativistic velocities, the Laser Sail is highly inefficient. However, at relativistic velocities with $\beta \sim 1$, $\eta_i \sim 1$. Thus, photon propulsion is highly efficient at relativistic velocities. One way of overcoming the low efficiency at low velocities in using photon propulsion is to use multistage approaches, in which at low spacecraft velocities, particle-based propulsion systems, such as electrical thrusters, are used. [25]

Here, the effort is focused on making the propulsion system simple such that the spacecraft carries only minimal propellant for attitude control and maximal equipment for crew habitation and safety environment. One of the best ways to achieve such a goal is overcoming the inherent inefficiency in producing thrust of the photon thruster by amplifying the momentum transfer of photons by recycling photons between two high reflectance mirrors. The simplest recycling scheme is a Herriot cell with multi-bouncing laser beams between two high reflectance mirrors without forming a resonant optical cavity. This Herriot cell type approach was first proposed by Meyer [12] followed by Simmons and McInnes, [11] Their study was in depth analyzed by Mertzger and Landis [26] recently. This approach requires highly focused laser beam spots on each mirror to avoid the beam interference that may induce optical resonance in the cavity, which works against thrust amplification as explained below.

Photon thrust amplification based on Herriot cells seems to be straightforward, however, the implementation of the concept turned out to be not. As the cavity length and the number of photon bouncing increase in Herriot cells, the focal spot diameter projected on mirrors increases, requiring extremely large mirrors to avoid the laser beam interference. [15] Once the laser beam starts to interfere, the non-resonant cavity becomes a passive resonant cavity that is shown below to be impractical for photon propulsion amplification. In fact, the first experimental attempt on photon thrust amplification in a non-resonant Herriot-cell type optical cavity was performed by Gray et al. [27] They could obtained amplified photon thrust of ~0.4 μN with a 300-W laser and a photon thrust amplification factor of ~2.6, which was much smaller than the anticipated amplification factor greater than 50. The much lower-than-expected amplification factor obtained by Grey et al. [27] revealed the above mentioned technical difficulties in the Herriot cell concept. [15]

Meyer et al. [14] proposed to overcome the challenge posed by Herriot cell type photon amplification, and published elaborate calculations on the energy efficiency of recycling photons in a passive resonance optical cavity, in which a laser system is located outside of the optical cavity. The passive resonant optical cavity, Fabry-Perrot optical resonator, has been extensively used in high-sensitivity optical detection methods, such as the cavity ring down spectroscopy. [28] In the cavity ring down spectroscopy, typically laser pulses are injected through the first mirror and bounced between two mirrors as many as tens of thousand times. The current off-the-shelf technological limit of the system reported is obtained with super mirrors used for the cavity ring down spectroscopy with the reflectance of 0.99995 with the photon bounce number of 20,000. [28] This experiment clearly demonstrated that thrust amplification in optical cavity by orders of magnitude (tens of thousands) is feasible in principle.

However, the passive resonant optical cavity for photon thrust amplification turned out to be unsuitable for propulsion applications, because it is highly sensitive to the small changes in the distance between the mirrors and mirror deterioration. This sensitivity was observed in the gravitational detection system (LIGO) with such high-Q passive optical cavities, in which even one nanometer perturbation in cavity length sets the system out of resonance and nulls the photon thrust. [29] In addition, the high-Q passive resonant optical cavity requires near single-frequency lasers to efficiently inject the laser through the input mirror. Typically such single-frequency lasers have poor power-to-photon conversion efficiency. Therefore, it was concluded that the passive resonant cavity photon thruster is unsuitable for photon thrust amplification. [15]

The author initially proposed PLT mainly to overcome the difficulties in injecting sufficient laser power in high-Q optical cavities for the usage of precision formation flying in which the mirrors of the optical cavity are in near static conditions. [14] In PLT, a laser cavity is formed between two space platforms with the laser gain media located between them as illustrated in Fig. 2, in contrast to the previously proposed multiple reflection laser photon propulsion concepts that use passive optical cavities with the laser amplification located outside of optical cavity.

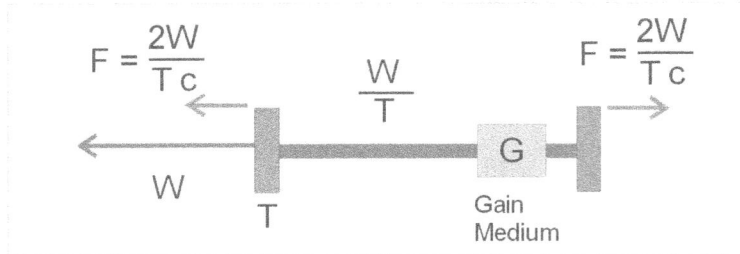

Figure 2. Schematic diagram of the Photonic Laser Thruster (PLT), which is based on the active resonant optical cavity approach to photon thrust amplification.

Under the auspice of NIAC/NASA, the author successfully demonstrated the proof-of-concept of a PLT. [3,16] In this demonstration, a PLT was built from off-the-shelf optical components and a YAG gain medium, and the maximum amplified photon thrust achieved was 35 μN for a laser output of 1.7 W with the use of a HR mirror with a 0.99967 reflectance. This performance corresponds to an apparent photon thrust amplification factor of ~3,000, and the actual amplification factor of 100. More importantly, in the experimental demonstration, the author accidentally discovered that the PLT cavity is highly stable against the mirror motion and misalignment unlike passive optical cavities. In fact, in the demonstration experiment by the author, the full resonance mode of the PLT was discovered to maintain even when one of the HR mirror was held, moved, and tilted by a hand to the author's surprise. In a more systematic experiment, the PLT cavity was systematically demonstrated to be highly stable against tilting, vibration and motion of mirrors. Subsequent theoretical analysis by the author showed that PLT can indeed be used for propulsion applications, and proposed Photonic Laser Propulsion (PLP), the propulsion with PLT. [15] The reason for the observed stability results from that in the active optical cavities for PLT and PLP the laser gain medium dynamically adapts to the changes in the cavity parameters, such as mirror motion, vibration and tilting, which does not exist in the passive optical cavities.

Recently, under the auspice of NASA Innovative Advanced Concepts, PLT has been demonstrated for photon thrusts up to 3.5 mN (100 times thrust scaling up) and enhancement factors up to 1,500 (15 times efficiency improvement) using a 1-kW thin disk laser system, and to successfully propel, slow, and stop a 1U cubesat in laboratory environment first time. The laboratory demonstration was performed on a 2-m air track that minimizes the friction. This represent the first time demonstration of moving a macro object of a satellite with laser propulsion in history. Fig. 3 shows the photograph of the actual demonstration of propelling the 1U cubesat.

Figure 3. Infrared photo of propelling a 1U cubesat with Photonic Laser Thruster in laboratory environment with a photon thrust of 3.0 mN and the amplification factor was 1,000.

Fig. 4 schematically shows the energy transfer efficiency from the photon energy to the spacecraft kinetic energy as a function of $\beta=v/c$ in the Laser Sail.

Photon Propulsion Efficiency

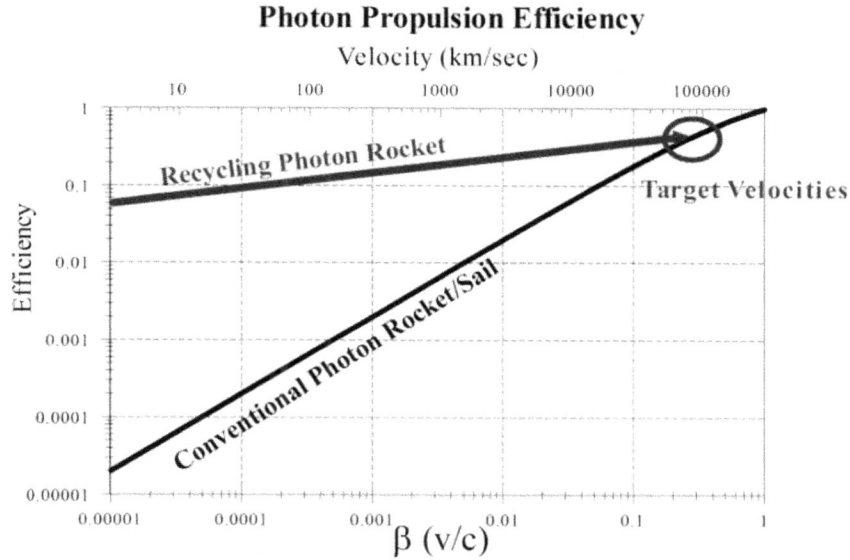

Figure 4. Energy transfer efficiency from the photon energy to the spacecraft kinetic energy as a function of β=v/c. As the spacecraft velocity approaches the relativistic velocity, the need of photon amplification decreases owing to the increasing energy to thrust conversion efficiency of photon propulsion.

Fundamentally, photons transfer their energy to the spacecraft by redshifting due to Doppler shift upon reflection, thus the higher the spacecraft speed is, the higher the efficiency is. It is interesting to note that at speeds near c nearly 100% of light energy is converted to the spacecraft kinetic energy, as if the spacecraft acts like a black hole in the moving direction. The lower solid curve in Fig. 4 represents the efficiency of conventional photon rocket and sail with photon recycling. The upper solid line represents schematically an example the efficiency of recycling photon rocket, PLT. At low β, the PLT can have a very high thrust amplification factor (in this example, ~3,000), however, it is expected that as β approaches 1, the PLT amplification factor should asymptotically converge to 1. Therefore, as the spacecraft velocity approaches the relativistic velocity, the need of photon amplification decreases owing to the increasing energy to thrust conversion efficiency of photon propulsion.

In PLT, the photon thrust, F_T, produced by a laser beam on each mirror is given by:

$$F_T = \frac{2}{c}\frac{W}{T}, \tag{11}$$

where W is the extracavity laser power through the mirror, and T is the cavity loss factor. More realistically, because PLT has the gain medium in the laser cavity, the circulating power in the cavity, W/T, can be estimated by [30]

$$\frac{W}{T} \approx \frac{G}{T'}\frac{I_{sat}}{2}A \tag{12}$$

where G is the unsaturated round-trip gain factor, I_{sat} is the saturation intensity of the gain medium, A is the effective lasing area in the gain medium, and T' is given by:

$$T' = T + a + s \tag{13}$$

where a is the roundtrip absorption coefficient and s is the roundtrip scattering coefficient. To have a high W/T, PLT should have high G and I_{sat}, but low T'. Examples of the maximum theoretical thrust as a function of the cross sectional area correlating with various laser powers with I_{sat} ~ 1.4 kW/cm², G~1, T'~0.001, are summarized in Table 2.

391

Table 2. The exemplary maximum theoretical thrusts of the photon thruster based on Nd:YAG with I_{sat} ~ 1.4 kW/cm2, G~1, and T'~0.001 (amplification factor of 1,000). The actual achievable thrust also depends on other parameters, such as thermal management capability. The large cross sectional area of gain media can be achievable either with a single crystal or by multiplexing numbers of smaller gain media.

Power Required in Intracavity Due to Loss Minimum Cross Sectional Area of Gain Medium

(Nd:YAG)	Maximum Intracavity Power	Maximum Theoretical Thrust	
1 k W	1.43 cm2	1 MW	6.7 mN
1 MW	1,430 cm2	1 GW	6.7 N
1 GW	143 m2	1 TW	6.7 kN
1 TW	143,000 m2	1,000 TW	6.7 MN

The PLT system for this estimation is based on Nd:YAG crystal. The actual achievable thrust also depends on other parameters, such as thermal management capability. Another important question is how large the cross sectional area of gain media can be constructed. With the use of the recently developed slab gain medium design, achieving the cross sectional area of 100 cm^2 is within reach with the current-state-of-the-art high power solid state lasers. However, the gain medium with greater than 100 m^2, is technologically extremely challenging. One approach to overcome the thermal management problem is to combine lasers beams from a number of small gain media on a grating employing the spectral beam combining technique. [31] Another approach is to use gas laser technologies, as in the Air Borne Laser (ABL). One interesting approach alternative to diode pumped lasers, which may be highly important to PLT development, is solar pumped lasers, [32,33] as was first envisioned by Forward. [2]

One of the factors that limit the maximum obtainable velocity of the accelerating mirror and its accommodating spacecraft is limited by the Doppler shift of the bouncing photons. Doppler shift effect on the active resonant cavity behavior is an extremely complicated issue, which is beyond the scope of the current paper. Eventually, this aspect should be studied with computer optical simulation. Optical gain in the laser cavity can only occur for a finite range of optical frequencies. The gain bandwidth is basically the width of this frequency range. For example, the gain bandwidth of the YAG laser system with the laser wavelength in the order of 1,000 nm is in the order of 0.6 nm, [34] which is ~ 0.06 % of the wavelength. For an order of magnitude estimation, we assume that PLT utilizing the YAG laser system will be limited by the gain bandwidth to the first order, then, theoretical maximum spacecraft velocity is ~1.8×10^5 m/sec (180 km/sec) that is 0.06 % of the light velocity, $c=3 \times 10^8$ m/sec. To overcome this redshift limitation, PLT, multiple PLTs with different wavelength gain media need to be used for the Photonic Railway.

Traditionally, the intracavity laser arrangement required for PLT operation had been operated in relatively short cavities less than 10 m long. Therefore, there has been a concern that the action distance of PLT may not be more than tens of meters. However, recently, Bohn [35] of the German Aerospace Center (DLR) reported that the German company Rheinmetall Defense demonstrated a 1-km long laser resonator similar to the PLT optical resonator in 1994-1995 with the use of a telescopic arrangement in the optical cavity, and that such long laser resonators can be scalable to 100 km with the usage of optics in the diameter of 70 cm. [35] These successful demonstrations promise that PLT can be operated beyond distances in the order of 100 km. Further studies should be performed whether PLT can be used for interstellar scales, but so far there is no show stopper on this issue.

One of key technological issues in implementing PLT is in the intracavity laser beam aiming, aligning, and tracking, which will be addressed more in depth in the discussion section. With the rapid advancement in laser weapons, the aiming, alignment, and tracking of laser beams on rapidly moving uncooperative targets over the distance greater than 100 km have become technologically feasible. Although the technical details of such aiming, alignment, and tracking system is grossly classified, the nut-shell of the technology is available in open literatures. Especially, the technology developed for ABL will play crucial role in PLT systems. Based on open literature, in ABL, the aiming, alignment, and tracking of the main laser rely on the scattered beam of the beacon laser (also diode pumped lasers at power level of a few kW). Similar to this, a small laser (power level of a few watts) in the mission vehicle can be used as a beacon laser. It seems that the aiming, aligning, and tracking system can be scaled to interstellar distances.

4. The Interstellar Photonic Railway

In this section, we consider the roadmap for developing the roundtrip manned interstellar flight based on PLT in a century with projected power production capabilities and technologies. Fig. 5 shows an example of a hypothetical projection of the required power for PLT as a function of year. The projected total world power production at 1.9 % yearly growth rate in TW (1012 W) as a function of year, which has recently analyzed by Millis [8] in depth, is plotted as the upper thick black line. Assuming an equivalent power of the total world power production (probably in a form of space solar power with SSP) can be dedicated to the ambitious interstellar flight, and assuming that the PLT laser has an energy conversion efficiency of 10%, the originally estimated power consumption by the Forward roundtrip interstellar flight would require about 80,000 TW (Forward, 1984), which is projected to be achievable well beyond the year 2500 as shown in Fig. 5. The 10 % of the total world power production as a function of year is plotted as the lower thick gray line.

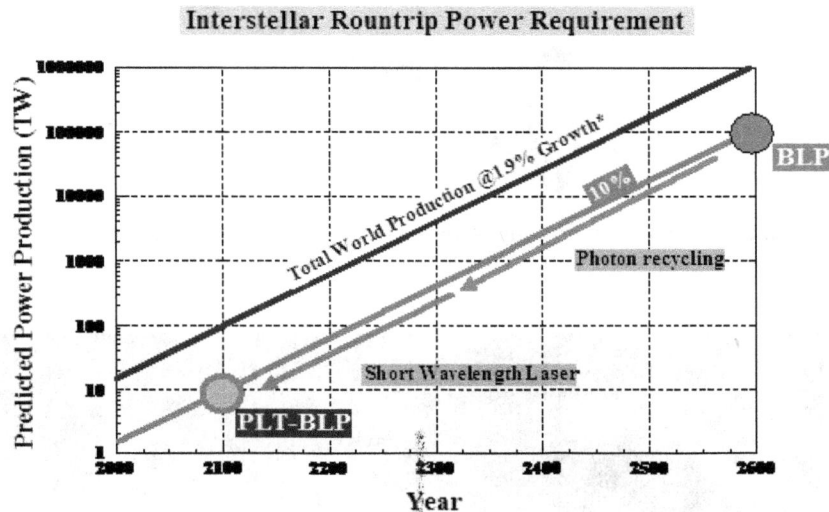

Figure 5. A projection of the required power for PLT-THE LASER SAIL as a function of year based on the world power production projection by Millis. [8]

In this example, it is estimated that the PLT would cut down the power requirement of BLP by a factor of 10 – 1,000 and that the incorporation of short wavelength laser by a factor of 10-100. In this case the total required photon power for BLP can be reduced by a factor of 100 – 100,000. In an optimal case with a reduction factor greater than 10,000, these two technological developments will be sufficient to make the PLT within reach by the year 2100. However, in less optimal case, suppose that PLT factor reduction is only 30 and the short wavelength laser 30, then an additional power reduction by a factor of 10 is required. Such power reduction can be achieved by spacecraft weight reduction with the rapidly developing material technologies, such as carbon nanotube materials, [36] and physiological technologies, such as minimal long-duration survival closed systems for crews in the future. Once PLT is implemented, the initial exploration flight would be performed, which will be followed by construction of the Photonic Railway as shown in Fig. 6. In this figure, a multitude of structural parts for constructing a large lens system and laser system will be transported from the Earth. In this case, the PLT system in the earth will be used for delivering the necessary components by both propelling and slowing down them.

Once the components start to arrive at the vicinity of the exoplanet, multiple PLT systems will be assembled and made to operate fully automatically by probably a sophisticated self-directing robotic system projected to be available by the end of the 21st century. Once the PLT system at the exoplanet becomes fully functional, it can be used for stopping the exoplanet-bound Space Train and for propelling the earth-bound Space Train that rides on the Photonic Railway. Fig. 7 illustrates a Photonic Railway consists of four PLTs: two for acceleration and two for deceleration. One important factor is that the Photonic Railway PLT needs to operate much shorter distance than the distance between the earth and the exoplanet. Typically, depending on the Space Train acceleration condition (the optimal case would be 1 g acceleration for maximum crew comfort), the system operation distance would be at least a factor of 3.2 shorter than the flight distance. Because of this, the Photonic Railway optical system can be at least a factor of 10 smaller in size than the PLT-BLP optical system.

393

Photonic Railway Construction

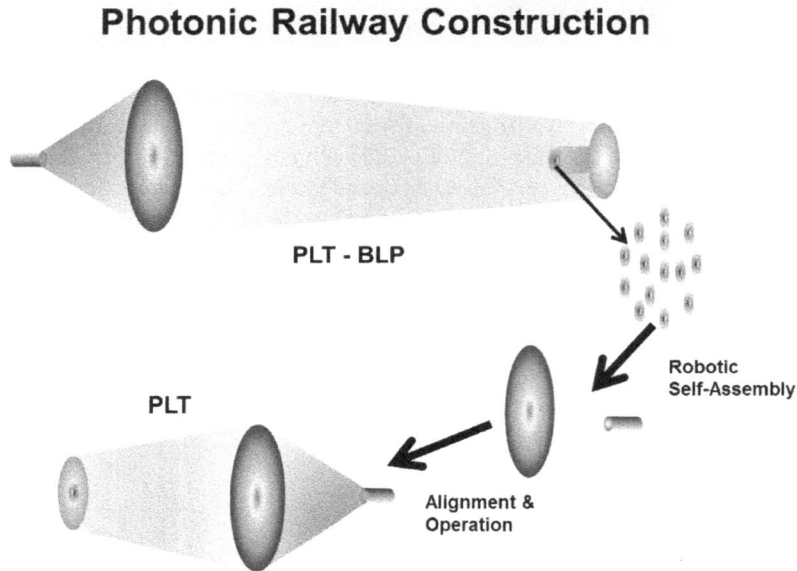

Figure 6. Construction of Photonic Railway with PLT-BLP with a multi-stage mirror sail. Focusing of the laser beam here was achieved with a large lens, however, a large mirror can be used as in Fig. 7.

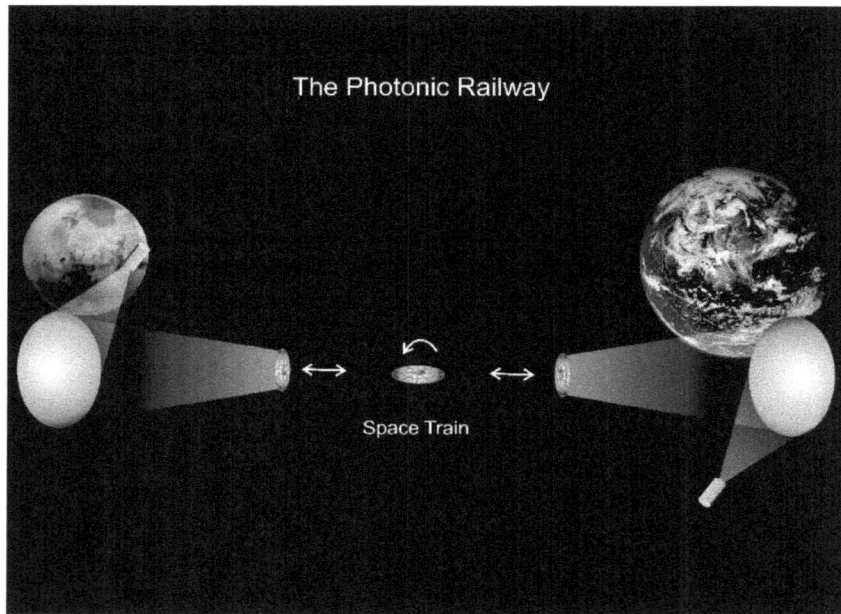

Figure 7. Illustration of a Photonic Railway that consists of four PLTs: two for acceleration and two for deceleration, which can generate artificial gravity. The Space Train will have small thrusters for attitude control and most of the onboard spacecraft resource will be dedicated to crew comfort and safety.

5. The Four-Phased Developmental Pathway Of The Photonic Railway

In the previous sections, it was shown that PLT has a potential to revolutionize future human endeavors in space, beyond that can be achieved with conventional rocketry. It was also proposed that the Photonic Railway, a permanent transport structure based on multiple PLTs, has a potential to enable routine interstellar commutes via

Space Trains. The Photonic Railway, as the transcontinental railway systems did, is projected to inspire sustainable economic interest and return investment, and to potentially achieve the goal: rountrip manned interstellar flight potentially within a century. [17] In this section, a four-phased developmental pathway of the Photonic Railway toward interstellar manned roundtrip is proposed: 1) Development of PLTs for satellite and NEO maneuvering, 2) Interlunar Photonic Railway, 3) Interplanetary Photonic Railway, and 4) Interstellar Photonic Railway. It is projected that these developmental phases will result in systematic evolutionary applications, such as satellite formation flying, NEO mitigation, lunar mining, and Space Solar Power (SSP), which is projected to generate sufficient sustainable economic interest and return investment to the development pathway.

5.1   Phase I: Near-Earth Objects (NEO) Maneuvering with PLT

The first phase in the developmental pathway towards implementing the interstellar Photonic Railway is maturing PLT technologies and systematic scaling up of its power/thrust and operation distance capabilities. PLT can provide the unprecedented capabilities in maneuvering spacecraft in near earth orbits, thus PLT is predicted to meet the needs of the next generation of space industry by enabling a wide range of innovative space applications near the earth. Examples of such unprecedented capabilities include propellantless operation, thrust and power beaming, and ultra-precision spacecraft maneuvering. [14,16,37] In this phase, which is predicted to evolve over the 5 – 30 year time frame, PLT would be capable of providing thrusts in the range of 1 mN – 1 kN, which requires the operation power of 100 W – 100 MW. The solar panel based space power currently can provide electrical powers on the order of 100 kW, therefore, the PLT capable of providing thrusts ~ 1 N can be readily implemented in the near future. Further scaling up of PLT with 1 kN thrust will require a large solar power system capable of providing power up to 100 MW, of which development and implementation may depend on the space solar power or space nuclear power development in the near future.

In Phase I, the operation distance of PLT is projected to reach ~ 100,000 km, which can cover a wide range of spacecraft maneuvering at LEO, MEO, and GEO. For example, once the spacecraft is in orbit, a 1 ton vehicle will take about 2.3 hours to cover 100,000 km via a 100 MW PLT with a thrust amplification factor of 1,000 for flying by. The diffraction limited size of the beaming mirror/lens should be on the order of 100 m and the spacecraft mirror diameter 2.44 m. The maximum thrust of such a system would be on the order of 1 kN. For rendezvous missions, such a spacecraft will take about 5 hours including deceleration time to traverse 100,000 km.

As PLT is successfully implemented in space and systematically scaled up, its economic interest is predicted to grow exponentially. In addition, PLT is predicted to reduce the use of toxic chemicals and minimizes the pollution in near earth space environment, which is becoming a growing concern as the number of space activities rapidly increase. Spacecraft formation flying is vital to the construction of next generation satellites with fractionated components and space solar power facilities for power beaming the harnessed solar power in space to the earth. Numerous commercial and defense applications of the formation flying technologies are projected to be potentially enabled by PLT, which include large space telescopes at GEO for real-time "Google" map and large low-cost space radars. Precision formation flying with PLT will play a vital role in constructing large scale space structures in GEO by significantly reducing the orbit raising cost from LEO. Examples of such space structures, which would need this technology, would be space solar power beaming facilities, large space radars, large surveillance optical satellite structure, space stations and space habitats. These applications will provide sustainable development of the Phase I PLTs.

The nanometer precision formation flying of satellite using a combination of PLTs and tether, the Photon Tether Formation Flight (PTFF), was proposed and investigated in detail by the author. [14] In an another approach, a group of spacecraft exploit relative positions and velocities so that differential gravity provides a force opposite that of the photon thrust from PLT in a way similar to what a tether might provide. [16,37] In such a scheme with two orbiting platforms, their positions to the center of the mass can be controlled by adjusting the two balancing forces: 1) the photon thrust from PLT, 2) the counterbalancing "virtual tug" that is generated by relative-orbital perturbation to create a capability for maneuvering spacecraft in earth orbit without using propellant and tethers. Fig. 8 illustrates examples of satellite formation using PLT pushing-out force and counterbalancing "virtual tug" generated by gravity gradient. [16,37]

A more general motions are possible in the case of a PLT system without tethers than that with tethers, [15] where the relative displacements are very limited. The PLT maneuvering system, however, could accomplish control of out-of-plane motions as shown in Fig. 8, thus would represent a breakthrough technology. [16,37] In addition, its propellantless performance prevents contamination, and saves considerable mass. PLT force acting in the along-track direction has the long-term effect of speeding up and slowing the pair of satellites between which it acts, which can be used for propellantless rendezvous of satellites. [37,38] In addition, PLT can be used for second-party propellantless stationkeeping. [38]

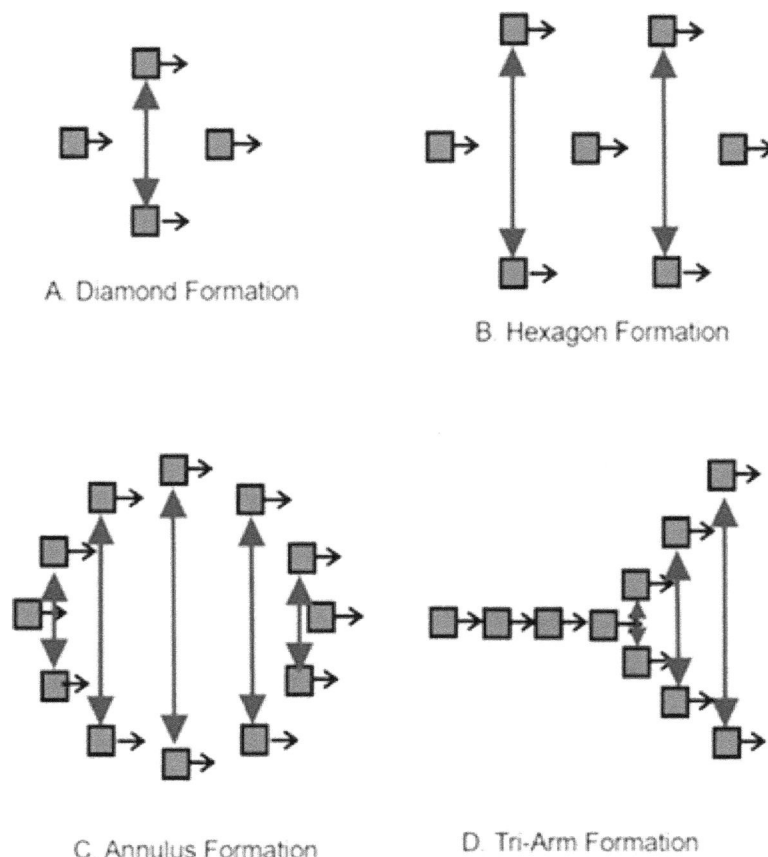

A. Diamond Formation

B. Hexagon Formation

C. Annulus Formation

D. Tri-Arm Formation

Figure 8. Examples of satellite formation using PLT pushing-out force and counterbalancing "virtual tug" generated by gravity gradient. [14,16,37]

PLT can be also used for second-party propellantless orbit-drag compensation. This PLT application will greatly reduce propellant requirement. Here, two spacecraft in very similar orbits with low inter-vehicle velocity are used for making the mission much more economical and reliable. [38] For example, a large resource space vehicle carries no payload of military value and conventional propellant, while relatively small multiple-mission vehicles carry a specific payload of interest but minimum conventional propellant. The replacement of the resource vehicle can be very faster and more economical than replacing the more important mission vehicles. This situation is similar to that of in-air refueling of fighter jets. PLT technology can be used for imparting ΔV to mission vehicles for making up orbit drag by a resource vehicle to extend its mission duration in the same way as a refueling tanker. Of course, the orbital energy of the resource vehicle is ultimately lost to the mission vehicles through energy exchange. To the extent that the mission vehicles are less readily replaced than the resource vehicle, this trade favors a PLT architecture in which orbital energy is beamed to the mission vehicles, allowing such a space system to persist in a LEO orbit. [38] In a similar way, it is projected that PLT can be used for mitigating or mining NEO or NEA.

5.2   Phase II: The Interlunar Photonic Railway

After Phase I PLT technologies and applications are fully developed and implemented, further scaling up of the PLT thrust and operation distance can enable maneuvering spacecraft and objects over the lunar distance of

384,500 km. In this phase, which is predicted to evolve over the 30 – 50 year time frame, PLT would be capable of providing thrusts in the range of 1 - 100 kN, which requires the operation power of 100 MW – 10 GW. The operation distance of PLT is projected to be up to 1,000,000 km, which can cover a wide range of spacecraft maneuvering over lunar-scale distances. For example, if PLT could be built on one of the earth orbits, such as GEO, and then used for transporting materials and robots for constructing PLTs either one of the earth-moon Lagrange points or directly on the moon to structure the Interlunar Photonic Railway as illustrated in Fig. 9.The diffraction limited size of the beaming lens should be on the order of 200 m and the spacecraft mirror diameter 50 m. The sizes the lens and mirror will decrease proportionally as the laser wavelength decreases. For example, a 100 time reduction in the laser wavelength will result in the lens diameter 20 m and the mirror diameter 5 m.

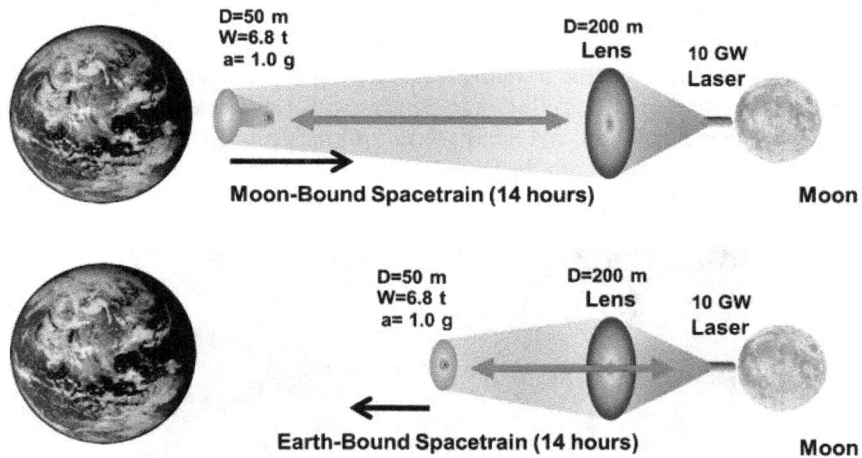

Figure 9. An Interlunar Photonic Railway consists of one PLT-BLP system located on the moon.

The Interlunar Photonic Railway with PLTs is predicted to meet the needs of the future generation of space industry market by enabling a wide range of innovative space applications involving the moon as a second step towards interstellar manned roundtrip commutes. For example, a 10 GW PLT with a thrust multiplication factor of 1,000, will generate a thrust of 66.7 kN, which can accelerate 6.8 ton Space Train at 1.0 g, a comfortable cruising acceleration as illustrated in Fig. 9. In this example, the Photonic Railway consists of one PLT-BLP system, which handles both Moon-bound and Earth-bound Space Trains. However, if the orbit issues are mitigated, a Photonic Railway system with four PLTs can be constructed similar to the one shown in Fig. 7: two PLTs for Moon-bound Space Trains and two for Earth-bound Space Trains. At this acceleration, lunar flyby will take about 6.8 hours, and for landing on moon about 14 hours.

In this example, only one PLT system located on the moon is used. For the earth-bound Space Train from the moon, the sail would be divided into two nested circular segments as shown in Fig. 9. The total vehicle mass would be, for example, 6.8 tons, including 3 tons for the crews, their habitat, and their exploration vehicles. The sail would be accelerated at 1 g by a 10 GW PLT system. Near the half way to the earth after about 7 hours later, the 10 m spacecraft with a built-in mirror detaches from the sail center and turns to face the 50 m diameter ring sail. The laser light from the moon reflects from the ring sail. The reflected light decelerates the spacecraft and inserts it into a near earth orbit. The total flight time would be 14 hours. For the moon-bound Space Train, in an earth orbit the 10 m diameter spacecraft with a built-in mirror detaches from the 50-m diameter sail center and accelerates from the face of the ring sail. The laser light from the moon reflects from the ring sail. As the return spacecraft approaches the moon 7 hours later, it is decelerated to a halt on the moon surface or inserted into a near lunar orbit by the laser light from the moon. The detailed orbit dynamics of the Space Train and Interlunar Photonic Railway is well over the scope of the present paper, and it will be presented elsewhere.

The Photonic Railway system can be built with very light materials using advanced materials and structures, which are predicted to be developed in Phase II time frame, and thus will have a large comfortable crew environment with small and light attitude control thrusters and electronics. Some of the applications of Interlunar Photonic Railway and Space Train include lunar mining and permanent lunar habitation. The moon is atmosphere free and has much less gravity (1/6 of the earth gravity), thus, it would be highly ideal place to form space launch station for traveling to other planets and stars. Therefore, it seems that the Interlunar Photonic Railway will be an important stepping stone towards Interstellar Photonic Railway, which is described in detail in the following sections.

5.3   Phase III: The Interplanetary Photonic Railway

The Phase III is predicted to evolve over the 50 – 70 year time frame, PLT will be capable of providing thrusts in the range of 100 kN - 10 MN, which will require the operation power of 10 GW – 1 TW. By this time frame, it is projected that high-power short wavelength laser will be fully developed for the required PLT power level. The operation distance of PLT is projected to be up to 10 billion km, the Earth-Pluto distance. One of the important milestone of this phase is the construction of Earth-Mars Photonic Railway. With the Earth-Mars distance of 225 million km, the diffraction limit sets the beaming lens diameter 2.5 km, and the spacecraft mirror diameter 220 m with 1 µm lasers. For the Earth-Pluto Photonic Railway with a distance of 7.3 billion km, the diffraction limit sets the beaming lens diameter 35 km, and the spacecraft mirror diameter 500 m with 1 µm lasers. A 1,000 times reduction in wavelength will reduce both the lens and mirror diameters by a factor of 32 respectively, and the lens and mirror diameters required for Earth-Pluto Railway will be 1 km and 16 m, respectively.

Figure 10. Large scale Photonic Rails extending beyond Pluto, via, for example, Mars and Jupiter PLTs.

However, the required PLT system size can be further decreased by using multiple PLT systems stationed near the planets between Pluto and the earth. Fig. 10 illustrates such a large scale Photonic Railway that extends beyond Pluto.

Let us compare the energy need to speed spacecraft for conventional rockets and that for PLT, in terms of specific energy (J/kg) that is the energy required for propelling a unit mass to a given velocity. The founding physics of this issue for non-relativistic cases was obtained by Meyer et al. [13] The specific energy of rockets, E_R is given by [13]

$$E_R = \frac{1}{2} m_f u^2 (e^{\frac{\Delta v}{u}} - 1) \tag{14}$$

Where mf is the mass of the payload, u is the velocity of the rocket engine jet, which is u = gI$_{sp}$, and Δv is the spacecraft velocity. For PLT, the specific energy, E_p is given by

$$E_P = \frac{1}{2M} m_f \Delta v c \tag{15}$$

where M is the thrust amplification factor of PLT. Fig. 11 shows examples of the specific energy, E_p, (J/kg) as a function of the spacecraft velocity (km/s) relevant to Mars Photonic Railway.

Specific Energy vs. Spacecraft Velocity

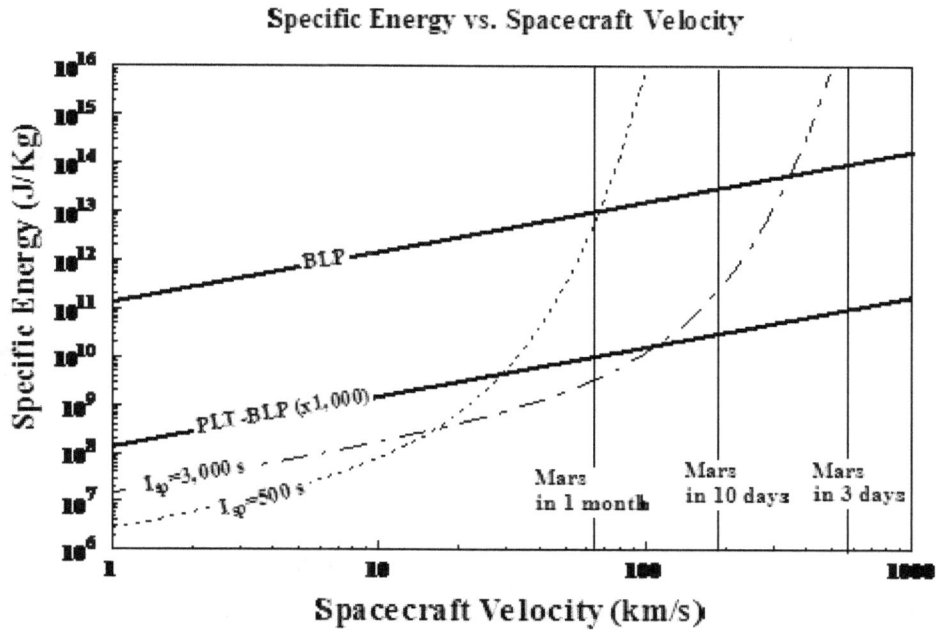

Figure 11. Specific energy as a function of spacecraft velocity relevant to Mars missions. The flight time to Mars is for flyby missions, and rendezvous missions would take more than twice longer.

Two curves represent the specific energies for rockets with I_{sp} = 500 s and 3,000 s respectively. The upper straight solid line represent the specific energy for the Laser Sail and the lower straight solid line for PLT with M=1,000. BLP without thrust amplification becomes more energy efficient than rockets with I_{sp}=500 s, if the travel time needs to be shorter than 1 month. BLP without thrust amplification becomes more energy efficient than rockets with I_{sp}=3,000 s, if the travel time needs to be shorter than a week. On the other hand, PLT-BLP with a thrust amplification factor of 1,000 becomes more energy efficient than rockets with I_{sp}=500 s, if the travel time needs to be shorter than 2 month. The BLP without thrust amplification becomes more energy efficient than rockets with I_{sp}=3,000 s, if the travel time needs to be shorter than two weeks. Eventually, when the flight time needs to be 3 days, for example, both BLP and PLT are much more energy efficient than rockets with I_{sp}=3,000 s. This estimate shown in Fig. 11 clearly demonstrates Photonic Railway based on PLT is potentially one of the most energy efficient ways to commute to planets in the solar system.

With the required Phase III technologies within reach, PLT can be built on one of the earth orbits, such as GEO, and then used for constructing PLTs either one of the Lagrange points of a planet of interest to structure an Interplanetary Photonic Railway. For Mars, the solar power is still strong, thus solar pumped PLT can be operated near Mars without too much disadvantages. However, planets farther away from the sun, such as Pluto, the solar pumping may not be efficient because of the reduced solar power at such a distance, therefore, the Photonic Railway would have two PLT systems near the Earth. The detailed orbit dynamics of the Space Train and Interplanetary Photonic Railway is well over the scope of the present paper, and will be presented elsewhere. The interplanetary Photonic Railway is predicted to meet the needs of the future space industry market by enabling a wide range of innovative space applications involving planets and asteroids. Some of the applications include mining and permanent habitation on other planets and asteroids. Once the permanent habitation on a planet is established, the planet can be used as a space station to go to other planets or exoplanets.

5.4   The Interstellar Photonic Railway

The Phase IV aims at the ultimate goal of human space exploration: manned roundtrip interstellar flight in one human generation. Once Phase III PLT technologies and applications are fully developed and implemented, further scaling up of the PLT thrust and operation distance will enable maneuvering and propelling the spacecraft and objects over interstellar distances. In this phase, which is predicted to evolve over the 70 − 100 year time frame, PLT is projected to be capable of providing thrusts greater than 10 MN, which requires the operation power of 1 - 100 TW. The operation distance of PLT is projected to be up to 100 trillion km, which can cover a wide range of spacecraft maneuvering over earth-nearby-star distance. For the Earth-ε-Eridani PLT with an operation distance of 10.8 ly (~100 trillion km), the diffraction limit sets the beaming lens diameter 1,000 km, and the spacecraft mirror diameter 252 km with 1 μm lasers as mentioned before in the previous sections. [2] Table 3 presents a comparison between the parameters for the original BLP by Forward (1984) and a hypothetical PLT with a thrust amplification factor of 1,000 and a 1 keV x-ray laser, which is projected to be available by the time frame of the Phase 4.

Table 3. Comparison between parameters for the original BLP by Forward [2] and a hypothetical PLT.

Original BLP Parameters By Forward (1984) Lens Diameter (1,000 km)				Interstellar Photonic Railway with PLTs Photon Recycling (x100) + Short Wavelength (x100) Lens Diameter (32 km)		
Phase	Laser Power	Spacecraft Weight	Sail Diameter	Laser Power	Spacecraft Weight	Mirror Diameter
Launch	75,000 TW	78,500 t	1000 km	10 TW	800 t	0.4 km
Rendezvous	17,000 TW	7,850 t	320 km	10 TW	800 t	0.4 km
Return	17,000 TW	3,000 t	100 km	10 TW	800 t	0.4 km
Stopping	430 TW	785 t	100 km	10 TW	800 t	0.4 km

The incorporation of the 1 keV x-ray laser relaxes the diffraction limit by a factor of 100. Millis [8] conservatively predicts that by the year 2100 the total world power production would be ~100 TW assuming 1.9 % average annual growth. Assuming that the same power is available for interstellar mission in a form of probably solar power and that the efficiency of the PLT from the energy to photon power conversion is about 12 %, the required 12 TW of photon power before amplification is projected to be met by the year 2100.

Once this is achieved, PLT can be built on one of the solar system planets or satellites, such as the Pluto, and then used for constructing PLTs either one of the Lagrange points of an exo-planet of interest to structure the Interstellar Photonic Railway. The detailed orbit dynamics of the Space Train and Interstellar Photonic Railway is well over the scope of the present paper, and it will be presented elsewhere. It is difficult to assess what would be a financial interest or value of interstellar roundtrip flight at that time, however, by the time of Phase IV, the rapidly growing space industries and science are predicted to discover advanced resources that can be harvested from exoplanets and hopefully the demand of such resources would inspire interstellar roundtrip flight. If such investment returns are present, the Interstellar Photonic Railway is predicted to meet the needs of the far-future space industry market by further enabling a wide range of innovative space applications involving exoplanets and other space objects beyond the solar system. Some of the applications would probably include exoplanet mining and permanent exoplanet habitation.

6. Discussions

Foretelling the development of technologies that are needed for the Photonic Railways over next 100 years is extremely challenging. Here based on extrapolation of the existing science and technology, it was predicted that PLT in conjunction with BLP would potentially revolutionize the way in which space missions and travels are executed. However, numerous daunting technological challenges exist in pursuing the development of the Photonic Railway. The present paper addresses some of critical issues, however, many other potential technological challenges are anticipated and the scope of these challenges covers an extremely wide range of science and engineering. For example, one of the fundamental challenges in applying PLT for the interstellar missions lies in the use of the astronomically long resonant optical cavities. A major technological difficulty exists in such a system: alignment of the optics. The precision in the unit of radian in selected missions are shown in Table 4.

In terms of available aiming accuracy of lasers, for example, typical ABL operation requires the operation distance on the order of 1,000 km and the target irradiation size of 10 cm (10^{-4} km). The reflected laser signal return from the moon requires the operation distance 4×10^6 km and the reflection mirror size on the order of 1 m (10^{-3}

km), which requires the aiming accuracy of 2.5×10^{-10} rad. Such an operation requires the angular aiming accuracy on the order of 10^{-10} rad. Therefore, the existing aiming accuracy can meet the accuracy required for missions up to the interplanetary mission.

Table 4. The precision in angular alignment required for exemplary missions.

Destination	Typical Distance (km)	Typical Spacecraft Mirror Diameter (km)	Angular Aiming Precision (Rad)
Near Earth	10^5	10^{-3}	10^{-8}
Moon	4×10^6	4×10^{-2}	10^{-8}
Mars	2×10^8	2×10^{-1}	10^{-9}
Pluto	7×10^9	1	1.5×10^{-10}
ε Eridani	10^{14}	1	$<10^{-14}$

Recent researches on the space telescope metrology propose systems that can achieve the angular aiming accuracy on the order of 10^{-12} rad. However, the aiming accuracy required for interstellar missions is more than 2 - 4 orders magnitude smaller than the currently used ones, and by the time the present propulsion system is applied for interstellar missions, which is 70 – 100 years from now, such a technology is predicted to be available.

Another important issue is in the feedback mechanism required for maintaining such high relative angular accuracy. For interstellar missions, the feedback signals in the PLT optical cavities would take years to arrive at the sensors for adjusting the aim. Thus, by the time the angular adjustment is performed to offset the misalignment, the spacecraft mirror/sail would be in an unpredictable angular and spatial position. Therefore, the angular aiming accuracy for interstellar mission will require to ability to sustain absolute angular accuracy without relying on feedback mechanisms. These topics are for future studies.

Finally, the present analyses primarily depended on the existing photonic technologies. Recently, there have been some very exciting breakthrough photonic technologies that can further reduce the size and weight of PLT, thus the Photonic Railway, are on horizon. For example, Bose-Einstein Condensation (BEC) of photons was recently demonstrated in a microcavity to drastically decrease laser beam divergence well below that predicted by the conventional diffraction theory. [39] The use of such photon BEC for long distance applications as in PLT and the Photonic Railway will require extensive studies on quantum electrodynamics of optical resonance cavities and PLT, which are still in infancy.

7. Conclusions

The present paper aimed to establish a roadmap that the interstellar manned roundtrip flight could be developed on in a century within a frame of exiting scientific principles, under the assumption that the required existing technologies can be further developed. The perspective on photon propulsion presented here results in a conclusion that mastering photon propulsion that uses photon momentum directly is the key to overcoming the limit of the current propulsion technology based on conventional rocketry and potentially opening a new space era. [15,17] A unified rocketry theory that can be applied for both particle and photon propulsion was developed to explain the general behavior of the specific thrust as a function of I_{sp}, which is a measure of how efficient a thruster is in generating thrust at a given input energy. The unified theory shows that at relativistic I_{sp}, there is no distinction between photons and particles as propellants.

In addition to its highest I_{sp} physically allowed, the chief advantage of photon propulsion stems from its capability in separating the energy source from the spacecraft as in Photonic Laser Thruster (PLT), which can avoid the exponentially increasing onboard fuel mass as a function of I_{sp}. PLT further permits the reusable power beaming structures that are similar to railway structures, the minimal weight of spacecraft, and generation of artificial gravity over the course of space travel. Thus, for the missions requiring high I_{sp}, PLT was shown to be far superior to conventional propulsion based on particles. The emerging science and technologies, such as high power lasers, solar sails, precision optics, ultra-light large scale space optics and telescopes, and detailed information of solar system planets, NEO, and exoplanets, now provide a fertile ground for a full-scale development of photon propulsion in space.

A roadmap towards the interstellar flight was proposed here that the permanent energy-efficient interstellar transportation structure, the Photonic Railway, similar to transcontinental railway systems, is necessary to attract sustainable economic interest and reinvestment over a century to the development pathway. [17] The technological foundation of the Photonic Railway lies on PLT, of which proof-of-concept was originally demonstrated by the author. [3,4,14-17] Recently, under the auspice of NASA Innovative Advanced Concepts, PLT has been demonstrated for photon thrusts up to 3.5 mN and enhancement factors up to 1,500 using a 1-kW thin disk laser system, and to dynamically propel, slow, and stop a 1U cubesat in laboratory environment in the author's lab. The new PLT demonstration proves that PLT development can ride on the Moore's curve, in which the thrust and the power capacity of the pump laser system can exponentially increase as a function of year. Therefore, our PLT demonstrations pose the first step towards the interstellar Photonic Railway.

PLT is projected to further advance its development by incorporating the anticipated x-ray laser, Bose-Einstein photon condensation, and advanced material science and technologies so that its power and engineering requirements for interstellar manned roundtrip flight can be achievable probably within a century by reducing its size and power requirement by orders of magnitude. Once the PLT is successfully implemented, multiple PLTs are proposed to be used to construct a more permanent and energy-efficient transportation structure. Such a structure would allow the interstellar commute in comfortable and safe, yet light Space Train that is mainly composed of crew habitation, safety environment, and navigation systems.

A four-phased evolutionary developmental roadmap or pathway of the Photonic Railway towards interstellar manned roundtrip travel is described: 1) Development of PLTs for satellites and NEO manipulation, 2) Interlunar Photonic Railway, 3) Interplanetary Photonic Railway, and 4) Interstellar Photonic Railway. It is projected that these developmental phases will result in systematic evolutionary applications, such as satellite formation flying, NEO mining/mitigation, and Space Solar Power, which will provide sufficient sustainable economic interest and return investment to self-sustain the development pathway. Once successfully implemented, the Photonic Railway is projected to bring about a quantum leap in the human economic and social interests in space from explorations to terraforming, mining, and permanent habitation in planets, asteroids, moons and exoplanets.

Acknowledgements

The author acknowledges the discussions and criticisms of many fellow rocket scientists and engineers, especially Prof. Mason Peck and Dr. Claude Phipps. In addition, the technical contributions of Dr. Injeyan, Mr. Williams, Mr. Harkenrider, and Dr. Neuhaus, in PLT lab demonstration are greatly appreciated.

References

1. Tsander, K., "From a Scientific Heritage", *NASA TTF-541* (1967) (quoting a 1924 report from Tsander).

2. Forward, R. L., "Roundtrip Interstellar Travel Using Laser-Pushed Lightsails", *Journal of Spacecraft and Rockets*, 21, 187–195 (1984).

3. Bae, Y. K., "Photonic Laser Thruster (PLT): Experimental Prototype Development and Demonstration", AIAA 2007-6156-318, *Space 2007 Conference Proceedings* (2007).

4. Bae, Y. K., "Photonic Laser Propulsion: Proof-of-Concept Demonstration", *AIAA Journal of Spacecraft and Rockets*, 45:, 153-155 (2008).

5. Takeuich, H. et al., "VLBI Tracking of the Solar Sail Mission IKAROS", *URSI 30th Symposium Proceedings*, 1-4 (2011).

6. Forward, R. L., "Advanced Space Propulsion Study", *AFRL Final Report*, F04611-86-C-0039 (1987).

7. Forward, R. L., "Advanced Propulsion Systems", I*n: Space Propulsion Analysis and Design,* ed. R. W. Humble, G. N. Henry, and W. J. Larson, McGraw-Hill Co., New York (1995).

8. Millis, M. G., "Energy, Incessant Obsolescence, and the First Interstellar Missions", *61st International Astronautical Congress*, IAC-10-D4.2.7, Prague, CZ (2010).

9. Marx, G., "Interstellar vehicle propelled by terrestrial laser beam", *Nature*, 211, 22–23 (1966).

10. Redding, J.L. "Interstellar vehicle propelled by terrestrial laser beam", *Nature*, 213, 588–589 (1967).

11. Simmons, J. F. L., McInnes, C.R., "Was Marx right? Or how efficient are laser driven interstellar spacecraft?", *Am. J. Phys.* 61, 205–207 (1993).

12. Meyer, T. R. et al., "The Laser Elevator: Momentum Transfer Using an Optical Resonator", In: *38th IAF Conference*, Brighton, UK, October 11-17 (1987).

13. Meyer, T. R., "Laser Elevator: Momentum Transfer Using an Optical Resonator", *Journal of Spacecraft and Rockets*, 39, 258-266 (2002).

14. Bae, Y. K. "A Contamination-Free Ultrahigh Precision Formation Flying Method for Micro-, Nano, and Pico-Satellites with Nanometer Accuracy", *Space Technology and Applications International Forum*, ed. M.S. El-Genk, AIP Conf. Proc. AP813, 1213-1223 (2006).

15. Bae, Y. K., "Photonic Laser Propulsion (PLP): Photon Propulsion Using an Active Resonant Optical Cavity", AIAA 2007-6131-818, *Space 2007 Conference Proceedings* (2007).

16. Bae, Y. K., "Photon Tether Formation Flight for Distributed and Fractionated Architectures", AIAA 2007-6084-275, *Space 2007 Conference Proceedings* (2007).

17. Bae, Y. K., "Prospective of photon propulsion for interstellar flight," *Physics Procedia* 38, 253 – 279. (2012).

18. Forward, R. L., "Pluto—The Gateway to the Stars," *Missiles and Rockets*, 10, 26-28 (1962).

19. Meyer, T. R. et al., "Rapid Delivery of Small Payloads to Mars", In: *The Case for Mars II, Vol. 62, Science and Technology Series*, Univelt, Inc., San Diego, CA, 419–431 (1985).

20. Matloff, G. L., "Interstellar Solar Sailing: Consideration of Real and Projected Sail Material", *JBIS*, 37, 135–141 (1984).

21. Forward, R. L., "Laser Weapon Target Practice with Gee-Whiz Targets", *Proceedings SDIO/DARPA Workshop on Laser Propulsion*, Vol. 2, ed. J. T. Kare, CONF-860778, Lawrence Livermore National Lab., Livermore, CA, 41–44 (1987).

22. Landis, G.A.. "Advanced Solar- and Laser-Pushed Lightsail Concepts", *Final Rept. for NASA Inst. for Advanced Concepts*, Ohio Aerospace Inst., Brook Park, OH (1999).

23. Fink, Y. et al., "A Dielectric Omnidirectional Reflector", *Science*, 282, 1679–1682 (1998).

24. Ritter, J. et al., "Photonic Muscles: Optically Controlled Active Optics", *Proc. SPIE 5894*, pp. 58941, (2006).

25. Kellett, B. J. et al., "Space Polypropulsion", *High-Power Laser Ablation VII, Proc. of SPIE 7005, 70052W* (2008).

26. Metzger, R. A. and Landis, G., "Multi-Bounce Laser-Based Sails", *STAIF Conference on Space Exploration Technology*, Albuquerque NM, AIP Conference Proceedings, 552, pp. 397-402 (2001).

27. Gray, A. P. et al., "Photon Flux Amplification for Enhancing Photonic Laser Propulsive Forces", AIAA 33 rd. Plasmadynamics and Lasers Conference, Maui Hawaii (2002).

28. Romanini, D. et al., "CW Cavity Ring Down Spectroscopy", *Chem. Phy. Lett.,* 264, 316-322 (1997).

29. Sheard, B. S. et al., "Observation and Characterization of an Optical Spring", *Phys. Rev. A, 69, 051801(R)* (2004).

30. Siegman, A. E., "Lasers", University science Books, Sausalito, California (1986).

31. Daneu, V. et al., "Spectral beam combining of a broad-stripe diode laser array in an external cavity", *Opt. Lett., 25*, 405-407 (2000).

32. Landis, G. A., "New Approaches for a Solar-Pumped GaAs Laser", *Optics Communications*, 92, 261-265 (1992).

33. Tsidulko, I. M., "Semiconductor Laser Pumped by Solar Radiation", *Soviet Journal of Quantum Electronics*, 22, 463-466 (1992).

34. Yariv, A., "Quantum Electronics", John Wiley & Sons, New York (1975).

35. Bohn, W. L., "Novel Aspect of Laser Propulsion", *High-Power Laser Ablation VII*, ed. Claude R. Phipps, Proc. of SPIE, 7005, pp. 70051C (2008).

36. Zhang, M. et al., "Strong, Transparent, Multifunctional, Carbon Nanotube Sheets", *Science*, 309:1215–1219 (2005).

37. Norman M. C. and Peck, M. A., "Orbit Maneuvers Through Inter-Satellite Forcing", *AIAA Conference Proceedings*, 2009-6097 (2009).

38. Peck, M. A.. private communications (2010).

39. Klaers, J., Schmitt, J., Vewinger, F., and Weitz, M., "Bose–Einstein condensation of photons in an optical microcavity", Nature 468, 545–548 (2010).

100 YEAR STARSHIP™

Interstellar Space, Stars, and Destinations

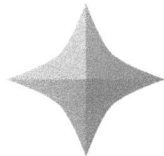

Chaired by Margaret Turnbull, PhD

Professor, Department of Physics and Astronomy

Track Description

Finding Earth 2.0 is rooted first and foremost in astronomy, planetology and astrophysics. Whether using ground based observation, earth orbiting technologies or deep space probes within or just outside of our solar system, understanding and advancing knowledge, instruments and theories of the history, formation, composition and evolution of the universe, galaxies, stars and planets is fundamental to finding earth analogues.

Presenters are asked to provide research, outline novel concepts, propose instrument designs and methods, review data and define capabilities, knowledge and mission parameters key to furthering the understanding: the composition of exosolar systems; the identification of exoplanets in the "goldilocks zone"; planets that are rocky; exoplanet atmosphere composition, size; as well as defining the interstellar medium and aspects of maps, navigation and guidance.

In addition, as our gaze is drawn many light years away, focusing on closer objectives as stepping-stones to deep space will be essential. Beyond Mars, what missions should be designed to eventuate successful travel to another star? How should potential destinations be evaluated?

Track Chair Biography

Margaret Turnbull, PhD

Lead Astronomer, NASA's New Worlds Observer Mission

Dr. Turnbull is an astrobiologist whose expertise is in identifying planetary systems that are capable of supporting life as we know it. She developed a Catalog of Habitable Stellar Systems for use in the search for extraterrestrial intelligence (SETI) and she has studied the spectrum of the Earth to identify telltale signatures of life. She is currently leading two teams to prepare for NASA's WFIRST mission, slated for launch in 2025. WFIRST will be the first mission to directly image planetary systems orbiting the nearest sunlike stars, and the first mission with the hope of determining the atmospheric composition and surface characteristics of those planets. When not thinking about alien worlds and missions to get us there, Maggie can be found keeping honey bees, raising monarchs, tapping sugar maples, and cross country skiing across the north woods with her dogs.

Testing the Boundaries of Planetary Habitability

Stephen Kane, Ph.D.

Fellow Stanford University and Maexeler Dataflow,

Testing the Boundaries of
Planetary Habitability

Stephen Kane

SAN FRANCISCO
STATE UNIVERSITY

ASTRONOMER

What my friends think I do — What my mom thinks I do — What society thinks I do

What the university thinks I do — What I think I do — What I really do

What is the "Habitable Zone"?

What the media (and public) think we mean ...

What is the "Habitable Zone"?

What the media (and public) think we mean ...

What is the "Habitable Zone"?

What astronomers think we mean ...

What is the "Habitable Zone"?

What we really mean ...

- The "Habitable Zone" is the region around a star where water **COULD** exist in a liquid state on the surface of a planet **IF** it has sufficient atmospheric pressure

- It does **NOT** comment on the presence of water

- It does **NOT** comment on habitability

- It does **NOT** comment on the presence of life

- Based on one data point (size/mass of planet also matters)

What is the "Habitable Zone"?

What we really mean ...

- The "Habitable Zone" is the region around a star where water **COULD** exist in a liquid state on the surface of a planet **IF** it has sufficient atmospheric pressure

- It does **NOT** comment on the presence of water

- It does **NOT** comment on habitability

- It does **NOT** comment on the presence of life

- Based on one data point (size/mass of planet also matters)

TARGET SELECTION!

What is the "Habitable Zone"?

Planet in the Habitable Zone ≠ Habitable Planet

"I THINK YOU SHOULD BE MORE EXPLICIT HERE IN STEP TWO."

Point #1:
We have not discovered any habitable planets
(planets that we know are habitable)

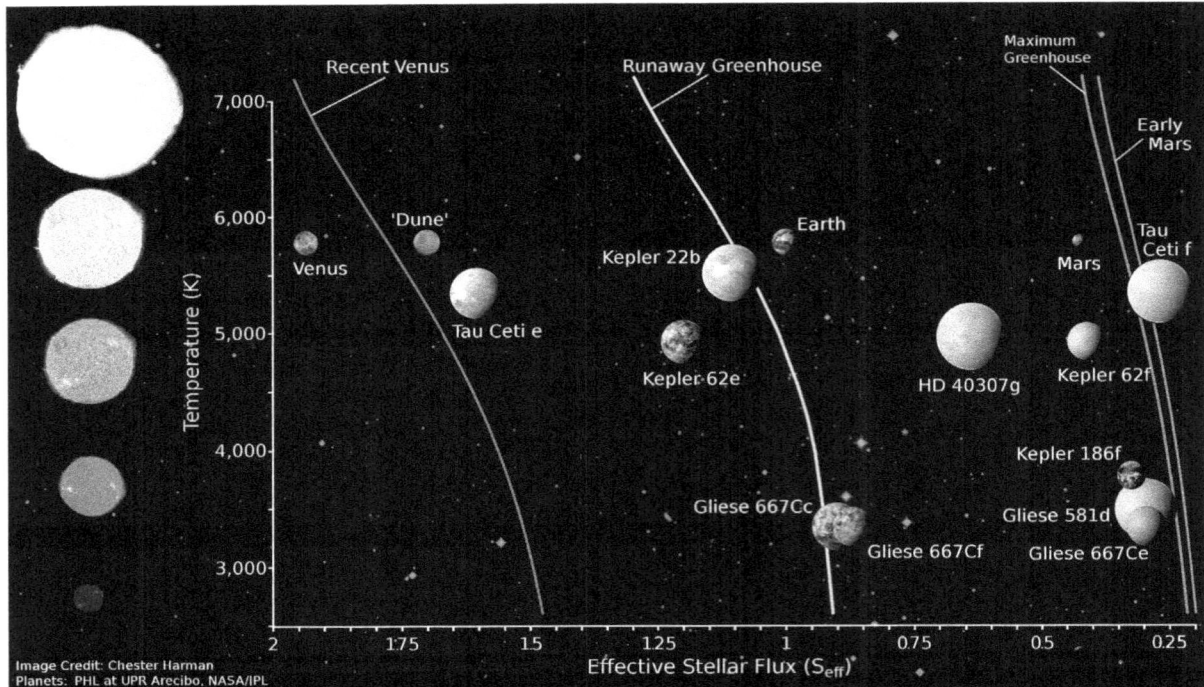

- **Conservative Habitable Zone: Runaway Greenhouse to Maximum Greenhouse**

- **Optimistic Habitable Zone: Recent Venus to Early Mars**

Stellar Parameters

Kopparapu et al. 2013, 2014

Habitable Zone Boundaries

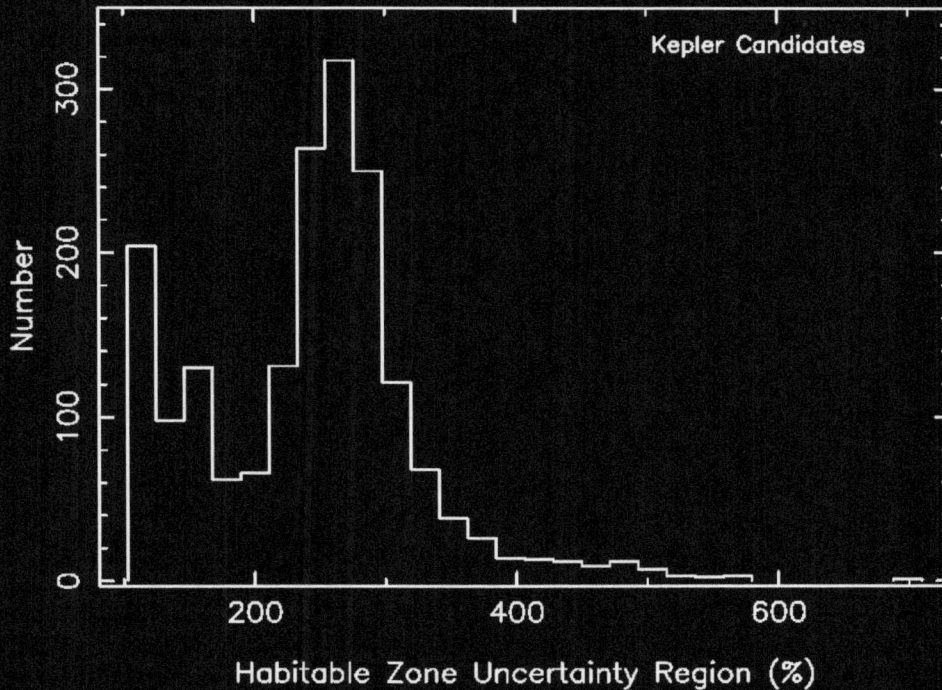

Point #2:
Habitable Zone boundaries
are uncertain
(sometimes by a lot!)

η(Earth) from Kepler

• Define $\eta_\oplus$ as fraction of stars with at least one terrestrial planet within the Habitable Zone

• For M stars: $\eta_\oplus$ = 0.15 +0.13/-0.06 (Dressing & Charbonneau 2013)
$\eta_\oplus$ = 0.48 +0.12/-0.24 (Kopparapu 2013)

• For Sun-like stars, $\eta_\oplus$ = 0.22 (Petigura et al. 2013)

• Kepler has a bias towards detecting transiting planets at small orbital periods

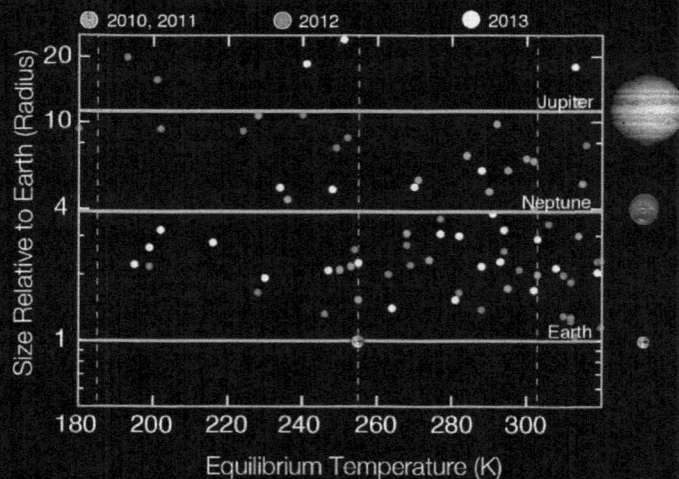

η(Earth) from Kepler

- Define $\eta_\oplus$ as fraction of stars with at least one terrestrial planet within the Habitable Zone

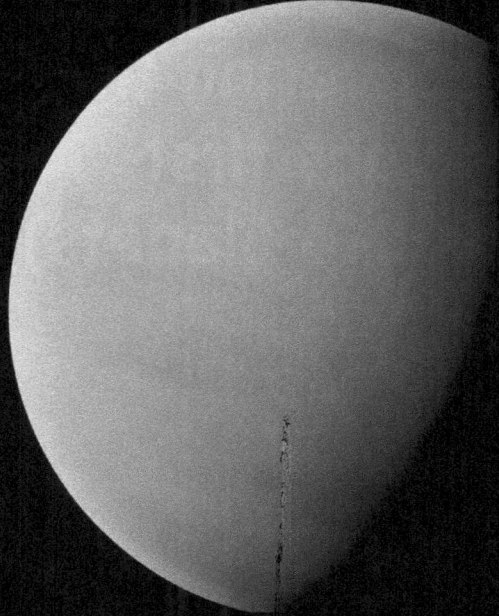

η(Venus) from Kepler

- Define η(Venus) as fraction of stars with at least one terrestrial planet within the Venus Zone

- For M stars: η(Venus) = 0.32 +0.05/-0.07
  For GK stars: η(Venus) = 0.45 +0.06/-0.09

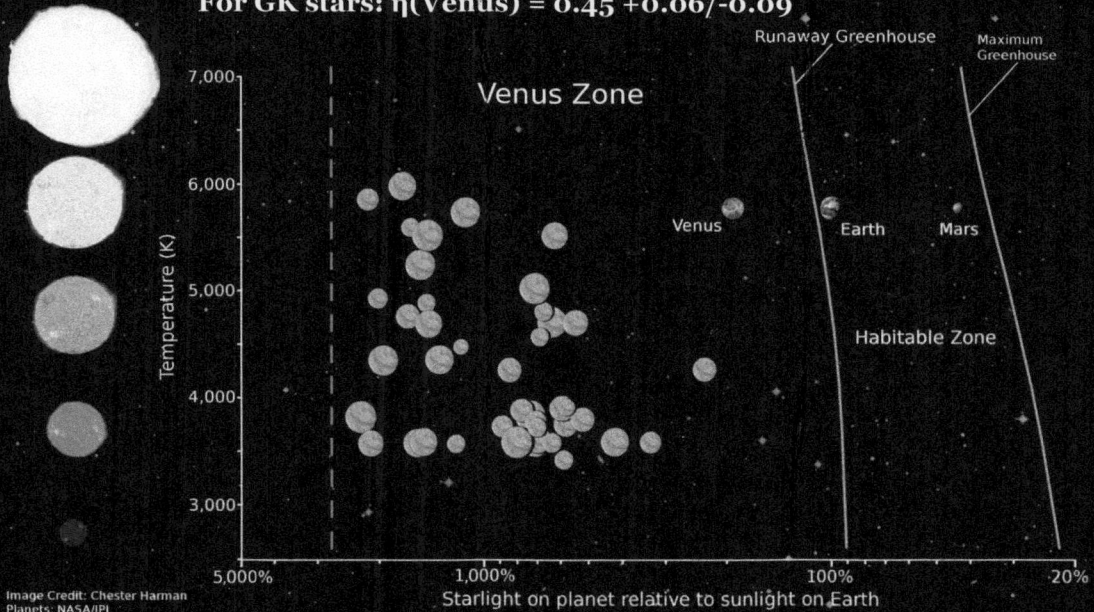

Image Credit: Chester Harman
Planets: NASA/JPL

Kane, Kopparapu, Domagal-Goldman. 2014, ApJ, 794, L5

Point #3:
Kepler is discovering the Venus analogs first
(Venus-like planets)

Sizes of exoplanets

• Our Solar System lacks super-Earths. Neptune is 3.8 Earth radii in size.

• Recent work show that the transition to a H/He dominated atmosphere occurs around 1.5 Earth radii (Lopez & Fortney 2013, Marcy et al. 2014).

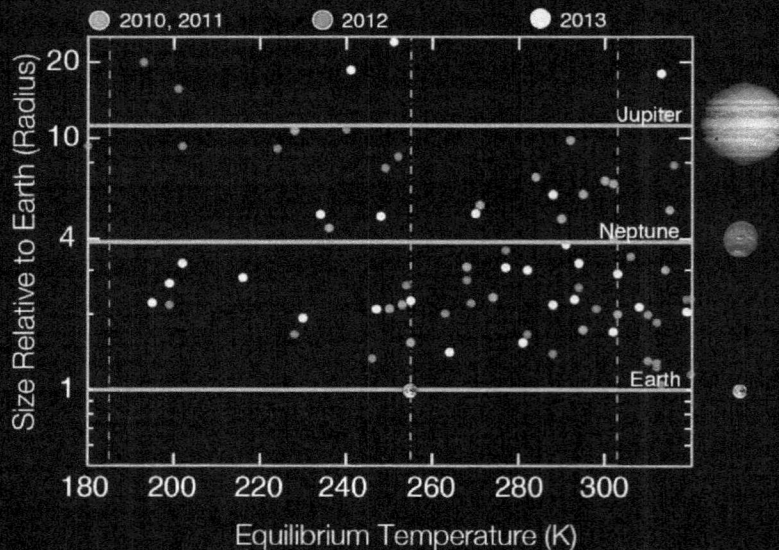

Sizes of exoplanets

- Our Solar System lacks super-Earths. Neptune is 3.8 Earth radii in size.

- Recent work show that the transition to a H/He dominated atmosphere occurs around 1.5 Earth radii (Lopez & Fortney 2013, Marcy et al. 2014).

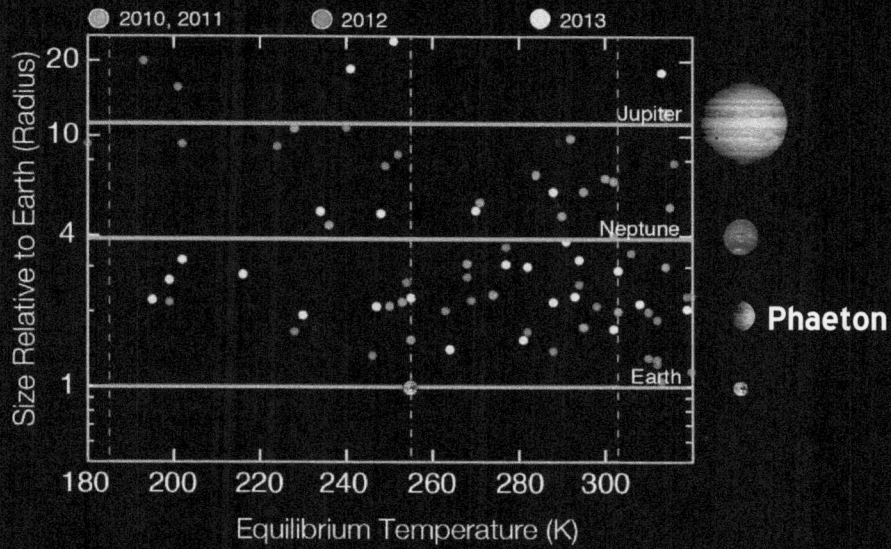

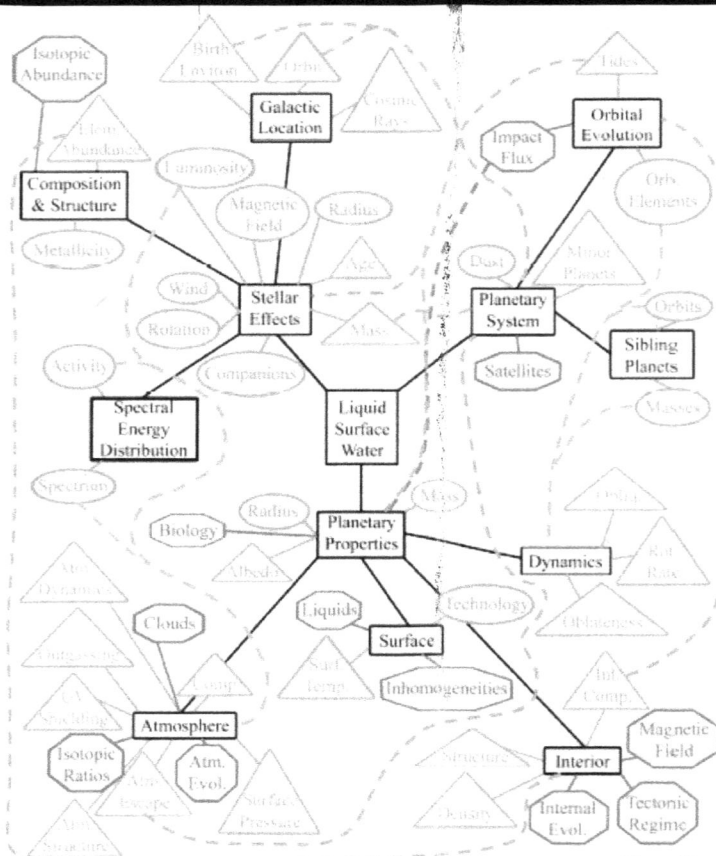

Point #4:
Super-Earth ≠ Earth

Rory Barnes & Vikki Meadows

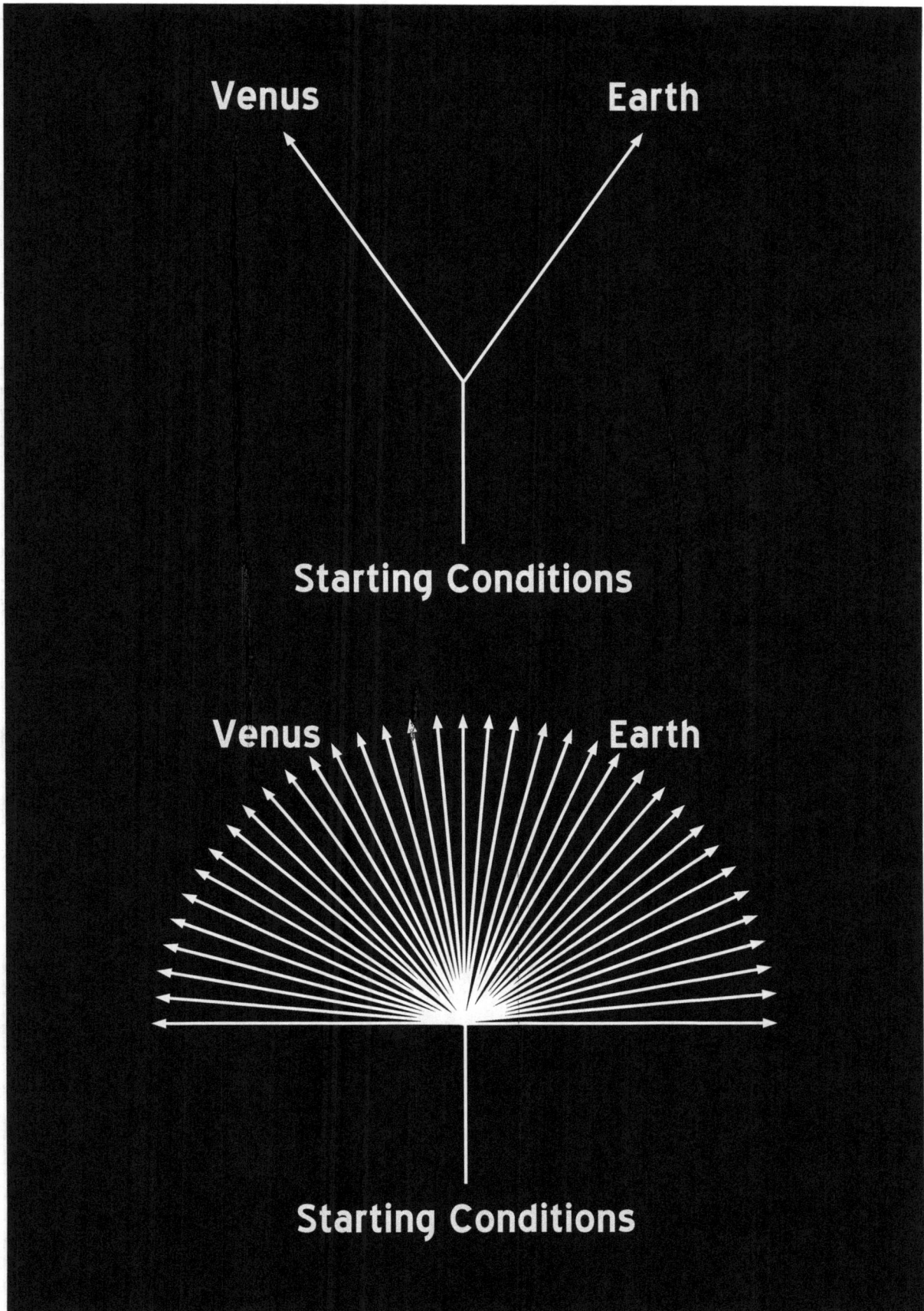

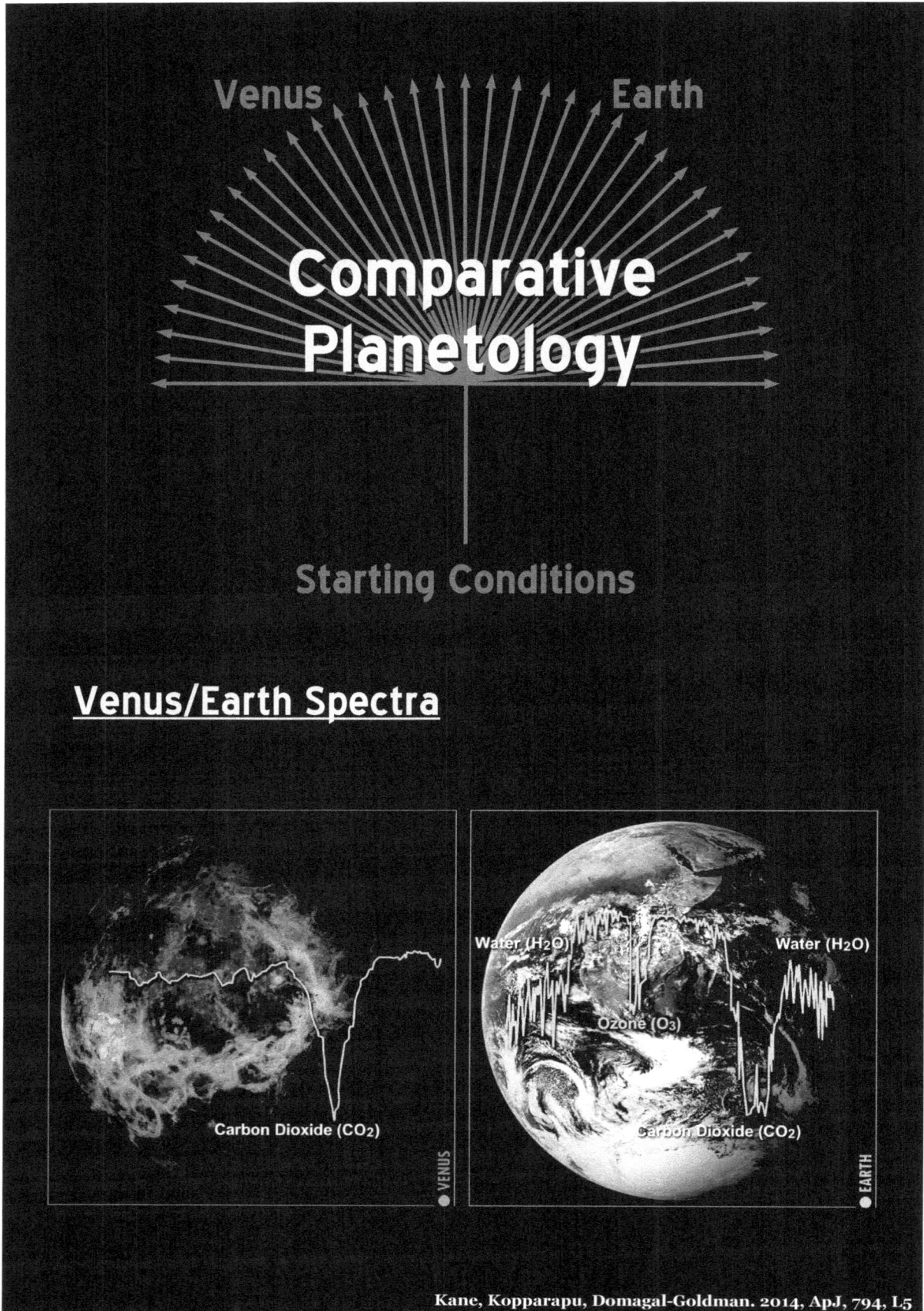

Kane, Kopparapu, Domagal-Goldman. 2014, ApJ, 794, L5

Conclusions

1. We have not discovered any habitable planets

2. Habitable Zone boundaries are uncertain

3. Kepler is discovering the Venus analogs first

4. Super-Earth ≠ Earth

1. Target selection for JWST and future TPF-type missions

2. Highly motivated to accurately determine stellar parameters (Huber et al. 2014)

3. Determining the frequency of Venus-like planets is important for understanding the Earth/Venus dichotomy

4. Super-Earths are common and may harbor moons

Goldilocks Zones— A Fine Grained Exoplanet Taxonomy

Patrick J. Talbot

Talbot Consulting, Jacksonville Beach, FL. 32250

patrick.talbot@hotmail.com

Dennis R. Ellis

Ellis Interstellar, Peyton, CO. 80831

bdellis@skybeam.com

Abstract

Research, engineering techniques, and data analytics identify Goldilocks zones as potential destinations. The taxonomy is based on available data, anticipated growth in the number of known exoplanets, and a flexible engineering methodology. The foundation of the taxonomy is a classic trade study to rank and score candidate attributes based on criteria derived from a set of exoplanet taxonomy requirements. A "critic", a sequential optimizer, and software-based tests find a minimum spanning taxonomy. Software assures that, to the extent practical, each object has a unique path through the taxonomy.

Why a taxonomy? It is a valuable teaching tool, it organizes and sorts a data set, and a taxonomy is structured as a hierarchical characteristics-based ontology. This knowledge structure is readily captured in a knowledge base for multi-strategy reasoning and data analytics. That is the true value of a taxonomy! The taxonomy defines the parameter space of exoplanet characteristics. While some work "inside the box", others work outside the box", a taxonomy defines the box.

Recent efforts produced an upper-level exoplanet taxonomy. This taxonomy doesn't span the orbital, physical, environmental, and stellar characteristics of interest. Nor does it provide a unique set of characteristics for each of ~ 2,000 exoplanets currently identified. The Goldilocks Taxonomy™ has 13 parameters chosen from a collection of potential characteristics. The Goldilocks Taxonomy™ provides fine-grained organization of exoplanet characteristics, a robust methodology to refine the taxonomy, and a suite of data analytics tools to discover interesting patterns.

Keywords

interstellar, Goldilocks zone, exoplanets, taxonomy,

1. Introduction and Summary

The proposed taxonomy is based on exoplanet insights insights from the literature and access to online exoplanet catalogs. These insights are the foundation for a classic trade study to rank and score candidate nodes based on criteria derived from missions and functional requirements. The nodes are further processed using a "critic", a sequential optimizer, and software-based tests to find a minimum spanning taxonomy. Software was written to assure that, to the extent practical, each object has a unique path through the taxonomy.

At first glance, a taxonomy is an important way of visualizing the widely varying set of attributes associated with an object. As such, it is a valuable teaching tool that provides people interested in space with an easy-to-understand view. Digging deeper, it provides a "world-view" of the domain that helps people understand objects in a more organized and holistic way. More practically, it organizes and sorts a data set, since all the attributes important to a taxonomy are not, today, available in a single data file. Building on the data set idea, a taxonomy is a hierarchical network and each node in the hierarchy has characteristics. This is the makings of a truly powerful knowledge structure, or ontology. The knowledge structure is readily captured in a knowledge base such as Protege [1]. The resulting hierarchical, characteristics-based ontology supports multi-strategy reasoning and learning, including simulation and planning [2] . That is the true value of a taxonomy!

2. Related Work

There have been multiple recent efforts to arrive at an exoplanet taxonomy [3],[4] supplemented with ongoing efforts and "conventional wisdom" that produce an upper-level taxonomy. These results are discussed as motivation to capture what's good and to avoid what's contrary to requirements for a fine-grained taxonomy.

3. Approach

- Define Requirements from mission analysis and functional analysis.
- Cite Assumptions and groundrules; for example, the taxonomy is limited to data in catalogs.
- Perform Trade Studies: Identify and characterize potential taxonomy nodes, citing rationale for accepting some object attributes and dismissing others. Assign weights to criteria using a pair-wise preference technique, citing rationale. Rank and score attributes, citing rationale.
- Sequential Optimization: Use a "critic" to add common sense and determine the optimum set of nodes, as the smallest number of nodes that produces less than 5% probability of one or more duplicate signatures.
- Software Development: Convert data files (for example, exoplanet.org) to a standard (.csv) format. Calculate discretized values for candidate taxonomy nodes from the integrated files. Assemble a single file with values for all candidate nodes. Analyze patterns, uniqueness, and requirements satisfaction of requirements using a uniqueness test, relative frequency, and sensitivity computations. Test the taxonomy using a Bayesian classifier to assure high probability of correct classification of an object.
- State Results: An exoplanet taxonomy that comprises current data sets integrated with additional sources. Show taxonomy resulting from the trade study and sequential optimization. Assure uniqueness, to the extent that it is practical, of each exoplanet signature, and explain optimization process. Provide a minimum description length representation of taxonomy attributes and states.
- Discussion: focus on an interactive taxonomy generator that provides a user centered taxonomy.
- Summarize: State results and cite advantages of the solution.

3.1 Define Requirements

Detailed description & requirements for an exoplanet taxonomy focus on the ability to view, understand, and predict the behavior of objects in orbit around the other stars. Two trade studies help top define the taxonomy:
- Mission Analysis maps potential nodes to mission requirements
- Functional analysis maps potential nodes to Challenge details & functional requirements.

3.2   Assumptions and Groundrules

The following are considered concerns:
* Missing Values: values for many exoplanet characteristics are unknown. The methodology and reasoning algorithms accommodate this aspect of the catalogs.
* Uniquely Described: the driving requirement for the minimal taxonomy is that the state space is large enough and that the states of characteristics are equally likely to guarantee a low probability that two objects have the same state. This is not the case because many states have unknown values. Hence the goal is to increase the number of characteristics and states thereof until we reach a point of diminishing returns. That is, when the percent of exoplanets with duplicate signature fails to diminish with increasing number of states.

3.3   Trade Studies: Potential taxonomy nodes are derived from Mission Analysis and Functional Analysis.

Mission Analysis: To "cast a wide net" and capture all potentially valuable nodes for describing exoplanets, the missions of potential interest are analyzed to identify important characteristics (Figure 1). Four attributes are identified: the extent to which an attribute facilitates exploration, the science value of the attribute, the cultural significance, and the extent to which the attribute is amenable to sensing technology. The characteristics are weighted by pair-wise comparison (top of Figure 1) and the potential taxonomy parameters are the scored based on the four attributes (bottom of Figure 1). The result is that from a mission perspective, the distance from our sun and the type of stat system (binary) ate rated most highly. Mission tasks will vary with project goals.

Mission Weighting

Mission Task	Exploration	Science	Culture	Sensing	Raw	Weight
Exploration	1	1	1	1	4	1
Science	1	1	1	1	4	1
Culture			1	1	3	0.75
Sensing		1		1	2	0.5

Scores indicate the degree to which a characteristic is expressive and distinctive

Characteristic	Exploration	Science	Culture	Sensing	Raw Score	% Score
Weights	1	1	0.75	0.5		
Distance from our Sun	10	10	10	10	32.5	100
Binary	10	10	10	10	32.5	100
Separation	10	10	9	10	31.75	98
Temperature	10	10	10	8	31.5	97
Period	10	10	9	9	31.25	96
Mass	9	10	8	7	28.5	88
Semi-Major Axis	10	9	7	8	28.25	87
Eccentricity	10	9	7	8	28.25	87
Inclination	10	9	7	8	28.25	87
Spectral Type	8	10	8	8	28	86
Physical Radius	9	10	8	6	28	86
Age	8	10	8	6	27	83
Spin Orbit Alignment	9	10	5	8	26.75	82
Stellar Metallicity	7	10	6	7	25	77
Discovery Method	7	8	8	8	25	77

Figure 1: Mission Trade Study

Functional Analysis: In similar fashion, a functional trade study, geared to identifying the utility and power of potential taxonomy characteristics, was completed. A classification scheme for exoplanets is required with these attributes:

* Explain: the exoplanet taxonomy (ET) shall be based on a set of characteristics that provide intuitive insights into these objects
* Discriminate: the ET shall identify differences among exoplanets
* Characterize: the ET shall comprise physical and dynamic characteristics, or a combination of both
* Group : the ET shall hierarchically organize and name objects based on similarity
* Predict: the ET shall allow the object trajectory to be accurately predicted
* Growth: the ET shall accommodate growth of the exoplanet catalog over at least two orders of magnitude.

The result of this trade study (Figure 2) is that temperatre is the most highly rated characteristic for inclusion into the taxonomy, while distance from our sum and star system type again are highly ranked.

Weighting Matrix

Functional Requirement	Explain	Discriminate	Characterize	Group	Predict	Growth	Raw	Weight
Explain	1	1	1	1	1	1	6	1
Discriminate	1	1	1		1	1	5	0.8
Characterize		1	1		1	1	4	0.7
Group	1	1	1		1		4	0.7
Predict		1			1	1	3	0.5
Growth			1			1	2	0.3

Characteristic	Explain	Discriminate	Characterize	Group	Predict	Growth	Raw Score	% Score
Weights	1	0.8	0.7	0.7	0.5	0.3		
Distance from our Sun	10	10	10	10	8	8	38.4	96
Binary	10	10	10	10	8	8	38.4	96
Separation	10	10	10	9	9	8	38.2	96
Temperature	10	10	10	10	10	10	40	100
Period	10	10	10	9	9	8	38.2	96
Mass	10	10	10	9	9	8	38.2	96
Semi-Major Axis	10	10	10	9	9	8	38.2	96
Eccentricity	10	10	10	9	9	8	38.2	96
Inclination	10	10	10	9	9	8	38.2	96
Spectral Type	10	9	9	10	10	6	37.3	93
Physical Radius	10	10	10	10	8	8	38.2	96
Age	10	9	9	10	10	6	37.3	93
Spin Orbit Alignment	10	10	10	6	7	5	34.5	86
Stellar Metallicity	10	9	8	9	8	7	35.2	88
Discovery Method	10	10	7	7	8	8	34.2	86

Figure 2: Functional Trade Study

3.4   Sequential Optimization

The candidate nodes of the taxonomy have scores for two trade studies: the mission tasks and the functional requirements trades. These are tabulated and averaged to produce a composite score for each candidate taxonomy node, and sorted according to rank (Figure 3).

Candidate Node	Mission Trade	Functional Trade	Composite Score	Rank
Temperature	97	100	98.5	1
Distance from our Sun	100	96	98	2
Binary	100	96	98	3
Separation	98	96	97	4
Period	96	96	96	5
Mass	88	96	92	6
Semi-Major Axis	87	96	91.5	7
Eccentricity	87	96	91.5	8
Inclination	87	96	91.5	9
Spectral Type	86	93	89.5	10
Physical Radius	86	96	89	11
Age	83	93	88	12
Spin Orbit Alignment	82	86	84	13
Stellar Metallicity	77	88	82.5	14
Discovery Method	77	86	81.5	15

Figure 3: Composite Ranking

Minimizing the Number of States: there are two ways to minimize the number of states in the taxonomy: reduce the number of nodes, or reduce the numbers of states in the nodes. A small number of states produces an exponentially large sample space; for example, if there are 13 nodes and each has six equally likely states, the sample space is 136 = 13,060,694,016. Given these simplifying assumptions, the probability that each object path through the taxonomy is not unique is given by the "Birthday Problem calculator [7]: for a catalog consisting of 15,065 objects, the probability of duplicates is < 1%.

Uniqueness: Many of the candidate nodes are continuous variables, for example, eccentricity. Such variables have an infinite number of states and must be discretized (Figure 4); for example, low, medium, and high eccentricities. This discretization produces a total of 96 states across 16 taxonomy nodes which results in 6*6*6*6*6*6*6*9*6*6*6*6*6 = 4,231,664,861,184 possible states. The optimum strategy is to bin the variables so that all values of the variable fit in a bin and each bin has about the same Figure 4: Discretization of Taxonomy Nodes number of instances. The probability that an object is not unique is computed using the Birthday Problem calculator. Let m=2,065 random selections (number of confirmed exoplanets) from n= 4,231,664,861,184 choices. The resulting probability that it is not unique is p = 2.6%. This appears suitable to guarantee uniqueness to the extent practical (unknown values being exceptions). It is tested on the taxonomy data set derived from the current catalog and the state space is perturbed to minimize the number of states while maintaining a unique taxonomy path for each object.

We see that 12 characteristics, with six or more states each are, in theory, sufficient to predict a vanishingly small probability, for a catalog containing 5,000 exoplanets, that two exoplanets share exactly the same characteristics. On the other hand, 15 characteristics, with the number of states shown, provide a near zero probability of duplicates for a 50,000 object catalog. Here, the curse of dimensionality" works in our favor: adding a small number of characteristics (three) greatly expands the cumulative number of states. Characteristics are grouped according to these categories. Optimization will later identify which, if any, characteristics can be dropped from the taxonomy.

Employing a Critic: Modifications to the composite ranking are required to impose knowledge of redundancy in parameters. Specifically, separation varies with position in orbit unless circular orbits are assumed (and the often are), while period and semi-major axis are related by Kepler's Third Law [5]. Separation is deleted in favor of the more intuitive period attribute. Period and semi-major axis are retained because the parameters are well known and discretized binning provides additional information.

Spectral type is not available in the catalog we are using [6], so this parameter is omitted.

Metallicity is chosen ahead of spin-orbit alignment. It better predicts planet type formation.

Discretizing States: Exoplanet catalogs [6] include tools to easily construct histograms. From these plots, rules are readily defined (Figure 5).

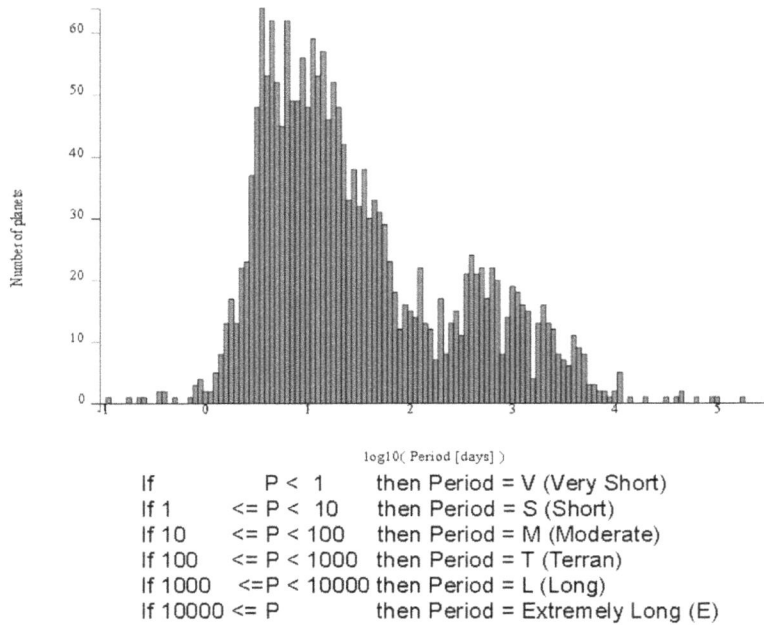

```
If          P <  1      then Period = V (Very Short)
If 1     <= P <  10     then Period = S (Short)
If 10    <= P < 100     then Period = M (Moderate)
If 100   <= P < 1000    then Period = T (Terran)
If 1000  <=P < 10000    then Period = L (Long)
If 10000 <= P           then Period = Extremely Long (E)
```

Figure 5: Rules Discretize Characteristics

Other rules, based on types of stars, discovery method, and a derived Earth Similarity Index [10] are:
- Spectral Type: O,B,A,F,G,K,M,L,T
- Discovery Method: R=Radial velocity, A = Astrometry, T = Timing, M = Microlensing, R = Transit, I = Imaging
- Earth Similarity Index (derived, but not included in this taxonomy)
  - **if .85 <= ESI   then ESI = E (Excellent)**
  - **if .80 <= ESI < .85 then ESI = V (Very Good)**
  - **if .75 <= ESI < .80 then ESI = G (Good)**
  - **if .65 <= ESI < .75 then ESI = M (Moderate)**
  - **if .50 <= ESI < .65 then ESI = P (Poor)**
  - **if   ESI < .50 then ESI = D (Dismal)**

Object Signature: A path through the taxonomy produces a unique 12 letter code that describes the object (Figure 6). A software application is executed to determine the percentage of objects that have unique signatures.

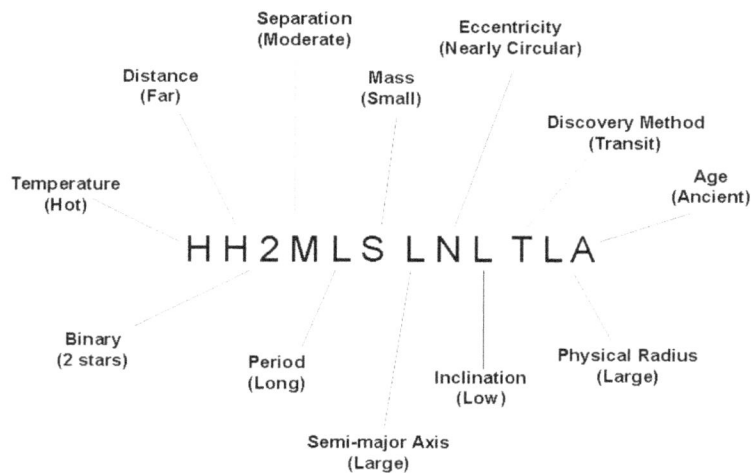

Figure 6: Sample Exoplanet Signature

Optimization Process: The challenge is to find the lowest number of representative states that provide the highest percentage of the exoplanets with unique signatures. An iterative processing technique (Figure 7) is employed. Three computed quantities are required: the object signature, the percentage of each state in each node, and the impact of attributes on uniqueness..

If the objects do not have unique signatures, states are added by increasing the granularity of populous states; for example, if the object type node has three states (payload, rocket body, and debris) and 80% of the objects are debris, then the debris state may be further discretized into small, medium, and large debris.

Once signatures are unique, the number of states is reduced by identifying minimally contributing states as possibly redundant and removing them; for example, if size=very small almost always is accompanied in a signature by radar cross section = very small, then the two states convey redundant information.

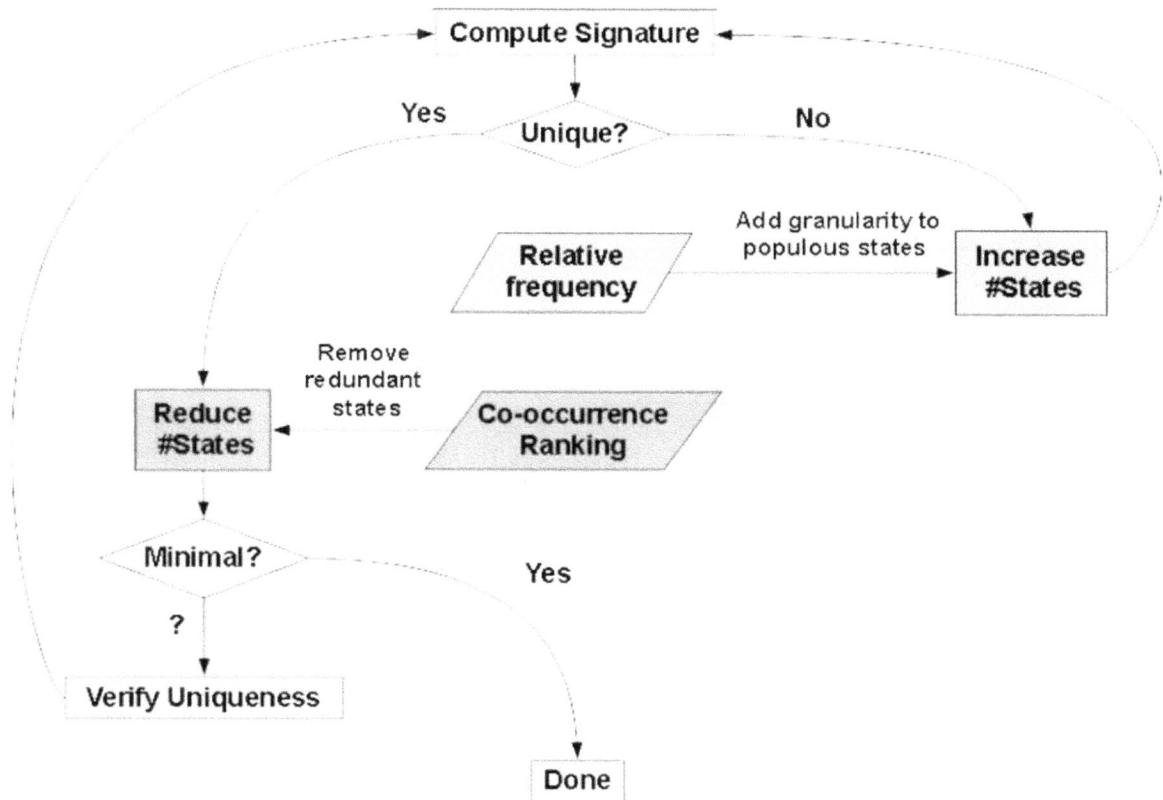

Figure 7: Optimization Process

3.6 Software Development

- Format: Convert an exoplanet catalog to a standard (.csv) format. Choose the exoplanet name as the pivot field: it is available in all data sets and uniquely identifies an object.
- Discretize: For each object in the catalog, discretize state values for candidate taxonomy nodes from raw data to produce an object "signature" (Figure 9).
- Attribute Bins: Compute the number of objects in each state of each attribute
- Signature Uniqueness: Calculate the percentage of signatures that are unique.
- Sensitivity Analysis: compute the percent duplicates with each of the attributes suppressed. This provides an indication of which attributes contribute most to uniqueness.

4. Result

The exoplanet taxonomy (Figure 8) is draw from earlier sections that provide rationale for a minimal set of taxonomy nodes that span the required range of missions and functions. The number of nodes is 12 and the number of states is 77. Many of the nodes have missing values, set the the "unknown" state. This lack of information precludes, for the present, the ability to form unique exoplanet signatures.

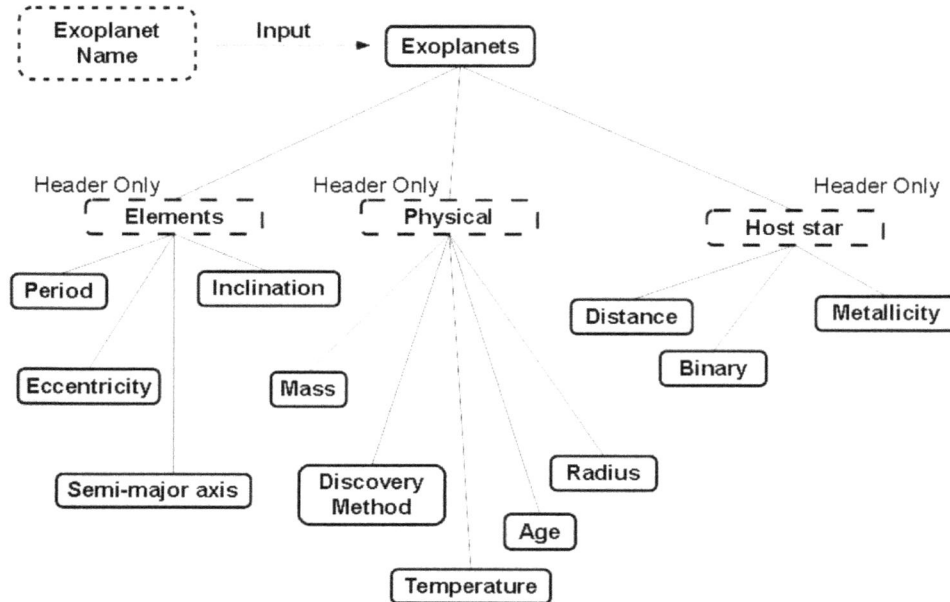

Figure 8: Exoplanet Taxonomy

The signatures of exoplanets were tested (Figure 9) to determine the percentage of unique signatures. Many duplicates (1166 or 56.5%) were identified. The rules provided earlier put objects in discrete bins for each attribute as shown in the table:

Iteration: 6   # states =77   #Objects: 2,065   Duplicate Signatures:1166
Attributes=12   # Possibilities =705,438,720   Probability of Duplicates: 56.5%

Attribute	States (#/state)
Temperature	Terran(0), Hot(0), Very Hot(0), Extreme(0), Intense(0), Sun Like(0), Missing(1275)
Distance	Very Near(184), Near(240), Moderate(121), Far(64), Very Far(64), Distant(1047), Missing(112)
Binary	Two Stars(1899), One Each(26), Two Binaries(138), CircumBinary(2), Other(0), Missing(0)
Period	Very Short(21), Short(741), Moderate(712), Terran(323), Long(172), Extreme(17), Missing(79)
Mass	Very Small(6), Small(35), Moderate(210), Terran(317), Large(476), Giant(2), Missing(947)
Eccentricity	Circular(964), Very Small(0), Small(0), Terran(0), Elliptical(0), Highly Elliptical(0), Missing(1101)
Inclination	Perp LOS (33), Nearly Perp (27), Prograde(0), Retrograde(0), Highly retro(0), Missing(1557)
Radius	Very Small(0), Small(0), Terran(991), Moderate(0), Large(0), Giant(0), Missing(730)
Age	Extra Young(31), Very Young(95), Sunlike(92), Young(52), Old(18), Ancient(9), Missing(1696)
Semi-Major Axis	Very Small(340), Small(362), Moderate(1), Terran(0), Large(0), Giant(0), Missing (846)
Metallicity	Extra Low(0), Very Low(7), Somewhat Low(45), Low(109), Neutral(231), Moderate(368), High(248), Abnormally High(25), Missing(1025)
Discovery Method	Radial Velocity(32), Transit(2023), Astrometry(0), Timing(0), Micro-lensing(0), Imaging(0), Other(0), Missing(10)

Figure 9: Testing for Missing Values

432

Sensitivity Analysis: to determine which attributes contribute most heavily to signature uniqueness, the percent duplicates was computed with each of the 12 attributes suppressed. The result (Figure 10) was that attributes, such as eccentricity, age, and inclination, that have many missing values do not influence signature uniqueness because they have values that are unknown or are not evenly distributed across states. In two other cases, Binary and Discovery method, the percent known is large, but the sensitivity to changing the number of duplicates is small because the majority of exoplanets have a dominant star system state (binary) and a dominant discovery method (transit).

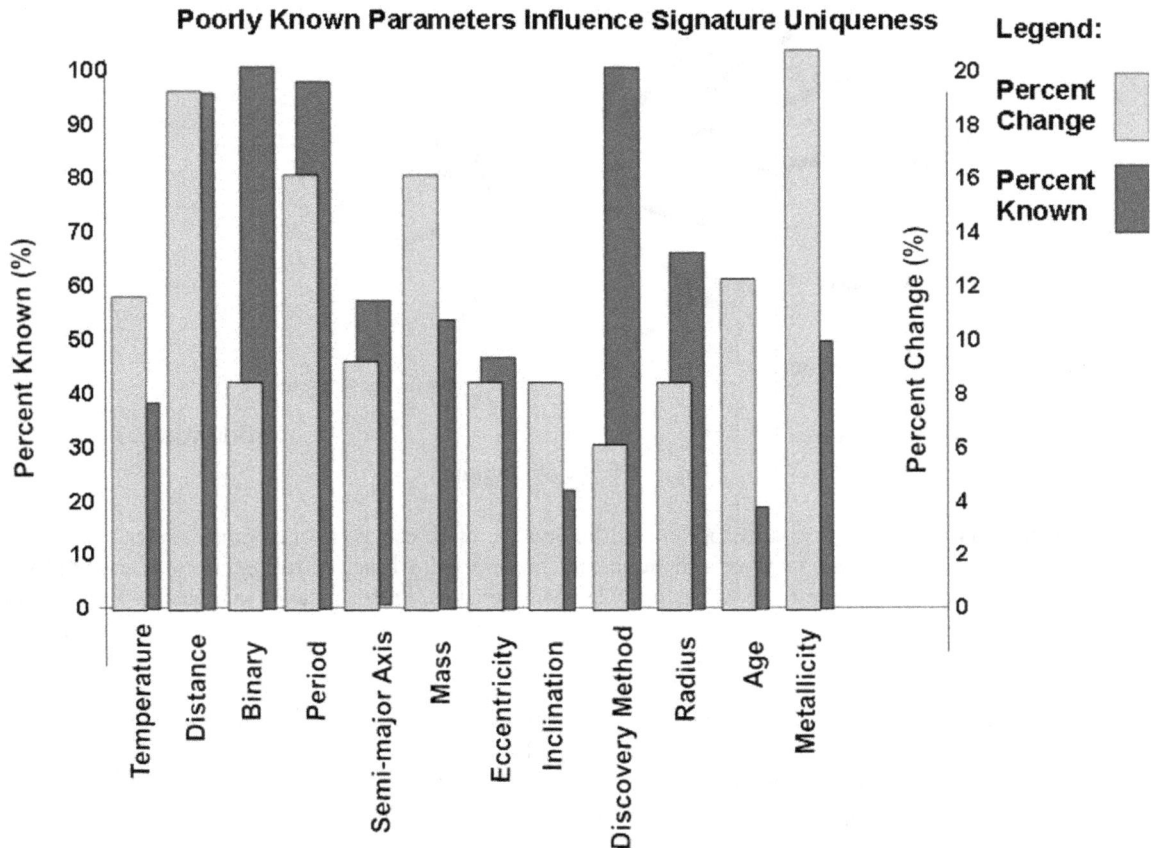

Figure 10: Signature Uniqueness Sensitivity

5. Discussion

The result of this paper is a fine-grained exoplanet taxonomy. However, the stakeholders in the astro-biology community have varied and sometimes conflicting interests. Knowing that there will be counter-proposals for an exoplanet taxonomy, and that users have varied needs, I propose that a "laundry list" of taxonomy nodes be made available. The user chooses those of interest, or chooses from among several pre-configured taxonomies. Underlying data, which will surely come from multiple databases, is assembled based on automated queries and integrated, with federated query technology used to automatically update the taxonomy data files.

Another reason to have a flexible ability to tailor a taxonomy "on-the-fly" is that various communities of stakeholders will operate from differing cross-disciplinarian perspectives. Finally, a user may want to deepen the taxonomy to include lower level detail.

Uses for the Taxonomy:
- Database query: use MySQL [12] to produce custom tables that sort exoplanets based on taxonomy nodes; for example, show a table with all exoplanets with earth-like diameter and temperature.

- Visualization: While the taxonomy hierarchy chart shows the overall knowledge structure, a Kiviat diagram (Figure 11) provides a great visual sense of all characteristics of a single exoplanet at a glance! Four quadrants (Physical, Orbital, Environmental, and Influential, are shown. A ring denotes the Goldilocks Zone. The exoplanet name and habitability score are shown in the center.

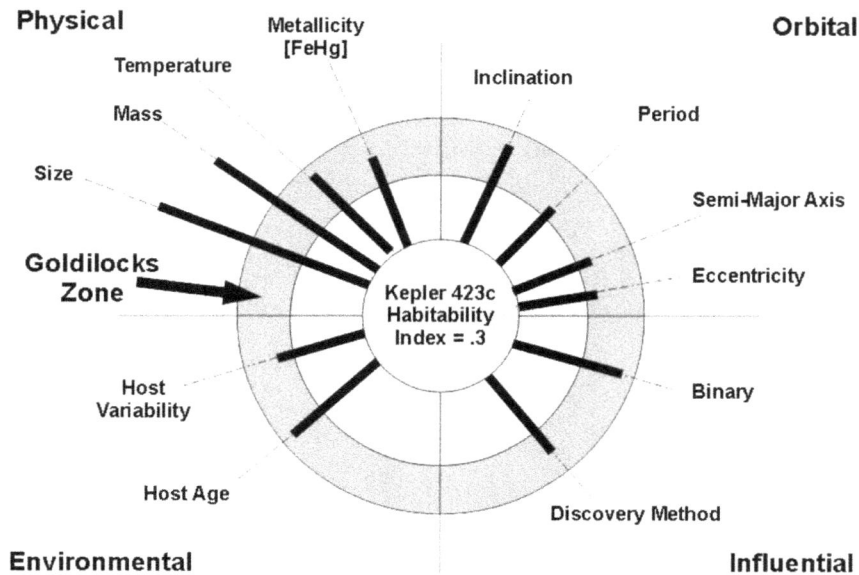

Figure 11: Kiviat Diagram

- Parameter Plots: use Excel to plot histograms of the number of objects versus taxonomy nodes
- Knowledge Base: use Protege to form a hierarchical, characteristics-based ontology. The taxonomy provides the tree-like structure and the node values provide characteristics. Protege allows knowledge base visualization and supports plug-in for automated reasoning; for example, GATE for information extraction, Weka for data mining, and SUBDUE for abductive reasoning.
- Pattern Discovery: use Weka rule induction to find patterns in the for of rules (Figure 12). The user identifies an independent variable (here, it is temperature) and the J48 rule induction algorithm automatically induces If,... AND,... AND,... THEN rules. For example, the longest rule shown is: If the exoplanet radius is less than .9 earth radii and the host star radius is greater than .13 sun radii and the exoplanet mass is less than .892 earth masses, then the exoplanet temperature is medium with two instances and no exceptions.

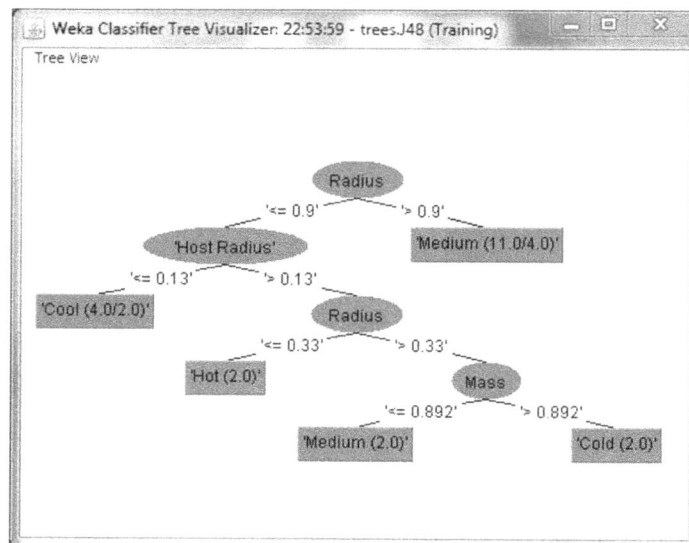

Figure 12: Weka Rule Induction

- Automated Discovery of Unknown Unknowns [13]: The knowledge base keeps track of what is known. As information arrives at the catalog, it is posted to an abductive reasoner that uses the knowledge base as the default structure, and automatically identifies previously unknown nodes and links.
- Additional Characteristics: two attributes that would be useful in filtering on habitable planets that are not yet available or have not been detected are the bio-signature (for example, rocky, tenuous atmosphere, significant oxygen, organic molecules, earth-like). Gemini Planet Imager [16] data has revealed that 51 Eri b, the recently discovered Jupiter-like exoplanet around the nearby star 51 Eridani has an atmosphere of methane and water, and likely has a mass twice that of Jupiter. The second is SETI [14] detection of electromagnetic energy (radar, infrared, gamma ray) that would signal the presence of intelligent life.
- Limitations: the goal of a unique set of attributes for each exoplanet was not met, due largely to that significant amount of unknown information about these objects.

6. Summary

- Elegant: a minimal set of attributes is easily understood
- Leverages domain understanding: mission and functional trade studies
- Provides a hierarchical structure for a knowledge base
- Provides compelling rationale for attribute selection via trade studies
- Satisfies constraints: spans problem space, meets requirements, is a minimal set
- Identifies commonsense limitations in the quest for uniqueness: sometimes it is important to understand that a some exoplanets are more the same than different
- We will revisit the exoplanet catalogs in a few years to update the taxonomy, hopefully with more information on attributes available.

Bibliography

1. http://protege.stanford.edu/, 07/28/2015.

2. http://www.amazon.com/Applications-Artificial-Intelligence-Decision-Making-Multi-Strategy/dp/1502907593, accessed 05/27/2015

3.   http://arxiv.org/abs/1106.0635, accessed 07/28/2015.

4. http://www.lpi.usra.edu/planetary_atmospheres/presentations/Taxonomies_plenary.pdf, accessed 11/06/2015.

5. https://en.wikipedia.org/wiki/Kepler%27s_equation, accessed 11/27/2015.

6. https://github.com/OpenExoplanetCatalogue/, accessed 11/27/2015.

7. https://lazycackle.com/Probability_of_repeated_event_online_calculator__birthday_problem_.html, accessed 05/03/2015.

8. https://en.wikipedia.org/wiki/Earth_Similarity_Index, accessed 11/07/2015.

9. https://en.wikipedia.org/wiki/Minimum_description_length, accessed 07/18/2015.

10. http://arxiv.org/ftp/arxiv/papers/1308/1308.0616.pdf, accessed 05/30/2015.

11. http://www.google.com/patents/US8078559, accessed 05/30/2015.

12. http://www.seti.org/, accessed 11/22/2015.

100 YEAR STARSHIP™

Poster Session

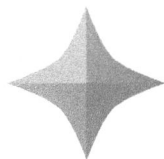

.

Chaired by Timothy Meehan, PhD

Saber Astronautics

Track Description

Great ideas arise through unique individual observations, from people of all ages and educational backgrounds. Students are especially encouraged to submit to this session.

The Poster Sessions are an opportunity to present snapshots of early concepts and experiments. Presentation in the poster format allows in-depth discussion in a small group setting. Presenters are welcome to present on any of the topics from the other technical tracks as well as other topics germane to the theme Finding Earth 2.0. Suitability is at the discretion of the Track Chair.

Track Summary

Tennessee Valley Interstellar Workshop's Virtual Science Competition

By Les Johnson, Joe Meany

The poster announced the Tennessee Valley Intersstellar Workshop sponsored worldship design competition. The competition is aimed primarily at university students. First place will be awarded $2500 and a travel stipend for the team lead to attend the TVIW meeting in 2017.

Commentary about Angelo Baymasecchi, S.J. (2015-2130), first Jesuit that went on a pilgrimage to the Dyson Sphere surrounding Rigil Kentaurus

By Óscar Garrido González

A description is presented of a fictional work based upon the author's personal experience during religious pilgrimages and volunteer work with refuge migrants. The micro story centers around Father Angelo Baymasecchi and weaves influences from a number of other novels such as A Case of Conscience, The Sparrow, and others.

A Roadmap for Interstellar Travel

By Abigail Elizabeth Sherriff, Chris Welch

This work presented a thorough and interdisciplinary roadmap to the development and launch of a worldship within the 100 year timeframe. Necessary enabling technological developments were identified and incorporated with existing or emerging technologies.

Gateway Space Station Foundation

By John Blincow, President

This unique video animation portrays the block construction method, which is similar to modern ship building methodologies, of the Gateway Spacestation. Manned pods and robotic construction pods would be utilized in the construction. The funding strategy uses a lottery system enabling participation from all economic levels.

Track Chair Biography

Timothy Meehan, PhD

Saber Astronautics

Timothy Meehan, PhD has extensive expertise in biomolecular analytical science and diagnostics. He has applied nanotechnology approaches to bioengineering through a collaborative effort with the Australian Stem Cell Centre in order to achieve large scale human synthetic whole blood production. Tim is a seasoned small business leader with experience in genetic diagnostics, microbial parasite detection and commercial New Space startup ventures. At Saber Astronautics he is working with NASA performing reduced gravity hardware flight tests and developing the next generation of autonomous fault detection and recovery solutions for spacecraft.

Commentary About Angelo Baymasecchi, S.j. (2015-2130),
First Jesuit that Went On a Pilgrimage to the Dyson Sphere Surrounding
Rigil Kentaurus and "His Holiness John Xxiv, About Father Angelo
Baymasecchi'S Diary"

Óscar Garrido González

hodosge@gmail.com

Editor's Note

The following was submitted full graphical representation to the Poster Session as well as a fictional piece to the Original Fiction category for the 2015 Canopus Awards, for which it was selected as a finalist. Due to the fictional nature of the piece, allowances in formatting are made. Titles are centered per the author's original intent. Hyphens are used were standard current stylistic guidelines would disallow them. There are additional variations from common formatting and punctuation. As this piece is intended to reflect an ecclesiastical reflection written far in the future, the formatting variations are intentional and an aspect of the fiction itself.

Abstract

At a glance it is a kind of tiny exercise on "exegesis" [the biggest component] of a fiction proposal ["His Holiness John XXIV, about Father Angelo Baymasecchi´s Diary", Finalist Text for Canopus Awards 2015] that takes a close look at an interreligious dialogue at the titanic 100YSS project. This work is based on a. The author´s own experience by making several pilgrimages to sanctuaries of Abrahamic Faiths, plus b. his readings on the history of the Society of Jesus, and c. his final stage (2015) at Red Cross-Spain – as a volunteer – by accompanying immigrants from Eastern Europe, Africa and Asia in their journey of suffering ("pilgrimage") to developed countries that usually do close their doors to them. The presenter has focused on himself as a pilgrim, going on with a micro-story about Father Angelo Baymasecchi, the character of a diary full of intimacy and spiritual beliefs, coming from the life of Saint Ignatius of Loyola, founder of the Jesuits in the XVI Century. Israel, Spain and Italy are starting points (cities such us Compostela, Jerusalem and Rome). The novels "A Case of Conscience" (1958), by James Benjamin Blish, and "The Sparrow" (1996), by Mary Doria Russell, are both remarkable influences, too, on this fiction narration. The books i. "Vast Universe: Extraterrestrials and Christian Revelation" (by Thomas O'Meara OP), and ii. "Christ and the Cosmos: a Reformulation of Trinitarian doctrine" (Keith Ward, August 2015), joined to iii. the talks of Robin Lovin and Guy Consolmagno, SJ (Library of Congress Astrobiology Symposium 2014), are the main foundations for this dissertation as a whole, tackling the frontier domain of "sacred and philosophical hermeneutics" applied to this imaginary plot, from the point of view of the presenter, a humble average citizen.

1.- Commentary (Prima Mentis Operatio, Secunda Mentis Operatio)

1.1. -   Regarding *Vast Universe Extraterrestrials and Christian Revelation* [1],

by Thomas O´Meara, O.P.

- **Carl Sagan should predict that *Baymasecchi still will be a fiction character in his time*** as the *challenges of interstellar travel are enormous – perhaps so enormous that its critics are right, and no civilization will ever be able to achieve it (...) It will not be who reach Alpha Centauri and the other nearby stars (...).*

- O´Meara writes religious phrases about ***"covenant" or "redemption" do not exhaust the ways*** by *which divine power can share its life of love and mercy*, this sentence would be enough answer to the end question of Father Angelo in his excerpt (number 9) *the salvation of the feasible souls outside of the Catholic Church* (…) also to the words of the Pope John XXIV at his approved rules: *we share the same God.*

- **In my story a discovering of other cultures at Cosmos has yet taken place, but O´Meara should note a controversy by reading the Bible**, he points out that *at no point in Christian Scriptures do we learn that there is another race of knowing corporeal beings in the universe*, also *the economy of salvation, a central phrase of Greek Christian theologians, could be wider than one divine plan for Earth.*

1.2 .- Regarding *Christ and the Cosmos. A Reformulation of Trinitarian Doctrine* [2],

by Keith Ward

- ***A n infinite being of generosity would tend to many incarnations*** rather than one, also *Aquinas made a marginal observation in his theology (...) he asked (...) whether a divine person could be incarnate in a further creature – someone other than Jesus of Nazareth –, and answered affirmatively*, this topic is connected with the fourth rule [376].

- **1st rule [373]: *We carry with us outside the Earth our own history of divine mercy***, it is explained by the author like this: *it is not that there is literally a human figure in a throne, surrounded by angels. This Son of Man image, the, would not describe an individual human person, but would be a symbol of one who has achieved the moral goal of humanity (...)*, by connecting also with values of new space cultures of my story: *Our cousins´ souls have the gift of the omni-knowledge, as a preternatural present from our mutual God*, and [375] Fifth rule*: that civilizations like the one at this Dyson Sphere, are peace-loving and respectful with our believes.*

- **My story should mention, as a nice way of meditation along the pilgrimage, the Prologue of John´s Gospel (1, 1-18), because Ward writes**: *the Prologue is now, anyway, part of the canonical text and can be taken as a genuine early reflection on the cosmic status of Jesus which became an important part of Christian tradition. It lives in a wholly different conceptual space from the other Gospels.*

- **First rule [373]: *Extraterrestrial cultures recently discovered deserve all our respect***, as we share the same God. Ward suggests**:** *Maybe, just as we should stop thinking that Jesus will literally rue the Earth one day and that all other religious beliefs will collapse except the Christian.*

- ***Votive objects and libations of the extraterrestrial cultures*** *recently discovered deserve all our respect, as we share the same God* (first rule [373]). Ward points out: *I am inclined to say that the representation of God as a male human is grossly inadequate, if not actually idolatrous. God must be beyond any finite and physical form.*

1.3.- Regarding the Seminar on Astrobiology 2014 at the Library of Congress.

Robin Lovin [3]

- *That civilizations like the one at this Dyson Sphere, are peace-loving* and *respectful with our believes* ([377] Fifth rule):, also, at Baymasecchi´s diary, I say: *It is obvious that they are our master teachers, and we – humans – their collaborators,* according to Lovin, *in the presumption of the intelligent life that we encounter, we must be very careful with the concepts of "dignity", "conquest", "conversion" and "explotation".*

- **Lovin points out about the responsibility in our terrestrial environment,** similar to the responsibility we notice at Baymasecchi: I *have been selected by God for being the first pilgrim to (...)*; endly we should anticipate for the transactions with extraterrestrials, as Pope John XXIV remarks; *for the true meaning that in the Pilgrimage of the Universal Church to the stellar system of Rigil Kentaurus we should have, these rules must be follow.*

1.4.- Regarding the Seminar on Astrobiology 2014 at the Library of Congress.

Guy Consolmagno, SJ. [4]

- *Would you baptize an alien?* If we say "No" we can consider that Christianity has not cosmic significance. Probably this was the mission of Carmelites in their trip to the Dyson Sphere of Rigil Kentaurus (as it is said at Baymasecchi´s diary), also one of the duties of Father Baymasecchi. But Consolmagno see other possible answers: a) *Direct meaning,* b) *Who are you to baptize an alien?* [376] Fourth rule: *To promote the meditative reading of several scholastic texts, above all the ones of Saint Thomas of Aquinas, who, at some marginal notes, wrote about the existence of other worlds;* according to Consolmagno: *it is important the cosmic meaning of the Gospel of Saint John,* same as Keith Ward at his book "Cosmic Christ". Consolmago also speaks about incarnations theories at universe, same commentary as the one of Ward.

- ***There is no doubt, for the history of space exploration,*** *that civilizations like the one at this Dyson Sphere, are peace-loving and respectful with our believes* [377, Fifth rule], also at the Diary: *Our cousins´souls have the gift of the omni-knowledge, as a preternatural present from our mutual God. It is obvious that they are our master teachers, and we – humans – their collaborators;* the brother Jesuit astronomer says; *love is what we have in common with the rest of the universe, besides, intelligence only has sense if there is anyone else to share with.*

2.- Commentary (Tertia Mentis Operatio)

2.1. Statistical Analysis

2.1.1.- Ordered sequencies (average value)

[(My Story – Robin Lovin) = 1,5], [(My Story – Guy Consolmagno) = 1,5)], [(Guy Consolmagno – Guy Consolmagno) = 1,5], [(My Story – *Christ and the Cosmos*) = 1,33], [(Keith Ward – *Christ and the Cosmos*) = 1,08], [(My Story – *Vast Universe*) = 1,07], [(Robin Lovin – Robin Levin) = 1].

2.1.2.- Classified Studied Texts (frequency)

Vast Universe (n=35), *Christ and the Cosmos* (n=30), Robin Lovin (n=14), Guy Consolmagno (n=11), total frequency (n=90).

2.1.3.- Praenotanda Rules, sorted by frequency

If (n_1 = frequency)- (n_2 = item/s), then.-

A   Vast Universe : [*Rule 1 - 373* = (4 – 3,6,13,14)], [*Rule 2 - 374* = (0)], [*Rule 3 - 375* = (0)], [*Rule 4 - 376* = (4 – 2 – 4,9)], [*Rule 5 -377* = (5 – 1- 10)], [*Rule 6 - 378* = (6, 0)].   *B   Christ and the Cosmos* : [*Rule 1 - 373* = (8 – 1,2,4,5,6,8,10,12)], [*Rule 2 - 374* = (0)], [Rule 3 - 375 = (1 – 5)], [*Rule 4 - 376* = (1 – 6)], [Rule 5 - 377 = (1 – 5)], [*Rule 6- 378* = (1 – 11)].   *C*   Robin Lovin : [Rule 1- 373 = (2 – 1, 3)], [Rule 2- 374 = (0)], [Rule 3- 375 = (0)], [Rule 4- 376 = (0)], [Rule 5- 377 = (1 - 4)], [Rule 6- 378 = (0)].   *D*   Guy Consolmagno : [Rule 1- 373 = (0)], [Rule 2- 374 = (0)], [Rule 3- 375 = (0)], [Rule 4- 376 = (1 - 2)], [Rule 5- 377 = (1- 3)], [Rule 6- 378 = (0)] .

Total : [Rule 1-373 = (14)], [Rule 2-374 = (0)], [Rule 3-375 = (1)], [Rule 4-376 = (4)], [Rule 5-377 = (4)], [Rule 6-378 = (1)] .

2.2. Baymasecchi is a copy of the author´s expectations in order to seek God. He has usually asked for a divine sign, many years, since his childhood; and a way to get it came from the religious pilgrimages to the Holy Land (year 1995, with a group of Jews, Muslims and Christians), Compostela (year 2003, accompanied by two seminarians and a prospective student of astrophysics in the Canary Islands) and Rome (2011, by myself, to speak with a Jesuit). My own pilgrimages have given me spiritual healing and peace. The Jesuit Reverend Angelo Baymasecchi [in memory of two great Jesuit scientists: Joseph Bayma (1816-1892), who taught mathematics at Jesuit college of Santa Clara, and the astronomer Pietro Angelo Secchi (1818-1892) has been also influenced by the author´s voluntary work with inmigrants at the Red Cross-Spain throughout 2015, from this event the author strongly requests more research into a theology that should study this field of refugees and immigration from a micro and macro viewpoint.

"His Holiness John XXIV, about Father Angelo Baymasecchi´s Diary" 6
by Óscar Garrido González

[371] [Annotation to the Spiritual Exercises of Saint Ignatius of Loyola, approved by the 274th Pope, His Holiness John XXIV, at the Feast of Saint Peter Claver, in the year 2141] For the true meaning that in the Pilgrimage of the Universal Church to the stellar system of Rigil Kentaurus we should have, these rules must be followed:

[372] The Father Angelo Baymasecchi, the first Jesuit that goes on a pilgrimage to the Dyson Sphere surrounding Alpha Centauri, entrusts his travel diary to the Society of Jesus; and after his own death in the year 2130, the Congregation for the Interstellar Divine Worship recommends and approves to attaching, in the Spiritual Exercises, the next praenotanda:

[373] 1st rule. The first one: To praise vestments and churches along the way. Votive objects and libations of the extraterrestrial cultures recently discovered deserve all our respect, as we share the same God. We carry with us outside the Earth our own history of divine mercy. The dominican preacher Thomas O´Meara even predicted clearly in his publication: *"Vaste Universe. Extraterrestrials and Christian Revelation"* (published at 2013*), how the Church should conduct itself relating to our cosmic relatives, a significant part of it, it is the following: "the power of a divine person is infinite and it cannot be limited to anything created. Could there not be other incarnations? Perhaps many of them and at same time? While the Word and Jesus are one, the life of the Jewish prophet and Earth hardly curtails the divine Word´s life"*.

[374] Second rule. The second one. To confess our sins with a priest if there is at least one in the supply starships. Also, to receive the Holy Host, by the end of human hibernation on return or non-return trips.

[375] 3rd rule. The third one. To praise the constitutions about fasts, vigils and penances. Commander of the starship is able to free Catholic pilgrimages from the obligations when gravitational harshness happens.

[376] 4th rule. The fourth one. To promote the meditative reading of several scholastic texts, above all the ones of Saint Thomas of Aquinas, who, at some side notes, wrote about the existence of other worlds.

[377] 5th rule. The fifth one. There is no doubt, for the history of space exploration, that civilizations like the one in this Dyson Sphere, are peace-loving and respectful towards our

beliefs. The theologian Hans Küng posited that the mystery of the created and uncreated light, and the mystery of God, they both go on unsolved.

[378] 6th rule. Lastly, we pray to the Blessed Virgin Mary that She accompanies us to the current pilgrimages and the ones in the near future to the Marian Sanctuary erected in 2135 and dedicated to Alpha Centauri Maris.

From the Vatican.

John XXIV

Diary (excerpts) of Father Angelo Baymasecchi. Year 2102, 10th September, 15 hours GMT+1 (Mother Earth). Fifteen days for the beginning of the Pilgrimage to Alpha Centauri Maris

[Sheet 1]

"*I have been selected by God* " (Arrupe, 2002:47) for being the first pilgrim to the Spring where H_2O flows without any scientific explanation but by the intercession of Alpha Centauri Maris, from the manned mission of the year 2089 and developed by the Catholic congregation of the Second Reformed Carmelites.

"*Each talent (or grace) has been given not to me, but for the Society of Jesus and the Church.*

Also the imperfections should be taken into account under this light and see that I must correct them and prevent their pernicious effects." (Arrupe, 2002:47)

1) The post of commander in this mission-pilgrimage means being an "*instrument, a representative and channel of God and His Graces for developing His plans through the organization most powerful organization of the Church*" (Arrupe, 2002:47).

"*Huge grace carries with great responsibility.*

2) *The safety of the existence of the grace is true. Our God has to help me but He demands my absolute loyalty to His directions and His graces. The union with Christ and His constant communication is absolutely necessary* "(Arrupe, 2002:47). On Our Lady depends the good of the pilgrimage to this Dyson Sphere where her miracle is repeated. *It is necessary to reach most perfect identification. Naturally it requires a great spiritual discretion in order for me not to be wrong and being inspired for what is not of my own spirit* (Arrupe, 2002:47).

"Tantum quantum

A principle that it is undoubtedly very clear " (Arrupe, 2002:54).

"To use all creatures" (without exception for our cousins) *"and arrange them in order they are able to serve for the Glory of God"* (Arrupe, 2002:54).

"The goal is God Himself, the created value is the glory of God, that glory is in particular the knowledge and love that humankind "(Arrupe, 2002:54) and "new" stellar civilizations have of God and for God. *"Therefore, every means must be means in order to result in more knowledge and love of God. In these means (or creatures) all of them are included without restriction of any kind: supernatural means and natural ones: persons"* (Arrupe, 2002:54), extraterrestrial beings *"and things: positive and negative ones: pleasant and unpleasant ones"* (Arrupe, 2002:54). Accordingly, the resources that there are in the Vatican Space Museum are a sign of the religious meaning of the civilizations that inhabited and inhabit the stellar system of Rigil Kentaurus (sharp tools with the overprinting "Arcos" that it alludes to a previous contact to our "discovering" with these universal cultures, and throughout diverse stages of human history; they suggest as well a development parallel to ours: sacrifices of animals to the gods are a proof of their first evolutionary period).

(N.B.- *"In this section we can get rid of the salvation of the souls outside the Catholic Church. In our work we can and in some way we must get rid of the final purpose, because this depends only and exclusively of the grace of God: I work and I do what I can for increasing that knowledge between Christians and pagans. What are the particular results?, an advance to the virtue?, a conversion?, a soul which is saved from the hell? I do not know: I must work in order to increase the glory of God through the most effective means"* (Arrupe, 2002:54). Our cousins´souls have the gift of the omni-knowledge, as a preternatural present from our mutual God. It is obvious that they are our master teachers, and we – humans – their collaborators. We praise Our Mother, that so much power and courage gave to our founder and first general – Saint Ignatius of Loyola - , at Storta, close to the *Eternal City.*

+ Angelo Baymasecchi, SJ

Acknowledgements

(a) James the Greater, son of Zebedee and Patron Saint of Spain. (b) Saint Francis Borgia, S.J., fourth Duke of Gandía, Grandee of Spain and third Superior General of the Society of Jesus, to whom I prayed September 23rd at the church-headquarters of the Society at Spain, in Madrid, just 50 meters in front of the US Embassy, where the Duchess of Lerma paid for an urn made of silver where the Superior´s relics can be visited; by the end of the day I opened an e-mail signed by Sir Jason Daniel Batt, announcing my unbelievable state of Finalist for Canopus Awards. I returned to pray at the same parish on April 12th , 2016, same day of the presentation of Breakthrough Starshot in New York, without my knowing of the celebration of the event (Starshot); my short story - focused on Alpha Centaury - was getting real! (c) Reverend Francisco Javier Avilés Jiménez [Catholic priest and theologian. Parish of Santo Domingo de Guzmán, Albacete – Spain] (d) My family and friends, specially to: (i) Catholic priest Juan Iniesta Sáez [Pontificio Colegio Español de San José, Rome], (ii) Pedro Tranque [neuroscientist, School of Medicine-Albacete, and Red Cross], (iii) Antonio Muñoz, (iv) Javier Mañas [Technology professor and vice president of the Astronomical Society Alba-5], (iv) Emilia Lara [last president of the Red Cross - Albacete Local] & (v) Marto Redondo [Red Cross].

This project has been supported by both Indiegogo -Life and Vorticex crowdfunding campaigns and a great diffusion help, voice and image: a report at a news programme of RTVE - Radio Televisión Española, the interviews at COPE radio, esRadio and Radio Chinchilla and at the newspapers La Tribuna de Albacete or La Cerca de Castilla- La Mancha, plus the kind call at the specialised magazine on fantasy literature, *Stardust*.

References

1. O´Meara, T. "Vast Universe Extraterrestrials and Christian Revelation". Liturgical Press, Collegeville, Minnesota (2012).

2. Ward, K. " Cosmic Christ. A Reformulation of Trinitarian Doctrine". Cambridge University Press, New York (2015).

3. https://astrobiology.nasa.gov/seminars/featured-seminar-channels/special-seminars/2014/9/18/nasalibrary-of-congress-astrobiology-symposium/

4. https://ac.arc.nasa.gov/p2bnkx7m970/?launcher=false&fcsContent=true&pbMode=normal

5. https://astrobiology.nasa.gov/seminars/featured-seminar-channels/special-seminars/2014/9/18/nasalibrary-of-congress-astrobiology-symposium/

6. https://ac.arc.nasa.gov/p575rvceanm/?launcher=false&fcsContent=true&pbMode=normal

7. My short story is based upon the documents:

   - Loyola, I. *"Spiritual Exercises of Saint Ignatius of Loyola"* . Sal Terrae, 10th Edition, Bilbao.

   - Arrupe, P. *"Here I am, Lord. Notes of the Spiritual Exercises (1965). Preface, transcription, and notes by Ignacio Iglesias, SJ"*. Mensajero. 2002, Bilbao. Pedro Arrupe, SJ. Pages 47, 54 y 55.

   - O´Meara, T. *"Vast Universe Extraterrestrials and Christian Revelation"*. Liturgical Press, Collegeville, Minnesota (2012).

A Roadmap to Interstellar Travel: The Societal Challenges [Poster]

Abigail Sherriff

International Space University

Abigail.Sherriff@community.isunet.edu

Chris Welch

International Space University

Chris.Welch@isunet.edu

Summary

This work presented a thorough and interdisciplinary roadmap to the development and launch of a worldship within the 100 year timeframe. Necessary enabling technological developments were identified and incorporated with existing or emerging technologies.

Editor's Note

Sherriff's and Welch's contribution was an intricate poster laying out a potential roadmap for a generational worldship. Due to the complexity of the design of the poster, it is displayed in two views in the next few pages. On the following page, rotated 90 degrees, the entire poster is presented. Then on the subsequent pages, the poster is provided, magnified, in parts so that the detail of the entry can be viewed and valued. Unfortunately, this does mean that certain words and lines are cut off in the magnified view. However, both views together provide an overall reproduction of their map.

We have also uploaded this poster online so that it can be viewed in color and in greater detail than printing reproduces. To access that PDF, go to: http://bit.ly/100yssSherriffWelchPoster.

Roadmap for Interstellar Travel

"An interdisciplinary roadmap for a nominal 2115 launch of a slower-than-light, multi-generational worldship for interstellar travel."

ASTRA PLANETA

INTERNATIONAL SPACE UNIVERSITY

Astra Planeta Team, astra-planet@isunet.edu.
International Space University, Strasbourg, France

INTRODUCTION

If humanity desires to traverse the cosmos and visit other stellar systems, it will have to overcome vast interstellar distances. One solution is creating a worldship to carry 100,000 people over multiple generations. This interdisciplinary roadmap presents key milestones for developing such a ship.

A lunar base would be an ideal choice to act as the center for resource extraction and construction activity. The worldship itself would be assembled at an Earth-Moon Lagrange point. In addition, the base would be useful to study the psychological and physiological effects of living in extreme environments over extended periods of time.

Shielding is required to protect the worldship and its occupants from the harsh environment of the interstellar medium (ISM), such as dust, galactic cosmic radiation and high energy, heavy-ions. Current proposed methods include both active and passive methods. However, further research and development is required.

The first torus constructed will serve as an engineering model used to test and validate the design. The lessons learned during the construction will benefit future construction. Furthermore, the torus can be used as an environment to test subsystems and run analog experiments.

Final construction will integrate all of the subsystems and will result in a populated and fueled worldship that is ready for launch. Construction would be staged with ecosystem establishment and habitation occurring after each stage is completed. This approach allows more time for ecosystem establishment and to develop societal norms prior to launch, reducing the risk of mission failure.

The International Interstellar Fund (IIF) would serve as the primary body for coordinating financing efforts for the worldship project. The payment requirements will be based on 'criteria' equality quota system, aiming for a $500 billion fund within 20 years of funding. The IIF will serve as a body that utilizes this pool to directly invest in the infrastructure and technological development needs of the project.

The lives of the worldship population rely on a closed-loop ecological system operating over hundreds of years. To date, a closed-looped ecological system has never been successfully engineered. It is estimated that development of a successful ecological engineering techniques would take approximately 80 years.

The selection of a population will be based on years of physiological and psychological experiments. Unlike traditional astronaut crew selection, a large population will be selected with a focus on diversity of skills, gender, and background.

A nuclear fusion drive is the most probable propulsion system. However, other technologies such as antimatter and interstellar ramjets should be considered as well. The selection and construction of such an engine is pushed to the end of the project to allow for sufficient technological development and testing.

CONCLUSION

The technical, societal, and logistical challenges of an interstellar worldship mission will require planning and organization spanning more than a century. The roadmap presented here is not meant to be a comprehensive schedule but rather outlines the key interdisciplinary developments required and identifies the critical path. It is intended to serve as a guide in identifying the way forward to making interstellar travel a reality.

MORE INFO HERE — Final Report — Executive Summary

LEGEND

Project start (line) and flow (arrow)
Depending branches
WS Key Project
WS Major Milestone
Probe launch line

Navigation and communication
Power, propulsion, and shielding
Structure and construction
Humanities
Finance
Governance

ASTRA PLANETA TEAM

K. Acierno
J. Bendandi
G. Bratukhin
C. Chen
G. De Vos
J. Grunin
K. Guseva
A. Kione
M. Kione
F.C. Leah
O. Labarde
H. Nuwatayan
B. Bruno
J.F. Rotoon
A. Scott
A. Sherriff
A. Witter
P. Zan

(Timeline years: 2015, 2025, 2035, 2045, 2055, 2065, 2075, 2085, 2095, 2105, 2115, 2125, 2135, 2145, 2155, 2165, 2175)

ASTRA PLANETA

"An interdisciplinary roadma[p]

INTRODUCTION

If humanity desires to traverse the cosmos and visit other stellar systems, it will have to overcome vast interstellar distances. One solution is creating a worldship to carry 100,000 people over multiple generations. This interdisciplinary roadmap presents key milestones for developing such a ship.

② A lunar base would be an ide[al] the center for resource extra[ction] activity. The worldship itself [would] at an Earth-Moon Lagrange [point] base would be useful to stud[y] and physiological effects of l[iving] environments over extented

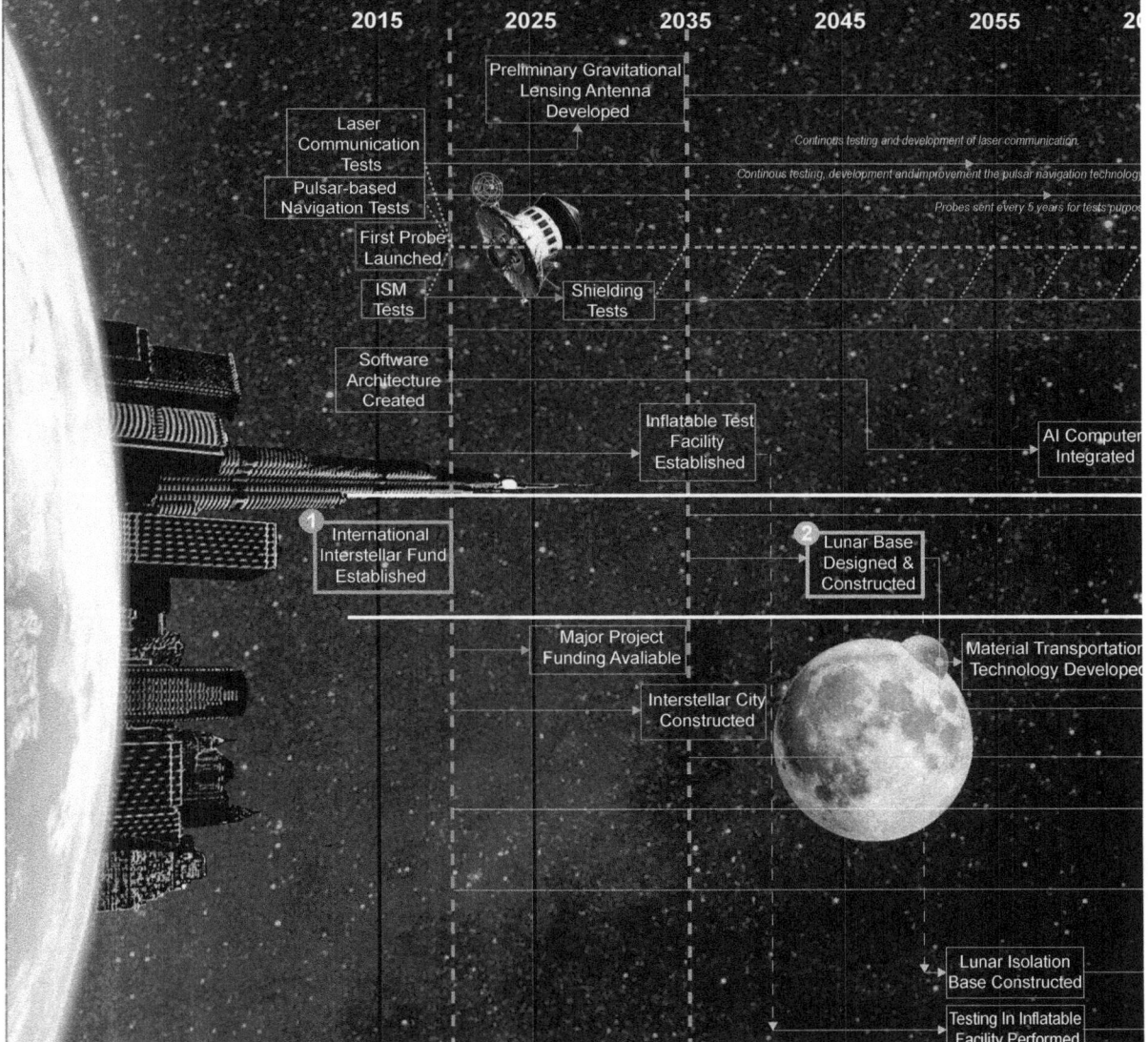

| 2015 | 2025 | 2035 | 2045 | 2055 | 2[065] |

Preliminary Gravitational Lensing Antenna Developed

Laser Communication Tests

Continous testing and development of laser communication.

Pulsar-based Navigation Tests

Continous testing, development and improvement the pulsar navigation technolog[y]

First Probe Launched

Probes sent every 5 years for tests purpo[ses]

ISM Tests

Shielding Tests

Software Architecture Created

Inflatable Test Facility Established

AI Computer Integrated

① International Interstellar Fund Established

② Lunar Base Designed & Constructed

Major Project Funding Avaliable

Material Transportation Technology Developed

Interstellar City Constructed

Lunar Isolation Base Constructed

Testing In Inflatable Facility Performed

Roadmap for Interstellar Trav

Astra Planeta Team, astra-planeta@isunet.edu,
International Space University, Strasbourg, France

oadmap for a nominal 2115 launch of a slower-than-light, multi-generation

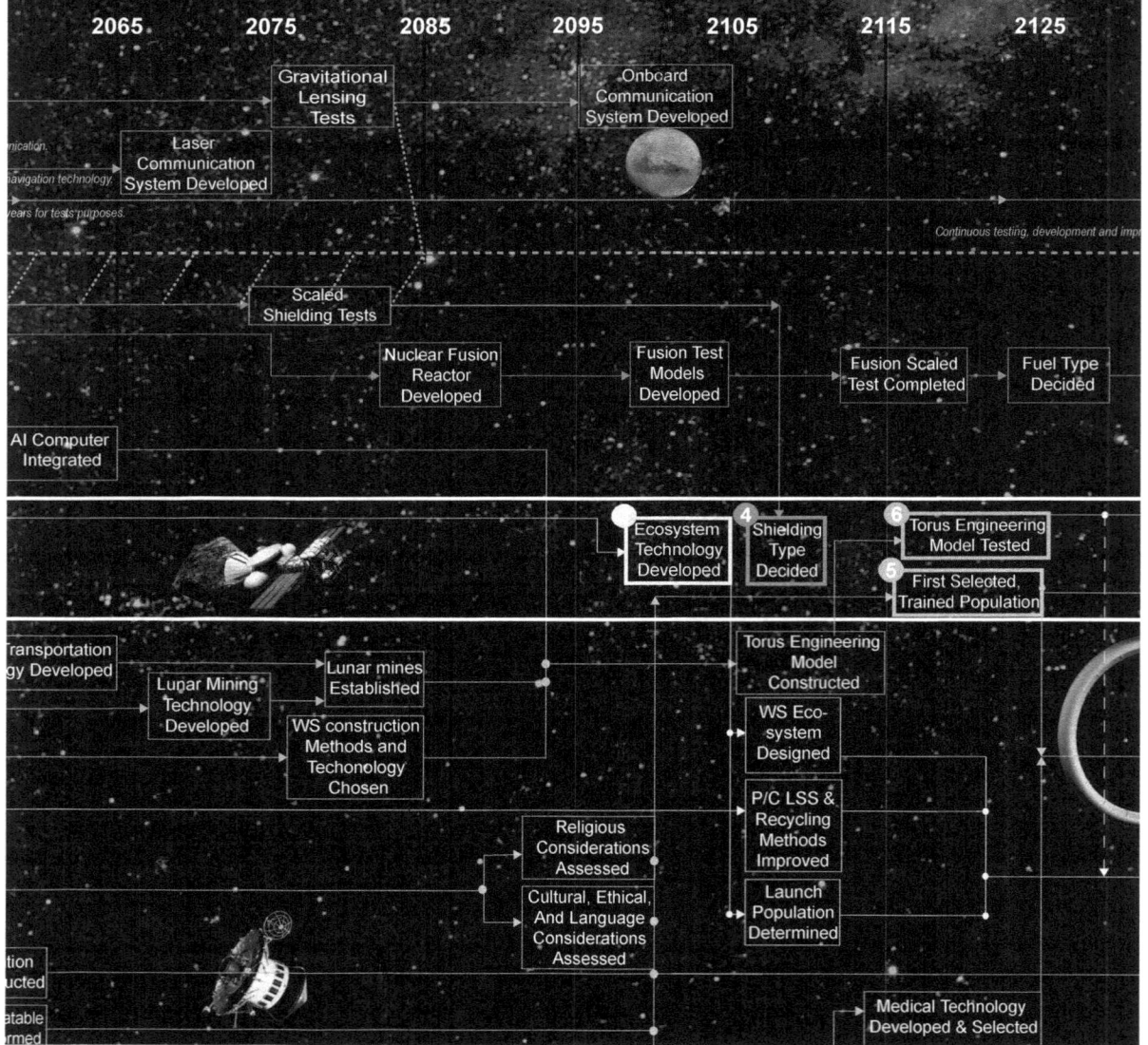

d be an ideal choice to act as
urce extraction and construction
ship itself would be assembled
Lagrange point. In addition, the
ful to study the psychological
effects of living in extreme
extented periods of time.

4 Shielding is required to protect the worldship and
its occupants from the harsh environment of the
interstellar medium (ISM), such as dust, galactic
cosmic radiation, and high energy, heavy-ions.
Current proposed methods include both active
and passive methods. However, further research
and development is required.

6 The first tor
engineering
the design.
constructio
struction. F
used as an
and run ana

| 2065 | 2075 | 2085 | 2095 | 2105 | 2115 | 2125 |

Gravitational
Lensing
Tests

Onboard
Communication
System Developed

Laser
Communication
System Developed

Scaled
Shielding Tests

Nuclear Fusion
Reactor
Developed

Fusion Test
Models
Developed

Fusion Scaled
Test Completed

Fuel Type
Decided

AI Computer
Integrated

Ecosystem
Technology
Developed

4 Shielding
Type
Decided

6 Torus Engineering
Model Tested

5 First Selected,
Trained Population

Transportation
gy Developed

Lunar Mining
Technology
Developed

Lunar mines
Established

WS construction
Methods and
Techonology
Chosen

Torus Engineering
Model
Constructed

WS Eco-
system
Designed

Religious
Considerations
Assessed

Cultural, Ethical,
And Language
Considerations
Assessed

P/C LSS &
Recycling
Methods
Improved

Launch
Population
Determined

Medical Technology
Developed & Selected

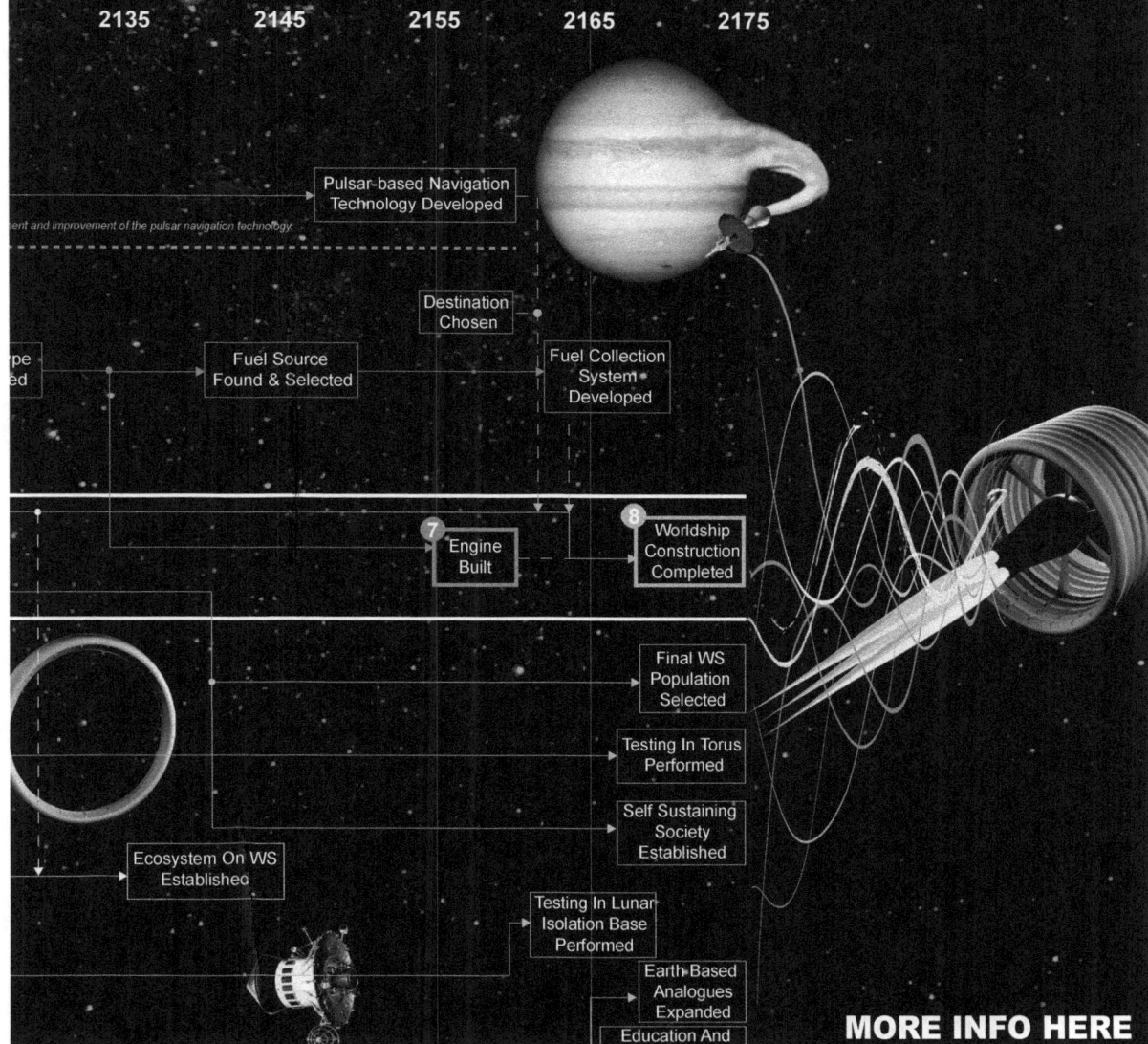

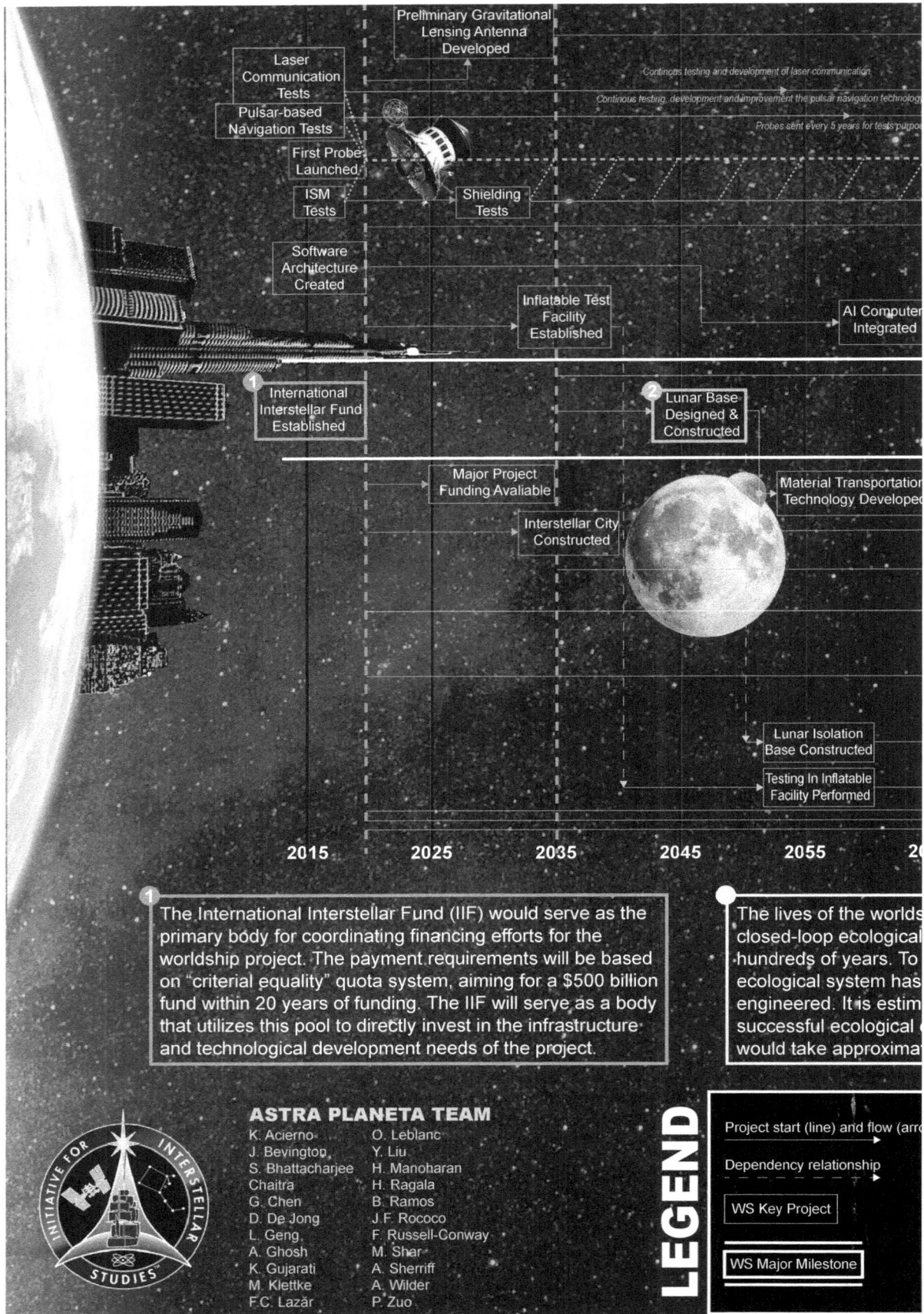

Preliminary Gravitational Lensing Antenna Developed

Laser Communication Tests

Pulsar-based Navigation Tests

First Probe Launched

ISM Tests

Shielding Tests

Software Architecture Created

Inflatable Test Facility Established

AI Computer Integrated

Continous testing and development of laser communication

Continous testing, development and improvement the pulsar navigation technology

Probes sent every 5 years for tests purpose

1 International Interstellar Fund Established

2 Lunar Base Designed & Constructed

Major Project Funding Avaliable

Interstellar City Constructed

Material Transportation Technology Developed

Lunar Isolation Base Constructed

Testing In Inflatable Facility Performed

2015   2025   2035   2045   2055   20

1 The International Interstellar Fund (IIF) would serve as the primary body for coordinating financing efforts for the worldship project. The payment requirements will be based on "criterial equality" quota system, aiming for a $500 billion fund within 20 years of funding. The IIF will serve as a body that utilizes this pool to directly invest in the infrastructure and technological development needs of the project.

The lives of the worlds closed-loop ecological hundreds of years. To ecological system has engineered. It is estim successful ecological would take approximat

ASTRA PLANETA TEAM

K. Acierno
J. Bevington
S. Bhattacharjee
Chaitra
G. Chen
D. De Jong
L. Geng
A. Ghosh
K. Gujarati
M. Klettke
F.C. Lazăr

O. Leblanc
Y. Liu
H. Manoharan
H. Ragala
B. Ramos
J.F. Rococo
F. Russell-Conway
M. Shar
A. Sherriff
A. Wilder
P. Zuo

LEGEND

Project start (line) and flow (arr

Dependency relationship

WS Key Project

WS Major Milestone

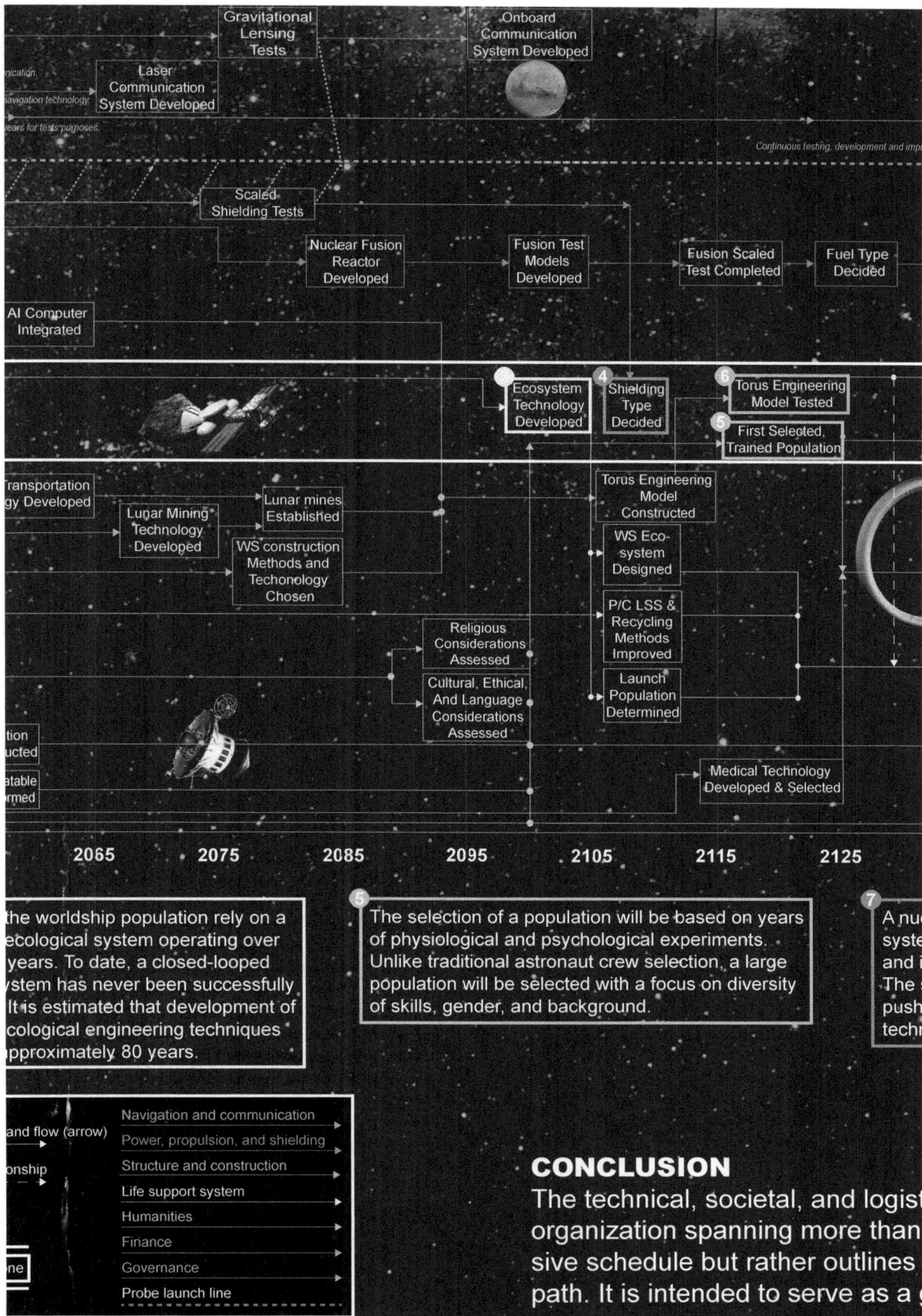

Gravitational Lensing Tests

Onboard Communication System Developed

...nication

Laser Communication System Developed

...navigation technology

...years for tests purposes.

Continuous testing, development and impr...

Scaled Shielding Tests

Nuclear Fusion Reactor Developed

Fusion Test Models Developed

Fusion Scaled Test Completed

Fuel Type Decided

AI Computer Integrated

Ecosystem Technology Developed

4 Shielding Type Decided

6 Torus Engineering Model Tested

5 First Selected, Trained Population

...ransportation ...gy Developed

Lunar Mining Technology Developed

Lunar mines Established

WS construction Methods and Techonology Chosen

Torus Engineering Model Constructed

WS Eco-system Designed

P/C LSS & Recycling Methods Improved

Religious Considerations Assessed

Cultural, Ethical, And Language Considerations Assessed

Launch Population Determined

...tion ...ucted

...atable ...rmed

Medical Technology Developed & Selected

| 2065 | 2075 | 2085 | 2095 | 2105 | 2115 | 2125 |

the worldship population rely on a ecological system operating over years. To date, a closed-looped ystem has never been successfully It is estimated that development of cological engineering techniques pproximately 80 years.

5 The selection of a population will be based on years of physiological and psychological experiments. Unlike traditional astronaut crew selection, a large population will be selected with a focus on diversity of skills, gender, and background.

7 A nu... syste... and i... The s... push... techn...

and flow (arrow)

...onship

...ne

Navigation and communication
Power, propulsion, and shielding
Structure and construction
Life support system
Humanities
Finance
Governance
Probe launch line

CONCLUSION
The technical, societal, and logist
organization spanning more than
sive schedule but rather outlines
path. It is intended to serve as a g

Pulsar-based Navigation
Technology Developed

...ent and improvement of the pulsar navigation technology.

Destination
Chosen

...pe
...ed

Fuel Source
Found & Selected

Fuel Collection
System
Developed

7 Engine
Built

8 Worldship
Construction
Completed

Final WS
Population
Selected

Testing In Torus
Performed

Self Sustaining
Society
Established

Ecosystem On WS
Established

Testing In Lunar
Isolation Base
Performed

Earth-Based
Analogues
Expanded

Education And
Outreach Program
Completed

2135 2145 2155 2165 2175

A nuclear fusion drive is the most probable propulsion
system. However, other technologies such as antimatter
and interstellar ramjets should be considered as well.
The selection and construction of such an engine is
pushed to the end of the project to allow for sufficient
technological development and testing.

MORE INFO HERE

Final Report

Execuitive Summary

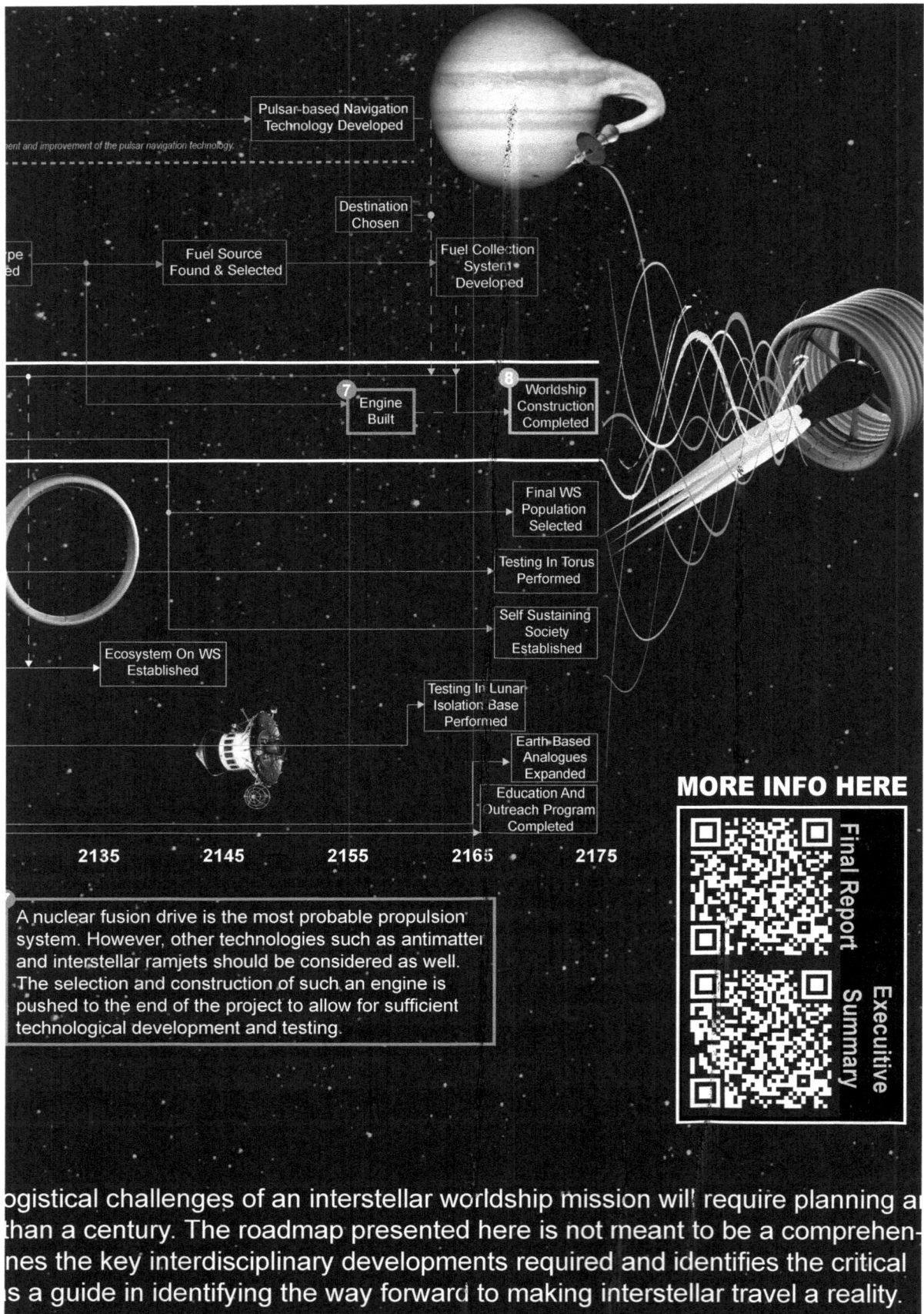

...ogistical challenges of an interstellar worldship mission will require planning a...
...than a century. The roadmap presented here is not meant to be a comprehen-
...nes the key interdisciplinary developments required and identifies the critical
...s a guide in identifying the way forward to making interstellar travel a reality.

Design & Layout

The wide variety of disciplines represented and information contained in the preceding papers provides unique design challenges. As in the previous year, the goal was to create a uniform presentation and easily accessible layout while still maintaining the individual author's needs and desires. The overall design is intended to be conducive to the experience of education and engagement. This year's improvements in design have focused on uniformity between the papers. Formulas, illustrations, and other non-text elements are more consistent (although not absolutely so). The challenge engaged and met is to provide a cross-discipline publication that highlights the variety of thought, research, and effort needed to achieve the audacious dream of interstellar travel.

Primary text is typeset in Adobe Caslon Pro. Headings are set in Helvetica Neue and League Gothic.

Design and layout by Jason D. Batt.

Editorial Board

Mae Jemison, MD
Pamela Contag, PhD
Kathleen Toerpe, PhD
Ron Cole
Alires J. Almon
Jason D. Batt

www.100YSS.org

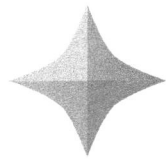

www.ingramcontent.com/pod-product-compliance
Lightning Source LLC
Chambersburg PA
CBHW082123210326
41599CB00031B/5852

9 780990 384021